PROGRESS IN HPLC
Volume 2

Related Titles of Interest

Books

Journals

PROGRESS IN HPLC
Volume 2

Electrochemical detection in medicine and chemistry

Editors
H. Parvez, M. Bastart-Malsot, S. Parvez,
T. Nagatsu and G. Carpentier

VNU SCIENCE PRESS

Utrecht, The Netherlands
1987

VNU Science Press BV
P.O. Box 2093
3500 GB Utrecht
The Netherlands

© 1987 VNU Science Press BV

First published 1987

CIP-DATA KONINKLIJKE BIBLIOTHEEK, DEN HAAG

Electrochemical detection in medicine and chemistry /
H. Parvez . . . [et al.]. — Utrecht : VNU Science Press.
— Ill. — (Progress in HPLC ; vol. 2)
ISBN 90-6764-062-X bound
SISO 542.1 UDC 541.13:[54+61]
Subject headings: electrochemical detection ; medicine
/ electrochemical detection ; chemistry.

Printed in Great Britain by J. W. Arrowsmith Ltd., Bristol

Preface

This second volume in the well received series Progress in HPLC bears witness to the rapid and continued growth of high-performance liquid chromatography as a powerful analytical technique. The reviews in this volume present a comprehensive overview of the current state of research in both the practical and theoretical aspects of electrochemical detection (ECD). Such detectors take full advantage of the electrical properties of many biological compounds and, due to their inherent sensitivity, now represent a substantial portion of all HPLC detectors.

The reviews are divided into three sections dealing with (1) instrument development, (2) examples of applications of ECD in medicine and chemistry (e.g. the analysis of drug metabolism) and (3) the analysis of biogenic amines and related enzymes.

It is hoped that this collection of reviews by many acknowledged experts will provide a valuable reference source to all workers in the field of HPLC, and will help to stimulate further research in all aspects of ECD.

Finally, the Editors and Publisher wish to express their gratitude to all contributors for the effort and care with which they prepared their manuscripts.

This book is dedicated to Salima, Natacha, and Alexandre

CONTENTS

INSTRUMENTAL DEVELOPMENTS IN ELECTROCHEMICAL DETECTION

Progress in HPLC, Vol. 2, pp. 3—36
Parvez *et al.* (Eds)
© 1987 VNU Science Press

Biochemical and pharmacological application of electrochemical detection in HPLC

TATSUO KURAHASHI* and HIROHITO NISHINO

Analytical Instrument Division, Yanagimoto Mfg Co., Ltd, Kyoto 612, Japan

SIMONE PARVEZ and HASAN PARVEZ

Unité de Neuropharmacologie, Université de Paris XI, 91405 Orsay, France

KOHICHI KOJIMA and TOSHIHARU NAGATSU

Laboratory of Cell Physiology, Department of Life Chemistry, Graduate School at Nagatsuta, Tokyo Institute of Technology, Yokohama 227, and Department of Biochemistry, Nagoya University School of Medicine, Nagoya 466, Japan

INTRODUCTION

Since 1906 when M. S. Twett employed the chromatographic technique for the first time to separate various plant pigments, chromatography has been applied as an effective analytical method in various fields. It has been remarkably developed in the last ten years or more, especially owing to the introduction of high-performance liquid chromatography (HPLC). For improving any analytical methods and instruments, much interest is taken in detection sensitivity, analytical time, operability and automation of data analysis. This tendency is also pronounced in HPLC; high sensitivity can be achieved by the development of new column packings, new liquid-flow methods, and new detectors. The analytical time, in which aspect liquid chromatography (LC) was far inferior to gas chromatography (GC), has been surprisingly shortened.

GC is very effective for separation of volatile samples, and about 20% of organic substances are volatile. However, it cannot be applied for compounds with molecular weight higher than 300, and for pyrolytic and ionic substances. For analyzing less volatile substances by GC, they must be derivatized or pyrolyzed. But such GC analysis involves many problems which often make it impractical. On the other hand, LC can be used for analysis of any samples provided that it is eluted with eluent. Thus LC can

---

* To whom correspondence should be addressed.

cover a wider application range than GC. LC has been widely applied to analyses of biological substances, pharmaceutical and xenobiotic (environmental) compounds. When examining biological samples such as body fluids (blood, urine, cerebrospinal fluid, saliva and amniotic fluid) and tissues, they are mixtures of complicated composition and the components to be identified are present frequently in trace amounts. In most cases, such biological samples are first treated by extraction, precipitation or adsorption—desorption (a clean-up procedure). Since a wide variety of compounds is analyzed by HPLC, a photometric or fluorometric detector cannot always provide satisfactory analysis in the aspects of sensitivity and selectivity. That is the reason why we have expected a more sensitive and selective detector for HPLC. Our expectation has been met by the development of an electrochemical detector (ECD) for HPLC using the principle of pholarography. The principle and applications of this useful detector are described below.

PRINCIPLE OF HPLC–ECD METHODS (OXIDATION)

Figure 1 shows an example of an electrochemical detector (Yanaco Model VMD-101A). This detector is composed of a three-pole potentiostat system. On the basis of the potential of a reference electrode (Ag/AgCl), constant potential electrolysis is given between the working electrode (glassy carbon) and the auxiliary electrode (stainless steel tube). Current occurs when an electrochemically active material contained in eluent is oxidized or reduced on the working electrode. This current is measured for finding the concentration of the electrochemically active material. Therefore, there are two modes in HPLC–ECD, oxidation mode and reduction mode. Oxidation mode is more frequently used than reduction mode, and examples of oxidation mode are mainly described in this chapter.

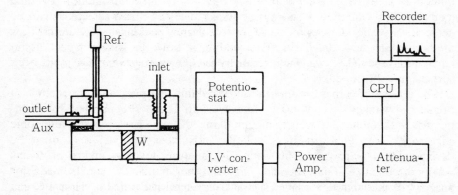

Figure 1. Construction and flow diagram of an electrochemical detector (VMD-101A).

Generally, limiting current in a flow system (i_L) is represented by Equation 1 (Fehér *et al.*, 1974).

$$i_L = knFCD^{2/3} U^x \qquad (1)$$

(k = constant kinetic viscosity coefficient depending on area of electrode, n = number of electrons contributing to electrode reaction, F = Faraday constant, C = concentration of a sample, U = flow rate, D = diffusion coefficient of a sample, x = exponential coefficient depending on structure of electrode). The logarithm of i_L is represented by the following equation:

$$\log i_L = \log K + x\log U \qquad (2)$$

($K = knFD^{2/3} C$). In this equation, the value of x can be obtained from the slope of the graph of $\log i_L$ against $\log U$. Generally, it is found to be one third in a cylindrical electrode, a half in a plate electrode and three quarters in a wall jet electrode.

Characteristics of the electrode system of the VMD-101A are shown in Figs 2 (linear dynamic range of current) and 3 (an example of analysis). As for dynamic range in analysis of catecholamines [norepinephrine (NE),

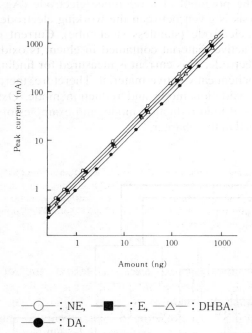

—○— : NE, —■— : E, —△— : DHBA.
—●— : DA.

Figure 2. Peak currents vs injected amounts of norepinephrine (NE), epinephrine (E), dihydroxybenzylamine (DHBA), and dopamine (DA). Column: Yanapak ODS-T (10 μm, 250 × 4.9 mm ID); mobile phase: phosphate buffer, pH 3.1; flow rate: 0.8 ml/min; and applied potential: +0.65 V vs Ag/AgCl.

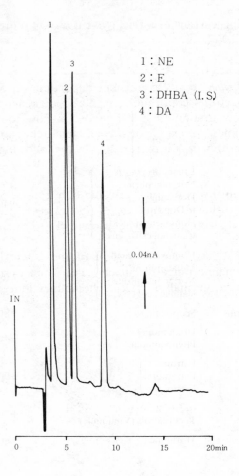

Figure 3. Analysis of catecholamines by HPLC—ECD. Samples: standards of norepinephrine (NE), epinephrine (E), dihydroxybenzylamine (DHBA), and dopamine (DA). Sample amount: each 100 pg; column: Yanapak ODS-A (250 × 4.6 mm ID); column temperature: 23° C; mobile phase: 0.1 M phosphate buffer, pH 3.1/ 10 μM Na_2EDTA; flow rate: 0.7 ml/min; applied potential: +0.8 V vs Ag/AgCl.

epinephrine (E), dihydroxybenzylamine (DHBA) and dopamine (DA)], linearity in peak height is assured in a wide range of 50 pg to 1 μg. As for detection limit (signal (S) to noise (N) ratio, $S/N = 2$), analysis of down to 5 pg is possible for NE. In addition to high sensitivity, this ECD exhibits good selectivity. It only responds to electrochemically active substances with functional groups. Also, by selecting a set potential depending on difference in oxidation or reduction potential among the substances, it can selectively detect only a certain kind of substances.

Typical substances to be analyzed by HPLC—ECD are shown in Table 1.

Table 1. Typical active substances of HPLC—ECD analysis

		mV (vs Ag/AgCl)
1. Aromatic Alcohols		
Phenols	Tyrosine, Tyramine, Thyroxine, Thyronine	+800—900
Catechols	Epinephrine, Norepinephrine, Dopamine, L-Dopa, Homogentisic acid, Catechol estrogens	+400—700
Methoxyphenols	Homovanillic acid, Metanephrine, Normetanephrine, Vanillic acid	+800—900
Hydroxycoumarins	Scopoletin	+800—900
Quinones	Ubiquinones, Phylloquinone	−400
Estrogens	Estron, Estradiol, Estriol	+900
Tocopherols	α-, β-, γ-, δ-Tocopherols (Vitamin E)	+700
Morphine	Morphine	+800
2. Aromatic Amines		
Anilines	Chloroaniline, Bromoaniline, p-Phenylenediamine, Benzidine	+900—1000
Sulfonamides	Sulfonamide (Sulfa drugs)	
3. Indoles		
Indole-3-derivatives	Tryptophane, Indolyl-3-acetic acid, Tryptamine, Melatonin	+800—900
5-Hydroxyindoles	Serotonine, 5-Hydroxyindoleacetic acid, 5-Hydroxytryptophane	+600—700

Table 1 (continued)

4. Phenothiazines		
	Chrolpromazine ⎫ Promethazine ⎬ Perphenazine ⎭	+900
5. Purines		
	Uric acid ⎫ Xanthine ⎟ Guanine ⎬ Theophylline ⎭	+800—1000
6. Others		
Ascorbic acid	Ascorbic acid	+800
Thiols	Cysteine ⎫ Penicillamine ⎬ Glutathione ⎭	+800
Anions	SCN^-, $S_2O_3^{2-}$, SO_3^{2-}, NO_2^-	

EXAMPLES OF BIOCHEMICAL AND PHARMACOLOGICAL APPLICATION OF HPLC—ECD METHODS (OXIDATION)

Examples of HPLC—ECD analysis (oxidation mode) for biochemical and pharmacological substances are shown in Table 2. Some typical chromatograms are described below.

Catecholamines, serotonin and their metabolites

Catecholamines and serotonin are generally known as neurotransmitters. Analysis of such monoamine neurotransmitters by HPLC—ECD is important for investigating the molecular mechanism and for biochemical and pharmacological studies on various diseases such as Parkinson's disease, neuroblastoma, mental diseases and hypertension. It is also applied to studies on the mechanism of drug actions. Figure 4 shows an example of measurement of catecholamines and serotonin in mouse brain. Figure 5 shows that of catecholamines in human plasma.

Drugs

Various drugs, e.g. L-dopa (Riggin et al., 1976), α-methyldopa, phenothiazin groups (Tjaden et al., 1976), penicillamine (Inoue et al., 1982), theophylline (Greenberg and Mayer, 1979), sulfa drugs, morphine (Wallace et al., 1980), etc., can be measured by HPLC—ECD. Figure 6 shows an example of measurement of theophylline, a drug for bronchial asthma.

Table 2. Examples of HPLC—ECD analysis

Substance	Sample	LC system			Detector		Reference
		Column*	Mobile phase	Flow rate (ml/min)	Voltage (V)	Sensitivity	
1. Aromatic alcohols							
1.1. Phenols							
Tryptophan, tyrosine, 5-hydroxyindoleacetic acid and homovanillic acid	Human CSF	μ-Bondapak C$_{18}$ 10 μm (300 × 3.9)	0.01 M Acetate buffer (pH 4.0)/methanol (17/3, v/v)	1.5	0.8	Tyr 60, Trp 400, 5HIAA 10, HVA 24 (pg)	Anderson et al. (1979)
Morphine and monoamines	Mouse brain	Ultrasphere-ODS 5 μm (250 × 4.6)	0.05 M Citrate buffer (pH 4.25)/THF (99/1, v/v)	0.85	0.725	Morphine 1, NE 0.2, DA 0.2, 5HT 0.5, Tyr 2, 3MT 0.5, Trp 2 (ng)	Ishikawa et al. (1982a)
p-Aminobenzoic acid	Human urine	Yanapak ODS-T 10 μm (250 × 4)	0.2 M Phosphate buffer (pH 3.5)/acetonitrile (8.75/12.5, v/v)	0.7	1.1		Ito et al. (1982)
Biogenic amines, precursor amino acids and related metabolites	Mouse brain	Nucleosil 7C$_{18}$ 7 μm (250 × 4)	0.1 M Sodium phosphate buffer (pH 2.2), 0.1 M citrate—NaOH buffer (pH 3.0), 0.1 M sodium phosphate	0.74 and 0.6	0.65, 0.73, 0.55 and 1.0	0.1—40 pmol	Oka et al. (1984)

Table 2 (continued)

Substance	Sample	LC system Column*	Mobile phase	Flow rate (ml/min)	Detector Voltage (V)	Sensitivity	Reference
			buffer (pH 6.5)/acetonitrile (96/4, v/v) and 0.1 M sodium acetate buffer (pH 5.2)				
1.2. Catechols							
Homogentisic acid	Serum and urine	Vydac-strong anion exchange (500 × 2.1)	0.5 M Acetate buffer (pH 4) or 0.75 M acetate buffer (pH 4)	0.2	0.45	1 ng/ml of serum or 100 ng/ml of urine	Zouthendam et al. (1976)
L-DOPA and dopamine	Serum	Vydac-bonded cation exchange (500 × 2)	10 mM H_2SO_4 and 40 mM Na_2SO_4	0.40	0.72	10 μg of DOPA/l of serum	Riggin et al. (1976)
Serotinin and dopamine	Mouse brain tissue	Zipax SCX (750 × 3)	Citrate—acetate buffer (pH 5.1)	1.3	0.6	0.1 pmol/ injection	Sasa and Blank (1977)
Biogenic amines	Mouse brain and rat urine	Lichrosorb RP-18 or Nucleosil C_{18} 10 μm (250 × 4.6)	Citrate—acetate buffer (pH 5.1) with 1 % THF	0.77	0.8 or 0.65	NE 5 DA 10 (pg)	Ikenoya et al. (1978)
Catecholamines	Plasma	Vydac CX (500 × 2)	Acetate—citrate buffer (pH 5.2)	40—50 (ml/h)	0.6	0.1 nmol/l	Hallman et al. (1978)
Catecholamines	Rat brain	μ-Bondapak C_{18} 10 μM (300 × 3.9)	Citrate—phosphate buffer with 0.3 mM SDS	1.0	0.72		Felice et al. (1978)

Compound	Sample	Column	Mobile phase			Detection limit	Reference
Methyldopa and metabolites	Human serum	Vydac SCX 30—44 μm (500 × 2)	0.02 M Phosphate buffer (pH 2.55) with 0.1 mM EDTA	0.4	0.54	$2-5 \times 10^{-7}$ M	Cooper et al. (1979)
DOPA	Rat and human brain	Yanapak ODS 5 μm (250 × 4)	0.1 M Potassium phosphate buffer (pH 3.5)	0.6	0.8	5 pmol	Nagatsu et al. (1979a)
Dopamine	Rat and human brain	Yanapak ODS (250 × 4)	0.1 M Potassium phosphate buffer (pH 3.5)	0.6	0.6	500 fmol	Nagatsu et al. (1979b)
Dopamine, DOPAC and HVA	Brain	Nucleosil C$_{18}$ 5 μm	Citrate buffer (pH 4.25)/methanol (92/8, v/v) with 1.7×10^{-2} M sodium hexyl sulfate	0.8	0.6		Magnusson et al. (1980)
Catecholamines and DOPAC	Human plasma	Ultrasphere-ODS (250 × 4.6)	0.1 M Sodium acetate—0.02 M citric acid buffer/methanol (90/10, v/v) with 100 mg/1 sodium octyl-sulfate and 50 mg/1 EDTA	1.0	0.6	5 pg	Mefford et al (1981)
Dopamine and serotonin	Animal serum	Yanapak ODS-T 10 μm (250 × 4)	0.1 M Phosphate buffer (pH 3.2)	0.6			Rahman et al. (1981)
DOPA	Rat, mouse and rabbit brain	Yanapak ODS-T (250 × 4)	0.1 M Phosphate/methanol (100/6.5, v/v)	0.7	0.6		Kato et al. (1981)
Norepinephrine	Human CSF, serum and brain	Yanapak ODS 10 μm (250 × 4)	0.1 M Potassium phosphate buffer (pH 3.0) with pentansulfonic acid (20 mg/100 ml buffer)	0.6	0.8	30 pmol	Matsui et al. (1981)

Table 2 (continued)

Substance	Sample	LC system			Detector		Reference
		Column*	Mobile phase	Flow rate (ml/min)	Voltage (V)	Sensitivity	
Catecholamines	Rat brain	Ultrasphere C$_{18}$ (250 × 4.6)	0.1 M Phosphate buffer (pH 2.0)/methanol (90/10, v/v) with 0.2 mM SDS and 0.1 mM EDTA	1.5	0.72		Hegstrand and Eichelman (1981)
Dopamine, HVA and DOPAC	Rat brain	Zorbax ODS 6—8 μm (250 × 2.1) and Zipax SCX 30—40 μm (1000 × 4.6)	Phosphate (0.1 M)—citrate (0.05 M) buffer (pH 4.3)/ methanol (90/10, v/v) and citrate (0.05 M)—acetate (0.1 M) buffer (pH 5.2) with 1 mM EDTA	0.4 and 0.6	0.61 and 0.85	DA 250 HVA 100 DOPAC 66 (pg)	Saraswat et al. (1981)
Norepinephrine and 3-methoxy-4-hydroxy-phenylethyleneglycol	Mouse brain and brain perfusate	μ-Bondapak C$_{18}$ 10 μm and (300 × 3.9) Lichrosorb RP-2 10 μm (250 × 4.6)	10 mM Phosphate buffer (ph 7.0)	1.0	0.8	MHPG 0.37 pmole	Towell and Erwin (1981)
Epinephrine	Rat brain	Nucleosil 7C$_{18}$ 7 μm (250 × 4)	0.1 M Sodium phosphate buffer (pH 2.6)/acetonitrile (99.5/0.5, v/v) with 5 mM sodium pentanesulfonate	0.9	0.6	0.5 pmol	Trocewicz et al. (1982a)
Epinephrine	Human brain	Nucleosil 7C$_{18}$ 7 μm (250 × 4)	0.1 M Sodium phosphate buffer (pH 2.6) with 5 mM sodium pentane sulfonate	0.9	0.6		Trocewicz et al. (1982b)
Monoamines and major metabolites	Rat CSF and brain	Lichrosorb RP-18 or RP-8 10 or 7 μm	0.1 M Phosphate buffer (pH 3.35)/methanol	1.0	0.9	less than 1.2—2.5 ng/	Wagner et al. (1982)

					injection	
	or Zorbax-ODS 5 μm (250 × 4.6)	(84/16, v/v) with 2.6 mM octane sulfonic acid (OSA), 0.1 M EDTA and 0.25 mM triethylamine (Et$_3$N) or 0.1 M phosphate buffer (pH 2.60)/methanol (80.5/19.5, v/v) with 2.75 mM OSA, 0.1 mM EDTA and 0.25 mM Et$_3$N or 0.15 M phosphate buffer (pH 2.70)/methanol (95/5, v/v) with 5 mM hexane sulfonic acid and 0.1 mM EDTA or 0.2 M phosphate buffer (pH 2.70)/methanol (89/11, v/v) with 2.5 mM heptane sulfonic acid and 0.1 mM EDTA				
Monoamines	Lichrosorb RP-18 10 μm (250 × 4.6)	0.05 M Phosphate buffer (pH 3.0) with 20 M EDTA and 1 mM sodium heptane—sulphonate	1.8	0.65	at least 200 pg	Warsh *et al.* (1982)
Catecholamines	Lichrosorb RP-18 5 μm or Nucleosil SA 5 or 10 μm (150 × 4.5)	Acetate (0.1 M)—citrate (0.04 M) buffer (pH 5.2)/methanol (90/10, v/v) with or without 3 mM 7,5-dimethylcyclohexyl sulfate	1.0	0.6–0.8		Eriksson and Persson (1982)
Catecholamines	Ultrasphere ODS 5 μm (250 × 4.6)	0.1 M Acetate—citrate buffer (pH 5.2)/methanol (78/22, v/v) with 5 mM sodium octane-1-sulphonate	1.2	0.6	30 pg/injection	Davies and Molyneux (1982)

Table 2 (continued)

Substance	Sample	LC system Column*	Mobile phase	Flow rate (ml/min)	Detector Voltage (V)	Detector Sensitivity	Reference
Free and conjugated catecholamines	Human urine and rat tissues	5 μm Octadecyl silica (250 × 4)	0.075 M Phosphate buffer (pH 2.8)/methanol (95/5, v/v) with 1 mM EDTA and 30 mg/1 sodium octyl sulfate	1.0—2.4	0.6—0.7		Elchisak and Carlson (1982)
DOPA, 5-S-cysteinyldopa and 5-S-glutathionedopa	Mouse melanoma and hair	Yanapak ODS-T 10 μm (250 × 4)	0.1 M Phosphate buffer (pH 2.1) at 45°C	0.7	0.5		Ito et al. (1983)
Catecholamines	Human plasma and urine	C₁₈ column	0.1 M phosphate buffer/methanol (89/11, v/v) with 20 mg/1EDTA (2Na) and 300 mg/1 HSS	1.0	0.8		Adachi and Mizoi (1983)
Monoamines and major metabolites	Rat brain	C₁₈ RCM 100 5 μm (8 mm ID)	0.1 M Acetate—0.1 citrate buffer (pH 3.7)/methanol (90/10, v/v) with 0.5 mM sodium octyl sulphate, 0.15 mM EDTA and 1 mM dibutylamine	1.0	0.8	NE 0.15 DA 0.28 5HT 0.43 MHPG 0.22 DOPAC 0.33 5HIAA 0.38 HVA 0.75 (pmols/injection)	Warnhoff (1984)
Catecholamines	Human plasma	Yanaco ODS-T 10 μm (250 × 4)	0.1 M Sodium citrate buffer 1.0 (pH 5.0)/acetonitrile (94/6) with 6 mM sodium	1.0	0.5 and 0.55	NE, E, DA 5 pg	Maruta et al. (1984)

Analyte	Sample	Column	Mobile phase	Flow rate	Potential	Detection limit	Reference
Neurotransmitters	Human CSF and rat nervous tissue	RP-18 Spheri 5 5 μm (350 × 2.1 and 100 × 4.6)	octanesulfonate and 2 mM disodium EDTA Sodium dihydrogen phosphate (0.1 M)/acetonitrile (92/8, v/v) with EDTA (0.08 mM) and n-octyl sodium sulphate (0.025 mM) (pH 3.5) and disodium hydrogen phosphate (6.7 mM)—citric acid (13.3 mM)/ methanol (87—85/13—15, v/v) with n-octyl sodium sulphate (2.5 mM) and EDTA (0.05 mM) (pH 3.3)	0.3 and 1.3	0.75 and 0.65		Honegger *et al.* (1984)
Biogenic amines and their metabolites	Rat brain	Ultrasphere ODS 3 μm or 5 μm (150 × 4.6 or 75 × 4.6)	0.1 M Monobasic sodium phosphate—citric acid (pH 4.0—4.35)/acetonitrile (100/8—11, v/v) with 1 mM disodium EDTA and 1 mM sodium octanesulfonic acid	1.0	0.8	20—150 pg/ 20 μl sample	Saller and Salama (1984)

1.3. Methoxyphenols

Analyte	Sample	Column	Mobile phase	Flow rate	Potential	Detection limit	Reference
3-Methoxy-4-hydroxyphenyl-ethyleneglycol	Human CSF	μ-Bondapak C$_{18}$ 10 μm (300 × 3.9)	Citrate—acetate buffer (pH 5.15)	2.0	0.75		Anderson *et al.* (1981)
3-Methoxy-4-hydroxyphenyl-ethyleneglycol and its sulfate conjugate	Human lumbar CSF	μ-Bondapak C$_{18}$ 10 μm (300 × 4.6)	0.1 M Phosphate buffer (pH 2.5) and methanol/ distilled water (3/2, v/v)	1.2	0.7 and 1.0	MHPG 50 pg	Krstulovic *et al.* (1981)

Table 2 (*continued*)

Substance	Sample	LC system Column*	Mobile phase	Flow rate (ml/min)	Detector Voltage (V)	Sensitivity	Reference
Amine metabolites	Human urine	Hypersil ODS 5 μm (150 × 4.6)	0.2 M Phosphate—citrate buffer (pH 2.7)/methanol (95 or 85/5 or 15, v/v)	1.0	0.72		Joseph et al. (1981)
Major monoamine metabolites	Primate brain	Hypersil ODS 5 μm (150 × 4.5)	0.1 M Phosphate buffer/methanol (92/8, v/v)		0.7	MHPG 120 HVA 250 (pg)	Cross and Joseph (1981)
3-Methoxy-4-hydroxyphenylglycol (free and conjugated)	Human urine	μ-Bondapak C$_{18}$ 10 μm (300 × 3.9)	0.009 M Acetate—citrate buffer (pH 5.1)/methanol (90/10, v/v)	1.0	0.8		Santagostino et al. (1982)
3-Methoxy-4-hydroxyphenyl-ethyleneglycol and 3-methoxy-4-hydroxy-mandelic acid	Plasma	Spherisorb-ODS 5 μm (250 × 3)	0.07 M Phosphate buffer (pH 2.5)/methanol (97/3, v/v) with 0.01 % EDTA	1.2	0.8	VMA 2 MHPG 1.2 (ng/ml)	Ong et al. (1982)
Normetanephrine, metanephrine and 3-methoxytyramine	Human urine	μ-Bondapak C$_{18}$ 10 μm (300 × 3.9)	0.02 M Phosphate—citrate buffer/methanol (100 or 90/0 or 10, v/v) with 2.5 mM sodium octylsulfonate and 50 μM EDTA	1.5	0.9		Jouve et al. (1983)
3-Methoxy-4-hydroxyphenylglycol	Human plasma and urine	Spherisorb ODS-2 3 μm (100 × 4.6)	0.01 M Acetate buffer (pH 4.0) with 1 mM EDTA	1.5	0.9	0.1 ng/injection	Shea and Howell (1984)

Compound	Sample	Column	Mobile phase			Detection limit	Reference
3-Methoxy-4-hydroxyphenylglycol sulfate ester	Rat brain	Lichrosorb RP-18 10 μm (250 × 4.6)	0.15 M Phosphate buffer (pH 3.5)/methanol (89/11, v/v) with 0.13 mM EDTA	1.0	0.85		Hornsperger et al. (1984)
3,4-Dihydroxy-phenylacetic acid, 5-hydroxyindole-3-acetic acid and 4-hydroxy-3-methoxy-phenyl acetic acid	Human plasma	Yanapak-ODS-A 7 μm (250 × 4.6)	0.1 M Phosphate buffer (pH 3.2)/methanol (82/18, v/v) with 10 μM EDTA (2Na)	1.2	0.6 and 0.75	DOPAC 200 5HIAA 200 HVA 800 (pg/ml)	Minegishi and Ishizaki (1984)
3-Methoxy-4-hydroxyphenyl-ethyleneglycol	Human urine	Ultrasphere-ODS 5 μm (250 × 0.46)	50 mM KH_2PO_4—H_3PO_4 (pH 2.5)/acetonitrile (88/12, v/v)	0.7	0.80		Shipe et al. (1984)
1.4. Estrogens Estriol	Human urine	Nucleosil C-18 10 μm (250 × 4.6)	$MeOH:H_2O:HClO_3$ (600:400:0.5) with 7.0 g of $NaClO_4 \cdot H_2O$	1.0	1.1	10 ng	Hiroshima et al. (1980)
2- and 4-hydroxylated estrone and estradiol	Human urine	μ-Bondapak C_{18} (300 × 3.9)	Acetonitrile/0.5 % ammonium dihydrogen phosphate (pH 3.0) (1/2.1, v/v)	1.0	0.8	1 ng	Shimada et al. (1981)
1.5. Tocopherols Tocopherols, ubiquinones and phylloquinone	Human serum and rat plasma	Nucleosil C-18 10 μm (250 × 4.6)	Ethanol/methanol (6/4, v/v)/70% $HClO_4$ (999/1, v/v) with 7 g of $NaClO_4 \cdot H_2O$ or methanol/pyridine (999/1, v/v) with 7 g of $NaClO_4 \cdot H_2O$ or ethanol/70% $HClO_4$ (999/1, v/v) with 7 g of $NaClO_4 \cdot H_2O$	0.44—1.56	0.9 or 0.7 or −0.3	TP 50 UQ 150—200 PQ 100 (pg)	Ikenoya et al. (1979)

Table 2 (continued)

Substance	Sample	LC system Column*	Mobile phase	Flow rate (ml/min)	Detector Voltage (V)	Sensitivity	Reference
1.6. Others							
β-Cetotetrine	Plasma or urine	Bondapak Phenyl Corasil 37—50 μm (2ft × 2 m ID)	0.025 M Phosphate buffer (93/7, v/v)	0.5—0.6	0.5	0.05—0.10 ng/ injection	Magic (1976)
Morphine	Human serum	Lichrosorb with octadecyl—silane bonded (300 × 4)	Methanol/0.01 M KH$_2$PO$_4$ (85/15, v/v)	1.0	1.0	1 ng/ml	Wallace et al. (1980)
Morphine	Mouse blood	Ultrasphere-ODS 5 μm (250 × 4.6)	0.075 M Citrate buffer (pH 3.75)/methanol/THF/acetic acid (81.5/10/2.5/6, v/v)	1.0	0.725		Ishikawa et al. (1982b)
3. Indoles							
3.1. Indolyl-3-derivatives							
Tryptophan and its metabolites	Human urine, serum and rat brain	μ-Bondapak C$_{18}$ 10 μm (300 × 4)	0.5 M Acetate buffer (pH 5.1)/methanol (85/15, v/v) and McIlwaine buffer (pH 4.0)/methanol (80/20, v/v)	1.0	0.50 and 1.0	5HIAA 1.0 / 5HT 1.0 / Trp 6.0 (ng/ml)	Koch and Kissinger (1979)
Tryptophan and metabolites	Rat brain and pineal tissue	Vydac 201 TP 10 μm (250 × 3.2)	0.1 M Sodium acetate— 0.1 M citric acid (pH 4.1)/methanol (100—75/0—25, v/v)	0.7	0.48— 0.91	5—20 pg	Mefford and Barchas (1980)

3.2. 5-Hydroxyindoles

Compound	Sample	Column	Mobile phase			Detection	Reference
5-Hydroxy-3-indoleacetic acid	Urine	Ultrasphere C$_{18}$ 5 μm (150 × 4.5)	0.01 M Phosphate buffer (pH 2.2)/methanol (83/17, v/v)	1.3	1.0		Shihabi and Scaro (1980)
Serotonin	Human serum and plasma	μ-Bondapak C$_{18}$ (300 × 4)	0.5 M Ammonium acetate buffer (pH 5.1)/methanol (85/15, v/v)	1.0	0.5	1.1 pg/ml or 6.5 pM	Koch and Kissinger (1980)
Serotonin	Rat and human brain and serum	Yanapak ODS-T 10 μm (250 × 4)	0.1 M Potassium phosphate buffer (pH 3.2)/methanol (90/10, v/v)	0.5	0.8	200 fmol	Rahman et al. (1980)
Serotonin	Human plasma	Spherisorb I ODS 5 μm (250 × 4.6)	0.1 M Sodium acetate trihydrate—acetic acid buffer (pH 4.2)/methanol (85/15, v/v)	1.2	0.6	250 fmol/ injection	Tagari et al. (1984)

4. Phenothiazines

Compound	Sample	Column	Mobile phase			Detection	Reference
Perphenazine and fluphenazine	Blood	Methyl-bonded silica (SI60) 6–7 μm (100 × 2.8)	Methanol/phosphate buffer [0.05 M, pH 7.40 (57/43, v/v) and pH 6.90 (53/47, v/v)] with 7 g/l of KCl			25 ng of fluphenazine/ ml of serum	Tjaden et al. (1976)
Promethazine and other phenothiazine compounds	Plasma and serum	MicroPak CN-10 10 μm (300 × 4)	0.02 M KH$_2$PO$_4$/aceto-nitrile (55/45, v/v)	2.0	0.9	0.2 g of promethazine/l	Wallace et al. (1981)
Chlorpromazine and levomepromazine	Human plasma and urine	Nucleosil C$_{18}$ 5 μm (150 × 4)	Pyridine/THF/acetonitrile/ 0.1 M acetate buffer (pH 3.5) (0.1/1.0/68.9/30.0, v/v) with 20 mM of NaClO$_4$	0.7	0.95	100 pg	Murakami et al. (1982)

Table 2 (continued)

Substance	Sample	LC system			Detector		Reference
		Column*	Mobile phase	Flow rate (ml/min)	Voltage (V)	Sensitivity	
5. Mercaptans							
Reduced glutathion	Guinea pig and rat tissue	Vydac CX (2.1 mm ID)	0.02 M H_3PO_4 (pH 2.1)	0.3	1.0	5×10^{-12} moles/injection	Mefford and Adams (1978)
Cysteine and homocysteine	Plasma and urine	Zipax SCX (500 × 2)	1% H_3PO_4 or phosphate—citrate buffer (pH 2.5)	0.5	0.1	cysteine 10^{-6} M	Saetre and Rabenstein (1978)
Penicillamine	Whole blood, plasma and urine	Zipax SCX 30 μm (300 × 4.1)	0.03 M Citric acid—0.01 M dibasic sodium phosphate buffer	2.5	0.1	10^{-7} M	Bergstrom et al. (1981)
6. Others							
Ascorbic acid	Urine and serum	Zipax SCX, Vydac SAX and Bondapak AX/ Corasil (500 × 2)	0.05—0.07 M Acetate buffer (pH 4.75)	0.30	0.7		Pachla and Kissinger (1976)
Uric acid	Amniotic fluid	Bondapak C_{18} 10 μm (300 × 4.6)	0.1 M K_2PO_4 (pH 2.50) and acetonitrile/distilled water (60/40, v/v)	1.2	0.8	picogram order	Krstulovic et al. (1979)
Ascorbic acid	Marine animals	Partisil 10 SAX (250 × 4.6)	60 mM Acetate buffer (pH 4.6)	2.9	0.75		Carr and Neff (1980)
Fatty acids, bile acids and prostaglandins (after derivatization)	Guinea pig plasma and human bile	Nucleosil C-18 10 μm (250 × 4.6)	Methanol/H_2O/$HClO_4$ (800—600/120—400/1) or methanol/H_2O/pyridine,	0.9—1.2	0.7 and 0.75	stearic acid 0.5 chenodeoxy-cholic acid 2	Ikenoya et al. (1980)

Compound	Sample	Column	Mobile phase			prostaglandin F$_{2\alpha}$ 2 (ng)	Reference
Sulfinalol hydrochloride	Human plasma and urine	μ-Bondapak NH$_2$ (300 × 3.9) or Zorbax NH$_2$ (250 × 4.6)	(900/100/1) with 0.05 M NaClO$_4$ Acetonitrile/0.2 M acetate buffer (pH 6.6) (93/7, v/v)	1.75	0.73	2.1 ng/ml	Park et al. (1981)
4,5-bis-(4-methoxyphenyl)-2-(2-hydroxymethyl-sulfinyl)imidazole	Human plasma and urine	Lichrosorb RP-18 10 μm (250 × 4.6)	Methanol/H$_2$O (1/1, v/v) with 0.01 M disodium hydrogen phosphate	2.0	1.0	10 ng/ml of sample	Krause (1981)
Bile acids (post-column derivatization)	Serum and bile	Radial-Pak A 10 μm (100 × 5)	0.3% Ammonium phosphate (pH 7.3)/acetonitrile/methanol (100/35/15 and 100/45/15)	1.0	0.1	20 pmol	Kamada et al. (1982)
Haloperidol and its reduced metabolite	Serum and plasma	μ-Bondapak CN 10 μm (300 × 3.9)	40 mM Phosphate buffer (pH 6.8)/acetonitrile (55/45, v/v)		0.9	0.5 ng/injection	Korpi et al. (1983)
o-Phthalaldehyde—mercaptoethanol derivatives of amino acids	Plasma and other biological materials	Hypersil ODS 5 μm (150 × 4.6)	0.1 M Sodium phosphate buffer (pH 7.0)/methanol (50:50, v/v), or 0.05 M sodium phosphate buffer (pH 5.5)/methanol (80/2) and (20/80)		0.4—0.7		Joseph and Davies (1983)
Coenzyme Q and Coenzyme Q H$_2$	Rat and guinea pig tissue	Nucleosil-C18 5 μm (150 × 4.6)	Methanol/ethanol/HClO$_4$ (500/500/1, v/v) with 0.7% NaClO$_4 \cdot$ H$_2$O	1.0	−0.3 and 0.7		Hiroshima et al. (1983)
Reducing carbohydrates	Serum and urine	LS212 or IEX220 (600 × 7.5, 1–3 columns)	Water	0.45 or 1.0	0.75	1 pmol (0.2 ng)	Watanabe and Inoue (1983)

Table 2 (continued)

Substance	Sample	LC system			Detector		Reference
		Column*	Mobile phase	Flow rate (ml/min)	Voltage (V)	Sensitivity	
Acetylcholine and choline	Rat brain	Bio-Sil ODS-5S (150 mm length)	0.01 M Sodium acetate–0.02 M citric acid (pH 5) with 4.5 mg/l sodium octyl sulfate and 1.2 mM tetra-methylammonium chloride	0.80	0.5	Ch 1 ACh 2 (pmol)	Potter *et al.* (1983)
Leucine- and methionine-enkephalin, cholecystokinin tetra-peptide and octa-peptide sulphate	Rat brain	RP-18 (150 × 2.1 or 125 × 4.6 or 150 × 3.0)	150 mM or 200 mM Phosphate buffer (pH 5.5)/ 1-propanol (88.5/11.5, v/v), or 40 mM phosphate buffer (pH 5.5)/1-propanol (84.5/ 15.1, v/v), or 300 mM phosphate buffer (pH 5.5)/ 1-propanol (85.5/14.5, v/v)	0.4 or 0.6 or 1.0	1.0	CCK-4 0.1 pmol	Sauter and Frick (1984)
3-Methoxy-4-hydroxy and 4-methoxy-3-hydroxy benzyl-amine	Rat tissues	Ultrasphere-ODS 5 μm (150 × 4.6)	0.1 M Sodium phosphate, 20 mM citric acid, 0.15 mM disodium EDTA, 2 mM sodium octane–sulfonic acid in 20% methanol (pH 3.2)	1.5	0.85	1 pmol/ injection	Nissinen and Männistö (1984)
o-Phthalaldehyde derivatives of amines	Rat brain	HS-3 C$_{18}$ 3 μm (100 × 4.6)	Gradient elution of citrate buffer and methanol	1.5	0.7	30–150 fmol	Allison *et al.* (1984)

* Name of stationary phase, particle size and column size (length × ID mm)

Abbreviations used in this table:

ACh = acetylcholine; CCK-4 = cholecystokinin-4; Ch = Choline; CSF = cerebrospinal fluid; DA = dopamine; DOPA = dihydroxyphenylalanine; DOPAC = 3,4-dihydroxyphenylacetic acid; E = epinephrine; EDTA = ethylenediamine tetraaacetic acid; 5HIAA = 5-hydroxyindoleacetic acid; 5HT = 5-hydroxytryptophan, serotonine; 5HTP = 5-hydroxytryptophan; HVA = homovanillic acid, 3-methoxy-4-hydroxyphenylacetic acid; MN = metanephrine; MHPG = 3-methoxy-4-hydroxyphenyl(ethylene)glycol (MOPEG); 3MT = 3-methoxytyramine; NE = norepinephrine; NM = normetanephrine; SDS = sodium dodecyl sulfate; THF = tetrahydrofuran; Trp = Tryptophan; Tyr = Tyrosine.

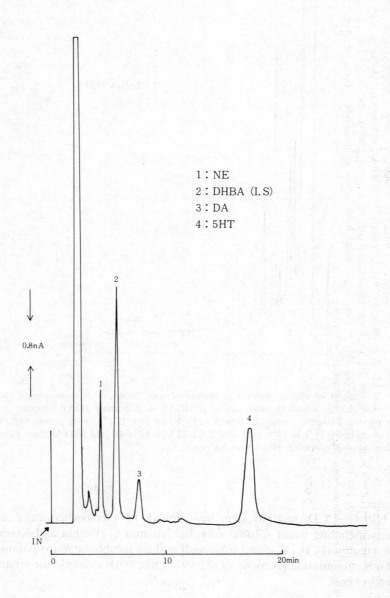

Figure 4. Analysis of catecholamine and serotonin in mouse whole brain. Peaks: 1. norepinephrine (NE); 2. dihydroxybenzylamine (DHBA, internal standard); 3. dopamine (DA); 4. serotonin (5-hydroxytryptamine, 5HT); column: Yanapak ODS-T (250 × 4.0 mm ID); mobile phase: 85% 0.1 M citric acid (3Na)—phosphoric acid buffer (pH 5.3)/14% methanol/1% THF/2.0 mM C_8SO_3Na/10 μM EDTA(2Na); flow rate: 1 ml/min; applied potential: +0.65 V vs Ag/AgCl.

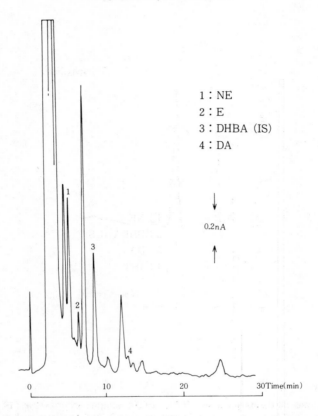

Figure 5. Analysis of catecholamines in human plasma. Peaks: 1. norepinephrine (NE); 2. epinephrine (E); 3. dihydroxybenzylamine (DHBA); 4. dopamine (DA). Sample: 3 ml of human plasma. Column: Yanapak ODS-A (250 × 4.6 mm ID). Mobile phase: 86% 0.1 M phosphate buffer (pH 3.3)/14% methanol/2.5 mM C_8SO_3Na/10 μM EDTA(2Na). Flow rate: 1 ml/min. Applied potential: +0.65 V vs Ag/AgCl.

Vitamins

The HPLC–ECD method can be applied to electrochemically active vitamins including water soluble vitamins [vitamin C (Pachla and Kissinger, 1976), vitamin B_2, B_6, B_{12}, and folic acid] and fat soluble vitamins [vitamin E, vitamin K, ubiquinone (Ikenoya *et al.*, 1979), etc.]. An example for vitamin E is given in Fig. 7.

Steroids

Estrogens such as estriol, estradiol, estrone and catecholestrogen are electrochemically active. Therefore, ECD can be effectively applied for monitoring the changes of estrogens during the process of pregnancy. Figure 8 shows data on estrogens in the blood of a 38-week pregnant woman.

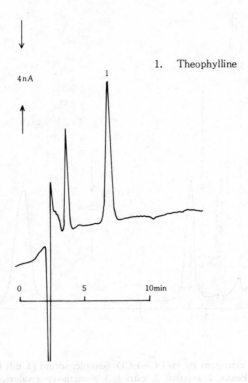

1.   Theophylline

Figure 6. Analysis of theophylline by HPLC—ECD. Sample: theophylline (10 ng). Column: Yanapak ODS—T (250 × 4.0 mm ID). Mobile phase: 85% 0.1 M citrate buffer (pH 4.5)/15% CH$_3$CN. Flow rate: 1 ml/min. Applied potential: +1.15 V vs Ag/AgCl.

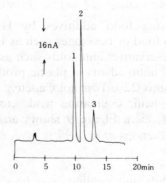

Figure 7. Analysis of tocopherols by HPLC—ECD. Sample: each 50 ng of tocopherols. Peaks: 1. δ-tocophenol; 2. γ-tocopherol and β-tocopherol; 3. α-tocopherol. Column: Yanapak ODS-T (250 × 4.0 mm ID). Mobile phase: 95% CH$_3$CN/5% 0.1 M NH$_4$Cl. Applied potential: +0.75 V vs Ag/AgCl.

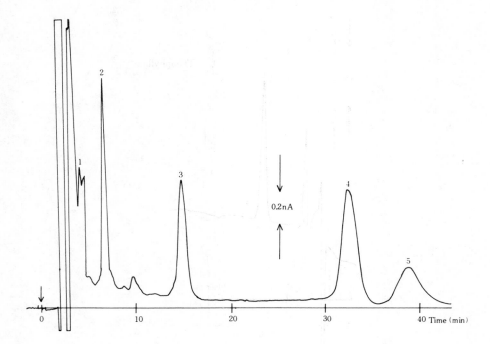

Figure 8. Analysis of estrogens by HPLC—ECD. Sample: serum (1 ml) from a woman at 38 weeks of pregnancy. Peaks. 1. estetrol; 2. estriol; 3. *p*-methoxy-*n*-valeric anilide; 4. estradiol; 5. estrone. Column: Yanapak ODS-T (250 × 4.0 mm ID). Mobile phase: 55% 0.1M phosphate buffer (pH 3.1)/30% CH_3CN/15% CH_3OH. Flow rate. 1 ml/min. Applied potential. +1.0 V vs Ag/AgCl.

Food additives

Analyses of the following food additives by HPLC—ECD have been reported: antioxidants in food or beverages such as BHA, BHT and vitamine E; *P*-oxybenzoate as preservative; antimolds such as OPP (*o*-phenylphenol); vanillin (flavour); phenol antioxidants in plastic products (BHT, Antioxidant ZKF, Antage W-300, Ionox 220, Topanol Ca, etc.); sterigmatocystin; phenol acids in beer (caffeine acid, coumarine acid, etc.); and others (toluene diamine, trichlorophenol, etc.). Figure 9 shows analytical data of BHA in fodder.

Amino acids and peptides

The following amino acids and peptides can be directly analyzed without derivatization: cysteine (with SH group) (Saetre and Rabenstein, 1978), tyrosine (with phenol group) and tryptophan (with indole ring); peptides (with cystein, tyrosine or tryptophan) such as glutathione and enkephalin. Analytical data for enkephalin is shown in Fig. 10.

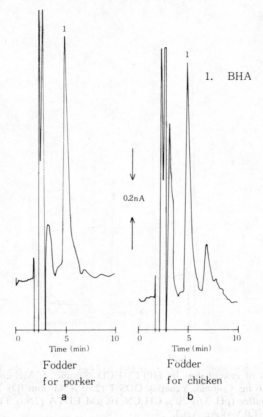

Figure 9. Analysis of tertiary-butyl-4-hydroxyanisole (BHA) by HPLC—ECD. Sample: 10 g of fodder. Column: Yanapak ODS-T (250 × 4.0 mm ID). Mobile phase: 30% 0.05 M phosphate buffer (pH 3.1)/70% CH₃CN. Flow rate: 0.8 ml/min. Applied potential: +0.8 V vs Ag/AgCl.

Indirect detection method of electrochemically inactive compounds by derivatization to electrochemically active compounds

Electrochemically inactive compounds can be analyzed by HPLC—ECD after their derivatization to electrochemically active compounds. Carboxylic acids such as fatty acid, bile acid and prostaglandins cannot be directly analyzed by HPLC—ECD, but can be analyzed with high sensitivity when labelled with *p*-aminophenol (Ikenoya *et al.*, 1980). In addition, various examples of analyses have been reported, including analysis by reacting a reducing sugar with a phenanthroline complex (Watanabe and Inoue, 1983), analysis of amino acids with their *o*-phthalaldehyde derivatives (Joseph and Davis, 1983) and a method in which bile acid is reacted with 3α-hydroxysteroid dehydrogenase in the presence of NAD to produce NADH (Kmada *et al.*, 1982). Derivatization can be carried out using enzymes. Acetylcholine and choline can be measured with acetylcholine esterase and choline oxidase as H_2O_2 (Potter *et al.*, 1983).

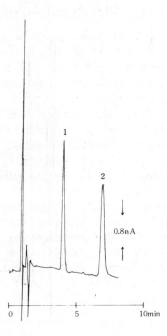

Figure 10. Analysis of enkephalins by HPLC—ECD. Sample: 1. Met-enkephalin 46 ng; 2. Leu-enkephalin 54.6 ng. Column: Yanapak ODS-T (250 × 4.0 mm ID). Mobile phase: 80% 0.1 M phasphate buffer (pH 3.6)/20% CH$_3$CN/10 μM EDTA (2Na). Flow rate: 1 ml/min. Applied potential: +1.0 V vs Ag/AgCl.

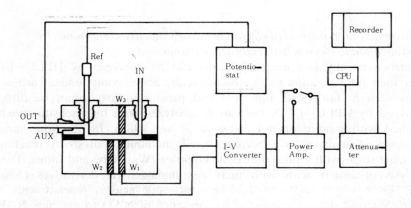

Figure 11. Construction of a twin-electrode electrochemical detector (VMD-501). W$_1$, W$_2$ and W$_3$ = glassy carbon working electrode (Ag/AgCl).

MULTI-ELECTRODE ELECTROCHEMICAL DETECTION

A detector with two or more working electrodes is more versatile than the three-pole detector. As a detector of this type, a block diagram of a Yanaco Model VMD-501 detector is shown in Fig. 11. This detector is provided with three working electrodes, two of which can be respectively set to a desired potential. Currents caused by electrode reactions on their surfaces can be recorded by a 2-pen recorder. Also, a signal obtained by addition or subtraction of the two currents can be recorded. Major advantages of the multi-electrode electrochemical detector are: higher sensitivity, improvement in detection specificity, simultaneous detection of oxidized and reduced substances, wider measuring range, and detection/measurement of products of electrode reaction.

Figure 12A shows signals obtained by applying +1200 mV to working electrode 1 (W_1) and +800 mV to working electrode 2 (W_2). When the oxidation potential is high as obtained with tyrosine, specific detection can be performed by subtraction of the signals as shown in Fig. 12B. Figure 13 shows an example of simultaneous analysis of oxidized and reduced substances. In the analysis, W_1 is set at -0.4 V and W_2 is set at $+1.0$ V with vitamin C, B_2 and B_6 as samples.

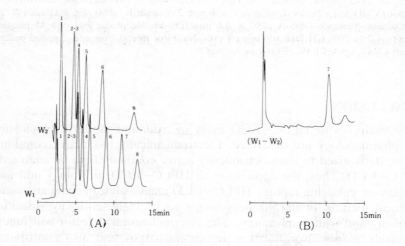

Figure 12. Analysis of catecholamines and the metabolities by HPLC with a twin-electrode electrochemical detector (VMD-501). Peaks: 1. norepinephrine (NE); 2. epinephrine (E); 3. dihydroxymandelic acid (DOMA); 4. dihydroxybenzylamine (DHBA); 5. normetanephrine (NM); 6. dopamine (DA); 7. tyrosine (Tyr); 8. vanillylmandelic acid (VMA). Samples: each 10 ng of standards. Column: Yanapak ODS-T (250 × 4.0 mm ID). Mobile phase: 0.1 M phosphate buffer (pH 3.1)/10 μM EDTA(2Na). Flow rate: 0.8 ml/min. Applied potential: W_1, +1200 mV vs Ag/AgCl; W_2, +800 mV vs Ag/AgCl. Sensitivity: 40 nA/ full scale.

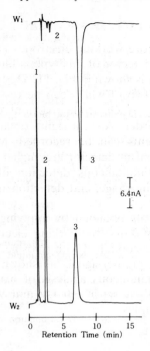

Figure 13. Analysis of vitamins by HPLC—ECD with a twin-electrode electrochemical detector (VMD-501). Peaks: 1. vitamin C (20 ng); 2. vitamin B_6 (160 ng); 3. vitamin B_2 (190 ng). Column: Yanapak ODS-A (250 × 4.6 mm ID). Mobile phase: 70% 0.1 M phosphate buffer (pH 3.1)/30% CH_3OH/400 mg/1 C_8SO_3 Na. Flow rate: 0.8 ml/min. Applied potential: W_1, −0.4 V vs Ag/AgCl; W_2, +1.0 V vs Ag/AgCl.

CONCLUSION

Some examples of HPLC—ECD assay in oxidation mode in biochemistry and pharmacology are described. Electrochemically less active compounds can be derivatized to electrochemically active compounds to be analyzed by HPLC—ECD. Thus, the application of HPLC—ECD in biology and medicine is now enlarging rapidly. HPLC—ECD analysis has become an essential chemical method in life sciences owing to its high sensitivity, specificity, simplicity and wide applications. The electrochemical detector will function effectively as a sensor for flow injection analysis and also contribute to development of a direct *in vivo* voltammetry.

REFERENCES

Adachi, J., and Mizoi, Y. (1983). Acetaldehyde-mediated alcohol sensitivity and elevation of plasma catecholamine in man. *Jpn J. Pharmacol. 33*, 531—539.
Allison, L. A., Mayer, G. S., and Shoup, R. E. (1984). *o*-Phthalaldehyde derivatives of amines for high-speed liquid chromatography/electrochemistry. *Anal. Chem. 56*, 1089—1096.

Anderson, G. M., Young, J. G., and Cohen, D. J. (1979). Rapid liquid chromatographic determination of tryptophan, tyrosine, 5-hydroxyindoleacetic acid and homovanillic acid in cerebrospinal fluid. *J. Chromatogr. 164*, 501—505.

Anderson, G. M., Young, J. G., Cohen, D. J., Shaywitz, B. A., and Batter, D. K. (1981). Amperometric determination of 3-methoxy-4-hydroxyphenylethyleneglycol in human cerebrospinal fluid. *J. Chromatogr. 222.* 112—115.

Bergstrom, R. F., Kay, D. R., and Wagner, J. G. (1981). High-performance liquid chromatographic determination of penicillamine in whole blood, plasma, and urine. *J. Chromatogr. 222,* 445—452.

Carr, R. S., and Neff, J. M. (1980). Determination of ascorbic acid in tissues of marine animals by liquid chromatography with electrochemical detection. *Anal. Chem. 52,* 2428—2430.

Cooper, M. J., O'Dea, R. F., and Mirkin, B. L. (1979). Determination of methyldopa and metabolites in human serum by high-performance liquid chromatography with electrochemical detection. *J. Chromatogr. 162,* 601—604.

Cross, A. J., and Joseph, M. H. (1981). The concurrent estimation of the major monoamine metabolites in human and non-human primate brain by HPLC with fluorescence and electrochemical detection. *Life Sci. 28,* 499—505.

Davies, C. L., and Molyneux, S. G. (1982). Routine determination of plasma catecholamines using reveersed-phase, ion-pair high-preformance liquid chromatography with electrochemical detection. *J. Chromatogr. 231,* 41—51.

Elchisak, M. A., and Carlson, J. H. (1982). Assay of free and conjugated catecholamines by high-performance liquid chromatography with electrochemical detection. *J. Chromatogr. 233,* 79—88.

Eriksson, B.-M., and Persson, B.-A. (1982). Determination of catecholamines in rat heart tissue and plasma samples by liquid chromatography with electrochemical detection. *J. Chromatogr. 228,* 143—154.

Fehér, Z., Nagy, G., Tóth, K., and Pungor, O. (1974). The use of precipitate based silicone rubber ion-selective electrodes and silicone rubber based graphite voltammetric electrodes in continuous analysis. *Analyst 99,* 699—708.

Felice, L. J., Felice, J. D., and Kissinger, P. T. (1978). Determination of catecholamines in rat brain parts by reverse-phase ion-pair liquid chromatography. *J. Neurochem. 31,* 1461—1465.

Greenberg, M. S., and Mayer, W. J. (1979). High-performance liquid chromatographic determination of theophylline and its derivatives with electrochemical detection. *J. Chromatogr. 169,* 321—327.

Hallman, H., Farnebo, L.-O., Hamberger, B., and Jonsson, G. (1978). A sensitive method for the determination of plasma catecholamines using liquid chromatography with electrochemical detection. *Life Sci. 23,* 1049—1052.

Hegstrand, L. R., and Eichelman, B. (1981). Determination of rat brain tissue catecholamines using liquid chromatography with electrochemical detection. *J. Chromatogr. 222,* 107—111.

Hiroshima, O., Ikenoya, S., Ohmae, M., and Kawabe, K. (1980). Electrochemical detector for high-performance liquid chromatography. III. Determination of estriol in human urine during pregnancy. *Chem. Pharm. Bull. 28,* 2512—1514.

Hiroshima, O., Ikenoya, S., Naitoh, T., Kusube, K., Ohmae, M., Kawabe, K., Ishikawa, S., Hoshida, H., and Kurahashi, T. (1983). Electrochemical detector for high-performance liquid chromatography. VI. Application of a twin electrochemical detector. *Chem. Pharm. Bull. 31,* 3571—3578.

Honegger, C. G., Burri, R., Langemann, H., and Kempf, A. (1984). Determination of neurotransmitter systems in human cerebrospinal fluid and rat nervous tissue by high-performance liquid chromatography with on-line data evaluation. *J. Chromatogr. 309,* 53—61.

Hornsperger, J.-M., Wagner, J., Hinkel, J.-P., and Jung, M. J. (1984). Measurement of 3-methoxy-4-hydroxyphenylglycol sulfate ester in brain using reversed-phase liquid chromatography and electrochemical detection. *J. Chromatogr. 306,* 364—370.

Ikenoya, S., Tsuda, T., Yamano, Y., Yamanishi, Y., Yamatsu, K., Ohmae, M., Kawabe, K., Nishino, H., and Kurahashi, T. (1978). Design and characterization of electrochemical detector for high-performance liquid chromatography and application to the determination of biogenic amines. *Chem. Pharm. Bull. 26.* 3530—3539.

Ikenoya, S., Abe, K., Tsuda, T., Yamano, Y., Hiroshima, O., Ohmae, M., and Kawabe, K. (1979). Electrochemical detector for high-performance liquid chromatography. II. Determination of tocopherols, ubiquinones and phylloquinone in blood. *Chim. Pharm. Bull. 27,* 1237—1244.

Ikenoya, S., Hiroshima, O., Ohmae, M., and Kawabe, K. (1980). Electrochemical detector for high-performance liquid chromatography. IV. Analysis of fatty acids, bile acids and prostaglandins by derivatization to an electrochemically active form. *Chim. Pharm. Bull. 28,* 2941—2947.

Inoue, K., Otake, T., Noda, K., Kyogoku, K., and Uchida, S. (1982). Determination of glutathione, D-penicillamine using a high performance liquid chromatography with a voltammetric detector. *Yakugaku Zasshi 102,* 659—665.

Ishikawa, K., Martinez, J. L. Jr., and McGaugh, J. L. (1982a). Simultaneous determination of morphine and monoamine transmitters in a single mouse brain. *J. Chromatogr. 231,* 255—264.

Ishikawa, K., McGaugh, J. L., Shibanoki, S., and Kubo, T. (1982b). a sensitive procedure for determination of morphine in mouse whole blood by high performance liquid chromatography with electrochemical detection. *Jpn J. Pharmacol. 32,* 969—971.

Ito, S., Maruta, K., Imai, Y., Kato, T., Ito, M., Nakajima, S., Fujita, K., and Kurahashi, T. (1982). Urinary p-aminobenzoic acid determined in the pancreatic function test by liquid chromatography, with electrochemical detection. *Clin. Chem. 28,* 323—326.

Ito, S., Homma, K., Kiyota, M., Fujita, K., and Jimbow, K. (1983). Characterization of structural properties for morphologic differentiation of melanosomes. III. Free and protein-bound dopa and 5-S-cysteinyldopa in B16 and Harding-Passey melanomas. *J. Invest. Dermatol. 80,* 207—209.

Joseph, M. H., Kadam, B. V., and Risby, D. (1981). Simple high-performance liquid chromatographic method for the concurrent determination of the amine metabolites vanillylmandelic acid, 3-methoxy-4-hydroxyphenylglycol, 5-hydroxyindoleacetic acid, dihydroxyphenylacetic acid and homovanillic acid in urine using electrochemical detection. *J. Chromatogr. 226,* 361—368.

Joseph, M. H., and Davies, P. (1983). Electrochemical activity of o-phthalaldehyde-mercaptoethanol derivatives of amino acids. Application to high-performance liquid chromatographic determination of amino acids in plasma and other biological materials. *J. Chromatogr. 277,* 125—136.

Jouve, J., Mariotte, N., Sureau, C., and Muh, J. P. (1983). High-performance liquid chromatography with electrochemical detection for the simultaneous determination of the methoxylated amines, normetanephrine, metanephrine and 3-methoxytyramine, in urine. *J. Chromatogr. 274,* 53—62.

Kamada, S., Maeda, M., Tsuji, A., Umezawa, Y., and Kurahashi, T. (1982). Separation and determination of bile acids by high-performance liquid chromatography using immobilized 3α-hydroxysteroid dehydrogenase and an electrochemical detector. *J. Chromatogr. 239,* 773—783.

Kato, T., Horiuchi, S., Togari, A., and Nagatsu, T. (1981). A sensitive and inexpensive high-performance liquid chromatographic assay for tyrosine hydroxylase. *Experientia 37,* 809—810.

Koch, D. D., and Kissinger, P. T. (1979). Determination of tryptophan and several of its metabolites in physiological samples by reversed-phase liquid chromatography with electrochemical detection. *J. Chromatogr. 164,* 441—455.

Koch, D. D., and Kissinger, P. T. (1980). Determination of serotonin in serum and plasma by liquid chromatography with precolumn sample enrichment and electrochemical detection. *Anal. Chem. 52,* 27—29.

Korpi, E. R., Phelps, B. H., Granger, H., Chang, W.-H., Linnoila, M., Meek, J. L., and Wyatt,

R. J. (1983). Simultaneous determination of haloperidol and its reduced metabolite in serum and plasma by isocratic liquid chromatography with electrochemical detection. *Clin. Chem. 29*, 624—628.

Krause, W. (1981). Determination of plasma and urine levels of a new antiinflammatory agent, 4,5-bis-(4-methoxyphenyl)-2-(2-hydroxymethylsulfinyl)imidazole, by high-performance liquid chromatography and electrochemical or ultraviolet detection. *J. Chromatogr. 222*, 71—79.

Krtulovic, A. M., Bertani-Dziedzic, L. M., Gitlow, S. E., and Lohse, K. (1979). Amnionic fluid uric acid levels determined by reversed-phase liquid chromatography with spectrophotometric and electrochemical detection. *J. Chromatogr. 164*, 363—372.

Krstulovic, A. M., Bertani-Dziedzic, L., Dziedzic, S. W., and Gitlow, S. E. (1981). Quantitative determination of 3-methoxy-4-hydroxyphenylethyleneglycol and its sulfate conjugate in human lumbar cerebrospinal fluid using liquid chromatography with amperometric detection. *J. Chromatogr. 223*, 305—314.

Magic, S. E. (1976). Determination of β-cetotetrine in plasma and urine using high-performance liquid chromatography with electrochemical detection. *J. Chromatogr. 129*, 73—80.

Magnusson, O., Nilsson, L. B., and Westerlund, D. (1980). Simultaneous determination of dopamine, DOPAC and homovanillic acid. Direct injection of supernatants from brain tissue homogenates in a liquid chromatography-electrochemical detection system. *J. Chromatogr. 221*, 237—247.

Maruta, K., Fujita, K., Ito, S., and Nagatsu, T. (1984). Liquid chromatography of plasma catecholamines with electrochemical detection after treatment with boric acid gel. *Clin. Chem. 30*, 1271—1272.

Matsui, H., Kato, T., Yamamoto, C., Fujita, K., and Nagatsu, T. (1981). Highly sensitive assay for dopamine-β-hydroxylase activity in human cerebrospinal fluid by high performance liquid chromatography-electrochemical detection: Properties of the enzyme. *J. Neurochem. 37*, 289—296.

Mefford, I., and Adams, R. N. (1978). Determination of reduced glutathione in guinea pig and rat tissue by HPLC with electrochemical detection. *Life Sci. 23*, 1167—1173.

Mefford, I. N., and Barchas, J. D. (1980). Determination of tryptophan and metabolites in rat brain and pineal tissue by reversed-phase high-performance liquid chromatography with electrochemical detection. *J. Chromatogr. 181*, 187—193.

Mefford, I. N., Ward, M. M., Miles, L., Taylor, B., Chesney, M. A., Keegan, D. L., and Barchas, J. D. (1981). Determination of plasma catecholamines and free 3,4-dihydroxyphenylacetic acid in continuously collected human plasma by high performance liquid chromatography with electrochemical detection. *Life Sci. 28*, 477—483.

Minegishi, A., and Ishizaki, T. (1984). Rapid and simple method for the simultaneous determination of 3,4-dihydroxyphenylacetic acid, 5-hydroxyindole-3-acetic acid and 4-hydroxy-3-methoxyphenylacetic acid in human plasma by high-performance liquid chromatography with electrochemical detection. *J. Chromatogr. 308*, 55—63.

Murakami, K., Murakami, K., Ueno, T., Hijikata, J., Shirasawa, K., and Muto, T. (1982). Simultaneous determination of chlorpromazine and levomepromazine in human plasma and urine by high-performance liquid chromatography using electrochemical detection. *J. Chromatogr. 227*, 103—112.

Nagatsu, T., Oka, K., and Kato, T. (1979a). Highly sensitive assay for tyrosine hydroxylase activity by high-performance liquid chromatography. *J. Chromatogr. 163*, 247—252.

Nagatsu, T., Yamamoto, T., and Kato, T. (1979b). A new and highly sensitive voltammetric assay for aromatic L-amino acid decarboxylase activity by high-performance liquid chromatography. *Anal. Biochem. 100*, 160—165.

Nissinen, E., and Männistö, P. (1984). Determination of catechol-O-methyltransferase activity by high-performance liquid chromatography with electrochemical detection. *Anal. Biochem. 137*, 69—73.

Oka, K., Kojima, K., Togari, A., Nagatsu, T., and Kiss, B. (1984). An integrated scheme for the simultaneous determination of biogenic amines, precursoramino acids, and related metabolites by liquid chromatography with electrochemical detection. *J. Chromatogr. 308*, 43—53.

Ong, H., Capet-Antonini, F., Yamaguchi, N., and Lamontagne, D. (1982). Simultaneous determination of free 3-methoxy-4-hydroxymandelic acid and free 3-methoxy-4-hydroxyphenylethyleneglycol in plasma by liquid chromatography with electrochemical detection. *J. Chromatogr. 233*, 97—105.

Pachla, L. A., and Kissinger, P. T. (1976). Determination of ascorbic acid in food stuffs, pharmaceuticals, and body fluids by liquid chromatography with electrochemical detection. *Anal. Chem. 48*, 364—367.

Park, G. B., Koss, R. F., O'Neil, S. K., Palace, G. R., and Edelson, J. (1981). Determination of sulfinalol hydrochloride in human plasma and urine by liquid chromatography with amperometric detection. *Anal. Chem. 53*, 604—606.

Potter, P. E., Meek, J., L., and Neff, N. H. (1983). Acetylcholine and choline in neuronal tissue measured by HPLC with electrochemical detection. *J. Neurochem. 41*, 188—194.

Rahman, M. K., Nagatsu, T., and Kato, T. (1980). New and highly sensitive assay for L-5-hydroxytryptophan decarboxylase activity by high-performacne liquid chromatography—voltammetry. *J. Chromatogr. 221*, 265—270.

Rahman, M. K., Nagatsu, T., and Kato, T. (1981). Determination of aromatic L-amino acid decarboxylase in serum of various animals by high-performance liquid chromatography with electrochemical detection. *Life Sci. 28*, 485—492.

Riggin, R. M., Alcorn, R. L., and Kissinger, P. T. (1976). Liquid chromatographic method for monitoring therapeutic concentrations of L-Dopa and dopamine in serum. *Clin. Chem. 22*, 782—784.

Saetre, R., and Rabenstein, D. L. (1978). Determination of cysteine in plasma and urine and homocysteine in plasma by high-pressure liquid chromatography. *Anal. Biochem. 90*, 684—692.

Saller, C. F., and Salama, A. I. (1984). Rapid automated analysis of biogenic amines and their metabolites using reversed-phase high-performance liquid chromatography with electrochemical detection. *J. Chromatogr. 309*, 287—298.

Santagostino, G., Frattini, P., Schinelli, S., Cucchi, M. L., and Corona, G. L. (1982). Urinary 3-methoxy-4-hydroxyphenylglycol determination using reversed-phase chromatography with amperometric detection. *J. Chromatogr. 233*, 89—95.

Saraswat, L. D., Holdiness, M. R., and Justice, J. B. (1981). Determination of dopamine, homovanillic acid and 3,4-dihydroxyphenylacetic acid in rat brain striatum by high-performance liquid chromatography with electrochemical detection. *J. Chromatogr. 222*, 353—362.

Sasa, S., and Blank, C. L. (1977). Determination of serotonin and dopamine in mouse brain tissue by high performance liquid chromatography with electrochemical detection. *Anal. Chem. 49*, 354—359.

Sauter, A., and Frick, W. (1984). Determination of neuropeptides in discrete regions of the rat brain by high-performance liquid chromatography with electrochemical detection. *J. Chromatogr. 297*, 215—223.

Shea, P. A., and Howell, J. B. (1984). High-performance liquid chromatographic method for determining plasma and urine 3-methoxy-4-hydroxyphenylglycol by amperometric detection. *J. Chromatogr. 306*, 358—363.

Shihabit, Z. K., and Scaro, J. (1980). Liquid-chromatographic assay of urinary 5-hydroxy-3-indoleacetic acid, with electrochemical detection. *Clin. Chem. 26*, 907—909.

Shimada, K., Tanaka, T., and Nambara, T. (1981). Studies on steroids CLXV. Determination of isomeric catechol estrogens in pregnancy urine by high-performance liquid chromatography with electrochemical detection. *J. Chromatogr. 223*, 33—39.

Shipe, J. R., Savory, J., and Wills, M. R. (1984). Improved liquid-chromatographic determination of 3-methoxy-4-hydroxyphenylethyleneglycol in urine with electrochemical detection. *Clin. Chem. 30*, 140—143.

Tagari, P. C., Boullin, D. J., and Davies, C. L. (1984). Simplified determination of serotonin in plasma by liquid chromatography with electrochemical detection. *Clin. Chem. 30*, 131—135.

Tjaden, U. R., Lankelma, J., Poppe, H., and Muusze, R. G. (1976). Anodic coulometric detection with a glassy carbon electrode in combination with reversed-phase high-performance liquid chromatography. Determination of blood levels of perphenazine and fluphenazine. *J. Chromatogr. 125*, 275—286.

Towell, J. F., and Erwin, V. G. (1981). Determination of the primary metabolite of central nervous system norepinephrine, 3-methoxy-4-hydroxyphenethyleneglycol, in mouse brain and brain perfusate by high-performance liquid chromatography with electrochemical detection. *J. Chromatogr. 223*, 295—303.

Trocewicz, J., Oka, K., and Nagatsu, T. (1982a). Highly sensitive assay for phenylethanol-amine *N*-methyltransferase activity in rat brain by high-performance liquid chromatography with electrochemical detection. *J. Chromatogr. 227*, 407—413.

Trocewicz, J., Oka, K., Nagatsu, T., Nagatsu, I., Iizuka, R., and Narabayashi, H. (1982b). Phenylethanolamine *N*-methyltransferase activity in human brains. *Biochem. Med. 27*, 317—324.

Wagner, J., Vitali, P., Palfreyman, M. G., Zraika, M., and Huot, S. (1982). Simultaneous determination of 3,4-dihydroxyphenylalanine, 5-hydroxytryptophan, dopamine, 4-hydroxy-3-methoxyphenylalanine, norepinephrin, 3,4-dihydroxyphenylacetic acid, homovanillic acid, serotonine, and 5-hydroxyindoleacetic acid in rat cerebrospinal fluid and brain by high-performance liquid chromatography with electrochemical detection. *J. Neurochem. 38*, 1241—1254.

Wallace, J. E., Harris, S. C., and Peek, M. W. (1980). Determination of morphine by liquid chromatography with electrochemical detection. *Anal. Chem. 52*, 1328—1330.

Wallace, J. E., Shimek, E. L. Jr., Stavchansky, S., and Harris, S. C. (1981). Determination of promethazine and other phenothiazine compounds, by liquid chromatography with electrochemical detection. *Anal. Chem. 53*, 960—962.

Warnhoff, M. (1984). Simultaneous determination of norepinephrine, dopamine, 5-hydroxy-tryptamine and their main metabolites in rat brain using high-performance liquid chromatography with electrochemical detection. Enzymatic hydrolysis of metabolites prior to chromatography. *J. Chromatogr. 307*, 271—181.

Warsh, J. J., Chiu, A., and Godse, D. D. (1982). Simultaneous determination of norepine-phrine, dopamine and serotonin in rat brain regions by ion-pair liquid chromatography on octyl silane columns and amperometric detection. *J. Chromatogr. 228*, 131—141.

Watanabe, N., and Inoue, M. (1983). Amperometric detection of reducing carbohydrates in liquid chromatography. *Anal. Chem. 55*, 1016—1019.

Zoutendam, P. H., Bruntlett, C. S., and Kissinger, P. T. (1976). Determination of homogentisic acid in serum and urine by liquid chromatography with amperometric detection. *Anal. Chem. 48*, 2200—2202.

Progress in HPLC, Vol. 2, pp. 37–51
Parvez *et al.* (Eds)
© 1987 VNU Science Press

Developments in electrochemical detection for liquid chromatography

SAM A. McCLINTOCK and WILLIAM C. PURDY

McGill University, Department of Chemistry, 801 Sherbrooke St. W., Montreal, Quebec, Canada, H3A 2K6

INTRODUCTION

The growth of high-performance liquid chromatography (HPLC) as an analytical technique has led to continued research into both the separation process itself and the development of more efficient detection systems. Due to recent advances in column technology the detection system is now often the limiting component in many HPLC analyses. Initial attempts to develop the universal detector have all but been abandoned in favor of developing specific detectors for specific classes of compounds (Vickery, 1983). Many biological compounds function *in vivo* as a result of electron transfer processes; it is therefore only a logical extension to assume that these same processes can be monitored by utilizing this 'electron transfer' process, i.e. measurement by electrochemical means. Electrochemical detectors (ECD) take advantage of the electrical properties of many substances and, due to their inherent sensitivity, now represent a substantial proportion of all HPLC detectors. The initial commercial success of these systems has prompted not only the investigation of other electrochemical strategies as suitable for HPLC detection but has revived an interest in electroanalytical techniques generally as a viable analytical alternative to existing procedures (Kissinger and Heineman, 1984). In this chapter we will first take a brief look at the development of electrochemical detectors as a means of establishing the state of commercially available systems. Having established a basis for comparison we will take what might be considered to be a somewhat subjective look at the more promising developments in electrochemical detection and speculate on their promise for expanding the capabilities of liquid chromatography.

Electrochemical detectors function in the following manner: a potential of sufficient magnitude is applied to a working electrode, usually made of carbon, over which the analyte passes as it exits from the HPLC column, forcing oxidation or reduction to take place. The resulting flow of electrons to or from the working electrode can be measured as a current, the magnitude of which is proportional to the amount of analyte passing through the detector. Attempts to optimize detector design have resulted in the development of exact solutions for the limiting current produced in the three main detector designs: the tubular electrode (Blaedel and Klatt, 1966), the wall-jet electrode (Yamada and Matsuda, 1973) and, the most commercially successful, the thin-layer electrode (Weber and Purdy, 1979). Recently, attention has been focused on the inclusion of more than one working electrode in the electrochemical detector (Roston et al., 1982). These have become known as dual-working — or multiple-working — electrode detectors. They can be classified into three distinct groups based on the position of the working electrodes in the cells as well as variations in the use of each type. However, the ultimate aim is either to improve the selectivity or sensitivity of a particular chromatographic measurement.

SERIES-DUAL-ELECTRODE DETECTORS

The most common type and original design (Blank, 1976), is the series-dual-electrode (ECD) in which one electrode is placed downstream from the other (Fig. 1). This configuration has been exploited in several ways, but mainly the first electrode acts as a screen for the second electrode removing unwanted species. A series coulometric—amperometric detector was utilized to remove interfering species that were more easily oxidized than the species of interest (Schieffer, 1980), effectively removing them from the stream before the amperometric cell. In a similar configuration, Hepler and Purdy (1980), maintained the potential of a coulometric cell low enough to prevent oxidation of the species of interest, but succeeded in oxidizing some of the interfering species in the chromatographic mobile phase reducing the background current and ultimately the detection limit for the species.

Most of the work in this area deals with the placement of two working amperometric electrodes in one conventional thin-layer cell. This technique has been used to advantage where the number and response of the electroactive species detected by the second electrode can be manipulated by the potential at which the first electrode is held. The number of species detected at the second electrode can be reduced because the products of the first electrochemical conversion must be electroactive and stable until they reach the second electrode. It is also possible to use the first electrode to produce or 'generate' a species that can be detected at the second electrode. This technique was used to convert disulfides to thiols at -1.1 V and then detect the thiols formed at a potential of $+0.15$ V (Allison and Shoup, 1983). This analysis was carried out at two gold amalgam electrodes and is reported to have increased the detection limit for the species due to the much lower background current produced at the second electrode. Series amperometric detection has also been used (Roston and Kissinger, 1982) to obtain current—potential data on eluting species. By changing the potential of each electrode in increments, anodic and cathodic current—potential curves can be constructed. This current—potential dependence can be used to assist in sample identification by comparing the response of a standard and an

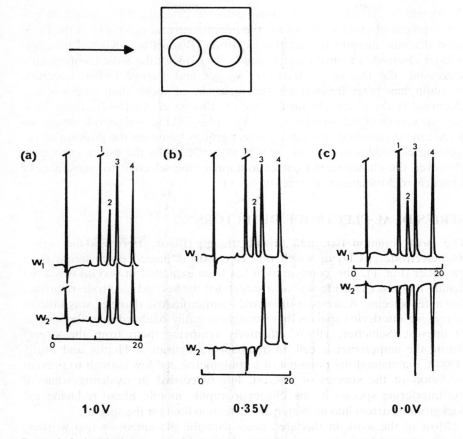

Figure 1. Series dual-electrode chromatograms for four phenolic acids: 1. 0.04 nmol of 4-hydroxybenzoic acid; 2. 0.02 nmol of vanillic acid; 3. 0.07 nmol of caffeic acid; 4. 0.06 nmol of 4-methylcatechol. Conditions: (a) W_1 = +1.10 V, W_2 = +0.95; (b) W_1 = +1.10 V, W_2 = +0.35 V; (c) W_1 = +1.10 V, W_2 = 0.0 V; 25 cm Biophase C_{18} column; 1 ml/min. (Reprinted with permission from *Anal. Chem.* **54**, 429—434, 1982)

unknown under similar conditions. Using retention times, electrochemical reactivity and dual-electrode collection efficiency Mayer and Shoup (1983) showed that what was thought to be a pure peak, was composed of two electroactive metabolites. At the second electrode only the metabolite of interest was detectable.

PARALLEL-DUAL-ELECTRODE DETECTORS

Electrochemical detectors have also been constructed by using two electrodes placed parallel and adjacent to each other (Fig. 2) in the flow stream (Roston and Kissinger, 1982). This configuration has been used to provide estimates of peak purity by using current ratios. However, it should be noted that

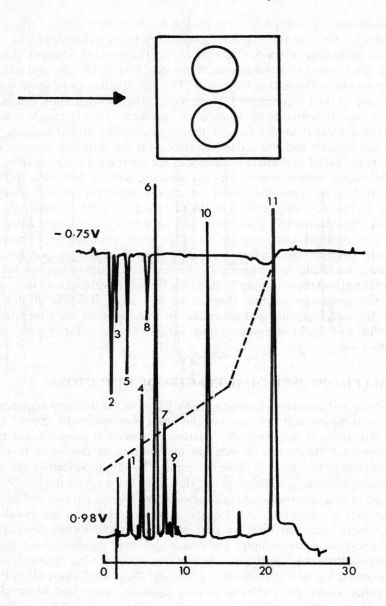

Figure 2. Simultaneous oxidation and reduction chromatograms of priority pollutant phenols using parallel-adjacent glassy carbon electrodes at −0.75 and +0.98 V vs Ag/AgCl. Peak identification: 1. phenol; 2. 2,4-dinitrophenol; 3. *p*-nitrophenol; 4. *o*-chlorophenol; 5. 4,6-dinitro-*o*-cresol; 6. 2,4-dimethylphenol; 7. 4-chloro-3-methylphenol; 8. *o*-nitrophenol; 9. 2,4-dichlorophenol; 10. 2,4,6-trichlorophenol; 11. pentachlorophenol. Conditions: linear gradient from 30% CH_3CN/70% 0.05 M $NaClO_4$, 0.005 M citrate pH 3.8 to 50/50% over 15 min, then to 80% CH_3CN/20% buffer at 20 min. Flow rate: 2 ml/min. Column: Biophase ODS 5 μm (250 × 4.6 mm). (Reprinted with permission from *Anal. Chem.* 54, 1417A–1434A, 1982.)

oxidation and reduction half-wave potentials are more indicative of a particular species. Current ratios can be affected by many parameters and should be used with care in species identification. The parallel-adjacent electrode design can be used to advantage in monitoring both oxidizable and reducible species in one solution simultaneously. This has similarities in some respects to the use of dual wavelength UV detectors. This configuration can also be used in the differential mode to enhance specificity. One electrode is held at a potential located at the foot of the hydrodynamic voltammogram for a particular species and the second electrode on the diffusion plateau. If the currents produced at these two electrodes are subtracted electronically, there will be signal enhancement for any species with a half-wave potential between these two potentials and a reduction in signal for species outside this region. In this way the potential of the two electrodes can be manipulated to increase the selectivity of a detector for a particular electroactive compound. This configuration also appears to offer some promise as a means of reducing baseline noise (Brunt *et al.*, 1981). The designs we have just discussed are available commercially; however, their utility has not yet been fully realized and their uses are limited only by the inventiveness of the user.

In the remainder of this chapter we will look at some of the more promising developments and speculate on their promise for expanding the capabilities of liquid chromatography with electrochemical detector in the future.

PARALLEL-OPPOSED DUAL-ELECTRODE DETECTORS

Microbore and capillary chromatography have placed stringent requirements on both pumping and detector systems for chromatography (Scott, 1984). In electrochemical detectors, the current produced is proportional to the flow-rate, and the current is substantially reduced at the lower flow rates encountered in this type of chromatography. Signal amplification associated with dual-electrode, parallel-opposed detectors helps offset this problem. In this type of detector two working electrodes are located parallel and opposite one another in a thin-layer cell. For this detector to function successfully, the species being detected must be part of an electrochemically reversible or quasi-reversible redox couple. The electrodes must be held close together and at potentials that are independent of one another. One electrode is held at a potential where oxidation takes place and the second electrode is held at a potential where the products of the oxidation at the first electrode are reduced to the starting material, making it available for re-oxidation. Then, as seen in Fig. 3, for each conversion that takes place, the current produced at each electrode will be inceased. This increase in current will depend on the number of conversions that take place; the thinner the flow cell and the slower the flow rate, the more efficient the conversion process will be.

This technique was developed, for stationary solutions in the early 1960s by Anderson and Reilley (1965a, b). It was shown that species could be ᶜorced to diffuse continuously between two electrodes in quiet solutions,

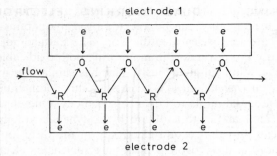

Figure 3. Electrode 1 is held at an anodic potential and Electrode 2 at a cathodic potential. Following the initial electrochemical conversion, the analyte can be converted back and forth from the oxidized to the reduced form producing current with each conversion. The more conversions, the greater the current produced for a given analyte concentration. (Reprinted with permission from *Internatl. Lab. 14*, 70—77, 1984.)

producing an infinite amount of current. This technique may still find some validity, using a stop-flow technique, for the detection of very low levels of species separated by chromatographic procedures.

The first attempts at using such a system as a detector for HPLC were of limited success (Fenn *et al.*, 1978). More recently, some predictions by Weber and Purdy (1978) based on a mathematical model suggested that construction of a functioning device was feasible. An amperometric device that showed current amplification of up to ten times, depending on the flow rates employed, has been developed and demonstrated in the analysis of catecholamines by McClintock and Purdy (1981, 1983) (see Fig. 4). An alternate cell design but similar in concept by Goto *et al.* (1983) has been shown to give current amplifications of up to 19 times for flow rates in the region of 1 µl/min. Since this device was operated at very slow flow rates, the coulometric model of Weber and Purdy was applied, and good agreement between theory and experiment was found.

As pointed out previously, the major criterion for such a detector design is that the electrodes be held as close to one another as possible so that the maximum number of conversions can be induced to take place. Advances in this area will undoubtedly come with the use of different insulating material to separate the electrodes. Sandwich-cell construction using insulating material such as silicon dioxide, and incorporation of the auxiliary and reference electrodes inside the flow-cell as alternate layers to help reduce iR drop, have shown promise. It should be possible with currently available technology to produce cells in the 2—5 µm thickness range, which would not only permit more conversions to take place than are presently possible but would also allow greater electrode areas to be used while maintaining small, fixed, flow-cell volumes. Cells with spacer thicknesses of less than 2 µm would probably have problems with clogging, contact between the electrodes or electrode 'cross-talk'.

ONE WORKING ELECTRODE

DUAL WORKING ELECTRODES

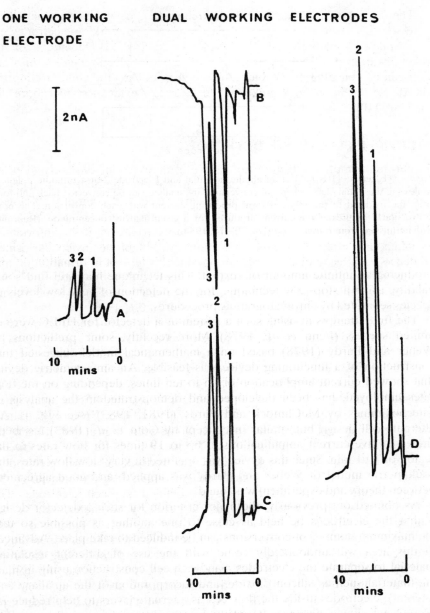

Figure 4. Chromatograms of a test mixture of: 1. 0.5 ng of norepinephrine; 2. 1 ng of epinephrine; 3. 1 ng of 3, 4-dihydroxybenzylamine. Flow rate: 1.0 ml/min; mobile phase: 95/5 0.1 M phosphate buffer pH 3.0/methanol with 90 mg/L sodium octyl sulfate; column C_{18} (150 × 4.6 mm). (A) Anodic current in the single-working electrode mode, $E_1 = 0.7$ V vs Ag/AgCl; (B) Cathodic current in the dual-working electrode mode $E_2 = -0.1$ V vs Ag/AgCl; (C) Anodic current in the dual-working electrode mode, $E_2 = +0.7$ V vs Ag/AgCl; (D) Total current in the dual-working electrode mode. (Reprinted with permission from *Anal. Chim. Acta 148*, 127–133, 1983.)

The flexibility of parallel-opposed, dual-working-electrode electrochemical detectors has been proven. The signal from the two working electrodes can be displayed separately, added together or subtracted. Depending on the potential applied to each electrode, various types of electrochemical information can be collected. In addition, these detectors have the ability to improve the detection limits for regular and to a greater extent for microbore and capillary HPLC.

MULTIPLE-ELECTRODE DETECTORS

In chromatographic detection not only are increases in sensitivity of prime concern, it is also important to be able to identify the analyte as it passes through the detector. This is particularly true with complex biological samples in which the identity of many of the components are unknown. Any information that can be used to identify the analyte, even if this information is taken under less than ideal conditions, would prove invaluable. Linear voltage-sweep techniques, which can provide this type of information in quiet solutions, are plagued by capacitative current in flowing streams. We will look at several approaches that are currently being investigated in an attempt to circumvent the problems of charging current and iR drop.

One approach to overcoming this problem is to extend the in-series dual-working-electrode concept to a multi-electrode system with as many as 32 small electrodes incorporated in a single flow cell. Dual-working-electrode electrochemical cells have been compared to dual-wavelength spectrophotometric detectors. Extending this analogy, multiple-electrode systems can be considered as the electrochemical equivalent of scanning-diode array detectors which are at present making a large impact on HPLC spectrophotometer detector design (Fell and Scott, 1983). It would seem logical that much of the work dealing with the manipulation of data from diode arrays is directly applicable to these systems. In multiple electrode devices each electrode is poised at some known potential different from its neighbor, ensuring that unique electrochemical information is obtained at each electrode. This information, when collected and reconstructed by computer, can be displayed as a simulated hydrodynamic voltammogram, a simulated cyclic voltammogram, or other simulated electrochemical technique, depending on the electrode potential pattern chosen. Two cell and controlling electronic designs (Roe and Ho, 1984; DeAbreu *et al.*, 1985), quite different in approach, although still in the developmental stages, show promise and may prove to be the electrochemical equivalent of the scanning-diode array spectrophotometers that are now becoming so popular.

PULSED POTENTIAL DETECTORS

Recently there has been renewed interest in detectors where the potential applied is a square-wave pulse superimposed on a slow linear ramp (Swartfzager, 1976; Osteryoung *et al.*, 1981), and on the design of controlling electronics for HPLC detectors (Eccles and Purdy, 1985). As with other pulse techniques, current is sampled just before a voltage pulse is applied and at some time near the end of the pulse, as charging current decays much more quickly than faradaic current the adverse effects of this charging current are minimized. Thus, if this square-wave pulse technique is superimposed on a linear voltage ramp, the ramp is repeated a number of times as a peak from the HPLC passes through the detector, and the information is summed. Then this information can be used not only to quantify but to identify the compound. It should also be possible to extend this square-wave concept to increase the sensitivity of single-working-electrode electrochemical detectors in the following manner. If a square-wave pulse of sufficient amplitude is applied to the electrode such that it covers the potential range in which both the oxidation and reduction potentials for a particular electrochemical reversible reaction are found, then the molecule can be made to oxidize and reduce in a cyclic manner many times before it exists the flow-cell. This technique is an alternative approach, using a conventional flow-cell, to the electrochemical amplification procedure described earlier using a parallel-opposed dual-electrode detector.

COULOSTATIC DETECTORS

Coulostatic HPLC voltammetric detectors have also shown promise in providing quantitation and also identification of HPLC effluents. Originally the coulostatic technique was developed in the early 1960s by Delahay (1962a, b) and Reinmuth (1962); a more recent review by van Leeuwen (1982) covers the theory and application of this technique. The coulostatic technique provides an alternate approach to discriminating against double-layer charging currents and iR drop. Coulostatics is based on the rapid injection of a charge pulse, more rapid than the electrode/analyte transfer rate, into the double-layer of the working electrode in an electrochemical flow cell. This charge pulse changes the double-layer capacitance. If the working electrode in the cell is disconnected after the charge injection, the only way for the charge to leave the double-layer is by a faradaic reaction. It follows that the charge will be removed more quickly for high concentrations of analyte and that the electrode will decay back to its equilibrium value more quickly. Because no current passes through the cell between charge injections, and because the double-layer charging occurs only during the time when the pulse is applied, coulostatic methods are free from the interferences of iR drop and double-layer charging current (Last, 1983; Reiss and Nieman, 1983; Barnes and Nieman, 1983). This offers certain advantages over existing electrochemical detectors.

Two major approaches have been used, single-potential coulostatic-pulse amperometry (CPA) and scanned-potential CPA. In single-potential CPA, small charge pulses are injected into the cell to maintain it at some equilibrium value. The total charge injected with time is current, which is proportional to the amount of electroactive species passing over the electrode surface; this is analogous to traditional d.c. amperometric techniques. In scanned-potential CPA, the potential is changed in small increments and the amount of charge required to bring the cell to this potential value is injected. In this case, a plot of the current versus potential yields a voltammogram free of charging current and iR drop. This voltammogram can be used not only for species identification but can also be used to resolve chromatographic peaks overlapped in the time domain. The associated instrumentation is, as one might expect, fairly complex and by necessity computer-controlled.

Systems capable of scanning the potential range at rates up to 3 V s^{-1} and of recording multiple channel chromatograms have been developed (Last, 1983). Although their present detection limits are approximately an order of magnitude poorer than those for conventional amperometric electrochemical detectors, these devices show promise for future development. This technique might be applied to the parallel-opposed thin-layer design discussed earlier using the single-potential mode by maintaining equilibrium potentials at two separate electrodes, both of which are supplying each other with electroactive products. This may be a way of overcoming the iR drop and the background current effects associated with this type of design, while at the same time increasing the detection limit of coulostatic voltammetric HPLC detectors.

A major application for devices of this type will be as detectors for normal phase chromatography (Mikkelsen *et al.*, 1985). Mobile phases in this type of chromatography are made up of non-conducting organic solvents and for this reason little work has been done to develop electrochemical detection methods. As the coulostatic technique does not depend on the conduction of current through the cell this type of detector may well prove useful as a detector for normal-phase HPLC.

ELECTRODE MATERIALS AND SOLVENTS

Although a number of electrochemical quantities may be measured using various techniques, the electrode material and the solvent in which the electrochemical measurement is carried out exert a powerful influence on detector behavior. The electrode material can determine and improve the electrode kinetics or lower the potential required to cause an electrochemical reaction to take place. This results in a more favorable signal-to-noise ratio and lowers detection limits for certain compounds in detectors of conventional design. Electrochemistry in nonaqueous solutions is a technique widely used by electro-organic chemists and a large body of knowledge exists on the electrochemical activity of many organic compounds in non-aqueous media (Mann and Barnes, 1070). Non-aqueous mobile phases, such as acetonitrile, DMF and DMSO, can be used with couloststic techniques or, in combination with alkylammonium salts, detectors can be operated in the conventional amperometric mode. The use of totally non-aqueous solvents in the conventional amperometric mode can provide a much wider electrochemical operating window, e.g. the useful operating range in aqueous solution is approximately +1.5 to −1.0 V while in dry acetonitrile the limits can extend from +3.0 to −3.0 V. Two major problems must be overcome, *iR* drop and the removal of water from the solvents. Even trace levels of water substantially reduce the useful potential window and increase background current fluctuations. It should be noted that due to the nature of the solution electrochemical reactions are often quite different from those in aqueous media. This could be used to advantage in the detection of molecules that have proven difficult to detect and/or separate by reversed-phase HPLC.

The most popular electrode material is carbon. In the original electrochemical cells carbon paste was used; this was followed by a series of other more mechanically stable carbon/binder combinations employing silicon rubber, polyethylene, polypropylene, Teflon and Kel-F. Various forms of solid carbon including glassy carbon, low temperature isotropic carbon, pyrolitic graphite, and reticulated vitreous carbon, have been used. A summary of carbon electrodes has been made elsewhere (Chesney *et al.*, 1981). Although carbon has been the most commonly employed material for oxidations, as its limitations are reached the use of specific electrode materials for specific purposes will become more widespread. For example, working electrodes of nickel oxide (Hui and Huber, 1982), and copper (Kok *et al.*, 1982), have been used for the detection of amines and amino acids, and other electrode materials may find application to a wide range of analytes. Although various metals may be employed as electrode materials the next big advance in electrode selectivity and sensitivity is likely to be achieved through chemical modification of electrode surfaces, a subject that has received a good deal of attention recently (Murray, 1980, 1984; and Faulkner, 1984). Preliminary studies in our laboratory have shown that carbon surfaces can be modified by polymeric films to facilitate the electrochemistry of certain drugs (Turk *et al.*, 1985).

POST-COLUMN REACTORS

Up to this point we have discussed possibilities for enhancing detection by manipulation of the electrochemical technique and the conditions under which the electrochemistry is carried out. An alternate approach is to take advantage of the inherent sensitivity of electrochemical detection by converting species that have no electrochemical activity under normal operating conditions into a product or products that can be seen by the ECD. To date most post-column reactions have been carried out by mixing the effluent from the HPLC with a second flow stream which contains the reactant. After the desired reaction takes place the products formed are then detected (Frei and Lawrence, 1981; Apffel *et al.*, 1983). This technique works well with electrochemical detection (Kissinger *et al.*, 1979; Shimada *et al.*, 1983) but has some disadvantages; the addition of the second liquid stream causes dilution of the peak of interest while at the same time causing turbulence in the stream. In electrochemical detection turbulence causes variation in

background noise and thus detector noise. Therefore both processes contribute to a reduction of the signal-to-noise ratio for the detector under these
operating conditions. Although interesting work has been carried out in this
area and commercial systems are currently available, post-column reactors
will only find widespread used when they are more convenient to use in
a general laboratory setting, i.e. requiring no direct involvement by the
operator.

A solid-phase post-column reaction system is one way to approach the
problem (Krull and Lankmayer, 1982; Krull et al., 1983). A small column
packed with material that can selectively react with the analyte is inserted
into the flow stream between the column and the detector. Band broadening
and peak distortion has been studied in these systems by Nondek et al.
(1983) and Nondek (1984) in an attempt to optimize their operating
efficiency. This approach has been extended to take advantage of the
selectivity of enzyme reactions (Bowers and Bostick, 1982); an enzyme is
bound to an inert matrix packed in a small post-column reactor. The use of
enzymes attached to a carbon electrode suface also shows promise (Heider
and Yacynych, 1984), in this case a post-column reaction is carried out at the
electrode surface. This has the advantage of introducing no post-column
band broadening and at the same time incorporating the selectivity of an
enzyme reaction. Although these types of electrodes will be useful for only
very specific and limited applications at this time, they hold promise for
being a major step in lowering the detection limit at least for certain classes
of compounds.

The recent introduction of post column on-line continuous photolytic
derivatization (Krull et al., 1984), will expand the use of HPLC even further.
A surprisingly large number of organic compounds are capable of forming
electroactive species after photolysis or photohydrolysis; they can then be
detected by conventional electrochemical detectors. In the simplest case
teflon tubing connecting the column and the detector is wound around a
broad spectrum mercury discharge lamp. As the analyte leaves the column
the photolysis takes place and if the tubing is of sufficient diameter and
length complete, conversion can be accomplished. Work has just been
initiated to establish the types of compound that can be analyzed by this
technique but organthiophosphates and organic nitro compounds have
already been shown to be amenable (Krull et al., 1984). It would seem that
further exploitation lies in the use of high-powered light sources or light
sources of specific wavelengths such as the laser or in the use of optically
transparent electrodes so that photolysis could be carried out right at the
electrode surface.

CONCLUSION

The wide variety of techniques that we have covered gives some indication of
the range of possibilities for electrochemical detection in liquid chromato-

graphy. However, the list is by no means limiting. It is hoped that this report will give some insights into promising areas of research and that it may further stimulate investigations in chromatographic detection.

REFERENCES

Allison, L. A., and Shoup, R. E. (1983). Dual-electrode liquid chromatography detector for thiols and disulfides. *Anal. Chem. 55*, 8—12.

Anderson, L. B., and Reilley, C. N. (1965a). Thin-layer electrochemistry: Steady-state methods of studying rate processes. *J. Electroanal. Chem. 10*, 295—305.

Anderson, L. B., and Reilley C. N. (1965b). Thin-layer electrochemistry: Use of twin working electrodes for the study of chemical kinetics. *J. Electroanal. Chem. 10*, 538—552.

Apffel, J. A., Brinkman, U. A. Th., and Frei, R. W. (1983). Use of non-segmented flow, post-column reaction detection with miniaturized HPLC systems. *Chromatographia 17*, 125—131.

Barnes, A. C., and Nieman, T. A. (1983). Coulostatic pulse amperometry for liquid chromatography/electrochemical detection. *Anal. Chem. 55*, 2309—2312.

Blaedel, W. J., and Klatt, L. N. (1966). Reversible charge transfer at the tublar platinum electrode. *Anal. Chem. 38*, 879—883.

Blank, C. L. (1976). Dual electrochemical detector for liquid chromatography. *J. Chromatogr. 117*, 35—46.

Bowers, L., and Bostick, W. D. (1982). *Chemical Derivatization in Analytical Chemistry. Vol. 2.* R. W. Frei and J. F. Lawrence (eds), Plenum Press, New York, Ch. 3.

Brunt, K., Bruins, C. H. P., and Doorbos, D. A. (1981). Comparison between a differential amperometric detector in the reductive mode and a UV detector in high-performance liquid chromatography with vitamin K3 as the test compound. *Anal. Chim. Acta 125*, 85—91.

Chesney, D. J., Anderson, J. L., Weisshaar, D. E., and Tallman, D. E. (1981) Evaluation of Kel-F Graphite electrodes as detectors for continuous flow systems. *Anal. Chim. Acta 124*, 321—331.

DeAbreu, M., McClintock, S. A., and Purdy, W. C. (1985). 'A multi-electrode amperometric detector'. Canadian Chemical Conference, Kingston, Ontario, Canada, 1—5 June 1984.

Delahay, P. (1962a). A new electroanalytical method: coulostatic or charge step method. *Anal. Chim. Acta 27*, 90—93.

Delahay, P. (1962b). Fundamentals of coulostatic analysis. *Anal. Chem. 34*, 1267—1271.

Eccles, G. N., and Purdy, W. C. (in press). Development of a pulse cyclic voltammetric instrument. *Anal. Lett.*

Faulkner, L. R. (1984). Chemical microstructures on electrodes. *Chem. Eng. News 62*, 28—45.

Fell, A., and Scott, H. P. (1983). Applications of rapid-scanning multichannel detectors in chromatography. *J. Chromatogr. 273*, 3—17.

Fenn, R. J., Siggia, S., and Curran, D. J. (1978). Liquid chromatography detector based on a single and twin-electrode thin-layer electrochemistry: application to the determination of catecholamines in blood plasma. *Anal. Chem. 50*, 1067—1073.

Frei, R. W., and Lawrence J. F. (eds) (1981). *Chemical Derivatization in Analytical Chemistry. Vol 1.* Plenum Press, New York.

Goto, M., Zou, G., and Ishii, D. (1983). Determination of catecholamines in ˙.uman serum by micro high performance liquid chromatography with micro precolumn and dual electro-chemical detection. *J. Chromatogr. 275*, 271—281.

Heider, G. H., and Yacynych, A. M. (1984). 'Characterization of an immobilized glucose oxidase chemically modified reticulated vitreous carbon flow-through electrode'. The Pittsburgh Conference on Analytical Chemistry and Applied Spectroscopy, Atlantic City, New Jersey, 5—9 March, 1984. Paper No. 536.

Hepler, B. R., and Purdy, W. C. (1980). The amperometric detection of thyroid hormones following reverse-phase high-performance liquid chromatography. *Anal. Chim. Acta 113*, 269—276.

Hui, B. S., and Huber, C. O. (1982). Amperometric detection of amines and amino acids in flow injection systems with a nickel oxide electrode. *Anal. Chim. Acta 134*, 211—218.

Kissinger, P. T., Bratin, K., Davis, G. C., and Pachla, L. A. (1979). The potential utility of pre- and post-column chemical reactions with electrochemical detection in liquid chromatography. *J. Chromatogr. Sci. 17*, 137—146.

Kissinger, P. T., and Heineman W. R. (eds) (1984). *Laboratory Techniques in Electroanalytical Chemistry*. Marcel Dekker, New York.

Kok, W. Th., Hanekamp, H. B., Bos, P., and Frei, R. W. (1982). Amperometric detection of amino acids with a passivated copper electrode. *Anal. Chim. Acta 142*, 31—45.

Krull, I. S., and Lankmayer, E. P. (1982). *Derivatization Reaction Detectors in HPLC*. American Laboratory, Fairfield, Connetticut. 18—31 May, 1982.

Krull, I. S., Xie, K. H., Colgan, S., Neve, U., Izod, T., King, R., and Bidlingmeyer, B. (1983). Solid phase derivatization reactions in HPLC polymeric reductions for carbonyl compounds. *J. Liq. Chromatogr. 6*(4), 605—626.

Krull, I. S., Ding, X.-D., Selavka, C., and Nelson, R. (1984). Electrochemical detection in HPLC, post-column, on-line, continuous photolytic derivatization for improved detection in HPLC—*hv*—EC. *LC Mag. 2*(3), 214—221.

Last, T. A. (1983). Coulostatic voltammetric liquid chromatography detector. *Anal. Chem. 55*, 1509—1512.

Mann, C. K., and Barnes, K. (1970). *Electrochemical Reactions in Non-aqueous Systems*. Marcel Dekker, New York.

Mayer, G. S., and Shoup, R. E. (1983). Simultaneous multiple electrode liquid chromatographic—electrochemical assay for catecholamines, indole-amines and metabolites in brain tissue. *J. Chromatogr. 255*, 533—544.

McClintock, S. A., and Purdy, W. C. (1981). A bi-potentiostat for four electrode electrochemical detectors. *Anal. Lett. 14*(B10), 791—798.

McClintock, S. A., and Purdy W. C. (1983). A dual-working electrode electrochemical detector for liquid chromatography. *Anal. Chim. Acta 148*, 127—133.

Mikkelsen, S. R., McClintock, S. A., and Purdy, W. C. (1985). *Single pulse coulostatic analysis*. Canadian Chemical Conference, Kingston, Ontario. 1—5 June, 1984.

Murray, R. W. (1980). Chemically modified electrodes. *Acc. Chem. Res. 13*, 135—141.

Murray, R. W. (1984). Chemically modified electrodes. In *Electroanalytical Chemistry. Vol. 13*. A. J. Bard (ed.), Marcel Dekker, New York, pp. 191—368.

Nondek, L. (1984). Band broadening in solid solid-phase derivatization reactions for irreversible first-order reactions. *Anal. Chem. 56*, 1192—1194.

Nondek, L., Brinkman, U. A., and Frei R. W. (1984). Band broadening in solid phase reactors packed with catalyst for reactions in continuous flow systems. *Anal. Chem. 55*, 1466—1470.

Osteryoung, R. A., O'Dea, J., and Osteryoung, J. (1981). Theory of square wave voltammetry for kinetic studies. *Anal. Chem. 53*, 695—701.

Reinmuth, W. H. (1962). Theory of coulostatic impulse relaxation. *Anal. Chem. 34*, 1272—1276.

Reiss, J. J., and Nieman, T. A. (1983). Quantitative and qualitative information from single coulostatic decay curves. *Anal. Chem. 55*, 1236—1240.

Roe, D. K., and Ho, I. P. (1984). Instrumentation for multi-electrode voltammetry. The Pittsburgh Conference on Analytical Chemistry and Applied Spectroscopy, Atlantic City, New Jersey. 5—9 March, 1984. Paper No. 532.

Roston, D. A., and Kissinger, P. T. (1982). Series dual-electrode for liquid chromatography. *Anal. Chem. 54*, 429—434.

Roston, D. A., Shoup, R. E., and Kissinger, P. T. (1982). Liquid chromatography/electrochemistry: Thin-layer multiple-electrode detection. *Anal. Chem. 54*, 1417A—1434A.

Schieffer, G. W. (1980). Dual coulometric—amperometric cells for increasing the selectivity of electrochemical detection in high-performance liquid chromatography. *Anal. Chem. 52*, 1994—1998.

Scott, R. P. W. (ed.) (1984). *Small Bore Liquid Chromatography Columns, their Properties and Uses.* John Wiley, New York.

Shimada, K., Tanaka, M., and Nambara, T. (1983). Sensitive derivatization reagents for thiol compounds in high performance liquid chromatography with electrochemical detection. *Anal. Chim. Acta 147*, 375—380.

Swartzfager, D. G. (1976). Amperometric and differential pulse voltammetric detection in high performance liquid chromatography. *Anal. Chem. 48*, 2189—2192.

Turk, D. J., McClintock, S. A., and Purdy W. C. (in press). The electrochemistry of amitriptyline at a chemically modified reticulated vitreous carbon electrode. *Anal. Lett.*

van Leeuwan, H. P. (1982). Coulostatic pulse techniques. In *Electroanalytical Chemistry*, A. J. Bard (ed.). Marcel Dekker, New York.

Vickery, T. M. (ed.) (1983). *Liquid Chromatography detectors. Chromatographic Science Series, Vol. 23.* Marcel Dekker, New York.

Weber, S. G., and Purdy W. C. (1978). The behaviour of an electrochemical detector used in liquid chromatography and continuous flow voltammetry. Part 1. Mass-transport limited current. *Anal. Chim. Acta 100*, 531—544.

Weber, S. G., and Purdy, W. C. (1979). The behaviour of an electrochemical detector used in liquid chromatography. *Anal. Chim. Acta 102*, 41—59.

Yamada, J., and Matsuda, H. (1973). Limiting diffusion currents in hydrodynammic voltammetry. *J. Electroanal. Chem. 44*, 189—198.

Progress in HPLC, Vol. 2, pp. 53—94
Parvez *et al.* (Eds)
© 1987 VNU Science Press

Electrochemical detectors for low-dispersion liquid chromatography

KAREL ŠLAIS

Institute of Analytical Chemistry, Czechoslovak Academy of Sciences, 611 42 Brno, Czechoslovakia

DEMANDS ON DETECTORS IN LOW-DISPERSION LIQUID CHROMATOGRAPHY

The contemporary trend in liquid chromatography (LC) towards packed columns with small diameters (microbore columns), all the time diminishing the size of the particles of the packings, as well as the study of open tubular (OT) liquid chromatography, lead to an ever decreasing volume of the eluted solute zone. This phenomenon is significant to such a degree that one can speak of a new branch, i.e. low-dispersion liquid chromatography. This technique intensifies demands on liquid chromatographs from the viewpoint of dead volume and the contribution of extra column broadening of the chromatographic zone. The strict instrumental demands are balanced, on the other hand, by the advantages gained, such as a lower consumption of the mobile phase, a high mass sensitivity (Kucera, 1984; Scott, 1980, 1983); and speeding up of the analysis (DiCesare *et al.*, 1981; Erni, 1983; Knox, 1980; Reijn *et al.*, 1983; Scott *et al.*, 1979), At present, it is anticipated (Hupe *et al.*, 1984) that that optimization of chromatographic analysis from the viewpoint of speed and miniaturization is not limited by the properties of the separation column, but by the possibilities of realizing chromatographic equipment with sufficiently small extra column contributions to the solute zone broadening.

The zone of the eluted solute can be simply characterized by the standard deviation, σ_V, expressed in volume units for the case of Gaussian profile in:

$$\sigma_V = t_R F_m/(n)^{1/2} \tag{1}$$

where t_R is the retention time of the solute, F_m the flow-rate of the mobile phase and n is the number of theoretical plates of the separation column. The relationship between the standard deviation expressed in volume, σ_V, and that expressed in time σ_t, is:

$$\sigma_V = \sigma_t F_m \tag{2}$$

Consider, for example, a packed microcolumn (Knox, 1980) of length $l = $ 150 mm and inner diameter ID = 1 mm, packed with particles of diameter d_p = 2.7 μm, having efficiency corresponding to n = 30000 at flow-rate F_m = 1.1 μl s^{-1}, the standard deviation of the zone of the unsorbed solute is then σ_V = 0.6 μl or σ_t = 0.5 s (see Table 1). Alternatively, consider an OT capillary column (Ishii and Takeuchi, 1983) of diameter ID = 30 μm and length l = 15 m, having a linear velocity for the mobile phase of u = 0.462 mm s^{-1}, i.e. volume flow-rate of 0.32 nl s^{-1}, the efficiency corresponding to the theoretical plate number n = 1150000, is σ_V = 6.5 nl or σ_t = 20 s.

Table 1. Examples of miniature columns in liquid chromatography and their demands on detector parameters (θ = 0.1; k = 0)

Column type	l (mm)	ID (mm)	d_p (μm)	n	F_m (nl s^{-1})	σ_V (nl)	σ_t (s)	References
Packed microbore	150	1	2.7	30000	1100	600	0.5	Knox (1980)
Open tubular capillary	15000	0.03	—	1150000	3.2	6.5	20	Ishii and Takeuchi (1983)

Chromatographic equipment must have properties such that the separation achieved on the column may be transferred onto the recording device without any distortion. Starting from the column outlet, this demand must be fulfilled successively by connections to the detection cell, the detection cell and, least but not least, the detector electronics. The effect of the extra column contributions has been studied intensively (Huber, 1978; Hupe *et al.*, 1984; Martin *et al.*, 1975; Reese and Scott, 1980; Sternberg, 1966). If θ^2 represents the maximal admissible relative decrease in column efficiency, then the maximal sum of the extra column contributions can be expressed as:

$$\sigma_{V,c}^2 + \sigma_{V,d}^2 \leqslant \theta^2 (F_m^2 \, t_R^2 / n) \tag{3}$$

where $\sigma_{V,c}^2$ is the volume variance of the broadening in the capillary connecting the column with the detector and $\sigma_{V,d}^2$ is the volume variance of the broadening of the solute zone in the detection cell.

For straight capillaries of a circular cross-section, $\sigma_{V,c}^2$ is proportional to the capillary length, L, and to the fourth power of the capillary diameter, d_c (Golay, 1958; Martin *et al.*, 1975; Scott and Kucera, 1971):

$$\sigma_{V,c}^2 = \frac{\pi L F_m d_c^4}{384 \, D} \tag{4}$$

(D is the diffusion coefficient of the solute in the mobile phase). For a packed microcolumn (assuming $D = 10^{-3}$ mm^2 s^{-1}), if a 1% decrease in the separation efficiency is admitted, a maximum length of L = 4 mm is obtained for a capillary of diameter $d_c = 0.1$ mm for the unsorbed solute using Equations 3 and 4. This means that commonly available stainless steel capillaries cannot be used. Provided that capillaries of internal diameter $d_c = 50$ μm (Takeuchi *et al.*, 1984) are available, the length of the connecting capillary can vary up to 64 mm, which permits certain variability in the positions of the detector and the column. For an adsorbed solute, the retention volume of which is twice that of an unsorbed solute, even a capillary of 0.1 mm diameter and 16 mm length can be admitted. However, optimum resolution is offered by designing the connection between the detector cell and column without connecting capillaries (Guiochon, 1980). This was realized with packed microcolumns, e.g. in connection with a UV detector (Erni, 1983; Kucera, 1980; Scott *et al.*, 1979) or an amperometric detector (Goto *et al.*, 1981; Šlais and Kouřilová, 1982, 1983). In the case of OT capillary columns, the design with an on-column detector is unavoidale, e.g. for a UV detector (Yang, 1981), an amperometric detector (Šlais and Krejčí, 1982), a coulometric detector (Knecht *et al.*, 1984), a potentiometric detector (Manz and Simon, 1983) and a fluorimetric detector (Guthrie and Jorgenson, 1984; Guthrie *et al.*, 1984). Due to the low flow-rates of the mobile phase in packed microcolumns and OT capillary columns, the increase in the radial dispersion caused by secondary flow in strained (Hofmann and Halász, 1979, 1980) and coiled (Katz and Scott, 1983; Tijssen, 1978) capillaries cannot be used for the reduction in volume variance of the connections.

Contributions to the distortion of the chromatographic zone owing to the detector originate from two sources: on the one hand is the influence of the time constant of the transfer function for the instantaneous concentration in the detection cell upon the resulting chromatographic analysis, and on the other is the influence of mixing in the detector cell. Equation 5 then holds for the maximal admissible time constant of detection, τ_d (Guiochon, 1980; Yang, 1982):

$$\tau_d \leqslant \theta \, \frac{t_R}{n^{1/2}} \tag{5}$$

Thus, if $\theta = 0.1$, then $\tau_d = 0.05$ s for the example of the packed micro-column for an unsorbed solute (see Table 1), which is the very strict demand of existing devices, recorders in particular. Recording devices commonly used nowadays, with a time constant of about 0.5 s even permit, e.g. with the microcolumn, a solute with a retention volume of 3.2-fold greater than that of unsorbed solute, the record showing a 10% decrease ($\theta^2 = 0.1$) in real column efficiency. When studying OT LC, the demands on the time constant of the recording device are rather different. The present state of the preparation technique for OT capillary columns, characterized by the example mentioned above (see Table 1), requires a time constant which is relatively great (2 s) due to a longer time of analysis (t_r of the unsorbed solute is 6 h). However, if a separation time comparable to that of pack columns is required, a time constant *c.* 10 ms is necessary (Yang, 1982).

For the case when the cell is considered to be a perfect mixer (Guiochon, 1980; Hupe *et al.*, 1984; Sternberg, 1966; Yang, 1982), the maximum admissible volume of the detector cell, V_d, can be expressed by the following relationship:

$$V_d \leqslant t_r F_m / (n)^{1/2} \tag{6}$$

For the example of the packed microcolumn (see Table 1), the highest admissible detection cell volume, V_d, becomes less than 60 nl for the unsorbed solute, and $\theta^2 = 0.01$. For a less exacting case, when $\theta^2 = 0.1$ and for a retention volume of the solute of three times that of the unsorbed solute, V_d is then 0.6 µl. It is obvious from the examples mentioned above that the detection cells of volumes of about 0.5 µl are a limited compromise in the optimization of the instrumentation for microcolumn chromatographs. OT capillary columns require detectors with cell volumes below 1 nl.

It follows from the above that the demands of the volume for the detection cell are a decisive factor in the use of the advantages of low-dispersion liquid chromatography. The time constant of the detection can be reduced for the majority of detectors by modifying their electronic sections, though at the expense of impairing the noise level, N (Guiochon, 1980). An adverse effect from the connections can be eliminated by designing the detection cell directly at the column outlet.

Besides the demands on the miniaturization of the cell volume, one must be aware of the main detector function, namely, the sensitive detection of solutes in the chromatographic column effluent. The solute concentration in the detector cell when the signal-to-noise ration (S/N) (i.e. the ratio of the measured quantity depending on the solute concentration to the noise of the same quantity uncorrelated with the change in the solute concentration) equals 2, which is usually considered to be the minimal detectable solute concentration, is thus particularly decisive. The reduction of the detection limit (defined as the minimal detectable quantity of the solute introduced to the column) is the main result of the miniaturization of the chromatographic equipment. For instance, at the minimal detectable concentration (ng ml^{-1}) and volume of the eluted peak (1 nl), the minimal detectable quantity of the solute is 10^{-15} g, or approximately 10^5 molecules (Poppe, 1983).

GENERAL ELECTRICAL DESCRIPTION OF THE CELL OF THE ELECTROCHEMICAL DETECTOR

The electrochemical detector uses the electrochemical properties of solutes for determining the flow of the liquid. These properties can be e.g. electrochemical reactivity at the electrode (amperometric, coulometric, polarographic or potentiometric detectors); ionic mobility in solution (conductivity detectors); permittivity in solution (permittivity detectors); or permittivity in the adsorbed state at the electrode (tensametric detectors). Sometimes several properties can appear at one time (electrokinetic detectors). A large group of electrochemical detectors (amperometric, coulometric, polarographic, conductivity, permittivity, tensametric) is represented by detectors which evaluate the current through the electrochemical cell at a controlled potential, with the change in the measured quantity being directly proportional to the change in the solute concentration in solution.

The cells of electrochemical detectors can be illustrated in a simplified way according to the classification mentioned above with the aid of a general diagram (Fig. 1). The reference electrode can be characterized as a source of voltage, E_r, with a capacity so great and an internal resistance so small that the impedance is negligible compared with the other impedances in the circuit. The resistance of the electrolyte in the cell, R_e, is given by the equation:

$$R_e = C/\kappa_e \tag{7}$$

and the capacity of the liquid in the cell, C_e, by the relationship:

$$C_e = \varepsilon_e/C \tag{8}$$

(κ_e is the conductivity and ε_e the permittivity of the liquid in the cell). The geometry of the electrochemical cell determines the so-called 'resistance constant' of the cell, C, defined as the ratio of the length of the liquid column between the electrodes to its cross-section.

Properties of the working electrode are determined by the type of electrochemical measurement. The electrode material, its size and geometrical arrangement are also particularly decisive. In general, the working electrode can be characterized by the leak resistance, R_w, the capacity, C_w, which is determined by the capacity of the electrode double-layer, C_d, and the electrode surface area according to the relationship:

$$C_w = AC_d \tag{9}$$

Capacity C_w varies due to the adsorption of the solute, i, at the working electrode and if the solute also undergoes a reaction at the electrode, the equilibrium potential of this reaction under the conditions of the solute

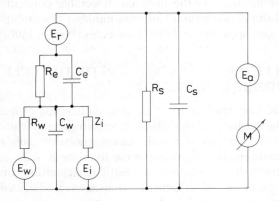

Figure 1. General equivalent electric diagram of the electrochemical cell.

solution in the mobile phase, E_i, and the polarization impedance of this reaction, Z_i, will be involved. The impedance can be characterized simply as the ratio of the difference between the electrode potential and the equilibrium potential, E_i, to the faradaic current produced by the electrochemical reaction of the solute, i, at the electrode; Z_i is, with reversibly reacting solutes, particularly dependent on the transport of the solute i to the working electrode.

It is also necessary to consider further stray resistance, R_s, and stray capacity, C_s, which are determined by the design and materials used for the construction of the cell and connections to the measuring device, M. The measuring device is usually equipped with a device for applying voltage, E_a, to the electrochemical cell.

The electrochemical property of the solute to be measured determines the selection of material and the design of the working electrode, cell geometry, applied voltage, E_a, the composition and the flow-rate of the mobile phase. At the same time, one is approaching the state when the measured response is a function of a single electrical quantity of the electrochemical cell, e.g. Z_i (amperometry, polarography), C_w (tensametry), R_e (conductivity measurement), C_e (permittivity measurement) and E_i (potentiometry). In some cases, however, this state cannot be reached (electrokinetic detection).

Classification by the electrochemical property of the solute, i.e. by the measured electrochemical quantity, is used for the characterization and evaluation of the possibilities of miniaturizing electrochemical cells of the detectors of various types.

AMPEROMETRIC, POLAROGRAPHIC AND COULOMETRIC DETECTORS

Faradaic current limited by diffusion

The faradaic current, i.e. the current produced by the electrochemical reaction of the solute on the working electrode, is measured at the amperometric, polarographic and coulometric detection. The working electrode is usually polarized by d.c. voltage, the magnitude of which is given by the sum of E_a and E_r (see Fig. 1). The effect of R_e on the current through the cell measured by the measuring device, M, can be substantially reduced by the addition of an electrolyte into the mobile phase, by the design and arrangement of the reference electrode, and by the potentiostatic connection (see Fig. 2). Electrolyte capacity, C_e, plays no role at the polarization of the electrochemical cell with both d.c. and a.c. voltage of low frequency, especially if the magnitude of R_e is kept at a low value. By careful selection of material for the working electrode (usually nobel metals or various forms of carbon) and its surface area, the highest possible leak resistance for the working electrode, R_w, and hence the least background current and thus also noise, N, can be achieved. Thereby, the influence of E_w on the measured current is also restricted, and the magnitude of E_w corresponds to the sum of E_a and E_r when the background current changes its sign. The capacity of the

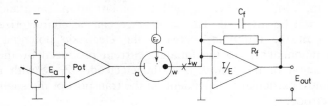

Figure 2. Simplified representation of the amperometric cell and the electronic circuits of the potentiostat and current amplifier: a, r, w — auxiliary, reference and working electrodes, respectively; Pot — potentiostat amplifier; I/E — current-to-voltage converter amplifier; R_f — feedback resistor; C_f — feedback capacitor for adjusting of low-pass filtering time constant τ_d $= R_f C_f$; E_{out} — output voltage, $E_{out} = I_w R_f$; I_w - working electrode current.

working electrode, C_w, is, according to Equation 9, dependent on the capacity of the double-layer, C_d. For solid electrode materials commonly used, C_d moves from 0.3 to 3 $\mu F\, mm^{-2}$ (van Rooijen and Poppe, 1983), for liquid mercury one can take into account $C_d = 0.1\ \mu F\, mm^{-2}$ (de Jong *et al.*, 1983a, b).

If solute i, subjected to reversible electrode reaction, is present in the mobile phase, by carefully choosing the applied voltage, E_a, the state can be reached when virtually all solute which will have diffused from the solution towards the electrode surface, will react at the electrode. Further, if the cell design and value of the mobile phase flow-rate, F_m, is such that when all the solute transported by convection (i.e. by the flow of the mobile phase) to the cell undergoes an electrode reaction, the detection is called 'coulometric'. Faradaic current can then be called 'coulometric current', I_c, and can be described simply by the following relationship:

$$I_c = zFc_0F_m \tag{10}$$

where z is the number of electrons exchanged in the electrode reaction, F is the faradaic charge and c_0 is the solute concentration in the mobile phase in the detector cell.

If only several per cents of the total quantity of the solute transported into the cell can diffuse towards the electrode, i.e. faradaic current is limited by the diffusion of solute towards the electrode, amperometric detection is involved. Provided that the working electrode of the liquid mercury is used, polarographic detection is in question. Faradaic current is then called 'limiting current' (diffusion current), I_d, the magnitude of which can generally be described by the following relationship (Poppe, 1978):

$$I_d = zFc_0\,\frac{DA}{\delta} \tag{11}$$

where δ is the thickness of the diffusion layer.

The dependence of I_d on the cell geometry, mobile phase flow-rate and other parameters is determined from relationship given below which was derived for amperometric and polarographic detectors (see Fig. 3) most

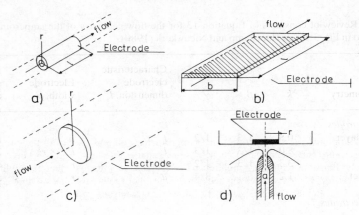

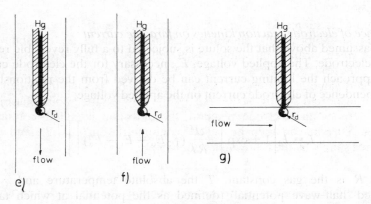

Figure 3. Types of flow geometry of amperometric and polarographic detectors: (a) tubular electrode; (b) thin-layer cell; (c) disc electrode; (d) wall-jet electrode; (e) parallel mercury and fluid flow; (f) opposite mercury and fluid flow; (g) mercury flow perpendicular to fluid flow. r — electrode radius; l — electrode length; b — electrode width; a — nozzle diameter; r_d — drop radius. Reproduced with permission from Hanekamp and Nieuwkerk (1980). © 1980. Elsevier Scientific Publishing Co.

frequently used (Hanekamp, 1981; Hanekamp and de Jong, 1982; Hanekamp and van Nieuwkerk, 1980):

$$I_d = kzFc_0DSc^\beta Re_x^\gamma b \tag{12}$$

(Re_x is Reynolds number based on l_x, Sc is the Schmidt number, b is the electrode width and k, γ and β are constants, the values of which are listed in Table 2 for the relevant types of the cell geometry. At the same time $Re_x = ul_x/\nu$ and $Sc = \nu/D$, where u is the mean linear velocity of the liquid in the cell, l_x is the characteristic electrode length and ν is kinematic viscosity of the liquid.

Table 2. Review of constants of Equation 13 for the flow geometry of the amperometric cells illustrated in Fig. 3. From Hanekamp and Nieuwkerk (1980).

Flow geometry	k	β	γ	Characteristic electrode dimension, l_x	Electrode width, b	Electrode surface area, A
Solid electrodes						
Thin-layer	0.8	1/3	1/2	l	b	$b \times l$
Tube	8.0	1/3	1/3	l	$l^{1/3} \times r^{2/3}$	$2\pi r \times l$
Disk	3.3	1/3	1/2	r	r	πr^2
Wall jet	1.2		3/4	a	r	πr^2
Liquid electrodes *stream:*						
Opposite	3.7	1/2	1/2		r_d	$4\pi r_d^2$
Parallel	7.9	1/3	1/3		r_d	$4\pi r_d^2$
Normal	4.3	1/2	1/2		r_d	$4\pi r_d^2$

Influence of electrode reaction kinetics on faradaic current

It was assumed above that the solute is subjected to a fully reversible reaction at the electrode. The applied voltage, E_a, necessary for the electrode current, I, to approach the limiting current can be derived from the relationship for the dependence of electrode current on the applied voltage:

$$I = I_d \left[1 + \exp \left[\frac{zF}{RT} (_{1/2}E_o - E_r - E_a) \right] \right]^{-1} \qquad (13)$$

where R is the gas constant, T the absolute temperature and $_{1/2}E_o$ is so-called 'half-wave potential' (defined as the potential at which faradaic current reaches the half of the value of the limiting current). If the diffusion and activity coefficients of the reduced and oxidized forms of the solute can be considered as equal, then $_{1/2}E_o$ is identical with the standard potential of electrode reaction, E_o, which is tabulated for a number of reactions. On the other hand, the relationship between $_{1/2}E_o$ and the equilibrium potential of the electrode reaction, E_i, is given by:

$$_{1/2}E_o = E_i + RT/zF(\ln c_o^{red}/c_o^{ox})$$

where c_o^{red}/c_o^{ox} are the concentrations of reduced and oxidized forms of the solute in the solution. This means that for amperometric detection, e.g. of oxidizable compounds when the major portion of the solute is present in the reduced form, E_i is much less than $_{1/2}E_o$ or E_o. In order to obtain I equal to at least 99%, of I_d, i.e. for achieving almost the limiting current, it is necessary for Relation 13 be valid for a two-electron oxidation:

$$E_a + E_r - {}_{1/2}E_o \gg 60 \text{ mV}$$

Organic compounds, however, react at the electrode to an irreversible extent, the relationship for the current of this irreversible reaction (e.g. oxidation) (Dvořák and Koryta, 1983a) being:

$$I = I_d \bigg/ \left[1 + \frac{D}{\delta k^o} \exp \left[\frac{(1-\alpha)zF}{RT} (E_i - E_r - E_a) \right] \right] \qquad (14)$$

where α is the coefficient of the charge transfer and k^o is the rate constant of the irreversible electrode reaction. It follows from Equation 14 that by increasing the transport of the solute towards the electrode, i.e. with a decrease in δ, it is ncessary to increase simultaneously (in the case of oxidation) the applied voltage E_a in order to reach the same ratio of I and I_d. Further, with the increasing irreversible character of the reaction, i.e. with a decreasing value of k^o, the applied voltage E_a further increases. Even though, e.g. in the detection of oxidizable compounds, E_i is less than $_{1/2}E_o$, to obtain the same I/I_d ratio E_a should be greater for irreversible than for reversible reactions.

Now consider, for example, the oxidation of hydroquinone at the platinum anode in a solution containing 1 mol l^{-1} KCl and 0.05 mol l^{-1} phosphate buffer of pH 6.65, when $\alpha = 0.63$ and k^o is c. 10^{-5} cm s^{-1} (Vetter, 1952). If detection proceeds in a thin-layer amperometric cell (see Fig. 3) of dimensions $b = 2$ mm, $l = 3$ mm, cell height $d = 0.05$ mm and surface area of the working electrode $A = 6$ mm^2, in connection with a microcolumn at flow-rate $F_m = 1$ mm^3 s^{-1}, the linear velocity in the cell is $u = 10$ mm s^{-1}. If $D = 10^{-3}$ mm^2 s^{-1} and $\nu = 10^{-6}$ m^2 s^{-1}, by using Equations 11 and 12, term $D/\delta k^o = 150$ is obtained. To provide for I reaching at least 99% of the value of I_d, it is ncessary that $E_a + E_r - E_i > 334$ mV should hold.

In the case of hydroquinone, the value of E_o is relatively low ($E_o = 254$ mV at pH 7 vs the standard hydrogen electrode) and o that of k^o relatively high since the quinhydrone electrode is usually considered as reversible. In this case E_a can be increased relatively easily with a decrease in δ. For a number of compounds detected amperometrically, e.g. monohydric phenols, the value of E_o is substantially greater and k^o much less so that not even at the potentials when decomposition of the mobile phase occurs can the limiting current not be reached (Armentrout et al., 1979). In such cases the improvement in the transport of the solute towards the working electrode can have only a slight or no effect on the increase in the faradaic current.

Optimization of the design of cells and electrode

To obtain low minimal detectable solute concentration the absolute value of signal, S, is not critical (i.e. in the case of the amperometric detector the value of faradaic current, I) but the ratio of this current to the fluctuation of the background current of the working electrode (denoted noise, N) is. The value of noise is affected to a considerable extent by the magnitude of the background current on one hand, and by the fluctuation of the applied voltage, namely by E_r (van Rooijen and Poppe, 1983), or the voltage of the operational amplifier of the potentiostat (see Fig. 2) (Hagahira *et al.*, 1983) on the other. For instance, the fluctuation of the applied voltage by 1 $\mu V\,s^{-1}$ on the electrode with the double-layer capacity $C_d = 1\ \mu F\,mm^{-2}$ will cause a fluctuation in the current of 1 $pA\,mm^{-2}$. In this context it is necessary to remember the temperature dependence of the reference electrode (van Rooijen and Poppe, 1983), e.g. the temperature coefficient of the Ag/AgCl electrode is $-1\ mV\,K^{-1}$ (Milazzo and Carolli, 1978), that of the calomel electrode is $-0.3\ mV\,K^{-1}$ (Ives and Janz, 1961).

From this discussion the question then arises about the efficiency of three-electrode (potentiostatic) arrangement of the amperometric or coulometric detection cell for liquid chromatography. In fact, two-electrode amperometric (Šlais and Krejčí, 1982; Šlais and Kouřilová, 1982, 1983) and coulometric (Knecht *et al.*, 1984; Blaedel and Wang, 1981) cells have been designed with comparable or better (Hagahira *et al.*, 1983) minimal detectable concentrations in comparison with the three-electrode cells.

The main functions of the three-electrode arrangement are:

1. to minimize the effect of the electrolyte resistance, R_e, on the measured current I and thereby permit a wide dynamic range of the response;
2. to ensure the definable potential;
3. to ensure the stable potential of the working electrode; and
4. to reduce the current density of the reference electrode in such way that polarization and hence the reduction of the linear range of the response may not occur.

The minimization of the influence of R_e is critical only at the measurement of higher solute concentrations. The amperometric detectors are particularly used for the detection of trace amounts so when higher value of R_e is attained a marked decrease in the potential of the working electrode does not result due to the passage of the solute through the cell. For instance, for one of the popular amperometric detectors — Model LC 4A, Bioanalytical Systems, West Lafayette, USA — the following interelectrode resistances were measured with a mobile phase composed of 0.01 $mol\,l^{-1}$ acetic acid in water (van Rooijen and Poppe, 1983):

the working electrode — the reference electrode: $3 \times 10^5\ \Omega$
the working electrode — the auxiliary electrode: $3 \times 10^5\ \Omega$
the reference electrode — the auxiliary electrode: $1.8 \times 10^4\ \Omega$.

It follows from both the measured values and the cell design that when the reference and the working electrodes are housed far from the measuring electrode the three-electrode arrangement cannot reduce effectively the influence of R_e. For a carbon electrode of surface area 7 mm^2, even in the absence of a solute one can expect a potential drop of up to several tens of mV on the resistance R_e owing to the influence of the background current. Sontag and Frank (1983) compared the two- and three-electrode arrangements of a Metroohm EA 1096 amperometric cell (Herisau, Switzerland) and pointed out that with the two-electrode arrangement the lower linear range of response occurred, peak tailing was increased and higher applied voltage at the electrode was required, although the minimal detctable concentration was not impaired. Negative phenomena were ascribed at the expense of the potential drop on the electrolyte resistance, R_e. Using a mobile phase composed of 0.01 mol l^{-1} NaNO$_3$ in water they measured the resistance between the working and the auxiliary electrode, $6 \times 10^4 \ \Omega$.

The potential drop on the electrolyte resistance can be reduced best by housing the auxiliary or the reference electrode in the wall opposite to the working electrode; this was implemented with both thin layer cells (Kissinger, 1983; Blaedel and Wang, 1981; Hagahira et al., 1983) and cells of the wall-jet design (Šlais and Kouřilová, 1982, 1983). Such an arrangement is particularly advantageous for the design of cells of thickness less than 50 μm.

The stability of the reference electrode potential contributes to the stability of the working electrode potential. For the correct reference function it is necessary to maintain a constant electrolyte composition in the vicinity of the reference electrode. With the currently used electrodes of the second type (calomel, Ag/AgCl electrode) the solution of the reference electrode is separated from the mobile phase with a diaphragm. At the isocratic elution in the liquid chromatography the reference electrode being washed with the mobile phase can be accomplished since the mobile phase composition is constant. In the two-electrode arrangement electrodes made of carbon (Blaedel and Wang, 1981) and stainless steel (Šlais and Kouřilová, 1982, 1983) were put into operation in this way.

The exactness of the working electrode potential permits reproducibility of the results under various conditions and facilitates use of the selectivity of amperometric detection. In liquid chromatography this fact is not so important since the selectivity is provided by separation on the column and the conditions optimal for detection should be found experimentally for every mobile phase — electrode material — solute combination. At the same time, the applied potential should be chosen so that, as far as possible, the limiting current of the solute may be reached (see discussion concerning Equation 14), otherwise the problem of increased tailing due to slow electrode reaction may occur.

Polarization of the reference electrode can be prevented by a large surface area with respect to the working electrode. This is easily done particularly with cells having miniaturized working electrodes.

It is obvious from the above discussion that to reach minimal detectable concentrations the three-electrode arrangement is not unavoidable. A simple two-electrode arrangement, old-fashioned though it may appear, can permit a lower noise level by up to one order of magnitude (Hagahira et al., 1983). The possibility of using a two-electrode arrangement is particularly significant when designing the cell for low-dispersion liquid chromatography when a simplified connection can markedly facilitate its realization.

From the analysis of Equations 12 and 13, from the above discussion of the origin of the working electrode noise and from earlier observations (Lanklema and Poppe, 1976b; Weber and Purdy) that in the case of a reversible reaction, enhancement of the signal-to-noise ratio (*S/N*) can be reached only by decreasing the thickness of the diffusion, layer, δ, or by improving the supply of the solute to the electrode by diffusion. Diminishing the characteristic dimension of the electrode to approximately less than 10 μm (Caudill *et al.*, 1982; Tallman and Weisshaar, 1983; Weisshaar *et al.*, 1981) is an analogous way of improving the *S/N* ratio, or the ratio of the electrode surface area to the thickness of the diffusion layer. For instance, with the disc electrode the electrode surface area diminishes with the square of the electrode diameter, whereas δ falls linearly with the electrode diameter. In addition to improving the *S/N* ratio, the independence of the magnitude of the limited current of the mobile phase flow-rate (Caudill *et al.*, 1982) can be anticipated with such electrodes.

Limits of the miniaturization of the cell and the working electrode
It follows from above that from the viewpoint of achieving maximal *S/N* ratio it is advantageous to keep the dimensions of both electrodes and cells as small as possible. This fact is advantageous for the miniaturization of the electrochemical cell; however, it cannot be use without limits. Problems associated with the measurement of small currents bring serious limits. This problem can be avoided partially by connecting a greater quantity of micro-electrodes in one cell (Caudill *et al.*, 1982) or by using a compound electrode material, e.g. carbon with a suitable insulator (Weisshaar *et al.*, 1981). This solution, however, leads unavoidably to a growth in the dmensions and thus volume of the detection cell.

For amperometric detectors currents of the order of magnitude of picoamperes with a time constant of about 1 s can easily be measured (van Rooijen and Poppe, 1983). Currents of about 0.1 pA at a time constant of about 0.1 s can be considered as the least measurable. The demand of the time constant is determined by the speed of the analysis; a fall in the measurable current below the above value is prevented by simultaneous section of a number of factors: noise currents of the input electronic circuits (operational amplifiers: ca 10^{-14} A), thermal noise of resistance (10^{-14} A or 1 μV for a resistance of 10^8 Ω at a band width of 1 Hz) and thermal coefficient of high-ohm resistances (0.1% K^{-1}) among others. The influence of stray capacities, C_s, and stray resistances, R_s, (see Fig. 1) can be suppressed efficiently by the arrangement commonly used today when the working electrode is connected to the virtual earth formed by the input of the measured current-to-output voltage converter (see Fig. 2).

In order to suppress the noise level, it is necessary to prevent the penetration of interfering currents by efficient screening. Since amperometric cells are currently designed from electrically insulating materials, it is necessary that, in addition to electric leads, the cell should also be screened. The electrical connection of the cell space with the surrounding due to the conductive mobile phase can also be a source of noise; this may be prevented by suitably designing the inlet capillary, particularly by diminishing its cross-section.

Starting from a minimal measurable current, then provided that the noise is c. 0.1% of background and current is of the order of 10 nA mm^{-2}, the minimum surface area of the working electrode is at 0.01 mm^2. Elbicki et al., (1984), who studied the conditions for the optimization of amperometric detectors, came to the conclusion that the minimum surface area of the working electrode should be 1 mm^2 if extra electrode noises are not to appear. The influence of the extra electrode interfering effects can be demonstrated by the following two examples:

Hagahira et al., (1983) using a coulometric detector with a working glassy carbon electrode of surface area 500 mm^2 in a two-electrode arrangement obtained a noise of 5 pA, at a working electrode potential of +0.8 V vs Ag/AgCl reference electrode and background current of 180 nA. However, Knecht et al., (1984) used a miniaturized coulometric detector with a working electrode of carbon fiber of length 0.7 mm and diameter 7 μm (surface area = 0.02 mm^2) and obtained noise of 5 pA at the same voltage on the working electrode and a background current of 400 pA.

The last cited authors assume a marked fall in the noise (by 1—2 orders of magnitude) with perfect screening of the cell and leads. It follows from the above examples that maintenance of the optimal minimum detectable concentration with miniaturization of amperometric and coulometric detectors is possible only with a perfectly designed electric arrangement.

Amperometric vs coulometric detection in miniaturized LC
For a complete view of the miniaturization of a cell designed for measuring Faradaic current it is necessary to compare the properties of amperometric and coulometric cells connected with miniaturized columns. From Equation 13 it follows that diminishing the electrode surface area causes both the surface area and noise to fall more rapidly than the conversion, i.e. the fraction of the reacted solute in the cell. This leads to the conclusion that amperometric detection, in comparison with coulometric detection, enable lower minimal detectable concentrations to be reached; this opinion is supported by a number of authors (van Rooijen and Poppe, 1983; Kissinger, 1983; Poppe, 1978; Weber and Purdy, 1978; Hanekamp and van Nieuwkerk, 1980). The opinions have, however, appeared recently (Curran and Tougas, 1984; Hagahira et al., 1983; Roe, 1983) which imply that the coulometric cell is more advantageous with respect to lower sensitivity towards changes in the flow-rate which contribute significantly to the noise level of the working electrode. In view of the requirements of minimal electrode surface area and low volume flow-rate passing through both packed micro- and OT capillary columns, diffusion transport towards the electrode approaches convection transport in the detection cell, and the cell then behaves more or less coulometrically. Examples of cells designed for packed microcolumns are, e.g. a thin-layer cell of 0.3 μl volume and 6 mm^2 surface area having a mobile phase flow-rate of F_m = 1.4 mm^3 s^{-1}, conversion of from 52% (p-aminophenol) to 76% (m-aminophenol) (Goto et al., 1981); and a wall-jet

cell of 20 nl volume with a working electrode surface area of 0.2 mm^2 having $F_m = 0.05$ mm^3 s^{-1} conversion of 10% (catechol) (Šlais and Kouřilová, 1982, 1983). Examples of the detectors for OT capillary columns are, e.g. a wall-ject cell of 1 nl volume with an electrode surface area of 0.01 mm^2, having a flow-rate of 0.5 nl s^{-1} and conversion of 13% (hydroquinone) (Šlais and Krejčí, 1982); and a single graphite fibre cell of 0.8 nl volume with an electrode surface area of 0.02 mm^2, having a flow-rate of 1.4 nl s^{-1} and conversion of 95% (catechol) or 73% (ascorbic acid) (Knecht *et al.*, 1984). The differences in conversions for various solutes under identical conditions can be explained by the differences in the kinetics of the electrode reaction (see discussion of Equation 14).

Electode cleaning
A progressive loss of sensitivity owing to contamination of the electrode surface with the products from the reaction of both solute and components of the mobile phase is an unavoidable phenomenon since the detectors are based on the electrochemical reaction at the measuring electrode. The design of the cell, however ingenious it may be, must therefore permit occasional re-standardizing of the electrode surface. The liquid mercury electrode used with polarographic detectors is advantageous in this respect, as the arrange-ment for liquid chromatography is such that a drop of mercury is always renewed after the chromatogram has finished (Lloyd, 1983a, b, c). Thus it is possible to renew the surface whilst obtaining a record without oscillations caused by mercury drops. With solid electrodes, occasional polishing of the surface with fine metallographic paper or the replacement of part of a paste electrode proved to be safest. The cell should be disassembled and reproducibly reassembled for this operation. Whereas in the case of detectors for packed microcolumns this can be performed relatively easily (Šlais and Kouřilová, 1982, 1983; Goto *et al.*, 1981), with the cells for OT LC this operation has been less exacting (Šlais and Krejčí, 1982) or almost un-realizable routinely (Knecht *et al.*, 1984) till now.

Electrochemical cleaning of the electrode based on periodic application of pulses of suitable polarity and magnitude (Štulík and Hora, 1976) appear imperfect. Somewhat more important is this procedure in cases when amperometric detection is made possible in this way, e.g. of sugars (Hughes *et al.*, 1981; Hughes and Johnson, 1981; Rocklin and Pohl, 1983) or amino acids (Polta and Johnson, 1983).

Examples of low-dispersion liquid chromatography with the amperometric, polarographic and coulometric detection
A modification of a 310 model of EG & G Princeton Applied Research, Princeton, USA with a supported mercury drop electrode (Lloyd, 1983a, b, c) (see Fig. 4) can serve as an example of a successfully miniaturized design for a polarographic detector. The volume of the mercury drop electrode can be adjusted to values from 0.2 μl upwards; and end of the inlet capillary is

68 K. Šlais

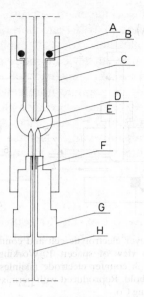

Figure 4. Cross-sectional diagram of the modified PAR flow cell: A — O-ring; B — polytetra-fluoroethylene sleeving which surrounds the mercury capillary and is separated and folded back at its upper to enable the O-ring to secure it in position; C — cell body; D — mercury capillary tip; E — eluent outlet jet; F — piece of tetrafluoroethylene tubing (1.59 mm outer diameter) holding the eluent capillary position; G — Omnifit end fitting; H — stainless steel capillary eluent line (0.51 mm outer diameter, 0.25 mm internal diameter). The distance between the cut-off points (broken lines) is 6 cm. Reproduced with permission from Lloyd (1983a). © 1983. Elsevier Scientific Publishing Co.

0.4 mm from the end of the capillary containing mercury and has the shape of an extending cone with the base 0.8 mm in diameter. The noise value of N = 10 pA at a background current of 1 nA was obtained with the minimum drop size, a time constant of the recording device of 0.3 s and a mobile phase composed of 0.025 mol l^{-1} phosphate buffer of pH = 0.3 in a methanol—water mixture (86:100 by volume) degassed by boiling, which flows through the detector at a flow-rate of F_m = 1 ml min^{-1}. The working electrode potential was −1.0 v vs Ag/AgCl reference electrode direct current (d.c. mode). The minimal detectable quantity of *m*-dinitrobenzene 4 pg was obtained on a 150 × 4.5 mm column. With suitable inlet capillaries, a polarographic cell of this type could be used successfully even in connection with packed microbore columns of inside diameter, 1 mm.

Great numbers of miniaturized amperometric cells have been constructed solid electrodes suitable for packed microcolumns: for thin-layer versions there are those of volume 0.3 µl (Goto *et al.*, 1981), 0.15 µl (Hirata *et al.*, 1980), 0.01—0.05 µl (Carlsson and Lundström, 1984); for wall-jet versions are those of volume 0.02 µl (Šlais and Kouřilová, 1982, 1983); and for tubular electrodes are those of volume 0.035 µl (Matysik *et al.*, 1981). A thin-layer cell designed by Goto *et al.*, (1981) (see Fig. 5) is an example of a

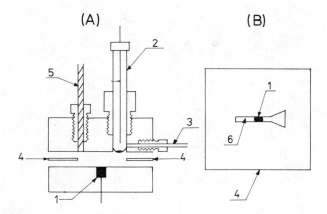

Figure 5. Construction of thin-layer electrolytic cell and connection with separation column. (A) Side view of cell; (B) top view of spacer. 1. Working electrode (glassy carbon); 2. reference electrode (Ag/AgCl); 3. counter electrode (stainless steel tube); 4. spacer (PTFE sheet); 5. separation column; 6. hold. Reproduced with permission from Goto *et al.* (1981). © 1981. Elsevier Scientific Publishing Co.

design which has already become classical (Kissinger *et al.*, 1973). The channel thickness is 50 μm and the volume is 0.3 μl; the working electrode is made of glassy carbon and has surface area of 6 mm^2. It is manufactured commerically as an electrochemical detector (Model 101-A by Yanaco company, Kyoto, Japan). In connection with a 0.05 × 145 mm microcolumn it has a detection limit of *c.* 10 pg (Goto *et al.*, 1981). A simple two-electrode detector working with a wall-jet cell (Šlais and Kouřilová, 1982, 1983) is manufactured with a minor modification (according to Fig. 6) (Šlais *et al.*, 1983) as an EMD 10 model by Laboratory Instruments Prague, Czechoslovakia). Connected with microcolumns, it permits a sensitive detection of a number of compounds (see Table 3). A negligible cell volume permits work with packed columns of 0.7 × 30 mm packed with particles of diameter (d_p) of 3 μm. Microcolumns of this type permit very rapid separations (see e.g. Fig. 7). Amperometric detectors are usually applied in reverse-phase liquid chromatography using suitable aqueous mobile phases. These detectors can, however, be applied successfully in straight phase systems, as in the example of phenothiazines (Fig. 8). To reduce the minimum detectable concentration, the technique of sampling in a non-eluting solvent, i.e. peak compression sampling (Guinebalt and Broquaire, 1981), is utilized. This technique is particularly advantageous in connection with microcolumns, as shown by the separation of chlorinated phenols (Fig. 9) (Krejčí *et al.*, 1984a). By connection of an amperometric detector with a microcolumn with a consumption of only 1 ml of the sample, the minimal detectable concentration is of the order of ng l^{-1} in magnitude of chlorinated phenols in water by a direct injection of the sample onto the microcolumn.

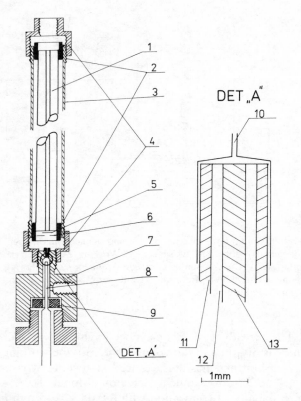

Figure 6. Construction of wall-jet amperometric cell and connection with compact glass cartridge microcolumn. 1. Glass microbore column; 2. metal ring of column; 3. metal jacket; 4. screw cap; 5. filter paper; 6. PTFE seal; 7. detector cell body; 8. working electrode; 9. silicon seal; 10. input capillary; 11. stainless steel working electrode jacket; 12. PTFE insulation; 13. platinum gold wire. Reproduced with permission from Šlais *et al.* (1983).

Detectors for OT capillary columns are likewise for OT capillary chromatography at the outset. At present only two examples of amperometric detectors with cell volume below 1 nl can be found (Šlais and Krejčí, 1982; Knecht *et al.*, 1984). The detector shown in Fig. 10 can serve as an example of the cell based on the principle of the wall-jet electrode. The detector was used to detect components in the effluent at the separation of eleven isomeric phenols on the open tubular capillary column (shown in Fig. 11). The total time of the separation did not exceed 15 min and the minimal detectable quantity of hydroquinone was 50 fg or 0.4 fmol (Šlais and Krejčí, 1982).

TENSAMETRIC DETECTORS

In tensametric detection a change in the capacity of the working electrode, C_w, caused by the adsorption of solute at the electrode surface (see Fig. 1) is evaluated. The alternating non-faradaic current caused by applying a suitable

Table 3. Detection limits for the use of 0.7 × 150 mm microcolumns, packed with Lichrosorb RP-18, d_p = 7 μ, (unless stated otherwise). Detector = EMD 10 (Laboratory Instruments, Prague, Czechoslovakia); Pt = working electrode; applied potential = +1.2 V; sample volume = 0.2 μl. From Vespalcová *et al.* (1984).

Solute	Mobile phase	Capacity ratio k	Minimum detectable Concen-tration ($\mu g\,l^{-1}$)	Quantity (pg)
Folic acid	Acetonitrile—water 3/97 (by volume) + 0.1 M NaClO₄	2.0	20	200
Vitamin A—acetate	Acetonitrile + 0.1 M NaClO₄	2.2	23	200
Quercetin	Acetonitrile—water—acetic acid 30/68/2 (by volume) + 0.1 M NaClO₄	3.2	2.3	30
Chlorpromazine	See Fig. 8	2.8	2.5	40
Prochlorperazine		11.5	3.5	100
Sulphanilamide	Acetonitrile—water 3/97	1.4	1.5	6
Sulphamethoxydiazine	(by volume) + 0.1 M NaClO₄	9.4	4	60
4-Chlorophenol	Acetonitrile—water 60/40	1.0	3	20
Pentachlorophenol	(by volume) + 0.1 M NaClO₄ + 0.001 M HClO₄	6.4	11	200
Rolitetracyclin	Acetonitrile—water 25/75 (by volume) + 0.1 M NaClO₄ + 0.001 M HClO₄	5.0	47	500
1, 2-Benzopyrene	Acetonitrile—water 80/20	9.0	1.5	40
20-methylcholanthrene	(by volume) + 0.5 M NaClO₄	15.5	2.5	90
N,N-Dimethyl-4-aminoazobenzene	Acetone—water 75/25 (by volume) + 0.1 M Mg(ClO₄)₂	3.0	3.3	20
2-Naphthylamine	Acetonitrile—water 60/40 (by volume) + 0.1 M NaClO₄ + 0.001 M NH₃	1.2	1.2	6

d.c. voltage modulated with a.c. voltage of several tens of mV amplitude (Jehring, 1974) is measured. With regard to the sensitivity towards adsorption, it is necessary to maintain a permanently well-defined, constant electrode surface state. Hence the material of the working electrode is limited practically exclusively to liquid mercury which can meet this demand in the polarographic arrangement. To prevent oscillations of the measured current and thus achieve minimum detectable concentrations, working with a static mercury drop electrode (SMDE) is unavoidable (de Jong *et al.*, 1983a, b; Bond and Jones, 1983). Consider the cell diagram of Fig. 1; in tensametric measurements there is usually no need to take into account C_e, R_s and C_s since their impedances are much higher in comparison with others and

72   K. Šlais

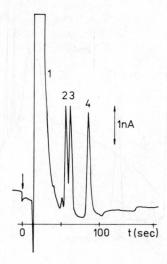

Figure 7. An example of fast separation of catecholamines on a microcolumn (0.7×30 mm) packed with Separon SI C-18 particle diameter, $d_p = 3$ μm). Mobile phase: 0.1 M NaClO$_4$, 0.001 M EDTA, 0.001 M HClO$_4$ in water, linear velocity $u = 2.5$ mm s^{-1}. EMD-10 detector with gold working electrode, applied potential $= 1.2$ V. Sample: 2. 1.0 μl containing 2.0×10^{-6} M noradrenaline; 3. 2.0×10^{-6} M DOPA; 4. 2.7×10^{-6} M adrenaline and 1. 5 wt % sodium naphthalenesulphonate stabilized with 2×10^{-4} M sodium bisulphite, acidified with HClO$_4$ to pH 2. Reproduced with permission from Kouřilová *et al.* (1984).

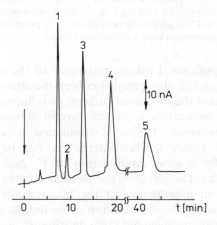

Figure 8. Separation of phenothiazines by straight-phase liquid chromatography with amperometric detection. Column: CGC 0.7×150 mm, packed with LiChrosorb SI 100, $d_p = 5$ μm. Mobile phase: acetonitrile $+ 0.001$ M NaClO$_4$ $+ 0.01$ M NH$_3$. Detector: EMD-10 with platinum working electrode, applied potential $= 1.2$ V. Sample: 1. Levopromazin (23 ng); 2. Clorprothixen (45 ng); 3. Chlorpromazin (18 ng); 4. Thioridazin (20 ng); 5. Prochlorperazin (19 ng) in mobile phase. Reproduced with permission from Krejčí *et al.* (1984). © 1984. Academia, Publishing House of the Czechoslovakian Academy of Sciences.

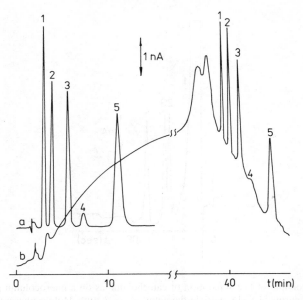

Figure 9. Separation of chlorinated phenols by reverse phase liquid chromatography with amperometric detection: (a) common sampling procedure, sample volume = 0.2 μl; (b) sampling by peak compression technique, sample volume = 1000 μl. Column: CGC 0.7 × 150 mm, packed with LiChrosorb RP-18, d_p = 7 μm. Detector: EMD-10 with platinum working electrode, applied potential = 1.2 V. Mobile phase: water—acetonitrile 40/60 (by volume) + 0.1 M NaClO$_4$, linear velocity is (a) 1.7 mm s^{-1}; samples 1. 4-chlorophenol (4.6 mg l^{-1}), 2. 2,4-dichlorophenol (6 mg l^{-1}), 3. 2,4,6-trichlorophenol (9 mg l^{-1}), 4. tetra-chlorophenol (3.5 mg l^{-1}), 5. pentachlorophenol (34 mg l^{-1}) in the mobile phase. (b) 1.4 mm s^{-1}; sample 1. 4-chlorophenol (1.6 μg l^{-1}), 2. 2,4-dichlorophenol (4 μg l^{-1}), 3. 2,4,6-trichlorophenol (6 μg l^{-1}), 4. tetrachlorophenol (1.6 g l^{-1}), 5. pentachlorophenol (15 μg l^{-1}) in water. Reproduced with permission from Krejčí *et al.* (1984a).

will not become significant. Leakage resistance of the electrode, R_w, has a particular influence on the phase angle between the alternating component of the applied voltage and the measured current; its influence can be suppressed by phase selective measurement of the current through the cell used in tensametry (de Jong *et al.*, 1983a, b; Bond and Jones, 1983; Lanklema and Poppe, 1976a). Hence the tensametric cell can be approximated in a simplified manner by a serial combination of C_w and R_e (Jehring, 1974; de Jong *et al.*, 1983a). The effect of the resistance of an electrolyte, R_e, will appear particularly in a reduction of the linear range of response. It is, therefore, best minimized by the addition of an indifferent electrolyte or by the potentiostatic cell arrangement. If a maximum of 1% deviation from linearity is admitted, the condition for maximum angular frequency, $\omega = 2\pi f$, of the applied voltage can be derived to be of the following form (de Jong, 1983a):

$$\omega \leqslant 0.1 \times 2^{1/2} \frac{\kappa_e}{r_d C_d} \qquad (15)$$

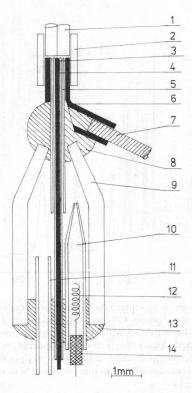

Figure 10. The amperometric cell and its connection to the open tubular capillary column. 1. capillary column; 2. PTFE tube; 3. platinum wire (diameter 0.1 mm); 4 and 5. glass capillaries; 6. stainless steel capillary; 7. connection wire for the auxiliary electrode; 8 and 13. epoxide resin; 9. glass vessel for the reference electrode; 10. reference electrode; 11. output capillary; 12. silver wire; 14. rubber stopper. Reproduced with permission from Šlais and Krejčí (1982). © 1982. Elsevier Scientific Republishing Co.

where r_d is the radius of the mercury drop electrode. Hence the necessity for a least radius of the drop and a least frequency for the applied modulated voltage; frequencies of $f = 20{-}100$ Hz are used in practice. At capacity of $C_d = 10$ $\mu F\, cm^{-2}$ and $r_d = 1$ mm measurement is possible with solutions of conductivities, $\kappa_e > 10^{-3}\,\Omega^{-1}\,cm^{-1}$.

The change in alternating current, caused by the effect on the capacity C_w, by solute adsorption, is proportional to the amount of solute adsorbed at the electrode surface. Under the conditions of tensametric detection in liquid chromatography, the time required for establishing the equilibrium surface concentration is substantially longer than that allowed by the maximal time of the limit drop with the time profile of the eluted peak. The value of the tensametric current is therefore controlled by solute diffusion to the drop time (de Jong, 1983a). From this observation (van Rooijen and Poppe, 1983; Weber and Purdy, 1978; Lanklema and Poppe, 1976b), that electrode noise is proportional to its surface, it was further derived (de Jong, 1983a) that to

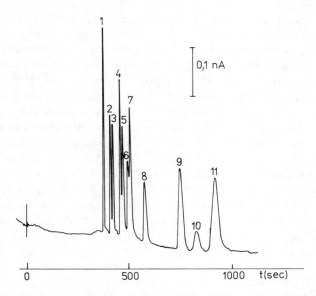

Figure 11. Separation of phenols by open tubular capillary liquid chromatography. Column: 2.8 m × 16 μm ID; Stationary phase: OV 101. Mobile phase: 10^{-3} M HClO$_4$ in distilled water; flow-rate: 1.7 nl s^{-1}. Pressure: 2.5 MPa. Peaks: 1. hydroquinone; 2. 4-methylphenol; 3. 2-methylphenol; 4. 3, 4-dimethylphenol; 5. 3, 5-dimethylphenol; 6. 2, 3-dimethylphenol; 7. 2, 4-dimethylphenol; 8. 2,6-dimethylphenol; 9. 2-methyl-4-ethylphenol; 10. 2-isopropylphenol; 11. 2,4,6-trimethylphenol. Reproduced with permission from Šlais and Krejčí (1982). © 1982. Elsevier Scientific Publishing Co.

obtain the maximum signal-to-noise ratio (*S/N*) it is advantageous to select the smallest possible drop radius, r_d. At the same time, the *S/N* ratio is, to a limited extent, dependent on neither the size nor the frequency of the alternating component of the applied voltage. In addition, the theoretical linear dependence of the tensametric current upon the square root of the mobile phase flow-rate was confirmed for a mobile phase flow-rate greater than 200 mm s^{-1} at a working electrode of $r_d = 0.736$ mm; for lower flow-rates the response is independent of the flow-rate (de Jong, 1983a).

In order to obtain minimum detectable concentration it is necessary that the compounds competing with solute sorption at the working electrode should not be present in the mobile phase. From this viewpoing it is necessary to evaluate the choice of non-polar component in the mobile phase in reverse-phase chromatography since the addition of a less polar component to the aqueous mobile phase decreases the response. It is therefore usually necessary to look for a compromise between the optimal separation and the sensitive detection (de Jong *et al.*, 1983b; Bond and Jones, 1983). Methanol (de Jong *et al.*, 1983b) and acetonitrile (Bond and Jones, 1983) proved best as organic components of the mobile phase.

The number of applications of tensametric detection to LC has been limited up to now, though the first notes on this technique appeared 20 years ago (Kemula *et al.*, 1965a, b). Aromatic hydrocarbons (Kemula *et al.*, 1965a, b), alkyl alcohols, alkylcarbonic acids, sulphonic acids, surface active substances (Lanklema and Poppe, 1976a), ethylene glycol ethers (Bond and Jones, 1983; Lanklema and Poppe, 1976a), cholanoic acids (Kemula and Kutner, 1981; Kutner *et al.*, 1982), benzoic acids (Kutner, 1982), glycosides (de Jong *et al.*, 1983b), steroids (de Jong *et al.*, 1983a, b) were detected tensametrically in LC. De Jong *et al.* (1983b) showed by example of Lynestrol steroid and glycosides that tensametric detection can be complementary to amperometric detection. Using a mobile phase containing 30—70 vol % methanol in water with 0.1 mol l^{-1} KNO$_3$ at a flow-rate of 2 ml min^{-1} with time constant of detection of 3 s, $r_d = 0.935$ mm, drop time of 2 s, applied voltage of 200 mV against an Ag/AgCl electrode modulated at a frequency of 20 Hz with amplitude of 10 mV they achieved the minimum detectable concentration for Lynestrol of 2×10^{-6} mol l^{-1} and a linear response range of two orders of magnitude. They achieved a minimum detectable concentration of 1×10^{-7} mol l^{-1} for digoxin under similar conditions.

None of the above examples considered the possibility of using tensametric detection in miniaturized versions of HPLC. The theoretical conclusions (de Jong *et al.*, 1983a) concerning both linearity and signal-to-noise ratio suggest advantageous properties from miniaturized design of the detector cell. One can therefore anticipate that by a miniaturized polarographic cell (Lloyd, 1983a, b, c) with a suitable connection to a microcolumn, tensametric detection can be used even in low-dispersion liquid chormatography.

The use of solid electrodes for tensametric measurements is not common. However, if the surface is regenerated by pulse oxidation with subsequent reduction, one can by a batch process at a spherical platinum electrode (5 mm^2 surface area) obtain minimum detectable concentrations, e.g. of toluene in water, down to 0.03 mg l^{-1} (Formaro and Trasatti, 1968). However, the total cycle of electrode surface regeneration and measurement lasted for about 2 min. If the regeneration and measurement process could be reduced to a fraction of a second, as with amperometric detection of sugars (Hughes *et al.*, 1981; Hughes and Johnson, 1981; Rocklin and Pohl, 1983) and amino acids (Polta and Johnson, 1983), it would be possible to consider using miniaturized cells with solid working electrodes for tensametric detection in liquid chromatography.

CONDUCTIVITY DETECTORS

Conductivity detectors have recently gained ever increasing popularity, particularly thanks to the extension of ion chromatography (Small *et al.*, 1975; Gjerde *et al.*, 1979, 1980). These are based on measurements of the changes in current through the detection cell caused by changes in the conductivity of the liquid in the cell. The relationship between the conductivity of the liquid and the resistance, R_e, which is the measured value of the detection cell, is given by Equation 7. The influence of electrode impedance on the measurement of R_e can be suppressed, e.g. by the application of an applied a.c. voltage, thus a relatively large capacity C_w with small impedance applies. The use of electrodes with large surfaces (i.e. reduction of R_w) and/or the use of increased applied voltage of up to several tens of volts are additional ways of suppressing the effects of electrical properties of electrodes on the measurement of R_e. Thereby electrolyte decomposition and reduction of electrode impedance are attained. Both these procedures assume the use of a cell with high cell constant, C. The effect of liquid permittivity or electrolyte capacity, C_e, on the magnitude of the measured current through the cell will not apply with d.c. or a.c. applied voltage, E_a. Provided that the fluctuation of conductivity of the measured liquid is the only source of noise, the signal-to-noise ratio and thus the minimal detectable concentration of ionized solute, is independent of the cell geometry (of cell constant C), magnitude and frequency of the applied voltage. A relative accuracy of measurement to about 0.1% (Pohl and Johnson, 1980; Cassidy and Elchuk, 1983) can be achieved by the usual measurement of resistance, R_e [i.e. bridge measurement (Mohnár et al., 1980; Scott and Simpson, 1982)], or the phase-selective detector (Hashimoto et al., 1978, 1981; Sato, 1982).

All the dissociated components of the mobile phase influence the conductivity of the electrolyte. The electrolyte concentration in the mobile phase must therefore be kept as low as possible (Small et al., 1975; Gjerde et al., 1979, 1980). The concentration of ions in the mobile phase is chosen to be less than 10^{-3} mol l^{-1} (Gjerde et al., 1979, 1980), which corresponds to a conductivity of below 10^{-4} Ω^{-1} cm^{-1} for common ions. With the precision of conductivity measurement mentioned above the estimate of minimum detectable ion concentration is less than 10^{-6} mol l^{-1}, or, at an ionic weight of 100, below 0.1 mg l^{-1}. The fact that conductivity is proportional to salt concentration for concentrations less than 0.01% means that a linear range of response of up to 4 orders of magnitude can be anticipated (Pohl and Johnson, 1980).

In order to reach optimal detectable concentrations, the cell of the conductivity detector (and usually also the inlets and chromatographic column particularly at a higher conductivity of the mobile phase) must be thermostatted since the coefficient of thermal dependence for electrolytic conductivity is about 2% of K^{-1}. Examples of the influence of temperature fluctuation on the detectability of ionized solutes are set by, e.g. Cassidy and Elchuk (1983), Glatz and Girard (1982), Jenke and Pagenkopf (1982) and Molnár et al. (1980). Differential connection, when two identical conductivity cells are connected in such a manner that one is connected upstream of the sampling device and the other after the column (Hashimoto et al., 1981; Sato, 1982), makes it possible to compensate for the mobile phase conductivity; however, it does not permit a decrease in the noise level.

The influence of stray capacities, C_s, which appears particularly at higher frequencies of the applied voltage and at low conductivities of the mobile phase, can be suppressed efficiently by the so-called 'bipolar pulse polarization' of the conductivity cell (Johnson and Enke, 1970; Keller 1981). With this technique, two identical rectangular pulses of opposite polarity are successively applied to the cell and the current is sampled at the end of the second pulse. Thereby a measurement precision for resistance R_e of up to 1:30000 can be reached, which enables the detection of changes in concentration of ions as low as 10^{-7} mol l^{-1} for an electrolyte concentration of c. 10^{-3} mol l^{-1} in the mobile phase.

The design of conductivity cells for LC usually has capillaries of electrically non-conductive materials (glass, PTFE), the ends of which are fitted to metallic (stainless steel) capillaries which form the inlet and outlet for the measured liquid on the one hand, and the cell electrodes on the other. This arrangement permits minor broadening of the chromatographic zone in the detector and easy realization of a cell of small volume. Examples of conductivity cells suitable for low-dispersion liquid chromatography are given in Table 4. Figure 12 presents a scheme for a miniaturized conductivity cell. An example of its application to the separation of ions on a microcolumn by suppressed ion chromatography is given in Fig. 13; the minimum detectable concentration of separated ions is about 10^{-6} mol l^{-1}. Separation of fatty acids on a microcolumn by reverse-phase chromatography using a miniaturized conductivity cell to which 30 V of d.c. was applied (Kouřilová *et al.*, 1983) is shown in Fig. 14. The minimum detectable concentration of acids is 6×10^{-7} mol l^{-1}, which is related to the minimum detectable amount 4×10^{-12} mole.

Table 4. Parameters of miniaturized conductivity cells

Cell volume (µl)	Cell constant (cm^{-1})	Application	References
0.33	300	Suppressed ion chromatography	Rokushika *et al.* (1983)
0.080	5.2	Measurement of extra column effects	Scott and Simpson (1982)
0.100	2500	RP chromatography of fatty acids	Kouřilová *et al.* (1983)
0.500	0.8	—	Tesařík and Kaláb (1973)

PERMITTIVITY DETECTORS

To determine changes in concentration of analysed compounds in the chromatographic column by permittivity detection, effluent variations in the liquid capacity of the cell, C_e, caused by the presence of solute in a given mobile phase are used. The dependence between C_e and liquid permittivity is given by Relationship 8. From the general scheme of Fig. 1, the effect of electrode impedance is suppressed by using a.c. applied voltage. The cell of the permittivity detector can then be viewed as a parallel combination of the capacity, C_e, and the resistance of the mobile phase, R_e. When measuring

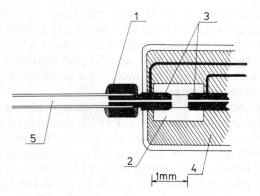

Figure 12. Schematic diagram of miniaturized conductivity cell. 1. stainless steel tube; 2. PTFE tube; 3. electrode; 4. epoxy resin; 5. Nafion tube (0.2 mm ID). Reproduced with permission from Rokushika *et al.* (1983). © 1983. Elsevier Scientific Publishing Co.

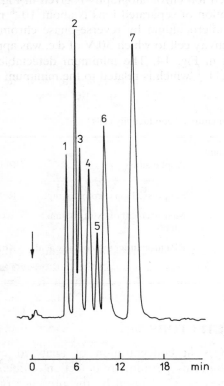

Figure 13. Chromatogram of inorganic anions separated by suppressed ion chromatography. Column: 190 μm × 47 cm, YEW AX-1 resin. Eluent: 4 mM Na_2CO_3 and 4 mM $NaHCO_3$ (pH 10.2). Temperature: 40° C. Flow-rate: 1.9 μl min^{-1}. Pressure: 4.8 MPa. Sample size: 20 μl. Splitting ratio = 70:1. Peaks: 1. F^- (1.4 ng); 2. Cl^- (2.8 ng); 3. NO_2^- (4.2 ng); 4. PO_4^{3-} (8.4 ng); 5. Br^- (2.8 ng); 6. NO_3^- (8.4 ng); 7. SO_4^{2-} (11.2 ng). Reproduced with permission from Rokushika *et al.* (1983). © 1983. Elsevier Scientific Publishing Co.

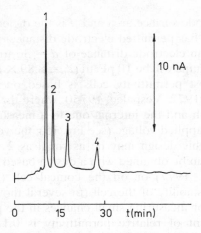

Figure 14. Chromatogram of monocarboxylic acids. Column: 750 × 0.5 mm ID, packed with separon SI C-18, d_p = 10 μm. Mobile phase: Methanol—water 40/60 by volume with 0.01% (by mass) of lauric acid. Flow-rate: 9 μl min^{-1}. Acids: 1. acetic; 2. propionic; 3. butyric; 4. valeric. Reproduced with permission from Kouřilová *et al.* (1983). © 1983 Academia, Publishing House of the Czechoslovakian Academy of Sciences.

capacity, it is necessary to suppress the influence of R_e, i.e. to reduce conductivity current of the cell in comparison with the capacitive current. Using the relationship for condenser capacitive current,

$$I_\varepsilon = E_a \omega C_e$$

the conductivity current

$$I_\kappa = E_a / R_e$$

and Equations 7 and 8, the I_ε / I_κ ratio is obtained as:

$$I_\varepsilon / I_\kappa = \omega R_e C_e = \omega \, \frac{\varepsilon_e}{\kappa_e} \tag{16}$$

It is obvious from Relationship 16 that the ratio of the capacitive to the conductivity current does not depend on the cell design and can therefore be increased only by increasing the applied voltage frequency. That is why high frequencies, usually in the range of 1—25 MHz, are used with permittivity detectors.

The demands on the small volume of a detection cell, governed by the need to minimize chromatographic zone broadening, also leads to a decrease in C_s such that even the stray capacity C_s usually plays a significant part. The detector cell design therefore aims at maximizing the C_e / V_d ratio or reaching the greatest possible cell capacity with the smallest possible volume. Consider the relationship for the capacity of a plate condenser, $C_e = \varepsilon_e A/d$, if the volume of the liquid dielectric is $V_d = Ad$, then the C_e / V_d ratio is

$$C_e / V_d = \frac{\varepsilon_e}{d^2} \tag{17}$$

where A is the electrode surface area and d is the dielectric thickness. Up to now the construction has permitted electrode distances down to several tens of a micrometer. At an electrode distance of $d = 30$ μm the ratio of the air capacity to the cell volume will be 10 pF/μl ($\varepsilon_0 = 8.89 \times 10^{-12}$ F m^{-1}).

The design of most permittivity cells is based on a coaxial condenser (Vespalec and Hána, 1972; Vespalec, 19750, where the external electrode is connected to the earth and the internal one to a measuring device M and a source of alternating applied voltage (see Fig. 15); the volume of the working space of the cell of this design may be as small as 2 μl (Vespalec, 1975). Comparable results can be obtained with a design based on the thin-layer cell (Poppe and Kuysten, 1977) or on the conical cell (Alder *et al.*, 1981). Maintaining thermal stability of the cell (to several thousandths of a degree K) is a prerequisite for measuring small changes in the permittivity since the temperature coefficient of relative permittivity is 0.1—0.5% K^{-1} (Slavík, 1978). In optimal cases, relative changes in permittivity of 7×10^{-7} (Vespalec and Hána, 1972; Vespalec, 1975), 1.5×10^{-7} (Poppe and Kuysten, 1977) and 1.6×10^{-6} (Alder *et al.*, 1981) can be indicated. Electronic compensation for temperature changes (Alder *et al.*, 1983) is less effective if compared with perfect thermostatting.

A serious disadvantage of permittivity detectors, in particular in comparison with refractometric detectors, is that they either cannot be applied at all (Vespalec and Hána, 1972; Vespalec, 1975; Poppe and Kuysten, 1977; Watanabe *et al.*, 1977; Benningfield and Mowery, 1981) or only with slight sensitivity (Alder *et al.*, 1981; Klatt, 1976; Mowery, 1982) to solute detection in mobile phases of high permittivities, especially in aqueous

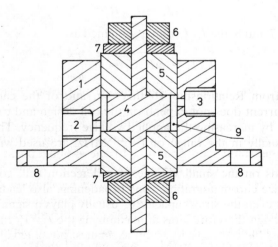

Figure 15. Schematic cross-section of the permittivity cell. 1. outer stainless steel earthed electrode; 2 and 3. openings for outlet and inlet tubings; 4. inner stainless steel electrode; 5. Teflon sealing ring; 6. nut; 7. support; 8. openings for fixing screws; 9. dielectric space: 30 μm thick, 10 mm diameter, air capacity 20 pF, volume 2.0 μl. Reproduced with permission from Vespalec (1975). © 1975. Elsevier Scientific Publishing Co.

organic mixtures used in reverse-phase chromatography. This can be avoided partially by measuring the phase shift (Klatt, 1976) or by using relatively high frequencies (Alder *et al.*, 1981).

The conductivity current was also used for measuring the cell time constant, i.e. the product of R_e and C_e (Hashimoto *et al.*, 1978), or for simultaneous measurements of permittivity and conductivity with using applied voltage at two frequencies (Alder *et al.*, 1984).

Permittivity detectors, despite their well-elaborated theory (Alder and Thoër, 1979; Alder *et al.*, 1981; Benningfield and Mowery, 1981; Haderka, 1974) are scarcely used. From the viewpoint of eventual cell miniaturization it is important to optimize the distance between electrodes, the influence of stray capacities and the thermostatting of the cell together with its inlets.

POTENTIOMETRIC DETECTORS

The equilibrium potential of the reversible electrode process, E_i, is measured at potentiometric detection. It is reached with practically zero current passage through the measuring cell, i.e. the measuring device has a considerable input resistance (ideally, infinitely great). The measuring device, M, gives a potential, E, equal to the sum of the reference electrode potential, E_r, and the potential of the electrode, E_i. Potential E_i corresponding to the equilibrium of chemical potentials of solute ions in the solvent and on the electrode, can be described by the Nernst equation. Since more components are usually present in the solution, the resulting mixed potential is defined by the following relationship (Dvořák and Koryta, 1983b; Manz and Simon, 1983; Schultz and Mathis, 1974):

$$E = E_r + E_i$$

$$= E_r + E_o + \frac{2.303\,RT}{z_i F}\,[\log(a_i + K_{ij}^{Pot} a_j^{z_i/z_j})] \qquad (18)$$

where a_i and z_i are the activity and charge of the solute ion in solution, respectively; K_{ij}^{Pot} is the selectivity factor or the measure of suppression of the interfering ion in relation to the solute ion (ideally 0). It follows from the last mentioned relationship that the measured potential is proportional to the logarithm of the solute concentration in the mobile phase and that this potential is not dependent on the geometrical design of the sensing element or on the volume of the measuring cell. The logarithmic dependence of concentration potential on the concentration at potentiometric detection in liquid chromatography shows peaks which are broader and have poorer resolution in comparison with detectors with a linear dependence on the signal from solute concentration (Manz and Simon, 1983).

Selectivity is an advantageous property for the majority of potentiometric sensing elements (see Equation 18). For LC detection the following were used: a silver electrode for the detection of halides (Deguchi *et al.*, 1978; Hershcovitz *et al.*, 1982; Suzuki *et al.*, 1983; Akaiwa *et al.*, 1982); electrodes with liquid membranes for the detection of inorganic anions (Fröbe *et al.*, 1983; Schultz and Mathis, 1974; Suzuki *et al.*, 1983), alkali metal cations and quarternary ammonium bases (Manz and Simon, 1983); and, finally, copper electrodes for the detection of amino acids (Alexander *et al.*, 1981; Alexander and Maitra, 1981). Detection selectivity is not unavoidable in liquid chromatography, and, therefore, non-selective potentiometric detection based on varying diffusivity of ions through an ion-exchange membrane (Deelder *et al.*, 1981) was also proposed.

Detection cells based on the use of ion-selective electrodes for batch measurements (Schultz and Mathis, 1974; Deguchi *et al.*, 1978) and some of the earlier designs (Hershcovitz *et al.*, 1982) are little suitable even for chromatography on columns of diameters of *c.* 4 mm since their volumes are large. Designs based on thin-layer cells which house the potentiometric electrode (Suzuki *et al.*, 1983; Deelder *et al.* 1981) or on the tubular electrode (Alexander *et al.*, 1981; Alexander and Maitra, 1981) are better.

The minimum detectable concentration is highly dependent on the type of detected ion and the electrode. For instance, for a halide electrode, minimum detectable concentrations are 10^{-5} mol l^{-1} for Cl^- ion, up to 10^{-7} mol l^{-1} for J^- ion. For an electrode with a liquid ion exchanger, minimum detectable concentrations of NO_2^-, NO_3^- and Br^- are *c.* 10^{-6} mol l^{-1} (Suzuki *et al.*, 1983). In the above instances, in contradiction with Equation 18, the dependence of the potential on the logarithm of the concentration has a nonlinear course, namely, at lower concentrations. In the case of the use of a non-selective ion-exchange membrane, the detection limit of quarternary ammonium bases was *c.* 7 nmol (Deelder *et al.*, 1981) using a mobile phase containing 0.05 mol l^{-1} phosphate buffer, 0.5 nmol l^{-1} dodecyl sulphate and 5% of isopropanol in water with the injection of 20 µl sample onto a 150 × 4.6 mm column.

The main factor limiting the use of the potentiometric detection is slow establishment of the equilibrium electrode potential. For instance, even with cell volume of 0.8 µl and mobile phase flow-rate of $F_m = 20$ µl s^{-1} (Deelder *et al.*, 1981) is the time detection constant $\tau_d = 1.3$ s, and at flow-rate of 2.5 µl s^{-1} is $\tau_d = 5.8$ s. This fact seriously restricts the use of this cell for fast analyses.

For OT capillary columns the demand on the speed of response is much less severe (see earlier). Manz and Simon (1983) and Fröbe *et al.* (1983) made use of this fact to design a potentiometric detector with an effective cell volume of several tens of picolitres; this was based on an ion-selective electrode with a liquid exchanger. Figure 16 illustrates the design of this detection device. Using a mobile phase containing 10^{-3} mol l^{-1} KNO_3 in water and a 0.225 × 600 mm OT capillary column, a detection limit of about 1 pmol was achieved for K^+ ions.

The disadvantage of a large time detection constant for ion-selective electrodes can be avoided partially by using a bipolar pulse conductivity measurement (Jenke and Pagenkopf, 1982; Johnson and Enke, 1970) proposed for commercial ion-selective calcium (Powley *et al.*, 1980) and

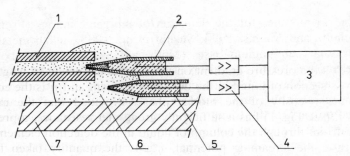

Figure 16. Schematic diagram of the ion-selective detection system for open tubular capillary columns. 1. Separation capillary; 2. ion-selective electrode; 3. voltmeter/recorder; 4. impedance converter; 5. reference electrode; 6. eluent drop; 7. glass plate. Reproduced from Manz and Simon (1983) by permission of Preston Publications, Inc.

fluoride (Powley and Nieman, 1982) electrodes. The establishment of equilibrium within several milliseconds, the avoidability of a real reference electrode, a response independent of the ionic strength of the solution and elimiation of the influence of the liquid potential are suggested as the main advantages of this procedure. The response then depends on the concentration, not on the activity of ions in solution. The bipolar pulse measurement with ion-selective electrodes was made also in a system with a flow-through thin-layer cell of volume, $V_d = 300$ μl, and thickness $d = 1.6$ mm (Powley and Nieman, 1983). At a flow-rate of 20 ml min^{-1} and pulse height of 2—5 V, the time necessary for the establishment of 95% of the full scale is shorter than 1 s, and at a flow-rate of 100 ml min^{-1} it is shorter than 0.5 s.

ELECTROKINETIC DETECTORS

Detection based on electrokinetic phenomenon cannot be simply classified according to the above scheme, since it is based on the transport of the diffusion portion of the electrode double-layer which is formed at the interface between the solid and liquid phases (Šlais and Krejčí, 1978; Krejčí *et al.*, 1978). Under the conditions of chromatographic analysis with straight phases, i.e. with the use of mobile phases of low permittivities, the streaming current I_s, is measured at electrokinetic detection, which is essentially the product of the mobile phase flow-rate and the charge density of the mobile phase which is in contact with the surface of the detection element. The relation between I_s and the potential between the solid phase and mobile portion of the diffusion electric double-layer i.e. zeta potential, ζ, can then be illustrated by the relationship

$$I_s = F_m \frac{\kappa_e}{D} \xi \tag{19}$$

For the conditions of reverse-phase chromatography, i.e. for the use of liquids of high permittivities, the magnitude of the streaming current, I_s, is given by the relationship (Terabe *et al.*, 1981)

$$I_s = \varepsilon_e \zeta \, \frac{\Delta P}{\Delta L} \, \frac{S}{\eta} \tag{20}$$

where ΔP is the pressure gradient on the detection element of length ΔL, S is the cross-section of liquid in the detection element and η is the coefficient of dynamic viscosity of the mobile phase. Provided that the measuring device, M, (see Fig. 1) has a sufficiently great inlet resistance, a preferential electric current through the column of liquid in the detection element occurs; in this case, the streaming potential, φ_s, is the quantity taken from the measuring device and is given by the Smoluchovský relationship (Kemula *et al.*, 1983b)

$$\varphi_s = \frac{\varepsilon_e}{\kappa_e} \, \frac{\Delta P}{\eta} \, \zeta \tag{21}$$

Equations 19—21 are the boundary relationships; practice usually approaches one of them more or less. The general theory of transport of charge and liquid in charged capillaries was described, e.g. by Sørensen and Koefoed (1974). For all cases of Equations 19—21 the influence of the solute on the measured magnitude is caused by the solute adsorption on the phase interface and thus by the change in zeta potential. The influence of the solute on I_s or φ_s by ε_s or κ_e is less significant. It further follows from Relationships 19—21 that the measured value is directly proportional to the mobile phase flow-rate or the pressure gradient on the detection element, and therefore F_m must be stabilized perfectly.

A detection element usually consists of one (Šlais and Krejčí, 1978; Kemula *et al.*, 1983a) or more capillaries (Šlais and Krejčí, 1978; Terabe *et al.*, 1981; Kemula *et al.*, 1983a), of PTFE (Šlais and Krejčí, 1978; Kemula *et al.*, 1983a) or stainless steel (Kemula *et al.*, 1983a). Additionally, the sorption bed shaped like a capillary of length 2 mm and diameter 1 mm (Šlais and Krejčí, 1978) or length 7 mm and diameter 0.5 mm (Terabe *et al.*,1981) packed with silica gel or alumina (Šlais and Krejčí, 1978; Terabe *et al.*, 1981), modified silica gel or TiO_2 (Terabe *et al.*, 1981) can be used as an electrokinetic detector. An example of this arrangement for measurement of the streaming current is given in Fig. 17. The possibility of directly sensing the electrochemical signal from the chromatographic column (Krejčí *et al.*, 1978, 1981) is of no little importance.

The possibility of constructing electrokinetic detection sensing elements of liquid volume below 1 μl in the detection element (Šlais and Krejčí, 1978; Terabe *et al.*, 1981; Kemula *et al.*, 1983a) is an important factor in miniaturization. In the case of signal sensing by chromatographic column, extra column detection volume is zero (Krejčí *et al.*, 1978, 1981). If the streaming current, I_s, is measured in the separation of solutes on the microcolumn, it is necessary, in agreement with Relationships 19 and 20, to take account of problems associated with the measurement of small electric currents (Krejčí *et al.*, 1978, 1981).

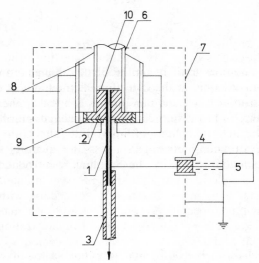

Figure 17. Schematic presentation of the electrokinetic detector. 1. Bundle of about 200 capillaries of 20 mm length and 20 μm diameter, prepared by drawing a glass tube packed with glass capillaries; 2. PTFE seal; 3. PTFE capillary for draining of the mobile phase; 4. connector; 5. electrometric amplifier; 6. chromatographic column; 7. shielding; 8. screw closure of the column; 9. metal washer for column closure; 10. filter paper. Reproduced with permission from Šlais and Krejčí (1978). © 1978. Elsevier Scientific Publishing Co.

However, long-term instability of the base line and reproducibility of the response are the main problems of electrokinetic detection. A certain improvement can be achieved by adding ionogenic surface active compounds into the mobile phase. This leads to impairment in reverse-phase chromatography (Kemula *et al.*, 1983b; Terabe *et al.*, 1982) and enhancement in straight-phase chromatography (Vespalec, 1981) for minimum detectable concentrations.

There have been only a few applications of electrokinetic detection in LC. The detection limits of bile acids and palmitic acid were about 20 ng (Terabe *et al.*, 1982), and 10 ng (Terabe *et al.*, 1981), respectively, for streaming current sensing in reverse-phase chromatography on a column of 250 × 4.6 mm. The minimum detectable concentration of propionic acid was 5×10^{-7} mol l^{-1} for streaming potential sensing under analogous conditions (Kemula *et al.*, 1983b). A linear range of response of three orders of magnitude can be reached in both direct (Kemula *et al.*, 1983b) and reverse-phase chromatography (Šlais and Krejčí, 1978). As for sensitivity, up to now electrokinetic detection has been capable of competing with none of the current detection modes. The simplicity and possible minimum extra column broadening of the eluted zone are amongst its advantages.

CONCLUSION

Electrochemical detectors represent a group of devices which can be used, according to the character of the compounds for detection, to solve a number of analytical problems. The suitability of a particular detection mode depends on the chromatographic system. To choose correctly and for the sake of optimization it is necessary to understand the principles of electrochemical instrumentation. Understanding of the chemical and physical processes to which a solute is subjected in the electrochemical cell is still more important. The present state and possibilities for the application of various types of detection in low-dispersion liquid chromatography can be summarized as follows.

Amperometric or coulometric detectors are the most frequently used, most sensitive and most highly investigated electrochemical detectors. Thanks to their simple design, they permit easy miniaturization of the electrochemical cell. This miniaturization in connection with packed microcolumns or open tubular capillary columns, leads to a mainly coulometric character of response from this detectors. Miniaturization together with optimization of cell geometry, leading to an enhancement of solute transport to the working electrode, gives rise to the possibility of speeding up the surface electrode reaction. With respect to the irreversible character of electrochemical reactions of organic compounds, however, this optimization may not lead to an improvement of the signal-to-noise ratio in all cases, particularly in the amperometric cell regime. A reduction in the surface area of the working electrode is unavoidably connected with problems arising from the measurement of small electric currents. The existing cell designs permit, by their hydrodynamic properties, a non-distorted record even if combined with open tubular capillary columns, at the expense of impaired minimum detectable concentration. To obtain a minimum detectable quantity of below 1 fmol, an arrangement is necessary for which the noise of an entire arrangement, with the exception of the working electrode, will be substantially less than 1 pA. It is, however, possible to anticipate a decrease in the noise by one order of magnitude at the maximum, below the last mentioned value.

Tensametric electrodes have, thanks to their principle, prerequisites for being universal electrochemical detectors. They have received relatively little attention up to now; the influence of interfering components in the mobile phase and the restriction of the working electrode practically exclusively to liquid mercury are complicating factors. This method of detection has not been used for low-dispersion LC yet, even though theoretical and instrumental prerequisites for its application to packed microcolumns are fulfilled. An intensive investigation, particularly from the viewpoint of using solid working electrodes, is necessary for further development.

Conductivity detection is, after amperometric, the most frequently used electrochemical method, primarily thanks to the development of ion chromatography. Low conductivity of the solution in the detection cell an cell thermostatting, which at best includes the chromatographic column, are assumptions for sensitive detection. The simple construction of miniaturized cells permits conductivity detection in connection with packed microcolumns; its use with open tubular capillary columns can be anticipated in the near future.

According to the theory, permittivity detectors are universal detectors and their sensitivity does not depend on cell volume. Construction stray capacities, which together with thermal stability determine minimum detectable solute concentrations, are the factors limiting miniaturization. The influence of mobile phase conductivity on the measurement of the capacity of the permittivity cell is the most serious restriction of the universal application of the permittivity detector. This effect can be removed neither by the cell design nor to a certain extent, by electric connection of the cell. Despite these limitations, realization of the permittivity detector can be anticipated for microbore packed columns.

Potentiometric detectors, or detectors based on the use of ion-selective electrodes, offer the possibility of cells of negligible detection volumes, as was shown by the realization of cells of several tens of picolitres in volume. Their wider application to low-dispersion liquid chromatography is prevented by a slow nonlinear change in potential depending on the change of concentration of solute in the cell. In some cases, undesird phenomena can be removed by conductivity measurement with ion-selective electrodes. Additional study of this group of detectors may extend their use in liquid chromatography.

Electrokinetic detection has not been studied to such an extent that it might become a routine detection technique. The simplicity of design, the possibility of realizing detection cells of negligible volume and of sensing the signal from the chromatographic column are amongst its advantages. The complexity of the electrochemical process and the dependence up on the state of the interface has not made it possible for detection parameters to be influenced effectively in the desired manner. The place of electrokinetic detection amongst other techniques will only be shown by additional studies.

The miniaturization of the effective cell volumes of flow-through detectors is unavoidable for increasing speed, improving mas sensitivity as well as economizing whilst enabling the use of expensive special mobile phases are offered by low-dispersion liquid chromatography. These advantages are such that intensive research activities on mastering the problems associated with miniaturization of liquid chromatographic systems are worthwhile.

REFERENCES

Akaiwa, H., Kawamoto, H., and Osumi, M. (1982). Simultaneous determination of bromide and chloride in natural waters by ion-exchange chromatography and direct potentiometry with an ion-selective electrode. *Talanta 29*, 689—690.

Alder, J. F., Drew, P. K. P., and Fielden, P. R. (1981). Improved high-frequency permittivity detector for the chromatography of alcohols and other polar species. *J. Chromatogr. 212*, 167—177.

Alder, J. F., Drew, P. K. P., and Fielden, P. R. (1983). Microcomputer-based temperature corrected permittivity detector for liquid chromatography. *Anal. Chem. 55*, 256—262.

Alder, J. F., Fielden, P. R., and Clark, A. J. (1984). Simultaneous conductivity and permittivity detector with a single cell for liquid chromatography. *Anal. Chem. 56*, 985—988.

Alder, J. F., and Thoër, A. (1979). Simple theoretical description and practical evaluation of a permittivity detector for flowing stream monitoring. *J. Chromatogr. 178*, 15—26.

Alexander, P. W., Haddad, P. R., Low, G. K. C., and Maitra, C. (1981). Application of a copper tubular electrode as a potentiometric detector in the determination of amino acids by high-performance liquid chromatography. *J. Chromatogr. 209*, 29—39.

Alexander, P. W., and Maitra, C. (1981). Continuous-flow potentiometric monitoring of α-amino acids with copper wire and tubular electrodes. *Anal. Chem. 53*, 1590—1594.

Armentrout, D. N., McLean, J. D., and Long, M. W. (1979). Trace determination of phenolic compounds in water by reversed phase liquid chromatography with electrochemical detection using a carbon—polyethylene tubular anode. *Anal. Chem. 51*, 1059—1045.

Benningfield, L. V. Jr., and Mowery, R. A. Jr. (1981). A commercially available dielectric constant detector for liquid chromatography and its applications. *J. Chromatogr. Sci. 19*, 115—123.

Blaedel, W. J., and Wang, J. (1981). Symmetrical two-electrode pulsed-flow detector for liquid chromatography. *Anal. Chem. 53*, 78—80.

Bond, A. M., and Jones, R. D. (1983). Microprocessor-based tensammetric detection for liquid chromatography. *Anal. Chim. Acta 152*, 13—24.

Carlsson, A., and Lundström, K. (1984). A miniaturized amperometric detector. Abstracts — Poster 39. 'Symposium on Liquid Chromatography in the Biomedical Sciences', Ronneby, Sweden, 18—21 June, 1984.

Cassidy, R. M., and Elchuk, S. (1983). Dynamic and fixed-site ion-exchange columns with conductometric detection for the separation of inorganic anions. *J. Chromatogr. Sci. 21*, 454—459.

Caudill, W. L., Howell, J. O., and Wightman, R. M. (1982). Flow rate independent amperometric cell. *Anal. Chem. 54*, 2532—2535.

Curran, D. J., and Tougas, T. P. (1984). Electrochemical detector based on a reticulated vitreous carbon working electrode for liquid chromatography and flow injection analysis. *Anal. Chem. 56*, 672—678.

Deelder, R. S., Linssen, H. A. J., Koen, J. G., and Beeren, A. J. B. (1981). A potentiometric membrane cell as a detector in liquid chromatography. *J. Chromatogr. 203*, 153—163.

Deguchi, T., Kuma, T., and Nagai, H. (1978). Application of a chloride ion-selective electrode as a detector in gel chromatography. *J. Chromatogr. 152*, 349—355.

de Jong, H. G., Kok, W. Th., and Bos, P. (1983a). Some considerations on the design of tensammetric flow-through detectors. *Anal. Chim. Acta 155*, 37—46.

de Jong, H. G., Voogt, W. H., Bos, P., and Frei, R. W. (1983b). Tensammetric detection in high performance liquid chromatography. Application to lynestrenol and some cardiac glycosides. *J. Liq. Chromatogr. 6*, 1745—1758.

DiCesare, J. L., Dong, M. W., and Ettre, L. S. (1981). Very-high-speed liquid column chromatography. The system and selected applications. *Chromatographia 14*, 257—268.

Dvořák, J., and Koryta, J. (1983a). *Elektrochemie*. Academia, Praha, pp. 276—284.

Dvořák, J., and Koryta, J. (1983b). *Elektrochemie*. Academia, Praha, pp. 366—369.

Elbicki, J. M., Morgan, D. M., and Weber, S. G. (1984). Theoretical and practical limitations on the optimization of amperometric detectors. *Anal. Chem. 56*, 978—985.

Erni, F. (1983). The limits of speed in high-performance liquid chromatography. *J. Chromatogr. 282*, 371—383.

Formaro, L., and Trasatti, S. (1968). Capacitance measurements on platinum electrodes for the estimation of organic impurities in water. *Anal. Chem. 40*, 1060—1067.

Fröbe, Z., Richon, K., and Simon, W. (1983). Anion selective microelectrodes as femtolitre cell volume detectors for open-tubular column LC. *Chromatographia 17*, 467—468.

Gjerde, D. T., Fritz, J. S., and Schmuckler, G. (1979). Anion chromatography with low-conductivity eluents. *J. Chromatogr. 186*, 509—519.

Gjerde, D. T., Schmuckler, G., and Fritz, J. S. (1980). Anion chromatography with low-conductivity eluents. II. *J. Chromatogr. 187*, 35—45.

Glatz, J. A., and Girard, J. E. (1982). Factors affecting the resolution and detectability of inorganic anions by non-suppressed ion chromatography. *J. Chromatogr. Sci. 20*, 266—273.

Golay, M. J. E. (1958). Theory of chromatography in open and coated tubular columns with round and rectangular cross-sections. In: *Gas Chromatography*, D. H. Desty (ed.), Butterworths, London, pp. 36—53.

Goto, M., Koyanagi, Y., and Ishii, D. (1981). Electrochemical detector for micro high-performance liquid chromatography and its application to the determination of aminophenol isomers. *J. Chromatogr. 208*, 261—268.

Guinebault, P., and Broquaire, M. (1981). Large-volume injection of samples dissolvd in a noneluting solvent. *J. Chromatogr. 217*, 509—522.

Guiochon, G. (1980). Optimization in liquid chromatography. In: *High-Performance Liquid Chromatography. Advances and Persoectives, Vol. 2*, C. Horváth (ed.), Academic Press, New York, pp. 1—56.

Guthrie, E. J., and Jorgenson, J. W. (1984). On-column fluorescence detector for open-tubular capillary liquid chromatography. *Anal. Chem. 56*, 483—486.

Guthrie, E. J., Jorgenson, J. W., and Dluzneski, P. R. (1984). On-column helium cadmium laser fluorescence detector for open-tubular capillary liquid chromatography. *J. Chromatogr. Sci. 22*, 171—176.

Haderka, S. (1974). Permittivity and conductivity detectors for liquid chromatography. *J. Chromatogr. 91*, 167—179.

Hagihara, B., Kogoh, K., Saito, M., Shiraishi, S., Hashimoto, T., Tagawa, K., and Wada, H. (1983). Novel coulometric detector for high-performance liquid chromatography. *J. Chromatogr. 281*, 59—71.

Hanekamp, H. B. (1981). *Polarographic continuous-flow detection*. Academisch proefschrift, Vrije Universiteit te Amsterdam, Amsterdam.

Hanekamp, H. B., and de Jong, H. G. (1982). Theoretical comparison of the performance of electrochemical flow-through detectors. *Anal. Chim. Acta 135*, 351—354.

Hanekamp, H. B., and van Nieuwkerk, H. J. (1980). Theoretical considerations on the performance of electrochemical flow-through detectors. *Anal. Chim. Acta 121*, 13—22.

Hashimoto, Y., Asui, Y., Moriyasu, M., and Uji, A. (1981). Microdetermination of bile acids by high-performance liquid chromatography equipped with a high sensitive conductance detector. *Anal. Lett. 14*, 1483—1492.

Hashimoto, Y., Moriyasu, M., Kato, E., Endo, M., Miyamoto N., and Uchida, H. (1978). Capacitance—conductance detector for high-pressure liquid chromatography. Application to alkaloids and fatty acids. *Mikrochim. Acta (Wien) II*, 159—167.

Hershcovitz, H., Yarnitzky, Ch., and Schmuckler, G. (1982). Ion chromatography with potentiometric detection. *J. Chromatogr. 252*, 113—119.

Hirata, Y., Lin, P. T., Novotný, M., and Wightman, R. M. (1980). Small-volume electrochemical detector for microcolumn liquid chromatography. *J. Chromatogr. 181*, 287—294.

Hofmann, K. and Halász, I. (1979). Mass transfer in ideal and geometrically deformed open tubes. I. Ideal and coiled tubes with circular cross-section. *J. Chromatogr. 173*, 211—228.

Hofmann, K. and Halász, I. (1980). Mass transfer in ideal and geometrically deformed open tubes. III. Deformed metal and plastic tubes. *J. Chromatogr. 199*, 3—22.

Huber, J. F. K. (1978). The chromatographic apparatus from the viewpoint of system theory. In: *Instrumentation for High-Performance Liquid Chromatography*. J. F. K. Huber (ed.), Elsevier, Amsterdam, pp. 1—9.

Hughes, S., and Johnson, D. C. (1980). Amperometric detection of simple carbohydrates at platinum electrodes in alkaline solution by application of a triple-pulse potential waveform. *Anal. Chim. Acta 132*, 11—22.

Hughes, S., Meschi, P. L., and Johnson, D. C. (1981). Amperometric detection of simple alcohols in aqueous solutions by application of a triple-pulse potential waveform at platinum electrodes. *Anal. Chim. Acta 132*, 1—10.

Hupe, K.-P., Jonker, R. J., and Rozing, G. (1984). Determination of band-spreading effects in high-performance liquid chromatographic instruments. *J. Chromatogr. 285*, 253—265.

Ishii, D., and Takeuchi, T. (1983). Capillary columns in liquid chromatography. In: *Advances in chromatography, Vol. 21*, J. C. Giddings, E. Grushka, J. Cazes, and P. R. Brown (eds.), Marcel Dekker, New York, pp. 131—164.

Ives, D. J. G., and Janz, G. J. (1961). *Reference Electrodes. Theory and Practice*, Academic Press, New York.

Jehring, H. (1974). *Elektrosorptionsanalyse mit der Wechselstrompolarographie*, Akademie Verlag, Berlin.

Jenke, D. R., and Pagenkopf, G. K. (1982). Temperature fluctuations in non-suppressed ion chromatography. *Anal. Chem. 54*, 2603—2604.

Johnson, D. E., and Enke, C. G. (1970). Bipolar pulse technique for fast conductance measurements. *Anal. Chem. 42*, 329—335.

Katz, E. D., and Scott, R. P. W. (1983). Low-dispersion connecting tubes for liquid chromatography systems. *J. Chromatogr. 268*, 169—175.

Keller, J. M. (1981). Bipolar-pulse conductivity detector for ion chromatography. *Anal. Chem. 53*, 344—345.

Kemula, W., Behr, B., Borkowska, Z., and Dojlido, J. (1965a). Adsorption of several organic compounds on the dropping mercury electrode in the system NH_4SCN—dimethylformamide—γ-picoline. *Coll. Czech. Chem. Commun. 30*, 4050—4060.

Kemula, W., Behr, B., Chlebicka, K., and Sybilska, D. (1965b). Chromato-polariographic studies. XXIV. Application of a.c. polarography to the separation of methylnaphthalene isomers. *Rocz. Chem. 39*, 1315—1325.

Kemula, W., Głód, B. K., and Kutner, W. (1983a). Electrokinetic detection in reversed phase high performance liquid chromatography. Part I. Volatile fatty acids. *J. Liq. Chromatogr. 6*, 1823—1835.

Kemula, W., Głód, B. K., and Kutner, W. (1983b). Electrokinetic detection in reversed phase high performance liquid chromatography. Part II. Quaternary ammonium ion-pairs of some volatile fatty acids. *J. Liq. Chromatogr. 6*, 1837—1848.

Kemula, W., and Kutner, W. (1981). Alternating voltage polarographic detection for high-performance liquid chromatography and its evaluation for the analysis of bile acids. *J. Chromatogr. 204*, 131—134.

Kissinger, P. T. (1983). Electrochemical detectors. In: *Liquid Chromatography Detectors*. T. M. Vickrey (ed.), Marcel Dekker, New York, pp. 125—164.

Kissinger, P. T., Refshauge, C., Dreiling, R., and Adams, R. N. (1973). An electrochemical detector for liquid chromatography with picogram sensitivity. *Anal. Lett. 6*, 465—477.

Klatt, L. N. (1976). Universal detector for liquid chromatography based upon dielectric constant. *Anal. Chem. 48*, 1845—1850.

Knecht, L. A., Guthrie, E. J., and Jorgenson, J. W. (1984). On-column electrochemical detector with a single graphite fiber electrode for open-tubular liquid chromatography. *Anal. Chem. 56*, 479—482.

Knox, J. H. (1980). Theoretical aspects of LC with packed and open small-bore columns. *J. Chromatogr. Sci. 18*, 453—461.

Kouřilová, D., Šlais, K., and Krejčí, M. (1983). A conductivity detector for liquid chromatography with a cell volume of 0.1 µl. *Collect Czech. Chem. Commun. 48*, 1129—1136.

Kouřilová, D., Šlais, K., and Krejčí, M. (1984). Modified peak compression sampling in fast micro-HPLC. *Chromatographia. 19*, 297—300.

Krejčí, M., Kouřilová, D., and Vespalec, R. (1981). Electrokinetic detection at differnt points in a narrowbore glass column in liquid chromatography. *J. Chromatogr. 219*, 61—70.

Krejčí, M., Šlais, K., and Kouřilová, D. (1984). Mikrokolony v kapalinové chromatografii. *Chem. Listy 78*, 469—486.

Krejčí, M., Šlais, K., Kouřilová, D., and Vespalcová, M. (1984a). Enrichment techniques and trace analysis with microbore columns in liquid chromatography. *J. Pharm. Biomed. Anal. 2*, 197—206.

Krejčí, M., Šlais, K., and Tesařík, K. (1978). Electrokinetic detection in liquid chromatography. Measurement of the streaming current generated on analytical and capillary columns. *J. Chromatogr. 149*, 645—652.

Kucera, P. (1980). Design and use of short microbore columns in liquid chromatography. *J. Chromatogr. 198*, 93—109.

Kucera, P. (ed.) (1984). *Microcolumn high-performance liquid chromatography*, Elsevier, Amsterdam.

Kutner, W. (1982). Alternating-current polarographic detection for reversed-phase ion-pair high-performance liquid chromatography of some benzoic acids. *J. Chromatogr. 247*, 342—346.

Kutner, W., Behr, B., and Kemula, W. (1982). Detection of cholanic acids in high-performance liquid chromatography based on effects of double layer capacity changes at the dropping mercury electrode. *Fres. Z. Anal. Chem. 312*, 121—125.

Lankelma, J., and Poppe, H. (1976a). Design of a tensammetric detecto for high speed high efficiency liquid chromatography in columns and its evaluation for the analysis of some surfactants. *J. Chromatogr. Sci. 14*, 310—315.

Lankelma, J., and Poppe, H. (1976b). Design and characterization of a coulometric detector with a glassy carbon electrode for high-performance liquid chromatography. *J. Chromatogr. 125*, 375—388.

Lloyd, J. B. F. (1983a). Optimization of the operational parameters of the supported mercury drop electrode detector in high-performance liquid chromatography. *Anal. Chim. Acta 154*, 121—131.

Lloyd, J. B. F. (1983b). High-performance liquid chromatography of organic explosives components with electrochemical detection at a pendant mercury drop electrode. *J. Chromatogr. 257*, 227—236.

Lloyd, J. B. F. (1983c). Clean-up procedures for the examination of swabs for explosives traces by high-performance liquid chromatography with electrochemical detection at a pendent mercury drop electrode. *J. Chromatogr. 261*, 391—406.

Manz, A., and Simon, W. (1983). Picoliter cell volume potentiometric detector for open-tubular column LC. *J. Chromatogr. Sci. 21*, 326—330.

Martin, M., Eon, C., and Guiochon, G. (1975). Study of the pertinency of pressure in liquid chromatography. II. Problems in equipment design. *J. Chromatogr. 108*, 229—241.

Matysik, J., Soczewiński, E., Żminkowska-Halliop, E., and Przegaliński, M. (1981). Determination of *o*-diphenols by flow injection analysis with an amperometric detector. *Chem. Anal. (Warsaw) 26*, 463—468.

Milazzo, G., and Caroli, S. (1978). *Tables of Standard Electrode Potentials*. Wiley, New York.

Molnár, I., Knauer, H., and Wilk, D. (1980). High-performance liquid chromatography of ions. *J. Chromatogr. 201*, 225—240.

Mowery, R. A. Jr. (1982). The use of a dielectric constant detector with reversed-phased LC. *J. Chromatogr. Sci. 20*, 551—559.

Pohl, C. A., and Johnson, E. L. (1980). Ion chromatography — The state-of-the-art. *J. Chromatogr. Sci. 18*, 442—452.

Polta, J. A., and Johnson, D. C. (1983). The direct electrochemical detection of amino acids at a platinum electrode in an alkaline chromatographic effluent. *J. Liquid Chromatogr. 6*, 1727—1743.

Poppe, H. (1978). Electrochemical detectors. In: *Instrumentation for High-Performance Liquid Chromatography*, J. F. K. Huber (ed.), Elsevier, Amsterdam, pp. 131—149.

Poppe, H. (1983). The performance of some liquid phase flow-through detectors. *Anal. Chim. Acta 145*, 17—26.

Poppe, H., and Kuysten, J. (1977). Construction and evaluation of a thermostatted permittivity detector for high-performance column liquid chromatography. *J. Chromatogr. 132*, 369—378.

Powley, C. R., Geiger, R. F. Jr., and Nieman, T. A. (1980). Bipolar pulse conductance measurements with a calcium ion-selective electrode. *Anal. Chem. 52*, 705—709.

Powley, C. R., and Nieman, T. A. (1982). Bipolar pulse conductometric monitoring of ion-selective electrodes. Part 2. Studies with the fluoride-selective electrode. *Anal. Chim. Acta 139*, 83—96.

Powley, C. R., and Nieman, T. A. (1983). Bipolar pulse conductometric monitoring of ion-selective electrodes. Part 3. Studies withthe calcium and fluoride electrodes in a continuous flow system. *Anal. Chim. Acta 152*, 173—190.

Reese, C. E., and Scott, R. P. W. (1980). Microbore columns — design, construction, and operation. *J. Chromatogr. Sci. 18*, 479—486.

Reijn, J. M., Poppe, H., and van der Linden, W. E. (1983). A possible approach to the optimization of flow injection analysis. *Anal. Chim. Acta 145*, 59—70.

Rocklin, R. D., and Pohl, C. A. (1983). Determination of carbohydrates by anion exchange chromatography with pulsed amperometric detection. *J. Liq. Chromatogr. 6*, 1577—1590.

Roe, D. K. (1983). Comparison of amperometric and coulometric electrochemical detectors for HPLC through a figure of merit. *Anal. Lett. 16*, 613—631.

Rokushika, S., Yin Qiu, Z., and Hatano, H. (1983). Micro column ion chromatography with a hollow fibre suppressor. *J. Chromatogr. 260*, 81—87.

Sato, H. (1982). Practical applicability of ion exchange chromatography with high sensitive, differential conductivity meter — Japan. *Bunseki Kagaku 31*, T23—T28.

Schultz, F. A., and Mathis, D. E. (1974). Ion-selective electrode detector for ion-exchange liquid chromatography. *Anal. Chem. 46*, 2253—2255.

Scott, R. P. W. (1980). Microbore columns in liquid chromatography. *J. Chromatogr. Sci. 18*, 49—54.

Scott, R. P. W. (1983). Small-bore columns in liquid chromatography. In: *Advances in Chromatography, Vol. 22*, J. C. Giddings, E. Grushka, J. Cazes, and P. R. Brown (eds.), Marcel Dekker, New York, pp. 247—294.

Scott, R. P. W., and Kucera, P. (1971). The design of column connection tubes for liquid chromatography. *J. Chromatogr. Sci. 9*, 641—644.

Scott, R. P. W., Kucera, P., and Munroe, M. (1979). Use of microbore columns for rapid liquid chromatographic separations. *J. Chromatogr. 186*, 475—487.

Scott, R. P. W., and Simpson, C. F. (1982). Determination of the extra column dispersion occurring in the different components of a chromatographic system. *J. Chromatogr. Sci. 20*, 62—66.

Šlais, K., and Kouřilová, D. (1982). Electrochemical detector with a 20 nl volume for micro-columns liquid chromatography. *Chromatographia 16*, 265—266.

Šlais, K., and Kouřilová, D. (1983). Minimization of extra-column effects with microbore columns using electrochemical detection. *J. Chromatogr. 258*, 57—63.

Šlais, K., and Krejčí, M. (1978). Generation of electricity in low-conductivity liquids as a detection principle in liquid chromatography. *J. Chromatogr. 148*, 99—110.

Šlais, K., and Krejčí, M. (1982). Electrochemical cell with effective volume less than 1 nl for liquid chromatography. *J. Chromatogr. 235*, 21—29.

Šlais, K., Pavlíček, M., and Krejčí, M. (1983). Dvouelektrodový elektrochemický detektor pro kapalinovou chromatografii. *Czechoslovakian Patent* No. AO 226 985.

Slavík, V. (1978). Influence of the temperature field on a permittivity detector in liquid chromatography. *J. Chromatogr. 148*, 117—125.

Small, H., Stevens, T. S., and Bauman, W. C. (1975). Novel ion exchange chromatographic method using conductimetric detection. *Anal. Chem. 47*, 1801—1809.

Sontag, G., and Frank, E. (1983). Charakteristik eines amperometrischen Durchflußdetektors mit zwei bzw drei Elektroden. *Mikrochim. Acta (Wien) II*, 1—16.

Sørensen, T. S., and Koefoed, J. (1974). Electrokinetic effects in charged capillary tubes. *J. Chem. Soc. Faraday II 70*, 665—675.

Sternberg, J. C. (1966). Extracolumn contributions to chromatographic band broadening. In: *Advances in Chromatography, Vol. 2*, J. C. Giddings, and R. A. Keller (eds.), Marcel Dekker, New York, pp. 205—270.

Štulík, K., and Hora, V. (1976). Continuous voltammetric measurements with solid electrodes. — Part I. A flow-through cell with tubular electrodes employing pulse polarization of the electrode system. *J. Electroanal. Chem. 70*, 253—263.

Suzuki, K., Aruga, H., Ishiwada, H., Oshima, T., Inoue, H., and Shirai, T. (1983). Determination of anions with a potentiometric detector for ion chromatography — Japan. *Bunseki Kagaku 32*, 585—590.

Takeuchi, T., Ishii, D., and Nakanishi, A. (1984). Instrumentation for fast micro high-performance liquid chromatography. *J. Chromatogr. 285*, 97—101.

Tallman, D. E., and Weisshaar, D. E. (1983). Carbon composite electrodes for liquid chromatography/electrochemistry: Optimizing detector performance by tailoring the electrode composition. *J. Liq. Chromatogr.* 6, 2157—2172.

Terabe, S., Yamamoto, K., and Ando, T. (1981). Streaming current detector for reversed-phase liquid chromatography. *Can. J. Chem.* 59, 1531—1537.

Terabe, S., Yamamoto, K., and Ando, T. (1982). Application of the streaming current detector to the analysis of individual bile acids. *J. Chromatogr.* 239, 515—526.

Tesařík, K., and Kaláb, P. (1973). A conductometric detector with a working volume of less than 0.5 μl. *J. Chromatogr.* 78, 357—361.

Tijssen, R. (1978). Liquid chromatography in helically coiled open tubular columns. *Separ. Sci. Technol.* 13, 681—722.

van Rooijen, H. W., and Poppe, H. (1983). Noise and drift phenomena in amperometric and coulometric detectors for HPLC and FIA. *J. Liq. Chromatogr.* 6, 2231—2254.

Vespalcová, M., Šlais, K., Kouřilová, D., and Krejčí, M. (1984). Příklady použití skleněných mikrokolon s amperometrickým detektorem ve farmaceutické analýze. *Českoslov. Farm.* 33, 287—294.

Vespalec, R. (1975). Improvement of the performance of the capacitance detector for liquid chromatography. Effects of the cell volume on band spreading and sensitivity of detection. *J. Chromatogr.* 108, 243—254.

Vespalec, R. (1981). Effect of the components of the chromatographic system on the electro-kinetic streaming current generated in liquid chromatography columns. *J. Chromatogr.* 210, 11—24.

Vespalec, R., and Hána, K. (1972). Performance of the capacitance detector for liquid chromatography. *J. Chromatogr.* 65, 53—69.

Vetter, K. J. (1952). Uberspannung und Kinetik der Chinhydronelektrode. *Z. Elektrochem.* 56, 797—806.

Watanabe, N., Azuma, M., and Niki, E. (1977). Study of dielectric constant detector for high speed liquid chromatography — Japan. *Bunseki Kagaku 26*, 295—300.

Weber, S. G., and Purdy, W. C. (1978). The behaviour of an electrochemical detector used in liquid chromatography and continuous flow voltammetry. Part 1. Mass transport-limited current. *Anal. Chim. Acta 100*, 531—544.

Weisshaar, D. E., and Tallman, D. E. (1983). Chronoamperometric response at carbon-based composite electrodes. *Anal. Chem.* 55, 1146—1151.

Weisshaar, D. E., Tallman, D. E., and Anderson, J. L. (1981). Kel-F-graphite composite electrode as an electrochemical detector for liquid chromatography and application to phenolic compounds. *Anal. Chem.* 53, 1809—1813.

Yang, F. J. (1981). On-column detection using a fused silica column. *J. High Resol. Chromatogr. Chromatogr. Commun.* 4, 83—85.

Yang, F. J. (1982). Open tubular column LC: Theory and practice. *J. Chromatogr. Sci.* 20, 241—251.

Progress in HPLC, Vol. 2, pp. 95—130
Parvez *et al.* (Eds)
© 1987 VNU Science Press

Characteristics of an amperometric detector with a rotating disc electrode and its application in clinical drug analysis

B. OOSTERHUIS and C. J. VAN BOXTEL

Clinical Pharmacology Section, Departments of Medicine and Pharmacology, Academical Medical Center, Meibergdreef 9, 1105 AZ Amsterdam, The Netherlands

INTRODUCTION

During the last few years a great number of studies dealing with liquid chromatography/electrochemical detection (LCEC) have been published in the analytical literature. Several recent review articles provide a survey of the developments in this field (Krull *et al.*, 1983; Majors *et al.*, 1984; McClintock and Purdy, 1984; White, 1984). Much effort is still invested in the improvement of instrumentation and techniques. Electrochemical flow cells with dual or multiple working electrodes in several configurations have been developed for enhanced detector selectivity and sensitivity, or to extend the number of compounds that can be detected. New and improved electrode materials or pretreatment procedures for electrodes were also investigated, and have lead to electrodes that are more selective for certain compounds (Kok *et al.*, 1983). As an alternative to the simple potentiostat, sophisticated electronic control devices have been developed, e.g. to provide more electrochemical information, or to resolve overlapping peaks.

In spite of these developments, the methods that employ electrochemical detection still form a small minority in the continuously expanding number of HPLC applications in biomedical (trace) analysis. In many routine laboratories (e.g. for clinical drug analysis) there is no running LCEC system. Moreover, the vast majority of realistic LCEC applications is thus far performed with simple amperometric detectors having a single carbon working electrode, that is kept at a constant potential. Several reasons may be given for this discrepancy. One of them is the inevitable time lag between new developments and their application. Another reason seems to be the limited number of compounds for which electrochemical detection has distinct advantages. The matrix in which a compound must be determined (e.g. human plasma) and the non-ideal electrochemical behaviour of many analytes may be the reason that the theoretical advantages of certain sophisticated concepts do not hold in practice. Finally, an important impediment for the introduction of LCEC as a routine method in many laboratories is the fact that many chromatographists are not familiar with some electrochemical principles and voltammetric techniques. Hopefully, the contents of the present volume will help to overcome this limitation.

In this chapter we describe the electrochemical principles and operation of an amperometric detector that is easy to handle. Some difficulties will be discussed that are inherent to voltammetry at solid electrodes. Attention will be given to important prerequisites that should be fulfilled for the development of a reliable and reproducible LCEC assay, with emphasis on the analysis of drugs and metabolites in human plasma. Finally, the detailed methodology of several drug assays will be described and discussed.

THE AMPEROMETRIC DETECTOR WITH A ROTATING DISC WORKING ELECTRODE

Some theoretical considerations

The response of the working electrode in any amperometric detector to a given concentration of electroactive analyte will be limited by charge transfer, mass transfer or both, depending on the chosen electrode potential. In any case, the response is proportional to the analyte concentration (Bard and Faulkner, 1980). Generally, the electrode potential is chosen so that the reaction rate at the electrode surface (charge transfer) is infinitely fast compared with the rate of mass transfer. Under such diffusion-limited conditions it follows from Fick's first law that the rate of mass transfer is inversely proportional to the thickness of the diffusion layer. In ampero-metric detectors with a stationary electrode, the thickness of the diffusion layer mainly depends on the detector flow-cell geometry and on the flow rate through the detector. In the various detector designs that are at present most widely used, the dependence of response on flow rate has been described theoretically and practically (Brunt, 1980; Kissinger, 1977; Yamada and Matsuda, 1973). One important practical consequence is that flow-rate variations caused by piston pumps (HPLC) or peristaltic pumps (continuous flow analysis) give rise to oscillating background which often forms a limiting noise (Weber and Purdy, 1981). Additionally, a reduced sensitivity is anticipated at lower flow rates, unless detector dimensions are adapted (McClintock and Purdy, 1984). In order to overcome these limitations we designed an amperometric detector flow cell with a rotating disc electrode (RDE) (Oosterhuis *et al.*, 1980). At an RDE, the thickness of the diffusion layer depends mainly on the rotation speed (Bard and Faulkner, 1980). Thus, in such a flow cell it should be possible to maintain a minimal and constant diffusion layer, with elimination of the influence of the flow rate. The construction of this flow cell will be described in detail in the following paragraph. An outline of its hydrodynamic and electrochemical characteristics will be given with emphasis on those parameters that are important for practical operation.

Construction of the rotating disc electrochemical detector (RDED)

Several models have been produced that differ only in certain details (Bruins *et al.*, 1982; Brunt *et al.*, 1980; Oosterhuis *et al.*, 1980). Figure 1 shows the flow-cell dimensions of the model that has been used for the experiments and applications described in this chapter. The working electrode compartment of the flow cell is, in principle, a wall-jet construction with a rotating electrode. Via the inlet channel in the stainless steel inlay (a) (diameter 1.0 mm) the eluent enters the flow cell, and impinges normally on the RDE. The solution leaves the working electrode compartment by streaming upward between the RDE and the wall of the vessel through channel (b) (diameter 5 mm) to compartment (g) in which the reference and auxiliary electrode are located. The working electrode shaft is constructed from a Kel-F tube (c) (diameter 6 mm) with a core consisting of a stainless steel ro (d). The outer diameter of the lower part of the RDE (e) is 8.5 mm, and fits in the working electrode compartment of the flow cell (diameter 9.0 mm) leaving enough space for the eluent to stream upward. The hold (f) (diameter 7 mm) in the electrode is filled with carbon paste. The carbon paste surface acts as the working electrode. At the top of the stainless steel core a mercury contact enables connection of the working electrode with the potentiostat. A small piece of platinum wire is therefore inserted into the mercury. The holder of the RDE is a modification of conventional electrochemical equipment. A detailed description of a holder of this kind has been given by Coenegracht (1972).

Figure 2 shows a picture of the complete detector. The cylindrical bottom of the stationary part of the holder fits precisely in the top of the working electrode compartment. The height of the RDE is variable. By using a belt-drive pulley system of different diameters the electrode can be rotated at different speeds (approximately, from 20 to 50 rps) with an electromotor. The rotation speed of the electromotor itself is 33 rps (Heidolph, Type 101.40.3). The pulleys at the RDE holder are connected by a sliding contact with the grounded stainless steel shield behind the flow cell. Several types of commercially available reference electrodes may be used as their shape is not very critical, e.g. a saturated calomel or silver chloride electrode (Radiometer K401 or K801, respectively). The auxiliary electrode consists of a stainless

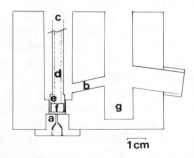

1cm

Figure 1. Schematic drawing of the RDED flow cell.

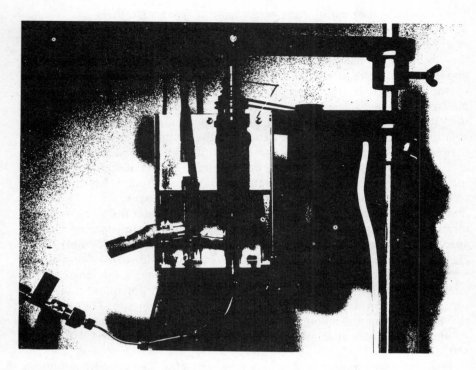

Figure 2. The assembled RDED flow cell.

steel wire and the detector is connected to a home-made potentiostat, comparable in performance to commercially available instruments. An electronic filter is used in the connection of the potentiostat with the power supply (Metrohm, E550 S). In Table 1 a survey is given of the composition and dimensions of several electrodes used for the experiments and applications described in this chapter. The graphite powder was purchased from Ringsdorff Werke, Bonn—Bad Godesberg (Spektralkohle Ringsdorff RW A). The glassy carbon electrode was purchased from Metrohm (Type AE 286-1) and was adapted to fit in the RDE holder.

Hydrodynamic and electrochemical characteristics of the RDED
The performance of the RDED, as described in the preceding section, may be affected by several variables, such as flow rate, rotation speed of the electrode and the height of the RDE in the flow cell above the eluent inlet (nozzle distance). The significance and influence of a number of these variables have been investigatd systematically (Bruins *et al.*, 1982; Brunt *et al.*, 1980). These studies were largely performed in combination with a continuous flow system, which means that the detector response was measured when the concentration of the test compound had reached a constant level in the flow cell.

Table 1. Composition and diameters of carbon electrodes.

Electrode material	Composition	Diameter of active surface (mm)
Silicone carbon paste	15.0 g graphite powder/8.0 g Dow Corning high vacuum grease	7
Nujol carbon paste	15.0 g graphite powder/9.0 g nujol	7
Vaseline carbon paste	15.0 g graphite powder/9.0 g vaseline	7
Glassy carbon (Metrohm AE 286-1)		4.8

The influence of flow rate and rotation speed on the response of the detector to varying concentrations of potassium hexacyanoferrate(II) was measured with a fixed potential (650 mV vs SCE) and a fixed nozzle distance (0.3 mm). The relationship between detector response and concentration was linear, independent of the choice of the other variables, provided the current was in such a range that the uncompensated resistance of the flow cell was negligible. The response surface of the detector to 2.35 µM potassium hexacyanoferrate(II) as a function of flow rate and electrode rotation speed is shown in Fig. 3. This figure illustrates how the detector response is increased by rotating the electrode. From the linear relationship between response and rotation speed it was concluded that, under the conditions given, a turbulent flow pattern exists in the flow cell (Levich, 1962). With a non-rotating electrode and using a wall-jet detector, the response is a power function of the flow rate. At rotation speeds of > 30 rps the response was independent of the flow rate when that exceeded 0.8 ml/min, whereas the dependence at lower flow rates could be attributed to the limited supply of electroactive species to the detector cell by the eluent.

The influence of nozzle distance and rotation speed on the response of the detector was investigated with ascorbic acid as the test compound (5 µg/ml) at a potential of 750 mV vs SCE. The resulting response surface is shown in Fig. 4. At a nozzle distance of 0.8 mm and rotation speeds of > 30 rps an almost linear relationship existed between response and rotation speed, possibly indicating a turbulent flow pattern in the flow cell. With increasing nozzle distance a linear relationship between response and the square root of the rotation speed was observed as long as the latter did not exceed 30 rps. A laminar flow pattern probably exists in the flow cell under these conditions. Varying the nozzle distance from approximately 0.8 to 4.0 mm and increasing the rotation speed above 20 rps hardly influenced the response in this experiment.

The performance of the detector and the influence of several variables was also tested in combination with HPLC when a detector has to fulfil more

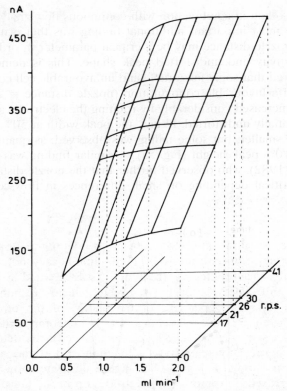

Figure 3. Response surface of the RDED as a function of flow rate and RDE rotation speed (rps = revolutions per second) at a constant concentration of 2.35 μM potassium hexa-cynanoferrate(II). [Reprinted from Brunt *et al.* (1980) with permission of Elsevier Science Publishers.]

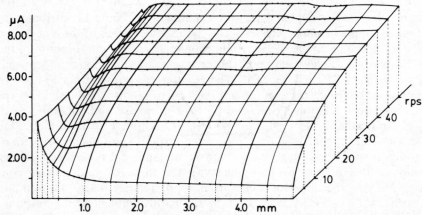

Figure 4. Response surface of the RDED in a continuous flow system as a function of nozzle height and electrode rotation speed. Concentration of ascorbic acid: 5 μg/ml; flow rate: 0.78 ml/min. [Reprinted from Bruins *et al.* (1982) with permission of Elsevier Science Publishers.]

stringent criteria than in combination with continuous flow analysis (Brunt *et al.*, 1980). The most important additional finding was that, with a rotating electrode, the nozzle distance may be a critical parameter in order to obtain maximum sensitivity and undistorted peak shape. This is demonstrated in Fig. 5. It appeared that some favourable and unfavourable cell configurations may exist. With a favourable configuration (nozzle distance = 2.4 mm) the sensitivity was increased considerably by rotating the electrode, and the peak shape was essentially undistorted. In fact, the peak width at 50% peak height became slightly smaller, but some tailing was observed, as quantified by the peak width at 10% peak height (Fig. 5a). A similar finding was reported by Feenstra *et al.* (1982), who observed further that the nozzle distance may be more or less critical depending on small differences in flow-cell geometry

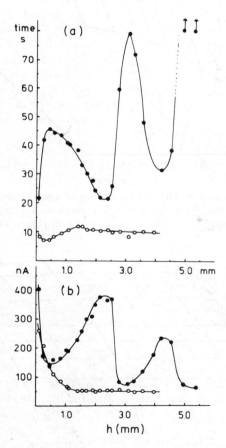

Figure 5. Influence of nozzle distance on the detector performance with a stationary (open circles) and a rotating (solid circles, 21 rps) electrode at a constant sample concentration of 105 ng of iron(II) in 20 μl samples. Peak width in seconds at 10% of height (a) and sensitivity of detector response (b) as a function of nozzle distance. [Reprinted from Brunt *et al.* (1980) with permission of Elsevier Science Publishers.]

(Feenstra, 1984). In a flow cell with a conical bottom the nozzle distance was apparently not critical (Westerink and Mulder, 1981).

The distinct difference between the plots for a stationary and rotating electrode in Fig. 5 indicates a completely different flow pattern under either of these conditions. This was verified by Bruins *et al.* (1982) who visualized the flow pattern of the sample in the detector flow cell by injecting coloured samples. By means of a motor-drive camera, series of photographs were taken of the sample flow pattern at a stationary as well as at a rotating electrode. Figure 6 shows pictures from these series a few seconds after the introduction of sample, with a stationary electrode (a) and a rotating electrode at two different rotation speeds (b, c). On entering the flow cell in the case of a stationary electrode, the sample moves straight towards the electrode, without mixing with the surrounding fluid. Subsequently, the sample spreads out radially over the electrode surface, developing a kind of mushroom shape. These phenomena have been observed and described by others who studied the characteristics of a wall-jet detector (Chin and Tsang, 1978; Yamada and Matsuda, 1973). From observations on the flow pattern in the case of a rotating electrode (Fig. 6b, c), it was concluded that there was reasonable agreement with theoretical hydrodynamic descriptions of the flow pattern between a stationary and rotating disc (Bruins *et al.*, 1982). An important detail is that the sample is apparently sucked towards the centre of the RDE, and then moves as a very thin layer over the electrode surface, due to centrifugal forces caused by the rotation of the electrode. In the case of a rapidly rotating electrode, the fluid entering the flow cell did not flow straight towards the centre of the RDE but developed a stable corkscrew pattern around the vertical axis of the flow cell (Fig. 6c). Whether the electrode is rotating or not, in both cases it is obvious from the visualized flow patterns that the effective cell volume of the detector is considerably smaller than the geometrical cell volume.

Practical operation of the RDED

Most of the difficulties that can be encountered in the operation and application of the RDED, and the precautions that should be taken, are similar to those with other electrochemical detectors (Weber and Purdy, 1981). Due to the sensitivity of electrochemical detection, the purity of mobile phase constituents is crucial. Particularly, the quality of the water, being the main constituent, requires careful consideration. Trace impurities from HPLC apparatus and column material may also contribute to background current and baseline instability, or may inactivate the working electrode. Therefore, when an LCEC system is started, it should be purged with at least 20—50 ml of fresh eluent before the potential is applied to the detector. The contribution of impurities in the eluent to the background current can be estimated with the RDED by comparing this current at a very low flow rate (e.g. 0.2 ml/min) and at a relatively high flow rate (e.g. 2.0 ml/min). The difference between these currents increases with increasing concentration of electroactive impurities.

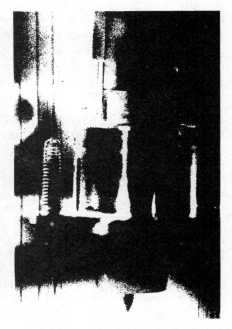

Figure 6. Sample flow patterns at different rotation speeds of the electrode: (a) stationary electrode; (b) slowly rotating electrode (2 rps); (c) rapidly rotating electrode (20 rps).

Under working conditions, the rotatory movement of the electrode (up to 40 rps) does not increase the noise in the detector signal, provided the pulleys at the top of the RDE are properly grounded as described earlier. If this is not the case, static electricity may accumulate on the pulleys, which causes an irregular noise and makes the detector sensitive to motions in its vicinity. When the detector was directly connected to the HPLC with stainless steel tubing, spikes were sometimes observed due to on-and-off switching of surrounding apparatus (e.g. refrigerators). These spikes could be eliminated when teflon tubing was used for the connection of the detector or when the stainless steel capillary was interrupted by a small piece of silicone tubing (see Fig. 2). Concerning the carbon paste electrodes used in the RDED, it was found that an extremely well-polished surface, as is recommended in thin layer cells, was not necessary. Malfunction of the working electrode is mostly due to the penetration of eluent between the electrode material and the wall of the electrode shaft, which gives rise to an increased and irregular background current. A very high background current, exceeding the offset capacity of the potentiostat, may occasionally be observed when the eluent comes in contact with the metal core of the electrode shaft. This current does not decrease when the flow rate is reduced. As a remedy, the shaft should be dried and the electrode material replaced. By having additional electrode shafts ready for use, the working electrode can easily and quickly be renewed or substituted by another type of electrode. This is also true for the reference electrode, and it is often useful to verify whether the latter is responsible for the observed noise in the detector signal. As was mentioned earlier, the nozzle distance may be more or less critical, depending on small differences in flow cell geometry. When the nozzle distance is properly tuned, the peak shape should be nearly identical whether the electrode is rotating or stationary, except for the difference in peak height.

GENERAL OBSERVATIONS CONCERNING THE APPLICATION OF LCEC FOR CLINICAL DRUG ANALYSIS

Some experimental conditions

Assay methods for several drugs that have been developed in our laboratory (Oosterhuis and van Boxtel, 1982, 1984; Oosterhuis et al., 1984) are described at the end of this section. These methods concern drugs from various therapeutic categories that also have different (electro)chemical properties. Thus, experience was obtained with certain aspects of amperometric detection that may be of value in the development and application of LCEC methods in general. Some of these aspects will be discussed in the present paragraph with reference to experiences reported in the literature. In addition, results of experiments that were performed in our laboratory will be presented to illustrate and support some of the revealed points. For these experiments, one representative compound was selected from each of three different therapeutic categories such that these three compounds could be separated in one chromatographic run. The three compounds and their chemcial structures are presented in Fig. 7. The compounds were separated on a short column (4.6 × 50 mm), packed with Spherisorb 5 μm ODS (Phase Sep, Queensferry, Clwyd, UK). The eluent consisted of 0.2 M sodium dihydrogen phosphate containing 7.5% 2-propanol and 10% methanol, and was delivered at a flow rate of 2.0 ml/min. The conditions of the electro-chemical detection were varied as indicated for the individual experiments. Unless otherwise stated, the electrode was rotating at a speed of 33 rps and the nozzle distance was 0.8 mm.

Electrochemical pretreatment of the working electrode

In many reports on voltammetric studies involving the use of carbon electrodes, electrochemical pretreatment procedures have been described that were found necessary to observe reproducible and well-defined electro-chemical behaviour (Blaedel and Jenkins, 1975; Engstrom, 1982; Rice *et al.*, 1983; van Rooijen and Poppe, 1981). Several pretreatment procedures have been proposed, but always pre-anodization of the electrode at a voltage greater then 1.5 V vs SCE is included, often in combination with pre-cathodization. The effect of pretreatment is dependent on the electroactive species and the composition of the solution under investigation, but in general the half-wave potential of the voltammetric wave is decreased, while

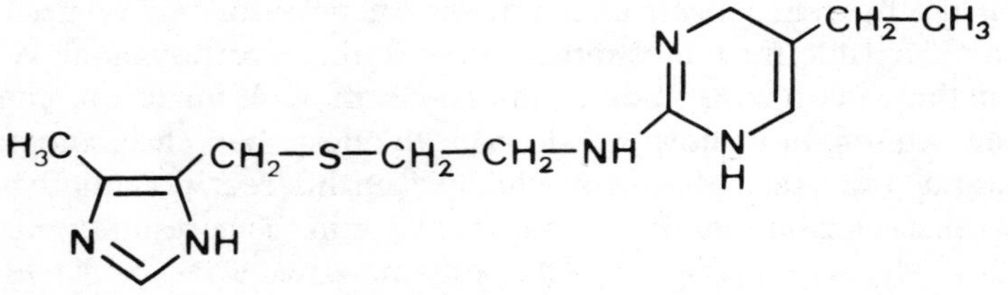

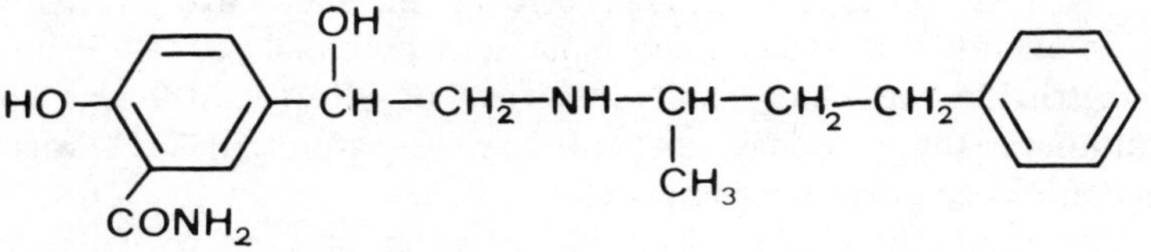

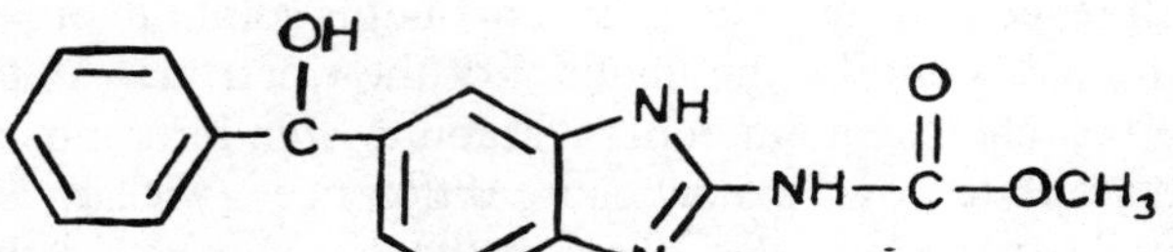

Figure 7. Structural formulas of 1. SK&F92909; 2. labetalol; and 3. hydroxymebendazole, being the test compounds for several experiments.

the shape of the wave becomes steeper and more reproducible (Engstrom and Strasser, 1984). It has also been shown that pretreatment enhances the performance of both glassy carbon and carbon paste electrodes in ampero-metric detection (Ravichandran and Baldwin, 1983, 1984).

In our laboratory it has become routine to initially apply a potential of >1500 mV for 15 min to the working electrode each time the chromato-graphic system is started while eluent is flowing through the detector. Subsequently, the potential is adjusted to the desired value in the particular system. Immediately after this adjustment, a large negative current is observed, which rapidly increases to a maximum and then decreases to some steady state value within about 15 min. According to our experience, this pretreatment provides several advantages. First, at a given potential, the detector response to many compounds is considerably improved. Second, the day-to-day variation in the electrode response and the variation in response between individual electrodes is reduced, particularly for carbon paste electrodes. As an additional advantage, the described pretreatment shortens the conditioning period of new electrodes, especially when a high detection potential has to be chosen. An alternative pretreatment procedure, including pre-anodization followed by a pre-cathodization step before setting the final potential, was also tested. With the compounds and systems used thus far, this alternative procedure yielded no advantages, but resulted in extended conditioning periods of the working electrode.

The consequences of pre-anodization for the performance of a silicone carbon paste electrode were illustrated by measuring hydrodynamic voltam-mograms of the three compounds mentioned earlier. A mixture containing 25 ng of each compound was injected under the described chromatographic conditions, and peak heights were determined at different potentials before and after pre-anodization at 1900 mV for 15 min. The resulting voltam-mograms are shown in Fig. 8. The voltammetric waves of hydroxy-mebendazole and labetalol became steeper and shifted over approximately 50 mV to a lower potential After pretreatment, the wave of hydroxy-mebendazole exhibited a well-defined plateau at potentials of >950 mV vs saturated AgCl. On the other hand, the wave of SK&F92909 was hardly affected, although the response at 1100 mV vs saturated AgCl was still considerably increased by pre-anodization.

Choice of electrode potential and influence of electrode rotation
The half-wave potential and the shape of voltammetric waves of many compounds is influenced by the composition of the solution in which the electrode reaction occurs and by the material of the electrode. As to carbon paste electrodes, the choice of binder,the graphite—binder ratio and even the source of the graphite powder may cause differences (Rice *et al.*, 1983). The history of an electrode may also have consequences in this respect. The detection potential in a given assay should therefore be determined with the selected working electrode under the established chromatographic condi-tions. Preferably, hydrodynamic voltammograms of the concerning analytes

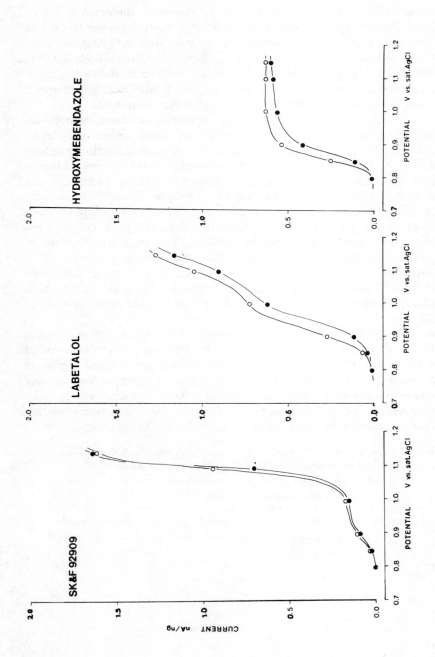

Figure 8. Hydrodynamic voltammograms of the three test compounds at a freshly prepared silicone carbon paste electrode (solid circles) and the same electrode after pre-anodization at +1900 mV during 15 min (open circles).

should be measured as described earlier. The detection potential is chosen at the plateau or the most horizontal part of the resulting voltammetric waves. In general, during use of an electrode, half-wave potentials tend to shift to more positive values. This is particularly the case with carbon paste electrodes, but also to some extent, with glassy carbon electrodes (Ravichandran and Baldwin, 1984). If the potential is chosen at or close to the steep part of the voltammetric wave of a particular compound, this shift will cause a considerable decrease in the detector sensitivity to that compound.

This was illustrated with an experiment employing the three compounds and chromatographic system described earlier. A nujol carbon paste electrode was used in the detector. After pre-anodization the detection potential was chosen at 1050 mV vs SCE (approximately 1100 mV vs saturated AgCl). As can be seen in Fig. 8, only the hydroxymebendazole wave is at a plateau at this potential, while particularly the SK&F92909 wave is very steep at this potential. A mixture containing 25 ng of each compound was injected at regular intervals over a period of 2.5 h, and peak heights were measured. The results are shown in Table 2. The detector response to hydroxymebendazole hardly changed, while the response to the other compounds decreased with about 10% (labetalol) and more than 50% (SK&F92909). Taking into account the voltammograms in Fig. 8, these reduced sensitivities correspond to positive shift of the voltammetric wave of about 25 mV.

These results also afford some conclusions about the use of an internal standard in combination with amperometric detection. Apart from other reasons for its use, there is a need to include an internal standard to correct for the instability of the detector response. This is valid as long as any variations in detector sensitivity affect equally the compound that is to be quantified and the internal standard. The results in Table 2 demonstrate that this is not necessarily the case. Ideally, the shape of the voltammograms of

Table 2. Decay of detector response towards a constant mixture of SK&F92909, labetalol and hydroxymebendazole (25 ng of each compound per injection).

Time after first injection (min)	Peak height (mm)		
	SK&F92909	Labetalol	Hydroxymebendazole
0	68	96	76
12	57	95	72
30	54.5	90.5	75
39	49	88	74
48	49	91	73
84	41	80.5	72
122	28	84	71
130	29	88	75
138	26	82	73

the compound to be quantified and the internal standard should be identical, but these voltammograms should at least have a common potential region that corresponds with a plateau in the voltammetric waves.

In the case of the rotating disc electrochemical detector, choosing the potential at the voltammetric wave plateau offers yet another advantage. As the plateau current is completely limited by diffusion, rotating the electrode will maximally increase the sensitivity under such conditions. This could be demonstrated with the three test compounds and the above chromatographic conditions. (Fig. 9). The detector response was measured with a stationary (S) and a rotating (R) electrode. In Fig. 9A a vaseline carbon paste electrode was used at 1200 mV vs saturated AgCl, and in Fig. 9B a silicone carbon paste electrode was used at 1100 mV vs saturated AgCl. According to what should be expected theoretically, the sensitivity for hydroxymebendazole was increased most (about fourfold) by rotation, whereas the difference was minimal in the case of SK&F92909. In addition, Fig. 9 illustrates that rotating the electrode does not significantly contribute to the detector noise.

Pretreatment of serum or plasma samples for LCEC

Pretreatment of a biological samle is usually required to some extent before it can be injcted into an HPLC, especially in combination with amperometric detection. Obviously, a more complicated and extensive clean up will be necessary when a lower detection limit and/or a higher detection potential is required. Drug assays for clinical research and therapy control are most often performed in plasma or serum. A widely used approach for the

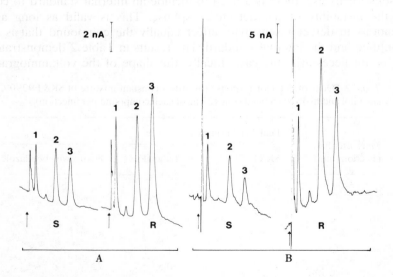

Figure 9. Influence of electrode rotation on baseline noise and on the peak height of: 1. SK&F92909; 2. labetalol and 3. hydroxymebendazole, as illustrated with (A) a vaseline carbon paste electrode and (B) a silicone carbon paste electrode; S = stationary electrode, R = rotating electrode (33 rps).

isolation of analytes from serum or plasma is liquid extraction from the appropriately buffered sample with an organic solvent. Generally, the more hydrophylic the analyte, the more polar the extraction solvent should be, and the more potentially interfering compounds are co-extracted.

Figure 10 shows the chromatograms of extracts from 0.2 ml alkalized plasma with ethyl acetate (Fig. 10A) and dichloromethane (Fig. 10B). The chromatographic conditions were as described earlier and the potential of the vaseline carbon paste electrode was 1000 mV vs SCE. As illustrated by Fig. 10, it appeared from our experience that an extensive additional sample clean-up is often necessary to obtain sufficient selectivity. Unfortunately, many manipulations appear to introduce new interferences. A useful alternative is the application of column switching HPLC. Several reviews have

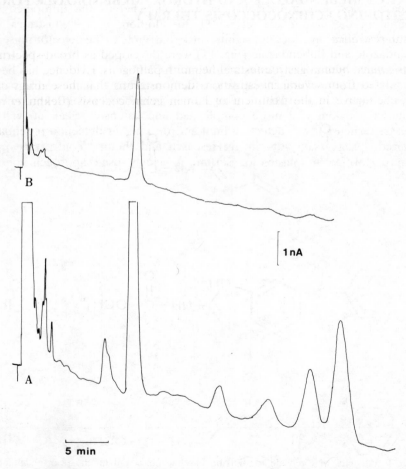

Figure 10. Chromatograms as obtained after the injection of an extraction residue corresponding with 250 μl alkalinated blank human plasma. Extraction was performed with (A) a polar (ethyl acetate) and (B) a relatively non-polar (dichloromethane) organic solvent.

been published describing the principles and applications of this concept (Freeman, 1981; Little *et al.,* 1984), and it has successfully been applied in LCEC methods (Edholm *et al.,* 1984). Column switching offers the possibility of improving the selectivity of a separation, as two or more different separation modes may be applied for the isolation of the analytes of interest from interfering compounds.

As a consequence of the improved selectivity, the internal standard for a particular analysis has to fulfil more stringent criteria with respect to its chromatographic properties. As these criteria should be combined with those that are related to the voltammetric properties, the selection of an appropriate internal standard has to receive careful attention.

ASSAY OF MEBENDAZOLE AND HYDROXYMEBENDAZOLE FOR MONITORING ECHINOCOCCOSIS THERAPY

Clinical relevance
Mebendazole and flubendazole (Fig. 11) were developed as broad-spectrum agents against human gastrointestinal helminth pathogens. Evidence has been accumulated from several investigations demonstrating that these drugs can also be effective in the treatment of human echinococcosis (Bekhti *et al.,*

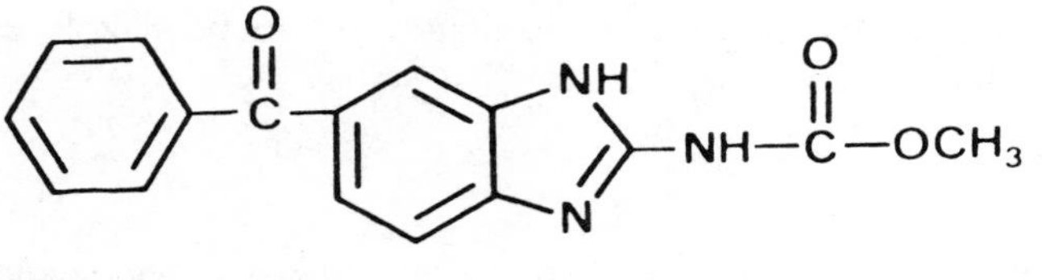

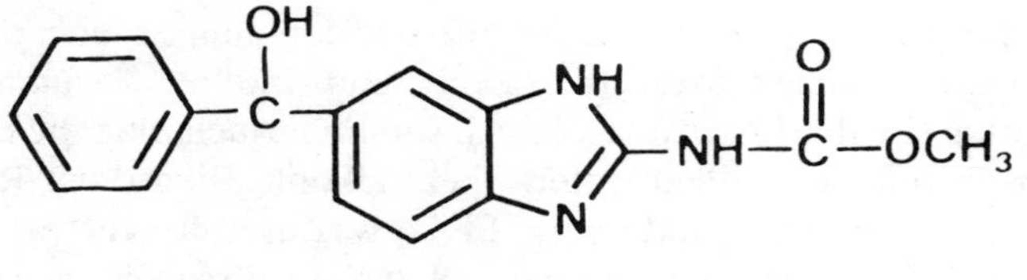

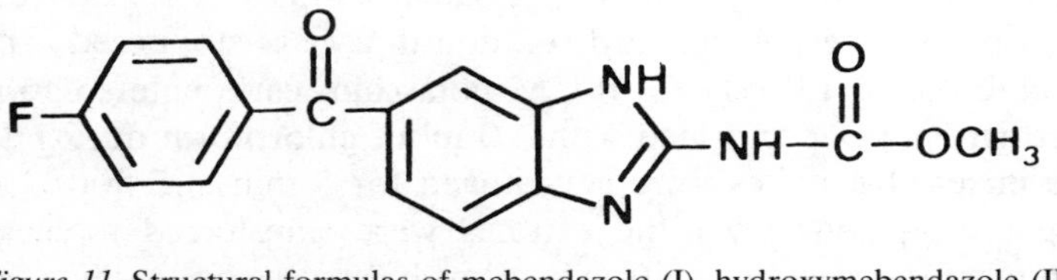

Figure 11. Structural formulas of mebendazole (I), hydroxymebendazole (II) and flubendazole (III).

1977). A very low and unpredictable systemic availability of mebendazole is observed when it is given orally. This is probably due to a combination of very low dissolution rate, limited absorption and high first-pass effect (Braithwaite *et al.*, 1982). Oral doses up to 5 g daily have to be administered to obtain mebendazole peak plasma levels exceeding the recommended 80 ng/ml (Witassek *et al.*, 1981). The main metabolites found in plasma are the aminometabolite, in which the carbamate has been hydrolysed, and the hydroxymetabolite (Fig. 11) which reaches the highest levels during chronic mebendazole therapy. To what extent these metabolites have any anthelmintic activity is unknown. A major elimination route for mebendazole and its metabolites is excretion in bile, almost entirely as conjugates (Witassek and Bircher, 1983). Thus far, several studies (Braithwaite *et al.*, 1982; Witassek and Bircher, 1983) have led to the conclusion that monitoring plasma levels of mebendazole and its metabolites is mandatory during therapy of echinococcosis, particularly when hepatic drug metabolizing capacity is impaired, or during cholestasis due to the frequently occurring cysts in the liver.

Analytical methods that are presently available for mebendazole monitoring are either rather insensitive (Allan *et al.*, 1980) or do not allow for quantification of metabolites (Alton *et al.*, 1979; Karlaganis *et al.*, 1979; Michiels *et al.*, 1978). The method described in this paragraph is sensitive and has the additional advantage of enabling the determination of mebendazole and a major metabolite simultaneously.

Methodology

Chemicals. Reference samples of mebendazole, hydroxymebendazole and flubendazole were kindly supplied by Janssen Pharmaceutica (Beerse, Belgium). Stock solutions (1 mg/ml) of these compounds were prepared in chloroform/formic acid (95:5 by vol). Dilute standard solutions were made in water/methanol/formic acid (80:19.5:0.5 by vol), one containing mebendazole (2 µg/ml) and hydroxymebendazole (4 µg/ml), and another containing 2 µg/ml of the internal standard flubendazole. All solutions, including the eluent were prepared with double-distilled water. Spektralkohle Ringsdorff RW-A (Ringsdorff Werke, Bonn—Bad Godesberg, FRG) was used for the preparation of carbon paste. All other chemicals were of analytical reagent grade and purchased from Merck (Darmstadt, FRG).

Extraction procedure. Plasma, cyst liquid, or saliva samples of 1.0 ml were spiked with 50 µl of the internal standard solution in a glass-stoppered 7 ml tube. After the addition of 1.0 ml of a 0.1 M potassium carbonate solution (pH = 11.5), extraction was performed with 4.0 ml of chloroform during 10 min in a tumble-mixer. The tubes were centrifuged for 5 min at 2.500 × g. After discarding the aqueous layer the extracts were transferred to clean glass tubes and evaporated under nitrogen. Residues were redissolved in 200 µl of eluent, and 100 µl or less was injected into the chromatographic system. Calibration curves for mebendazole (20—160 ng/ml) and hydroxy-mebendazole (40—320 ng/ml) were obtained by addition of 10—80 µl of the combined standard solution to 1.0 ml of blank human plasma samples using the described procedure. Peak height ratios of analytes versus internal standard were plotted against concentrations.

Chromatographic system. The chromatography column (4.6 × 50 mm) was packed in our laboratory with Spherisorb 5 µm ODS (Phase Sep, Queensferry, Clwyd, UK). Packing was performed with a 10% slurry in a 1:1 mixture (by volume) of chloroform/methanol employing the down-flow approach (Snyder and Kirkland, 1979). The mobile phase consisted of 20% 2-propanol in a pH 7.0 phosphate buffer (7.5 g KH_2PO_4 and 14.0 g $Na_2HPO_4 \cdot 2H_2O$ per liter). The eluent was delivered by a Kipp 9208 pump (Kipp Analytica, Emmen, The Netherlands) at a flow rate of 2.5 ml/min.

Electrochemical detection. The RDED as described earlier was equipped with a nujol carbon paste electrode. A saturated calomel electrode (SCE) (K401; Radiometer, Copenhagen, Denmark) was used as the reference electrode and a platinum wire as the auxiliary electrode. The potential was maintained at +950 mV SCE. Each time the system was started up, the pump and column were first flushed with at least 20 ml of eluent; the electrode was then initially oxidized at +1600 mV vs SCE for 15 min with the eluent flowing at 0.5 ml/min. About 15 min after setting the potential at +950 mV the system was ready for use.

Assay results

Typical chromatograms are shown in Fig. 12. No interfering peaks were observed in patient plasma, saliva, or cyst liquid samples, nor in fresh blank human plasma. In some cases a peak was observed immediately after the internal standard peak. Late eluting peaks were observed in some blank human plasma pools that had been stored frozen for a long time (1 year), making their use for calibration purposes impractical. The constancy of the test mixture response ratio (Fig. 12) in the detector was examined over several days using the same chromatographic column but different electrodes of varying ages. The response ratio of mebendazole versus the internal standard showed a coefficient of variation (CV) of 7.61% ($n = 10$); for hydroxymebendazole the CV was 6.14% ($n = 10$). Linear calibration plots were constructed. The mean equations with parameters ($\pm$S.D.) from five consecutive calibration plots of peak height ratio (y) versus concentration (x) were as follows:

mebendazole, $y = 0.0130(\pm 0.0008)x - 0.0062(\pm 0.018)$
hydroxymebendazole, $y = 0.0262(\pm 0.0018)x - 0.0196(\pm 0.0235)$

For all these curves the coefficient of correlation was > 0.999. The precision of the method was evaluated from duplicate measurements using the formula S.D. $= (\Sigma d^2/2n)^{1/2}$ where d is the difference between duplicates and n is the

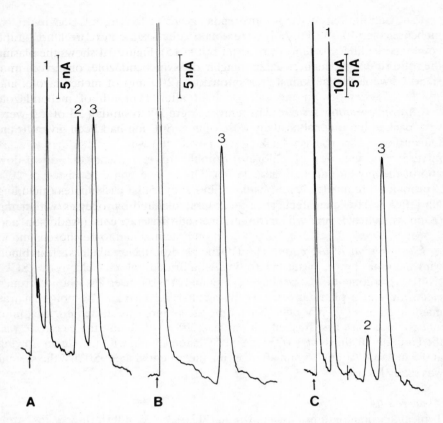

Figure 12. Chromatograms after injection of: (A) test mixture containing 20 ng hydroxy-mebendazole (1), 40 ng mebendazole (2) and 50 ng flubendazole (3); (B) 100 µl of the redissolved extraction residue of human blank plasma spiked with 100 ng/ml flubendazole; and (C) 100 µl of the redissolved extraction residue of patient plasma containing 125 ng/ml hydroxymebendazole and 14.3 ng/ml mebendazole. [Reprinted from Oosterhuis and van Boxtel (1984) with permission of Raven Press Publishers.]

number of duplicates. Thus, the CV was 9.02% ($n = 20$) for mebendazole at an average concentration of 33.0 ng/ml, and 9.91% ($n = 20$) for hydroxy-mebendazole at an average concentration of 181.5 ng/ml. The following recoveries were found for the different compounds:

mebendazole, $96.7 \pm 3.9\%$ ($n = 5$)
hydroxymebendazole, $68.4 \pm 7.9\%$ ($n = 5$)
flubendazole, $81.9 \pm 3.9\%$ ($n = 5$)

Considering a signal-to-noise ratio of 3:1 as a minimum, the limit of detection was about 5 ng/ml plasma for mebendazole and 2.5 ng/ml plasma for hydroxymebendazole. Fourteen plasma samples and one cyst liquid sample from several patients undergoing mebendazole therapy were analysed with the present method and with radioimmunoassay (RIA) for mebendazole (Michiels *et al.*, 1978) as employed by Janssen Pharmaceutica (Beerse, Belgium). The results are shown in Fig. 13. The regression equation for the correlation graph was $y = 1.0111x - 0.4792$ ($r = 0.9912$; $n = 15$). Figure 14 shows the plasma concentrations of mebendazole and hydroxymebendazole observed in a patient over a 6 h period after a pilot dose (200 mg) of mebendazole, and similar observations in the same patient after 1 month of mebendazole therapy. The second set of measurements were taken on day 23 of a 5-week period in which a constant daily dose of 3 × 800 mg had been given to the patient.

Discussion of methodology and results

Oxidative voltammetric investigations have shown that carbamates (including methyl 2-benzimidazole carbamate) exhibit oxidation processes within the practical positive potential limit of reversed-phase eluents (Anderson and Chesney, 1980). A relatively high detection potential had to be chosen, and in general this decreases the selectivity of the detection. Nonetheless, clean blank

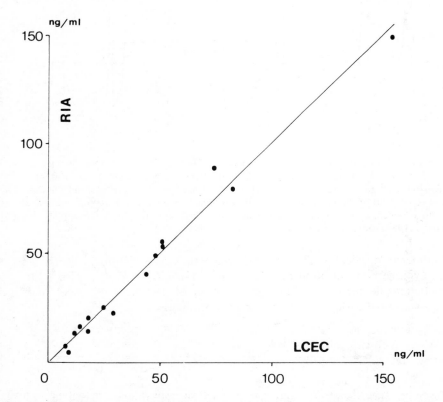

Figure 13. Correlation plot of the results of a radioimmunoassay (RIA) versus the results of the LCEC method in the analysis of 14 patient plasma samples and one cyst liquid sample for mebendazole. [Reprinted from Oosterhuis and van Boxtel (1984) with permission of Raven Press Publishers.]

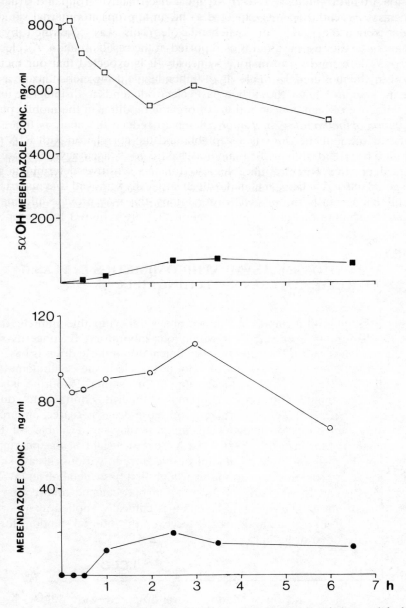

Figure 14. Time course of plasma concentrations of mebendazole (circles) and hydroxy-mebendazole (squares) in the same patient after the administration of a 200 mg pilot dose (solid symbols) and a morning dose of 800 mg (open symbols) during chronic mebendazole therapy. [Reprinted from Oosterhuis and van Boxtel (1984) with permission of Raven Press Publishers.]

chromatograms were obtained after a single chloroform extraction of alkalinated human plasma. Apparently, the majority of electroactive compounds in plasma are hydrophilic. Due to an identical electroactive group and structural similarity, flubendazole was expected to be an appropriate internal standard for mebendazole and hydroxymebendazole. This was confirmed by the variations in the response ratios as reported in the results section. As electrochemical detection is known for its linearity, it is expected that our method can easily be modified for the assay of flubendazole using mebendazole as the internal standard. Since the carbon paste electrode tends to disintegrate under the influence of high concentrations of organic modifier in the mobile phase, we chose a short chromatography column to reduce the analysis time. The reproducibility of the assay is acceptable and the correlation with RIA (Fig. 13) indicates that the present method has a good accuracy. The present method appears to be applicable also for the sensitive determination of mebendazole and hydroxymebendazole in surgically removed cyst material. In combination with parasitological observations, this may provide information on the concentrations at the site of action that are required for therapeutic effect.

DETERMINATION OF β-SYMPATHICOMIMETICS IN PLASMA WITH BIMODAL COLUMN SWITCHING HPLC

Clinical relevance

In Fig. 15 structural formulas are shown of several sympathicomimetic drugs that have thus far been analysed within our laboratory. It concerns sympathicomimetics with β2 selectivity. The clinical use of these drugs is based on the fact that they cause relaxation of smooth muscle tissues. Salbutamol and terbutaline are widely used in the treatment of asthma. Ritodrine is used mainly to terminate and prevent premature uterus contractions during pregnancy. Pharmacokinetic information on these drugs is scarce, apparently due to the limitations of analytical techniques thus far available for their determination (Martin *et al.*, 1976). An increased insight in the pharmacokinetics of these drugs could be of clinical relevance in various disease states, in relation with problems concerning their high first pass metabolism (Walker *et al.*, 1972), and for the stdy of the relation between the time courses of drug concentrations and effects. Bamethan is clinically applied as a peripheral vasodilator. In our methodology, it serves as the internal standard for the analysis of the other drugs in Fig. 15.

Methodology for salbutamol

Chemicals. All aqueous solutions were prepared with double-distilled water. Salbutamol sulphate and bamethan sulphate were kindly supplied by Glaxo and Boehringer Ingelheim, respectively. All other chemicals were of analytical reagent grade and obtained from Merck (Darmstadt, GFR). They were used as received. Aqueous solutions of standards were prepared con-

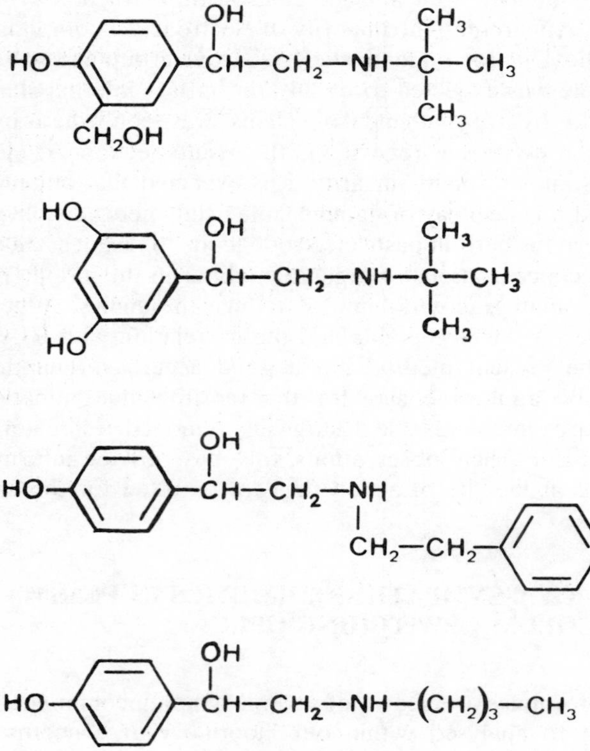

Figure 15 Structural formulas of salbutamol (I), terbutaline (II) ritodrine (III) and bamethan (IV).

taining 0.5 µg/ml salbutamol or 1.0 µg/ml bamethan. For purposes of preservation these solutions were slightly acidified with formic acid (±0.01 vol %). Sep-Pak cartridges were from Waters (Etten-Leur, The Netherlands). Spektralkohle Ringsdorff RW-A (Ringsdorff Werke, Bon—Bad Godesberg, GFR) was used for the preparation of carbon paste.

Sep-Pak manipulation. Sep-Pak cartridges were connected to 5 ml disposable PVC syringes, from which the plunger had been removed, to serve as an eluent reservoir. The outlets of the cartridges were fixed to 10 ml reagent tubes as recipients for the eluate, leaving a vent for the air to be displaced (Fig. 16). In this form twelve cartridges could be handled in a Sorvall refrigerated centrifuge in one run. The temperature in the centrifuge was kept between 5° and 10° C.

Chromatographic system. A schematic representation of the chromatographic system is given in Fig. 17. Eluents were delivered by two Kipp 9208 HPLC pumps (Kipp Analytica, Emmen, The Netherlands). Valve 1 was a

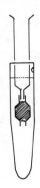

Figure 16. Construction for the handling of Sep-Pak cartridges in a centrifuge. [Reprinted from Oosterhuis and van Boxtel (1982) with permission of Elsevier Science Publishers.]

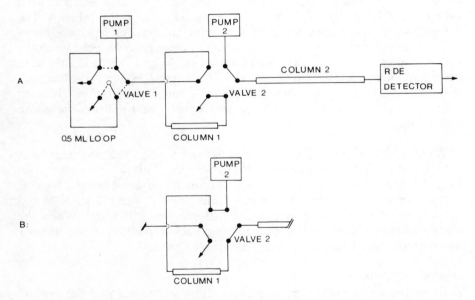

Figure 17. Schematic representation of the bimodal column switching HPLC system.

Rheodyne 7120 injection valve, equipped with a 0.5 ml loop. Valve 2 was a Rheodyne 7000 six-port rotatory valve. Column 1 (50 × 4.6 mm) was home-packed with Partisil SCX (10 μm). Packing was performed with a 10% slurry in water using the down flow approach. Column 2 (250 × 4.6 mm) contained LiChrosorb RP-2 (10 μm) and was obtained as a prepacked column from Brownlee Labs. (Santa Clara, CA, USA). The composition of the eluent solutions was as follows.

Eluent 1: phosphate buffer (pH 7.5) containing 2.58 g $Na_2HPO_4 \cdot 2H_2O$ and 0.3 g KH_2PO_4 per liter.

Eluent 2: sodium perchlorate (40 g/l) and 2-propanol (45 ml/l) in a phosphate buffer (pH 7) containing 14.0 g $Na_2HPO_4 \cdot 2H_2O$ and 7.5 g KH_2PO_4 per liter. The flow rate of pump 2 was 2.0 ml/min.

Electrochemical detection. The electrochemical detector (RDED) used in this method is described earlier. Nujol carbon paste was used for the working electrode, that was maintained at +950 mV vs SCE. Each time the system was started up, the electrode was initially conditioned at a potential of +1600 mV vs SCE for 15 min with eluent 2 flowing.

Sample preparation. In small glass tubes plasma samples (1.0 ml) were spiked with 20 µl of the internal standard solution. Before use Sep-Pak columns were washed with 5.0 ml methanol followed by 5.0 ml water. The spiked samples were transferred with pasteur capillary pipettes and passed through the prepared columns. Sample tubes and pasteur pipettes were rinsed with 1.0 ml water, which was then passed through the columns. The columns were then centrifuged at 1500 × g for 10 min. The aqueous eluates were discarded and the columns transferred to clean glass tubes to be eluted with 5.0 ml of a mixture of methanol—diethyl ether (25:75). The eluates were evaporated under nitrogen at about 40° C. Calibration curves for salbutamol in the range 2.5—20 ng/ml were obtained by adding 5—40 µl of the standard solution to 1.0 ml of blank human plasma samples and utilizing the above procedure.

Bimodal HPLC procedure. The residue from the Sep-Pak procedure was redissolved in 500 µl of eluent 1. Of this solution 250 µl were injected via valve 1 while valve 2 was in position A and the flow-rate of pump 1 was set at 3.0 ml/min. After about 5 min pump 1 was stopped. Valve 2 was switched to position B when the previous chromatogram was finished. When the solvent front appeared, valve 2 was switched back to position A and pump 1 was started again. A further sample could then be injected.

Results for salbutamol

Representative chromatograms of plasma samples obtained with the method for salbutamol are shown in Fig. 18. When the elution time for column 1 was chosen correctly no interfering peaks were observed. Blank human plasma from several pools was tested in this respect. The constancy of the detector response towards a test mixture containing salbutamol and bamethan was examined during a four-day period and a coeffiient of variation of 6.9% was found ($n = 8$). The average of seven calibration plots made during a nine-day period was described by the equation $y = 0.0769x + 0.0213$. The standard deviations of the slope and intercept were 0.00063 and 0.00022, respectively. The coefficient of correlation ranged from 0.9987 to 0.9999 with a mean of 0.9994. The standard deviation of the method was calculated from duplicate measurements in a similar way as described for mebendazole. Thus, the coefficient of variation was found to be 9.8% ($n = 13$) at an

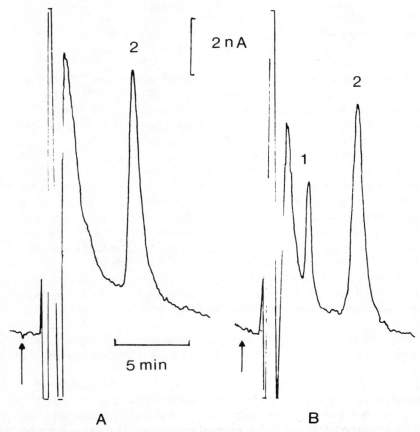

Figure 18. Chromatograms of a blank plasma sample (A) and a sample containing 5.8 ng/ml salbutamol (B). 1 = salbutamol, 2 = internal standard. [Reprinted from Oosterhuis and van Boxtel (1982) with permission of Elsevier Science Publishers.]

average concentration of 4.7 ng/ml. The signal-to-noise ratio indicates a detection limit of about 0.5 ng/ml plasma. The recovery of salbutamol and the internal standard was in all instances between 95 and 101%.

Ten plasma samples from a healthy volunteer after oral administration of 8 mg of salbutamol were analysed with the present method and a newly modified gas chromatographic—mass spectrometric (GC—MS) method in which salbutamol was converted to its trimethylsilyl derivative (Martin *et al.*, 1976). The correlation plot of the GC—MS results (*y*) vs the results of the present method (*x*) is described by the equation

$$y = 1.0002x + 1.02 \qquad (r = 0.9925; n = 10)$$

the GC—MS method consistently giving somewhat higher values (Fig. 19). No interference was observed from the following drugs when they were given to patients whose salbutamol levels were monitored: theophylline, prednisone, beclomethason, furosemide, triamterene, atenolol, methyldopa,

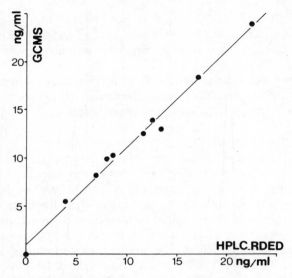

Figure 19. Correlation plot of the salbutamol levels as found with GC—MS versus the results of the method described in this chapter. Analysis of 10 plasma samples of a healthy subject obtained before and after a single oral dose of 8 mg salbutamol.

acenocoumarol and nitrazepam. As an illustration, the plasma concentration—time curve of a subject who received 2 mg salbutamol orally is shown (Fig. 20).

Adaptation of the methodology for terbutaline and ritodrine
The above-described sample pretreatment procedure, including the Sep-Pak step, is also applicable for terbutaline and ritodrine. The following modifications of the bimodal HPLC system were required for terbutaline: Column 2 (150 × 4.6 mm) was packed with Lichrosorb RP2 having a smaller particle size (5 μm). Composition of eluent 1 is 1.29 g of $Na_2HPO_4 \cdot 2H_2O$ and 0.15

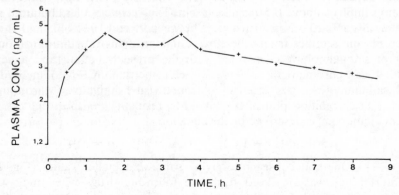

Figure 20. Time course of salbutamol plasma concentration in a subject after oral administration of 2 mg salbutamol. [Reprinted from Oosterhuis and van Boxtel (1982) with permission of Elsevier Science Publishers.]

g KH_2PO_4 per liter. Eluent 2 is sodium perchlorate (50 g/l) and 2-propanol (20 ml/l) in a phosphate buffer containing 14.0 g $NaH_2PO_4 \cdot 2H_2O$ and 7.5 g KH_2PO_4 per liter. The detection potential for terbutaline was +900 mV vs SCE.

For ritodrine, the only modification with respect to the methodology for salbutamol concerned eluent 2, which contained 20 g/l sodium perchlorate and 80 ml/l 2-propanol in the same phosphate buffer.

Discussion of the methodology for β-sympaticomimetics

In view of its chromatographic properties and the reported consistency of the salbutamol—bamethan response ratio, bamethan appeared to fulfil the requirements for an internal standard. At +900 mV vs SCE a similar reproducibility was obtained for the terbutalin—bamethan response ratio. A high electrode potential was required for detection. Thus selectivity had to be sacrificed for sensitivity. Since it appeared very difficult to gain adequate selectivity with conventional sample preparation techniques, including ion-pair extraction (Leferink, 1979), we have investigated the feasibility of mode-sequencing HPLC as discussed earlier. In the approach used, compounds of interest remain on the ion-exchange column during elution with eluent 1, due to their high capacity factor in this system. Elution focusing is achieved on transferring the analytes to the reversed-phase column as a result of the high ionic strength of eluent 2. Resolution of salbutamol and terbutaline from an interfering compound in plasma was increased by ion-pair formation of the analytes with the perchlorate ion. After approximately every 40 injections column 1 needed to be repacked. As salbutamol and terbutaline were eluted earlier than bamethan with eluent 1, their capacity factor on column 1 was determined at intervals by connecting the efflux of column 1 to the detector. The column-switching HPLC procedure can be easily automated with commerically available components. As sample pretreatment (column 1) and analysis (column 2) occur at the same time, five to seven samples can be analysed per hour.

The selectivity of the chromatographic system allowed direct injection of plasma samples after deproteinization with perchloric acid and neutralization with sodium bicarbonate. However, a poor recovery was obtained due to coprecipitation of the analytes with plasma proteins. Furthermore, in the Sep-Pak procedure not only proteins were removed, but also electrolytes that would cause band spread on the ion-exchange column. Evaporation of the organic eluate from the Sep-Pak columns was facilitated as the aqueous eluate was almost quantitatively removed by centrifugation. Sep-Pak columns were reused approximately 15 times with no measurable loss in performance.

PRELIMINARY METHOD FOR THE ANALYSIS OF RANITIDINE IN PLASMA USING BIMODAL COLUMN SWITCHING HPLC

Clinical relevance

Ranitidine (AH 19065) is a highly potent H2 receptor antagonist. Like other H2-antagonists (e.g. cimetidine), it is of value as an inhibitor of gastric acid secretion in the treatment of peptic ulcer disease. Within our laboratory, ranitidine plasma levels had to be determined in connection with a comparative study on the influence of ranitidine and cimetidine on the hepatic metabolism of other drugs. Ranitidine and several other H2-antagonists can be oxidized electrochemically, which is most probably due to the thioether function in their molecular structures (Fig. 21). Based on this observation we developed an LCEC method for ranitidine using SK&F92909 as the internal standard.

Methodology for ranitidine

Chemicals. Ranitidine hydrochloride was obtained from Glaxo (Hoofddorp, The Netherlands) and SK&F92909 was kindly supplied by Smith Kline & French. All solutions were prepared with double distilled water including stock solutions (1 mg/ml) and dilute solutions (5 μg/ml) of the standards. For the extraction procedure, a carbonate buffer (pH 9.5) was prepared containing 8 g sodium hydrogencarbonate and 9 g potassium carbonate per 100 ml. The other chemicals were of analytical reagent quality and purchased from Merck (Darmstadt, GFR).

Extraction procedure. In glass stoppered tubes, plasma samples of 1.0 ml were spiked with 60 μl of the dilute internal standard solution (= 300 ng).

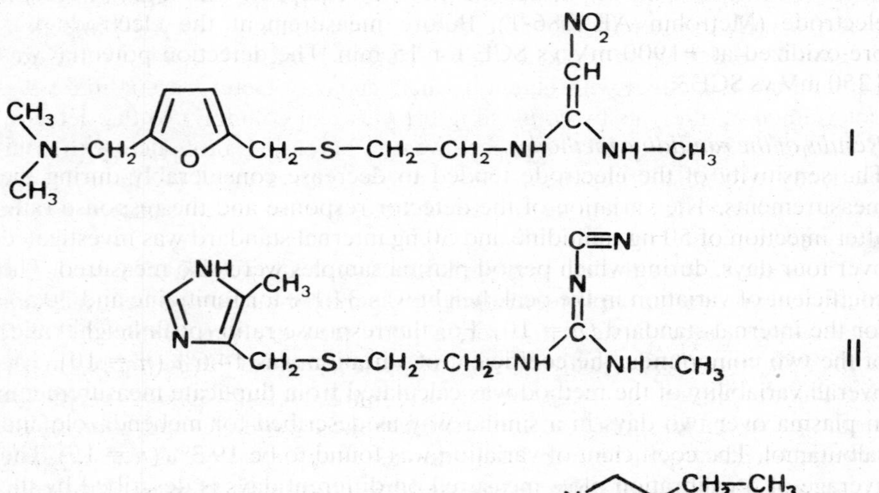

Figure 21. Structural formulas of ranitidine (I), cimetidine (II) and SK&F92909 (III).

After addition of 1.0 ml carbonate buffer the samples were extracted with 5 ml of a 2-propanol:dichloromethane (10:90) mixture. After extraction (10 min in a tumble mixer) the layers were separated by centrifugation for 5 min at $2500 \times g$ and the aqueous layers were discarded. The organic layers were transferred to clean glass tubes and evaporated at approximately 40° C under nitrogen. Residues were redissolved in 200 μl of eluent 1 and 50—100 μl was injected into the chromatographic system. For calibration curves (50—800 ng/ml) blank human plasma samples of 1.0 ml were spiked with 10—160 μl of the dilute ranitidine standard. Peak height ratios of analyte vs internal standard were plotted against concentrations.

Chromatographic system. The scheme of the bimodal HPLC system was as described earlier for salbutamol. Column 1 (30 × 4.6 mm) was packed with LiChrosorb-Si 60 (5 μm). Column 2 (100 × 4.6 mm) was packed with Nucleosil 10-SA (10 μm). Eluent 1 was a 0.01 M sodium hydrogenphosphate solution. Eluent 2 consisted of 0.2 M sodium hydrogenphosphate:methanol (80:20). The flow rate of pump 2 was 2.0 ml/min. The column switching procedure was analogous to the procedure for salbutamol. After injection at column 1 (valve 2 in position A) the sample was eluted with 6 ml eluent 1 (e.g. pump 1 at 3 ml/min for 2 min). Subsequently, valve 2 was switched to position B for the final separation and detection (Fig. 17).

Electrochemical detection. The RDED was equipped with a glassy carbon electrode (Metrohm AE 286-1). Before measurement the electrode was pre-oxidized at +1900 mV vs SCE for 15 min. The detection potential was 1250 mV vs SCE.

Results of the ranitidine method
The sensitivity of the electrode tended to decrease considerably during the measurements. The variation of the detector response and the response ratio after injection of 50 ng ranitidine and 50 ng internal standard was investigated over four days, during which period plasma samples were also measured. The coefficient of variation in the peak height was 34.1% for ranitidine and 30.5% for the internal standard ($n = 10$). For the response ratio (peak height ratio) of the two compounds, the coefficient of variation was 14.6% ($n = 10$). The overall variability of the method was calculated from duplicate measurements in plasma over two days in a similar way as described for mebendazole and salbutamol. The coefficient of variation was found to be 19.3% ($n = 15$). The average of 4 calibration plots measured on different days is described by the following equation:

$$y = 0.0053x - 0.1098$$

The standard deviations of slope and intercept were 0.002 and 0.072, respectively ($n = 4$). The coefficient of correlation ranged from 0.9966 to 0.9994, the mean value being 0.9979. The recovery of the extraction procedure ranged from 60 to 70%. Representative chromatograms of a blank human plasma and plasma containing ranitidine and internal standard are shown in Fig. 22. Considering the signal-to-noise ratio, the lower limit of detection is 10—20 ng/ml when extracts corresponding with 0.5 ml plasma are injected. Column 1 could be used for several days and no loss of ranitidine or internal standard during the first elution step was observed. The analytical column (2) had a relatively short lifetime of about two weeks. Due to considerable batch-to-batch differences of the Nucleosil, the composition of eluent 2 had to be adapted to a particular batch, while some batches appeared to be inappropriate.

In Fig. 23 the concentration—time course is shown for ranitidine, as measured in a human volunteer after an oral administration of ranitidine (300 mg).

Discussion of the ranitidine assay

Due to the relatively small area of the glassy carbon electrode, a high potential could be chosen without overloading the offset capacity for the background current. Nonetheless, the potential of 1250 mV SCE was still not at a plateau in the voltammetric waves of ranitidine and internal standard. This was apparent from the gradual decrease of the detector response to both

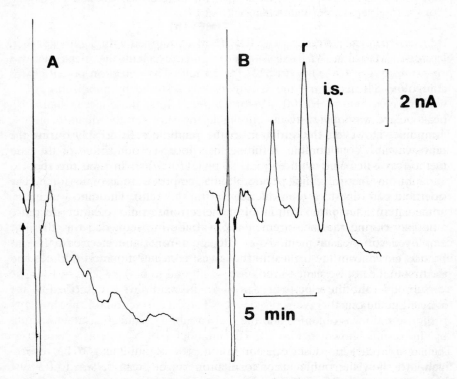

Figure 22. Chromatograms of blank plasma and a plasma sample containing 176 ng/ml ranitidine as obtained with the described methodology (1/4 of extraction residue injected).

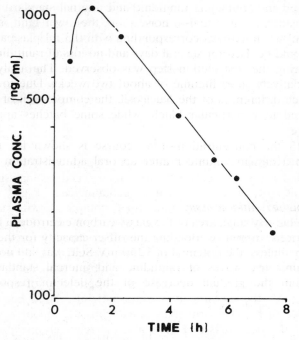

Figure 23. Time course of ranitidine plasma concentrations in a healthy subject after administration of 300 mg ranitidine at $t = 0$.

compounds. If necessary, the sensitivity was restored by anodization of the working electrode at $+1900$ mV for 15 min. Under these circumstances the peak height was obviously an unreliable measure for the quantification of ranitidine. However, the variation in the ranitidine—SK&F92909 response ratio was also considerable, resulting in a poor reproducibility of the assay method as calculated with formula $S = (\Sigma d^2/2n)^{1/2}$. In fact, the reported variation in the ranitidine—SK&F92909 response ratio was, for a considerable part, due to a reproducible shift in this ratio. The ratio was highest immediately after preoxidation of the electrode and thereafter decreased gradually during the measurements. The following sequence was therefore employed for measurement. After preoxidation of the electrode the calibration curve from the highest to the lowest concentration and samples from the first to the last were measured once.

Subsequently, the electrode was again preoxidized and duplicates were injected in the exactly reversed sequence. Thus, a correction for the shift in the response ratio was obtained and the real variability reduced, as demonstrated by the results shown in Fig. 23. The bimodal HPLC system is based on a similar approach as described for salbutamol. Ranitidine and the internal standard show the rather exceptional property of being retained at a silica column during elution with aqueous eluent. This property is utilized in the first elution step where interfering compounds from the plasma extraction residue are not retained and are thus eliminated. The combination of liquid extraction and clean-up obtained by the first elution step yields good selectivity, as no interfering peaks were observed from blank plasma in spite of the high detection potential. The relatively short lifetime and poor batch-to-batch reproducibility of the Nucleosil (column 2) was also notified by Edholm *et al.* (1984) in their terbutaline method. This seems to be a common feature of silica based ion-exchange materials for HPLC (Snyder and Kirkland, 1980).

CONCLUSIONS

The LCEC methods for several drugs as described in this chapter illustrate the potentials of electrochemical detection for clinical drug analysis.

For mebendazole and the β-sympathicomimetics, LCEC is the most likely method of choice. For ranitidine, compared with a previously published HPLC method using UV detection at 320 nm (Carey and Martin, 1979), the present method is more complicated without improved sensitivity. Nevertheless, when an expensive variable wavelength UV detector is not available the present LCEC method may be a viable alternative.

It was demonstrated that the successful application of electrochemical detection requires the investment of time and interest to become familiar with some electrochemical principles and technique. In our opinion this is worthwhile as the potentials of LCEC and the relatively inexpensive equipment that is required makes electrochemical detection an attractive technique for clinical drug analysis.

ACKNOWLEDGEMENTS

We are indebted to Dr K. Brunt and Dr P. M. J. Coenegracht who provided us the originals for several of the presented figures.

Ms M. van den Berg is also acknowledged for her excellent technical assistance in the experimental work within our laboratory that is described in this chapter.

REFERENCES

Allan, R. J., Goodman, H. T., and Watson, T. R. (1980). Two high performance liquid chromatographic determinations for mebendazole and its metabolites in human plasma using a rapid Seppak C18 extraction. *J. Chromatogr. 183*, 311—319.

Alton, K. B., Patrick, J. E., and McGuire, J. L. (1979). High-performance liquid chromatographic assay for the anthelmintic agent mebendazole in human plasma. *J. Pharm. Sci. 68*, 880—882.

Anderson, J. L., and Chesney, D. J. (1980). Liquid chromatographic determination of selected carbamate pesticides in water with electrochemical detection. *Anal. Chem. 52*, 2156—2161.

Bard, A. J., and Faulkner, L. R. (1980). *Electrochemical methods. Fundamentals and applications.* John Wiley & Sons, New York, p. 283.

Bekhti, A., Schaaps, J. P., and Capron, A. (1977). Treatment of hepatic hydatid disease with mebendazole: preliminary results. *Br. Med. J. 2*, 1047—1051.

Blaedel, W. J., and Jenkins, R. A. (1975). Study of the electrochemical oxidation of reduced nicotinamide adenine dinucleotide. *Anal. Chem. 47*, 1337—1343.

Braithwaite, P. A., Roberts, M. S., Allan, R. J., and Watson, T. R. (1982). Clinical pharmacokinetics of high dose mebendazole in patients treated for cystic hydatid disease. *Eur. J. Clin. Pharmac. 22*, 161—169.

Bruins, C. H. P., Doornbos, D. A., and Brunt, K. (1982). The hydrodynamics of the amperometric detector flow cell with a rotating disc electrode. *Anal. Chim. Acta 140*, 39—49.

Brunt, K. (1981). Electrochemical detectors for HPLC and flow analysis systems. In: *Trace Analysis, Vol. 1*, J. F. Lawrence (ed.), Academic Press, New York, p. 47.

Brunt, K., Bruins, C. H. P., Doornbos, D. A., and Oosterhuis, B. (1980). The response surface of an amperometric detector based on a rotating disc electrode for HPLC and continuous flow analysis. *Anal. Chim. Acta 114*, 257—266.

Carey, P. F., and Martin, L. E. (1979). A high performance liquid chromatography method for the determination of ranitidine in plasma. *J. Liq. Chromatogr. 2*(9), 1291—1303.

Chin, D. T., and Tsang, C. H. (1978). Mass transfer to an impinging jet electrode. *J. Electrochem. Soc. 125*, 1461—1470.

Coenegracht, P. M. J. (1972). Precieze semi automatische registrerende amperometrishe titrator. *Pharm. Weekbl. 107*, 769—782.

Edholm, L. E., Kennedy, B. M., and Bergquist, S. (1982). Automated analysis of terbutaline (Bricanyl) in human plasma with liquid chromatography and electrochemical detection using column switching (multidimensional chromatography). *Chromatographia 16*, 341—344.

Edholm, L. E., Kennedy, B. M., and Bergquist, S. (1984). Multidimensional column liquid chromatography with electrochemical detection for the analysis of terbutaline in human plasma. *Eur. J. Resp. Dis. 65, suppl. 134*, 33—40.

Engstrom, C. E. (1982). Electrochemical pretreatment of glassy carbon electrodes. *Anal. Chem. 54*, 2310—2314.

Engstrom, R. C., and Strasser, V. A. (1984). Characterization of electrochemically pretreated glassy carbon electrodes. *Anal. Chem. 56*, 136—141.

Feenstra, M. G. P. (1984). Dopamine receptor agonists: Neuropharmacology and bioanalytical evaluation. Thesis, University of Groningen.

Feenstra, M. G. P., Homan, J. W., Dijkstra, D., Mulder, Th. B. A., Rollema, H., Westerink, B. H. C., and Horn, A. S. (1982). Reversed-phase liquid chromatography with amperometric detection of lipophilic dopamine analogues and determination of brain and serum concentrations after sample clean-up on small Sephadex G-10 columns *J. Chromatogr. 230*, 271—287.

Freeman, D. H. (1981). Review: ultraselectivity through column switching and mode sequencing in liquid chromatography. *Anal. Chem. 53*, 2—5.

Karlaganis, G., Munst, G. J., and Bircher, J. (1979). High-pressure liquid-chromatographic determination of the anthelmintic drug mebendazole in plasma. *High Resol. Chromatogr. Chromatogr. Commun. 2*, 141—144.

Kissinger, P. T. (1977). Amperometric and coulometric detectors for high-performance liquid chromatography. *Anal. Chem. 49*, 447A—456A.

Kok, W. Th., Brinkman, U. A., and Frei, R. W. (1983). Amperometric detection of amino acids in HPLC with a copper electrode. *J. Chromatogr. 256*, 17—26.

Krull, I. S., Bratin, K., Shoup, R. E., Kissinger, P. T., and Blank, C. L. (1983). LCEC for trace analysis. Recent advances in instrumentation, methods and applications. *Am. Lab. 15*, 57 –65.

Leferink, J. G. (1979). Terbutaline. Analytical, clinical pharmacological and toxicological aspects. Thesis, University of Utrecht.

Levich, V. G. (1962). *Physicochemical Hydrodynamics*, Prentice Hall, Englewood Cliffs. Chaps. 2 and 3.

Little, C. J., Stahel, O., Lindner, W., and Frei, R. W. (1984). Column switching techniques in modern HPLC. *Int. Lab. 14*, 26—34.

Majors, R. E., Barth, H. G., and Lochmueller, C. H. (1984). Column liquid chromatography. *Anal. Chem. 56*, 300R—349R.

Martin, L. E., Rees, J., and Tanner, R. J. N. (1976). Quantitative determination of salbutamol in plasma as either its trimethylsilyl or *t*-butyldimethylsilyl ether using a stable isotope multiple ion recording technique. *Biomed. Mass. Spectrom. 3*, 184—190.

McClintock, S. A., and Purdy, W. C. (1984). Liquid chromatography/electrochemical detection: A review. *Int. Lab. 14*, 70—77.

Michiels, M., Hendriks, R., Thijssen, J., and Heijkants, J. A. (1978). Sensitive radio-immunoassay for mebendazole and flubendazole. *Janssen Pharmaceutica Clinical Research Reports.* R17635/11 and R17889/9.

Oosterhuis, B., Brunt, K., Westerink, B. H. C., and Doornbos, D. A. (1980). Electrochemical detector flow cell based on a rotating disc electrode for continuous flow analysis and high performance liquid chromatography of catecholamines. *Anal. Chem. 52*, 203—205.

Oosterhuis, B., and van Boxtel, C. J. (1982). Determination of salbutamol in human plasma with bimodal HPLC and a rotated disc amperometric detector. *J. Chromatogr. 232*, 327—334.

Oosterhuis, B., and van Boxtel, C. J. (1984). Analysis of sympathicomimetics in man with HPLC using mode sequencing and electrochemical detection. *J. Resp. Dis. 65, suppl. 135*, 153—156.

Oosterhuis, B., Wetsteyn, J. C. F. M., and van Boxtel, C. J. (1984). Liquid chromatography with electrochemical detection for monitoring mebendazole and hydroxymebendazole in echinococcosis patients. *Ther. Drug Monit. 6*, 215—220.

Ravichandran, K., and Baldwin, R. P. (1983). Liquid chromatographic determination of hydrazines with electrochemically pretreated glassy carbon electrodes. *Anal. Chem. 55*, 1782—1786.

Ravichandran, K., and Baldwin, R. P. (1984). Enhanced voltammetric response by electro-chemical pretreatment of carbon paste electrodes. *Anal. Chem. 56*, 1744—1747.

Rice, M. E., Galus, Z., and Adams, R. N. (1983). Carbon paste electrodes. Effect of paste composition and surface states on electron transfer rates. *J. Electroanal. Chem. 143*, 89—102.

Snyder, L. R., and Kirkland, J. J. (1979). *Introduction to Modern Liquid Chromatography.* John Wiley & Sons, New York, p. 212.

van Rooijen, H. W., and Poppe, H. (1981). An electrochemical reactivation method for solid electrodes used in electrochemical detectors for HPLC and flow injection analysis. *Anal. Chim. Acta. 130*, 9—22.

Walker, S. R., Evans, M. E., Richards, A. J., and Paterson, J. W. (1972). The clinical pharmacology of oral and inhaled salbutamol. *Clin. Pharmacol. Ther. 13*, 861—867.

Weber, S. G., and Purdy, W. C. (1981). Electrochemical detectors in liquid chromatography. A short review of detector design. *Ind. Eng. Chem. Prod. Res. Dev. 20*, 593—598.

Westerink, B. H. C., and Mulder, T. B. A. (1981). Determination of picomole amounts of dopamine, noradrenaline, 3,4-dihydroxyphenylalanine, 3,4-dihydroxyphenylacetic acid, homovanillic acid and 5-hydroxyindoleacetc acid, in nervous tissue after one step purification on Sephadex G-10, using HPLC with a novel type of electrochemical detection. *J. Neurochem. 36*, 1449—1462.

White, P. C. (1984). Recent developments in detection techniques for HPLC. Part I: Spectroscopic and electrochemical detectors. A review. *The Analyst 109*, 677—697.

Witassek, F., and Bircher, J. (1983). Chemotherapy of larval echinococcosis with mebendazole: microsomal liver function and cholestasis determinants of plasma drug level. *Eur. J. Clin. Pharmacol. 25*, 85—90.

Witassek, F., Burkhardt, B., Eckert, J., and Bircher, J. (1981). Chemotherapy of alveolar echinococcosis. Comparison of plasma mebendazole concentrations in animals and man. *Eur. J. Clin. Pharm. 20*, 427—433.

Yamada., J., and Matsuda, H. (1973). Limiting diffusion currents in hydrodynamic voltammetry. III. Wall jet electrodes. *J. Electroanal. Chem. 44*, 189—198.

Progress in HPLC, Vol. 2, pp. 131–178
Parvez *et al.* (Eds)
© 1987 VNU Science Press

Some theoretical and practical aspects of electrochemical detection in high-performance liquid chromatography

KAREL ŠTULÍK* and VĚRA PACÁKOVÁ
Department of Analytical Chemistry, Charles University, Albertov 2030,
128 40 Prague 2, Czechoslovakia

INTRODUCTION

At present, some applications of high-performance liquid chromatography cannot be imagined without the use of voltammetric or polarographic detection. Electrochemistry offers other methods for detection of substances in flowing liquids, mainly coulometry, equilibrium potentiometry, conductometry and high-frequency impedance measurements, but these methods have much more limited application. In fact, most authors use the term 'electrochemical' as a synonym for 'polarographic' and 'voltammetric'.

Attempts to electrochemically detect solutes in effluents from chromatographic columns were made long before the advent of HPLC; the first papers describing polarographic detectors were published by Drake (1950) and Kemula (1952). However, detectors based on electrolysis of substances eluted from columns have only become really important with the wide application of high-performance techniques in the seventies. Very many authors have elaborated on all aspects of electrochemical detection in HPLC and their work has been summarized in numerous reviews [the most recent review articles involve, e.g. Heineman and Kissinger (1980), Rucki (1980) and Štulík and Pacáková (1981a, 1984)].

In the present work we summarize our own results obtained in the study of electrochemical detection in HPLC and give references to other works only for the sake of comparison; for a general survey of the field, the reader is referred to the above reviews. Our work has been devoted to the construction, evaluation and application of certain types of voltammetric and polarographic detectors.

The construction of an electrochemical detector requires a solution of three main problems:

(a) The working electrode material must be chosen and the working

---

To whom correspondence shoud be addressed.

electrode constructed, so that the accessible potential range suffices for the given purpose, the residual signal and noise are sufficiently low and constant, the surface activity of the working electrode is constant and reproducible for as long a time as possible and the kinetics of the solute electrode reactions are favourable, to achieve a high sensitivity and reproducibility of measurement.

(b) The measuring cell must have a very small working volume and favourable hydrodynamics to avoid post-column band broadening and distortion. This requirement becomes especially stringent when using microbore and capillary columns.

(c) A suitable measuring technique must be selected, from the point of view of measuring sensitivity, reproducibility, selectivity and ease of signal handling.

The performance of detectors is evaluated and various detectors compared in terms of their operational parameters, which primarily include:

1. The sensitivity, S, defined as the slope of the dependence of the signal on the solute concentration or amount.

2. The detection limit, c_d, defined as a multiple of an estimate of the standard deviation of the noise s_V (in volts)

$$c_d = ks_V/S \tag{1}$$

where S is the sensitivity and k is usually 2—4, depending on the statistical significance level chosen. The detection limit is also often defined as twice or three times the absolute, peak-to-peak, noise value.

3. The linearity of the response function, expressed in terms of the linear correlation coefficient of the calibration curve.

4. The linear dynamic range, which is usually defined as that portion of the response function, R,

$$R = Sc^x \tag{2}$$

where c is the solute concentration, for which it holds that $0.98 < x < 1.02$.

5. The absolute values of the background signal and noise, and their stability in time.

6. The stability and reproducibility of the response.

7. The time constant, usually defined as the time required for the signal to attain 63.2% of the maximum value.

8. The contribution of the detector to the elution curve broadening, expressed usually in terms of the time or volume standard deviation, as a criterion of the width of a Gaussian peak.

For a more detailed discussion of these parameters see e.g. Štulík and Pacáková (1984) and the references cited therein.

DESIGN AND EVALUATION OF VOLTAMMETRIC AND POLAROGRAPHIC DETECTORS

Voltammetric detectors

Of the many types of voltammetric detectors [for a survey see e.g. Štulík and Pacáková (1984)], we have dealt with tubular, thin-layer and wall-jet systems and with a system of cylindrical fibres in a narrow channel. The main hydrodynamic characteristics are summarized in Table 1.

The first detector we constructed (Štulík and Hora, 1976) used a tubular system with platinum electrodes (see Fig. 1). This detector operates very reliably and the electrodes are readily removed for reactivation by mechanical polishing if necessary; however, its working volume is rather large for HPLC (close to 10 μl), and thus the detector is better suited for flow-injection analysis, where the requirements on solute zone broadening are less stringent.

To decrease the working volume of the detector, we later modified the construction (Štulík and Pacáková, 1980) by using a cylindrical counter electrode and placing it inside the tubular working electrode (see Fig. 2), thus obtaining a working volume of *c.* 2 μl while retaining the ruggedness of the original construction.

These detector types have, however, two limitations. Firstly, the reference electrode is placed far from the working electrode, which may lead to a large uncompensated voltage drop in the circuit, especially when mobile phases with low electrical conductivity are used. Secondly, the choice of working electrode materials is limited to those that can be appropriately shaped. Therefore, we turned our attention to the most common hydrodynamic systems used for the construction of voltammetric detectors, i.e. the thin-layer and wall-jet systems.

From the electrochemical point of view, an optimum arrangement of the electrodes should minimize the voltage drop between the working and reference electrodes by placing them as close together as possible and enable homogeneous charge distribution on the working electrode (uniform polarization), which is best achieved by placing the working and counter electrodes against one another in the channel through which the mobile phase flows. However, it is simultaneously required that the geometric volume of the detector cell should be as small as possible, i.e. the channel thickness must be very small (*c.* 0.02—0.1 mm). With such a small distance between the surfaces of the working and counter electrodes, the products of the electrochemical reaction on the counter electrode might interfere in the reaction on the working electrode. For these reasons we consider as optimal the arrangements shown in Fig. 3a for the thin-layer system and in Fig. 3b for the wall-jet system (Štulík *et al.*, 1981), i.e. with the counter electrode placed downstream from the working electrode and the reference electrode opposite to the working and counter electrodes. It is of advantage when the bottom part of the detector cell, containing the working and counter electrodes, can be turned by 180°, thus permitting easy change from the thin-layer to the

Table 1. Most important hydrodynamic electrode systems

Electrode	Equation for limiting current	Notes	References
Spherical	$I_1 = 4\pi r_0 nFDc + kn r_0^2 D^{2/3} f^{1/2} c$	Turbulent flow; constant k must be determined empirically	Levich (1942, 1944a, b)
Dropping mercury; parallel flow	$I_1 = 0.0605\, nFcD^{2/3}(mt_1)^{4/9} u^{1/3}[1 + 1.86 D^{1/3}(mt_1)^{-1/9} u^{-1/3} + 0.00332 D^{-1/3} m^{4/9} t_1^{-5/9} u^{-2/3}]$		Kimla and Štráfelda (1964)
Dropping mercury; opposite flow	$I_1 = 0.0154\, nFcD^{1/2}(mt_1)^{1/2} u^{1/2}[1 + 39.3 D^{1/2}(mt_1)^{-1/6} u^{-1/2} - 2154 D(mt_1)^{-1/3} u^{-1} + 0.0710 m^{1/3} t_1^{-2/3} u^{-1}]$		Štráfelda and Kimla (1965)
Dropping mercury; perpendicular flow	$I_1 = 0.0178\, nFcD^{1/2} u^{1/2}(mt_1)^{1/2}[1 + 13.8 D^{1/2}(mt_1)^{-1/6} u^{-1/2} + 0.134 m^{1/6} t_1^{-1/3} u^{-1/2}]$		Okinaka and Kolthoff (1957)

Planar	$I_1 = 0.68nFD^{2/3}cbl^{1/2}u^{1/2}\nu^{-1/6}$	Laminar flow	Levich (1947)
Tubular	$I_1 = 2.01nF\pi D^{2/3}l^{2/3}u_k r^{2/3}c$	$k = 0.33$ for laminar flow; $k = 1$ for turbulent flow	Levich (1962)
Cylindrical	$I_1 = 3.22nFc[\phi(a)]^{1/3}[\pi^{2/3}l^{2/3}(r_2 - r_1)D^{2/3}w^{1/3}/\{2(r_2 - r_1)\}]^{1/3} \times (r_2^2 - r_1^2)^{1/3}$ $$\phi(a) = \frac{1-a}{a}\left\{\left[\left[0.5 - \frac{a^2}{(1 - a^2)}\ln(1/a)\right]\right]\right/ \left[\left(\frac{1+a^2}{1-a^2}\right)\ln(1/a) - 1\right]\right\}; \ a = r_1/r_2$$	Laminar flow	Lown et al. (1980) Ross and Wragg (1965)
Conical	$I_1 = 0.77nFAcD^{2/3}u^{1/2}\nu^{-1/6}l^{1/2}$	Laminar flow	Jordan et al. (1958)
Wall-jet	$I_1 = knFAcD^{2/3}\nu^{-1/6}(u/L)^{1/2}$ $I_1 = 1.60knFcD^{2/3}\nu^{-5/12}w^{3/4}a^{-1/2}R^{3/4}$	Turbulent flow; k is an empirical constant; L is the characteristic dimension of the disc electrode	Matsuda (1967) Yamada and Matsuda (1973)

Symbols: A = electrode surface area; c = concentration; D = diffusion coefficient; F = Faraday constant; k = proportionality constant; m = mercury flow-rate; n = number of electrons exchanged; t_1 = mercury electrode drop time; u = linear flow-rate; w = volume flow-rate; v = kinematic viscosity; f = rotation frequency (rps).

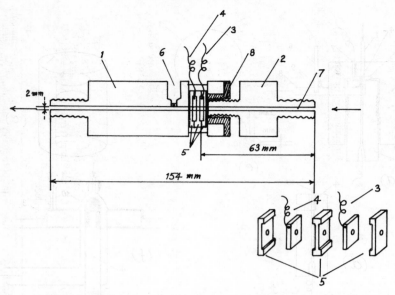

Figure 1. Detector with tubular electrodes. 1. cylindrical PTFE cell body with an opening for the placement of electrodes; 2. PTFE part which is screwed into the cell body and fixes the electrodes in place; 3. working electrode (Pt foil, 1 mm thick); 4. counter electrode (Pt foil, 1 mm thick); 5. PTFE insulation; 6. reference electrode liquid bridge; 7. channel for solution passage; 8. brass thread. From Štulík and Hora (1976).

wall-jet system and vice versa (Podolák *et al.*, 1982) (see Fig. 4). We have used these detectors with electrodes made of glassy carbon or consisting of classical graphite paste with nujol diluent.

A comparison of the operating parameters of these detectors (Štulík and Pacáková, 1981b) yielded the results summarized in Table 2. A number of conclusions can be drawn on the basis of this table.

1. All the electrode materials have similar anodic potential limits (about +1.2 V (SCE) at a pH of *c.* 4.5 for a current of 1 μA), except for graphite paste in the wall-jet system where the limit is somewhat lower. The negative potential limit is least favourable with platinum and the most negative for graphite paste; from this point of view, graphite paste is the most suitable electrode material.

2. From the point of view of residual current and noise, graphite paste is by far the best electrode material as its residual current is more than an order of magnitude lower than with glassy carbon and platinum, and exhibits a very low regular noise. The residual current and noise at platinum and glassy carbon electrodes strongly depend on the quality of the electrode surface polishing; glassy carbon is especially sensitive to polishing and is easily damaged by the passage of high currents. The noise with platinum electrodes is regular and increases uniformly with increasing sensitivity of measurement, while the noise with glassy carbon electrodes is low and increases suddenly when currents below *c.* 10^{-7} A are measured.

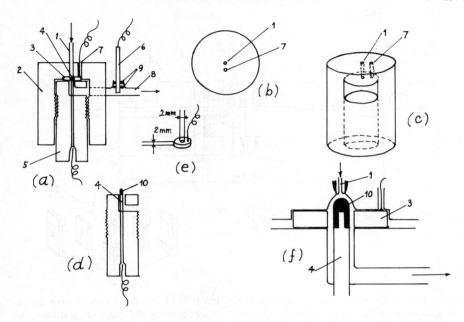

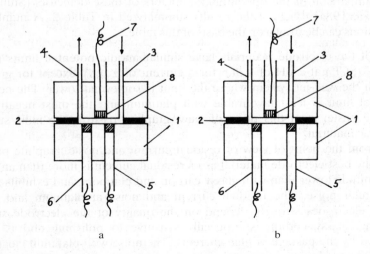

Figure 2. Tubular detector. (a) cross-section through the assembled detector; (b) top view; (c) side view of the outer part; (d) the inner part; (e) working platinum tubular electrode; (f) exploded view of the working — counter electrode systems. 1. Inlet (stainless steel capillary, 0.23 mm ID); 2. PTFE body, outer part; 3. tubular platinum working electrode; 4. cylindrical platinum counter electrode; 5. PTFE internal part; 6. reference electrode; 7. channel for the lead from the working electrode; 8. glass tube to waste; 9. PTFE O-rings; 10. PTFE insulating cap. From Štulík and Pacáková (1980).

Figure 3. (a) Thin-layer and (b) wall-jet detectors. 1. PTFE blocks; 2. 0.05 mm PTFE spacer; 3. inlet; 4. outlet; 5. carbon paste or glassy carbon working electrode; 6. glassy carbon counter electrode; 7. reference electrode; 8. porous glass. From Štulík *et al.* (1981).

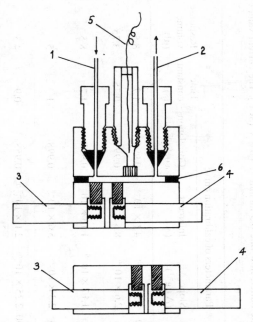

Figure 4. Thin-layer/wall-jet detector. 1. inlet; 2. outlet; 3. glassy carbon working electrode; 4. glassy carbon counter electrode; 5. reference electrode; 6. spacer. From Podolák *et al.* (1982).

On the other hand, platinum and glassy carbon electrodes, when polished well, yield reproducible values for a long time (one month or more in the absence of surfactants), and virtually the same absolute current values are obtained after repolishing. Graphite paste electrodes require frequent replacement (every day or, at best, every two or three days), as their response deteriorates quickly due to bleeding of the diluent from the paste and to mechanical damage to the electrode, especially at high flow-rates and in the wall-jet system. Graphite paste electrodes cannot be used in non-aqueous media that dissolve the paste diluent. After replacement of the paste, recalibration must always be carried out as absolute current values are not reproducible.

The current values stabilize almost instantaneously after potential adjustment at graphite paste electrodes, whereas at glassy carbon and platinum electrodes they drift for tens of seconds or even several minutes.

Thus it can be concluded that for measurements in aqueous systems graphite paste electrodes are usually most suitable. Glassy carbon and noble metal electrodes are useful for measurements in non-aqueous and mixed media, where the background current and noise are much lower than in aqueous solutions, apparently due to passivation of the electrode surface by the organic components of the solution. The rate of the electrode reaction of the solutes on graphite paste electrodes in any medium and on glassy carbon and noble metal electrodes in non-aqueous solvents are slower than the corresponding rates on glassy carbon and noble metal electrodes in aqueous

Table 2. Operating parameters of the detectors tested. (Adrenaline; citrate–phosphate buffer, flow-rate 0.3 ml min^{-1}; +0.8 V (SCE). The current values are scaled to an electrode surface area of 7.07 mm^2.)

Detector No.	Type	Potential range (V)	Residual current (A)	Detection limit (ng)	Linear dynamic range (ng)	Parameters of calibration curve			Time constant (s)	Response volume (µl)
						Slope (A µg^{-1})	Intercept (A)	Correlation coefficient		
1	Tubular (Pt)	−0.5 to +1.2	3×10^{-6}–1×10^{-5}	0.3	0.3–2500	1.3×10^{-6}	4.1×10^{-8}	0.998	1.7	8.5
2	Thin-layer (glassy C)	−0.8 to +1.2	2.4×10^{-6}–5.6×10^{-6}	0.5–1.0	1–5000	2.6×10^{-6}	8.6×10^{-8}	0.999	0.75	3.75
3	Thin-layer (C paste)	−1.5 to +1.2	1×10^{-7}–2×10^{-6}	0.5–1.0	1–5000	1.17×10^{-6}	1.13×10^{-7}	0.997	1.7	8.5
4	Wall-jet (glassy C)	−0.8 to +1.2	1×10^{-5}	0.3	0.3–3000	2.3×10^{-6}	5.6×10^{-8}	0.998	1	5
5	Wall-jet (C paste)	−1.8 to +1.0	1.8×10^{-7}	0.03–0.1	0.03–3000	2.25×10^{-6}	2.1×10^{-7}	0.958	0.9	4.3

solutions, but, due to decreased residual current and noise, the signal-to-noise ratios are most favourable.

To eliminate the main disadvantages of graphite paste, i.e. poor mechanical stength and bleeding of the diluent, various carbon electrodes with polymeric matrices have been devised [for a survey see e.g. Štulík and Pacáková (1984)]. We have used two polymeric matrix electrodes (Štulík *et al.*, 1981). One is prepared by dissolving *c.* 0.010 g polyvinyl chloride (PVC) in 3 g tetrahydrofuran, adding a weighed amount of finely pulverized graphite, mixing thoroughly and allowing excess solvent to evaporate (about 1 h at laboratory temperature). The viscous (but not dry) mixture is pressed into the depression in the detector body containing a platinum contact and left to solidify completely (about 1 h). The electrode thus obtained is polished with a fine metallographic paper. The other electrode is prepared by mixing a 1:1 mixture of chloroprene rubber and an alkylphenol resin (the commercial product 'Alkapren' by Matador, Czechoslovakia) with a weighed amount of finely pulverized graphite, pressing the mixture into the depression in the detector, allowing it to polymerize with catalysis by atmospheric moisture (about 4 h) and polishing as above.

The matrix-to-graphite ratio strongly affects the properties of these electrodes, but with optimal ratios (which are 1:17 and 1:1.27 for PVC and Alkapren, respectively) the electrodes retain all the advantages of graphite paste electrodes (see Table 2), are mechanically strong, can be polished to a mirror-like finish, resist certain organic solvents and last at least one month of continuous measurement.

3. The detection limits, calculated by us as twice the absolute noise value, was generally found to be lower for the wall-jet systems than for the thin-layer systems and the lowest value was attained for graphite paste in the wall-jet system. The detection limit obtained with the tubular system is between the values found for the wall-jet and the thin-layer systems.

4. All the studied detectors exhibit broad linear dynamic ranges, with upper limits dependent virtually only on the properties of the electronic circuitry (currents too high to be measured). The calibration curves obtained by the least-squares method have good linearity (see the correlation coefficients of the regression straight lines in Table 2), except for detector No. 5, with which linearity is poorer for large injected amounts. The detectors with glassy carbon electrodes generally have a higher sensitivity than those with graphite paste and platinum electrodes (see the slopes in Table 2). All the calibration curves have a positive intercept, caused by adsorption of the test solute, adrenaline, in the measuring system.

5. The time constants, determined as the time required for attaining current $I = 0.632 I_{max}$ on a step change in the solute concentration, are similar for all the detectors studied. The response volumes, obtained as the products of the time constants and the volume flow-rate, are somewhat greater than the geometrical volumes of the cells, indicating that the cells do not act as ideal mixing chambers, but that diffusion plays a role in the overall hydrodynamics of the systems.

The reproducibilities of the peak heights, half-widths and areas are similar for all the studied detectors and strongly depend on the state of the electrode surface. Typical values of the relative standard deivations for these parameters are c. 1% for amounts greater than 500 ng, 3% between 100 and 500 ng, 5% for 20—100 ng and 10% for lower amounts.

The detector response strongly depends on the spacer thickness. The thinner the spacer, the higher the linear flow-rate and the higher the sensitivity; however, the noise also increases, so that the detection limit is not changed. The noise value depends primarily on the quality of polishing of the cell walls; the better polished they are, the thinner the cell can be without an increase in the noise value and the lower is the detection limit. With decreasing thickness of the cell, the electric impedance increases and greater demands are placed on the potentiostat used. A typical value of the resistance between the electrodes is $10^4 \, \Omega$ for a cell thickness of 0.05 mm.

When the theoretical response of the studied detectors is calculated from the equations given in Table 1, it can be seen that the experimental sensitivities are somewhat lower than predicted (see Fig. 5). This is caused by several factors, the most important of which are:

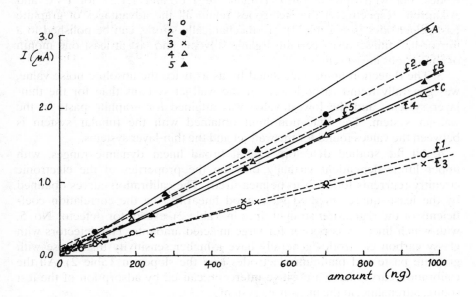

Figure 5. Experimental and theoretical calibration curves for voltammetric detectors. 1. tubular detector (Pt electrode); 2. thin-layer detector (glassy carbon electrode); 3. thin-layer detector (graphite paste electrode); 4. wall-jet detector (glassy carbon electrode); 5. wall-jet detector (graphite paste electrode). (A) Theoretical curve for the thin-layer detector; (B) theoretical curve for the wall-jet detector; (C) theoretical curve for the tubular detector. Conditions: +0.8 V (SCE); flow-rate, 0.3 ml min^{-1}; adrenaline in a citrate—phosphate buffer; the experimental curves were not corrected for adsorption of adrenaline in the measuring system (the positive intercept). From Štulík and Pacáková (1981b).

(a) uncertainty in the values used for the calculation, mainly the diffusion coefficients and empirical constants;

(b) specific effects of the electrode material causing deviations from the behaviour of an ideal, inert redox electrode;

(c) passivation of the working electrode;

(d) the effect of a very small working space (in the derivation of the equations it is assumed that the thickness of the hydrodynamic boundary layer is negligible compared with the cell dimensions, which is not fulfilled with the present cells);

(e) deviations from laminar flow.

The contribution from factors (d) and (e) strongly depends on the quality of polishing of the electrodes and cell walls, and on the precision with which the cell has been made. The wall-jet detectors exhibit a better agreement with the theory, because the turbulent flow, which is assumed by the theory, is apparently the predominant parameter. On the whole it can be concluded that there is little difference between the practical thin-layer and wall-jet detectors; the flow in thin-layer detectors exhibits turbulence to a greater or lesser degree and the behaviour of the wall-jet detectors is actually an intermediate between the thin-layer and wall-jet systems.

We have obtained especially good results with a detector consisting of a strand of glassy carbon fibres placed in a plastic capillary (Štulík *et al.*, 1984, Fig. 6). The capillary is placed in a macroscopic overflow vessel containing counter and reference electrodes immersed in the mobile phase. The fibre

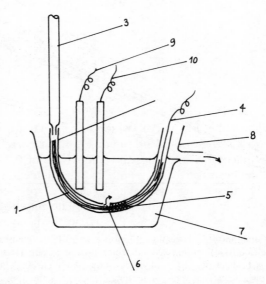

Figure 6. Carbon fibre detector. 1. working electrode (a strand of carbon fibres); 2. PTFE tube 0.6 mm ID; 3. column; 4. electrical contact; 5. silicone rubber seal; 6. outlet; 7. base electrolyte; 8. overflow vessel; 9. counter electrode; 10. reference electrode. From Štulík *et al.* (1984).

diameter is 4—7 μm. This detector is marked by extreme simplicity and cheapness. Its working volume can be varied by varying the effective length of the fibres. In Table 3, the geometric volumes, response times and response volumes are given as functions of the fibre effective length. It can be seen that the shortest possible detector should be used from the point of view of the response time and volume. A very short working length is also advantageous because the charge distribution along the working electrode is more homogeneous than with long fibres (a shorter distance from the counter and reference electrodes). The response volumes are not much larger than the geometrical volumes (determined as the volumes of distilled water required to fill the working space), indicating that the detector behaves as a good mixing chamber.

Table 3. Dependences of the geometric volume, response time and response volume on the fibre detector working length

		Flow-rate (ml/min)					
		0.1		0.2		0.3	
Detector length (mm)	Geometric volume (μl)	Response time (s)	Response volume (μl)	Response time (s)	Response volume (μl)	Response time (s)	Response volume (μl)
82	12.0	11.1	18.5	5.6	18.7	4.0	20.2
76	7.0	7.6	13.0	4.9	16.2	3.3	16.4
61	5.8	5.8	9.7	3.6	12.2	2.5	12.6
15	1.5	1.5	2.6	1.0	3.3	0.6	3.2
8	0.4	0.4	0.7	0.3	1.0	0.2	1.0

The noise value is virtually independent of the detector length and the mobile phase flow-rate and varies from 2×10^{-11} to 8×10^{-11} A. Typical parameters of the calibration curve (which are similar for all the detector lengths tested) are a slope of 3.73 nA/ng and a linear correlation coefficient of 0.9967 (for a detector 8 mm long and a flow-rate of 0.2 ml/min). Hence the detector is highly sensitive (cf. Table 2). The detection limits obtained with this detector are 20 pg adrenaline or even 3 pg benzidine and the calibration curves are linear down to the lowest solute amounts.

The detector has two drawbacks. Firstly, its hydrodynamics is poorly defined and may be changed by changing the position of the plastic capillary; however, reproducible results are readily obtained when the capillary position is maintained constant. Secondly, the detector linear dynamic range is narrower than that obtained with thin-layer and wall-jet detectors; the calibration curve ceases to be linear at injected solute amounts of about 50 ng; therefore, the detector is primarily useful for trace analysis and for microbore columns.

As the active surface area of the working electrode is quite large due to the carbon fibres, we tested the possibility of using it in the coulometric mode. However, when the theoretical charge required to electrochemically convert the injected amount of the solute was compared with the corresponding peak area, it was found that the degree of conversion is low, 0.44% and 2.8% for flow-rates of 0.2 and 0.1 ml/min, respectively (adrenaline, detector working length, 82 mm). Therefore, coulometric measurement would only be possible with extremely low flow-rates.

An additional advantage of the detector is a small dependence of the current signal on the mobile phase flow-rate; at flow-rates greater than $c.$ 0.2 ml/min the signal is virtually independent of the flow-rate.

A great problem in working with solid electrodes is working electrode passivation, which is especially important with this detector as the working electrode cannot be reactivated mechanically but only by chemical or electrochemical means. However, no passivation has so far been encountered, and several detectors of this type have reliably worked for more than one year without the need of any reactivation, even when the test compounds included substances that strongly passivated a rotating disc electrode in voltammetric measurements, e.g. phenols. The main reasons apparently involve the extremely low amounts of solutes in the HPLC measurements and the fact that the working electrode is in contact with the solute for only a short time interval during the passage of the particular solute zone, and for most of the time is washed by the pure mobile phase which leads to gradual desorption of any passivating film.

The reproducibility of measurement is similar to that obtained with other voltammetric detectors. Typical values of the relative standard deviation for five parallel measurements and $\alpha = 0.05$ are 2.7% for solute amounts above 50 ng, 4.7% for amounts around 1 ng and about 10% for less than 0.1 ng (tested with adrenaline in an aqueous buffer mobile phase).

A polarographic detector

Many polarographic detectors have been designed [for a survey see e.g. Štulík and Pacáková (1984)]. Our own construction (Štulík *et al.*, 1983) is similar to that of the commerical detector from Princeton Applied Research, USA, i.e. the eluate is fed to macroscopic dropping mercury electrode placed in an overflow vessel containing counter and reference electrodes. In contrast to the commercial detector, in which the eluate is fed to the electrode vertically, we use a horizontal jet (Fig. 7), as we feel that the system is thus better defined hydrodynamically (i.e. is closer to the model of Matsuda (1967) where a sphere is placed in a laminar stream of liquid, with the depolarizer impinging on the spherical surface) and should exhibit the highest measuring sensitivity [the liquid flowing perpendicularly to the mercury flow (Hanekamp and van Nieuwkerk, 1980)].

This detector was tested under various measuring conditions and was found to work reliably. From the point of view of measuring reproducibility it is advantageous to work with a hanging mercury drop; however, electrode

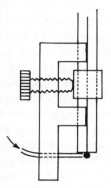

Figure 7. Working electrode of a polarographic detector. From Štulík *et al.* (1983).

passivation may create a serious problem. When working with a dropping electrode, current-sampled techniques must be used to avoid large oscillations of the signal.

The experimental results compare well with the theoretical values (for the equations see Table 1) and the signal dependence on the mobile phase flow-rate is also in agreement with the theory (an exponent of 0.5). An important parameter is the distance of the jet orifice from the surface of the mercury drop. It has been found that a variation of this distance within 0.5 and 3 mm had little effect on the time constant, but an increase in the distance caused a decrease in the peak height and especially a peak distortion with pronounced tailing.

Typical values of the time constant and the response volume are 3.7 s and 18.5 μl, respectively (0.3 ml/min with picric acid as solute), indicating that the effective volume of the polarographic detector is substantially larger than those of the voltammetric detectors discussed above and that diffusion plays a substantial role in the detector response. Therefore, the danger of elution curve post-column broadening is much greater than with the above voltammetric detectors.

Measuring techniques

The question of an optimal measuring technique has often been discussed [for a survey see e.g. Štulík and Pacáková (1984)]. The simplest direct current amperometry, when the solute limiting current is monitored at a constant working electrode potential, is used most often. It has sometimes been claimed that the use of pulse voltammetric techniques, especially differential pulse voltammetry (DPV) and square-wave voltammetry (SWV) leads to an increase in the measuring sensitivity, an improvement in the selectivity, suppression of the working electrode passivation and elimination of the signal dependence on the mobile phase flow-rate.

We have tested various measuring techniques on both solid and mercury electrodes and arrived at the following conclusions.

Solid electrodes (Štulík and Pacáková, 1980, 1981b). When a pulse technique is applied to a solid electrode, the signal is subject to interference from a large and poorly reproducible background current associated with a large noise. This effect is caused by the presence of oxygen-containing groups on the surface of solid electrodes [see e.g. Majer *et al.* (1973)] that undergo electrochemical reactions producing slowly decaying currents whose effect cannot be suppressed by current sampling, as can be done with normal charging current. Therefore, the sensitivity of measurement and its repro-ducibility are poorer than with the d.c. method. A suppressed effect of the flow-rate on the signal has also not been observed. Therefore, the only advantage of pulse measurement with solid electrodes is an increased selectivity which, however, rquires careful selection and maintenance of the working electrode potential. The signal dependences on the pulse amplitude and frequency are the same as in normal DPV measurements.

Hence, for detection with solid electrodes classical d.c. measurement can be recommended, using a filter with a time constant of about 2 s to suppress high-frequency noise. To alleviate electrode passivation in some systems it is useful to use electrode polarization by the waveform shown in Fig. 8 (Štulík and Hora, 1976). The working potential value is alternated with cleaning pulses of a suitable, empirically determined amplitude, at which the pas-sivating film is desorbed. The current is sampled at the end of the measuring pulses to decrease the charging current effect. The pulse frequency is of the order of tenths of a Hertz. The method is, of course, by no means generally applicable.

Mercury electrodes (Štulík *et al.*, 1983). In contrast to solid electrodes, mercury electrodes, which are characterized by a homogeneous surface without electroactive functional groups, bear a greater promise for the use of pulse techniques. Therefore, we studied the effect of various measuring techniques on the response of the polarographic detector depicted in Fig. 7, with both a stationary mercury drop and a dropping mercury electrode. The

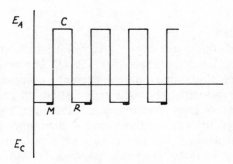

Figure 8. Electrode polarization by alternating measuring and cleaning pulses. The pulse length is variable, from 0.1 s to infinity. *M* — measuring interval; *R* — relaxation interval; *C* — cleaning interval. From Štulík and Hora (1976).

methods involved d.c. voltammetry (DCV), damped DCV, sampled DCV (tast), normal pulse voltammetry (NPV) and differential pulse voltammetry (DPV). The results are summarized in Tables 4 and 5 and are compared with the appropriate theoretical values calculated from the equations given in Table 6. It can be seen that the DCV data compare well with theory (Equation 1 and 2 in Table 6; Equation 3 gives values one order of magnitude lower than the experimental values). On the other hand, the DPV measurements do not agree well with the equations available that have been derived for reversible electrode reactions in quiescent solution (Equation 4 and 5) and for steady-state currents at a rotating disc electrode (Equation 6). On the whole, DCV measurement exhibits a better signal-to-noise ratio than pulse measurements and should be preferred in most measurements. Measurement with a stationary drop electrode is preferable to the use of a drop-

Table 4. Characteristics of the calibration curves and the detection limits. (The detection limit was obtained as twice the noise value. Method A: direct solute injection into the cell; method B: injection through the long capillary; flow-rate, 0.3 ml/min; large static drop; drop time (t_1) = 1 s, f = sampling interval.)

Method		Slope ($\mu A/\mu g$)	Correlation coefficient	Detection limit (ng)
DCV, static drop, A		16.00	0.9970	6
DCV, static drop, B		6.18	0.9974	22
DCV, dropping electrode, A		—	—	13
Damped DCV, static drop, A		6.60	0.9991	5
Damped DCV, static drop, B		5.37	0.9995	16
Damped DCV, dropping electrode, A		—	—	13
Sampled DCV, static drop, A	$f = 1.0$ s	6.90	0.9950	6
	$f = 0.2$ s	6.80	0.9970	6
	$f = 2.0$ s	3.70	0.9960	11
Sampled DCV, dropping electrode, A		13.10	0.9950	12
Theoretical DCV calibration curve, A,	Equation 1	15.60	—	—
	Equation 2	14.60	—	—
Theoretical DCV calibration curve, B,	Equation 1	6.50	—	—
	Equation 2	6.40	—	—
NPV, static drop, A, pulse = 100 mV, $f = 1$ s		10.1	0.9870	4.8
NPV, dropping electrode, A, pulse = 100 mV, $f = 1$ s		13.5	0.9989	120
DPV, dropping electrode, B, pulse = 100 mV, $f = 1$ s		0.30	0.9856	4
Theoretical DPV calibration curve, B, pulse = 100 mV, $f = 1$ s				
	Equation 4	2.3	—	—
	Equation 5	0.33	—	—
	Equation 6	0.42	—	—

Table 5. Noise values. (flow-rate, 0.3 ml/min; large drop; drop time $(t_1) = 1$ s, sampling interval $(f) = 1$ s, ΔE — pulse amplitude)

Method	Noise (A)	Note
DCV, static drop	6×10^{-8}	
DCV, dropping electrode	1.6×10^{-7}	Regular oscillations, 6.2×10^{-6} A, peak-to-peak
Damped DCV, static drop	4.6×10^{-8}	
Damped DCV, dropping electrode	1.6×10^{-7}	Regular oscillations, 3.4×10^{-6} A, peak-to-peak
Sampled DCV, static drop; various f	2.5×10^{-8}	Slight baseline drift; noise increases with decreasing f
Sampled DCV, dropping electrode; various f	$4 \times 10^{-8}—1.2 \times 10^{-7}$	Noise increases with decreasing f
DPV, static drop, various ΔE and f	$1—6 \times 10^{-9}$	Noise increases with decreasing f and increasing ΔE
DPV, dropping electrode, various f and ΔE	$2 \times 10^{-9}—1 \times 10^{-8}$	Noise increases with decreasing f and increasing ΔE
NPV, static drop	$1—2 \times 10^{-8}$	Small effect of ΔE
NPV, dropping electrode	8×10^{-7}	Small effect of ΔE

Note: There is a very small decrease in the noise with decreasing flow-rate and a pronounced decrease with decreasing electrode surface area; the latter may be expected from the theory.

ping electrode, provided that there is no electrode passivation. If a dropping electrode must be used, the current sampled technique must be employed to eliminate large current oscillations. In contrast to measurements with solid electrodes it has been found that with pulse techniques the signal is virtually independent of the mobile phase flow-rate, which is an advantage, together with the advantage of the selectivity of measurement. However, DCV measurement exhibits better precision than the pulse techniques, especially with a stationary electrode.

SELECTED APPLICATIONS OF VOLTAMMETRIC AND POLAROGRAPHIC DETECTION

In practical application of voltammetric and polarographic detectors, two main advantages can be gained:

1. The measuring sensitivity is higher and the detection limit lower than with other detection techniques available.

2. The selectivity of the electrochemical measurement helps in resolving mixtures of solutes that cannot be resolved satisfactorily in the column.

Below, we demonstrate these two advantages on the results of our own works. The determination of plasma catecholamines, indole amines and

Table 6. Equations for maximum current, I_p

No. Equation	Reference
1 $\quad I_p = k_1 nFAD^{2/3} \nu^{-1/6} u^{1/2} L^{-1/2} c$	Matsuda (1967)
2 $\quad I_p = 4\pi r nFDc + k_2 nr^2 D^{2/3} f^{1/2} c$	Levich (1942, 1944a, b)
3 $\quad I_p = 230 nD^{1/2} [m^{2/3} t_1^{1/6} + 103 D^{1/6} (mt_1)^{1/3} + 7.45 u^{1/2} (mt_1)^{1/2}] c$	Okinaka and Kolthoff (1957)
4 $\quad \Delta I_p = -\dfrac{n^2 F^2}{4RT} (A\Delta E D^{1/2} \pi^{-1/2} t_1^{-1/2} c)$	Parry and Osteryoung (1965)
5 $\quad \Delta I_p = nFAD^{1/2} \pi^{-1/2} t_1^{-1/2} \left[\left(\dfrac{\sigma - 1}{\sigma + 1} \right) \right] c;$ $\quad \sigma = \exp\left[\dfrac{\Delta E}{2} \right] \left(\dfrac{nF}{RT} \right)$	Parry and Osteryoung (1965)
6 $\quad \Delta I_p = \dfrac{nFAD}{\delta} \left[\left(\dfrac{\sigma - 1}{\sigma + 1} \right) \right] c; \quad \sigma = \exp\left[\dfrac{\Delta E}{2} \right] \left(\dfrac{nF}{RT} \right);$ $\quad \delta = \dfrac{1.6116 D^{1/3} \nu^{1/6}}{\omega^{1/2}}$	Myers *et al.* (1974)

Symbols: A = electrode surface area (cm^2); c = solute concentration (mol cm^{-3}); D = solute diffusion coefficient (5.07×10^{-6} cm^2 s^{-1}); ΔE = pulse amplitude (V); f = stirring frequency (rpm) for which linear flow-rate was substituted; F = Faraday constant (96 487 C mol^{-1}); I_p = maximum current (A); k_1 = 0.85; k_2 = empirical const. (k_2 = 1.717 $\times$ 10^7 A $cm^{5/6}$ $s^{7/6}$ mol^{-1}); L = characteristic dimension of the electrode ($L = 2/3r$); m = mercury flow-rate (g s^{-1}); n = number of electrons exchanged (n = 18); r — electrode radius (cm); R = gas constant (8.314 J K^{-1} mol^{-1}); t_1 = drop (sampling) time (s); T = temperature (K); u = linear flow-rate (cm s^{-1}); ν = kinematic viscosity (0.0106 St); π = Ludolf number; ω = angular frequency ($\omega = 2\pi f$, the linear flow-rate, u, was substituted for f).

carcinogenic aromatic amines are examples of the superior sensitivity of voltammetric detection. Indeed, the concentrations of these substances in most practical samples are so low that they cannot be determined in any other way. On the other hand, the determination of pyrimidine derivatives and tricyclic neuroleptics described below are examples of determinations where electrochemical detection is equally or even less sensitive than common UV photometric detection, but where the selectivity of electrochemical detection substantially contributes to the resolution of complex mixtures and to identification of the test compounds. This determination also demonstrates the usefulness of using several detectors connected in series, in this case the UV photometric, voltammetric and polarographic detector.

The examples given below also illustrate other advantages of voltammetric and polarographic detection, mainly a good measuring precision, even at very low solute concentrations and good linearity of the calibration curves, down to the lowest solute amounts.

Determination of plasma catecholamines

The most important application of voltammetric detectors is the determination of catecholamines in body fluids. These substances are readily oxidized at a potential as low as +0.6 V and thus are detected with high sensitivity at solid electrodes, as illustrated by a great number of publications dealing with this topic. As this field has been investigated in detail, both theoretically and with the view of application to the determination of catecholamines and their metabolites in cerebrospinal fluid, serum, urine and various tissues, especially the brain tissue, we will not discuss it and refer the reader to pertinent literature, e.g. the reviews by Kissinger (1977); Kissinger *et al.* (1977, 1981); Krstulovic (1979). We will describe only one application of electrochemical detection to the determination of adrenaline and noradrenaline in the plasma of obese children before and after physical stress. The values of plasma catecholamines provide a short-term index of the response of an organism to a stimulus, e.g. an effect of a drug or psychic or physical stress on the sympathetic nervous system. On the other hand, the urine content represents an average value over a relatively long time.

The chemicals used were noradrenaline bitartrate, adrenaline monotartrate (Fluka, Switzerland), sodium octyl sulphate (Merck, GFR). All other chemicals used were of p.a. purity (Lachema, Czechoslovakia).

Procedure. Sample pretreatment was carried out along the lines described in *Application Note No. 14* (1981). To a 5 ml conical reaction vial were added 2 ml plasma, 25 μl 3,4-dihydroxybenzylamine standard solution containing 100 ng/ml and 50 mg acid-washed aluminium oxide. Then 1 ml Tris buffer was added and the mixture was shaken 5—10 min on a reciprocating shaker. After settling, the alumina was separated, washed three times with water and dried. An amount of 200 μl 0.1 M $HClO_4$ was added, the mixture was centrifuged and filtered. The filtrate was injected onto the chromatographic column.

HPLC separation was carried out using an LC—XP liquid chromatograph with a Partisil ODS column (10 μm, 25 × 0.3 cm, Pye Unicam, England) and a carbon fibre voltammetric detector. The mobile phase consisted of 14.14 g monochloracetic acid, 4.68 g NaOH, 0.75 g Na_2EDTA and 25—30 mg sodium octyl sulphate in 1 l of distilled water. Other operating conditions were: flow-rate, 1.0 ml/min; injected amount, 50 μl; applied potential, +0.65 V; temperature, 20°C.

Sample calculations were made against a standard. For calibration purposes, a synthetic sample consisting of 2 ml phosphate buffer, 50 μl adrenaline and noradrenline standard solution (25 ng/ml adrenaline, 75 ng/ml noradrenaline) and 25 μl 3,4-dihydroxybenzylamine standard solution was assayed. Peak height ratios (relative to the internal standard DHBA) for unknown plasma were compared to those for this syntetic standard whose original concentrations were known:

$$\text{conc.}[NA_{unknown}] = \frac{[NA/DHBA]_{unknown}}{[NA/DHBA]_{known}} \times \text{conc.}[NA_{known}]$$

Results and discussion. Samples of plasma were taken from children of 12—17 years of age before and after physical stress. The noradrenaline and adrenaline concentrations were determined and are given in Table 7. The values found correlated well with the values given in the literature. The error of determination varied from 10 to 50% depending on the solute concentration.

Table 7. Determination of noradrenaline (NA) and adrenaline (A) in plasma before (B) and after (C) physical stress

Patient No.	NA (pg/ml)		A (pg/ml)	
	B	C	B	C
1	500	875	95	118
2	670	810	231	317
3	500	667	95	110
4	545	428	95	unmeasurable
5	400	187	75	70
6	586	461	111	57
7	546	444	59	63
8	727	667	103	125
9	857	900	215	unmeasurable
10	506	439	81	57

The changes in the NA and A values before and after physical stress correlated well with the values found by parallel fluorescence determinations in urine. The average values, in nanomoles per litre, were as follows:
Before physical stress the contents of NA were 2.95 ± 3.44 in plasma and 151.9 ± 69.42 in urine; the corresponding values after stress were 3.44 ± 1.29 and 225.5 ± 139.5. The contents of A before stress were 0.53 ± 0.31 in plasma and 162.9 ± 54.3 in urine; after stress, the corresponding values were 0.56 ± 0.51 and 179.8 ± 101.5.

Determination of indole amines with simultaneous UV photometric and voltammetric detection

Indole amine derivatives are interesting because of their hallucinogenic properties. They are formed in human organisms as metabolites of tryptophan (especially *N*-methyltryptamine, *N,N*-dimethyltryptamine and *N,N*-dimethyl-5-hydroxytryptamine) and are contained in some mushrooms (psilocybin, psilocin). Sensitive methods are required for their determination by psychiatry and forensic medicine.

A great variety of analytical methods have been applied to the determination of indole derivatives, but only gas chromatography has found wide use, as it is sufficiently sensitive and exhibits a high separation efficiency. However, these substances are poorly volatile and thus derivatization is required that makes the determination tedious and introduces an additional error. These drawbacks are removed in HPLC that has been used to determine indole amines in body fluids, tissues and mushroom extracts [see e.g. Christiansen *et al.*, (1981); Christiansen and Rasmussen (1982); Perkal *et al.*, (1980); Wagner *et al.*, (1982); White (1979); Wurst *et al.*, (1984); Yamaguchi *et al.*, (1982)]. While tryptophan derivatives, such as serotonin, tryptamine, etc.., are best separated on nonpolar chemically-bonded phases, silica gel and ion-exchangers have been successfully used to separate psilocin and psilocybin.

In addition to more universal UV photometric detectors, spectrofluorimetric and recently also electrochemical detectors have been used. We separated some hallucinogenic indole amines with detection by the tandem of a UV photometric and a voltammetric detector, and applied the procedure to the determination of these substances in cerebrospinal fluid and in some mushrooms [Kysilka *et al.* 1985].

The chemicals used were tryptophan (Kodak, USA), tryptamine (Calbiochem, USA), 5-hydroxytryptamine, 5-hydroxy-N,N-dimethyltryptamine, 5-methoxytryptamine, 5-hydroxy-N-acetyltryptamine, N-methyltryptamine, N,N-dimethyltryptamine, 5-methoxy-N,N-dimethyltryptamine (all from Sigma, USA), 4-hydroxy-N,N-dimethyltryptamine (psilocin), 4-phosphoryloxy-N-methyltryptamine (psilocybin) (Merck, GFR). All other chemicals were of p.a. purity, from Lachema (Czechoslovakia).

Procedures. A liquid chromatograph consisting of a 3B high-pressure pump, a sampling valve with a 10 μl loop, an LC-75 UV photometric detector and a Chromatographics 2 datasystem (all from Perkin Elmer, USA was used). The UV photometric detector was connected in series with an EDLC voltammetric detector (Laboratorní Přístroje, Czechoslovakia). The chromatographic column contained Partisil ODS (10 μm, 250 × 3.0 mm ID) and Separon SI C18 (10 μm, 250 × 3.2 mm ID). The mobile phase consisted of aqueous citrate—phosphate buffer of pH 3.8 (300 ml 0.1 M citric acid and 160 ml 0.1 M NaH_2PO_4) and various amounts of ethanol. The flow-rate was 1 ml/min

Sample pretreatment was carried out in the following manner.

1. Cerebrospinal fluid (CSF) was pretreated by a modified procedure of Yamaguchi *et al.*, (1982). To a 5 ml CSF sample, 70% $HClO_4$ was added to attain a concentration of 7% for the removal of proteins. The mixture was centrifuged for 30 min at 1000 g. The pH of the separated supernatant was adjusted to 12 with 45% KOH, while cooling in an ice bath. The $KClO_4$ precipitate formed was separated by a 5 min centrifugation, 1 g NaCl was added to the supernatant and the solution was extracted twice with 6 ml portions of redistilled dichloromethane (15 min shaking). The joint organic extracts were dried by adding 3 g anhydrous Na_2SO_4 and centrifuged for 5 min at 1000 g to remove the drying agent. The supernatant was transferred to a conical flask and the solvent was evaporated in vacuo at laboratory temperature.

The same procedure was repeated with a 5 ml CSF sample, to which standard methanolic solutions of 5-hydroxy-*N,N*-dimethyltryptamine, 5-methoxytryptamine, *N*-methyltryptamine, *N,N*-dimethyltryptamine and 5-methoxy-*N,N*-dimethyltryptamine were added, to obtain resultant concentrations of 10 ng/ml for these substances.

2. Mushrooms were pretreated by taking a 300 mg sample of the mushroom Psilocybe Bohemica Šebek, homogenizing and extracting for 24 h on a reciprocating shaker into 30 ml methanol. The mixture was then filtered, the solvent evaporated *in vacuo* and the residue dissolved in 3.0 ml methanol. A 10 μl aliquot was then injected into the chromatograph. The standard solutions of psilocin and psilocybin contained 1 and 5 μg of the substances in 5 μl methanol.

Determination of some clinically important indole amines
Optimal conditions for the separation and detection of the above indole amines have been studied. Various stationary phases were tested, the best results being obtained with Partisil ODS and Separon SIC. Of the mobile phases tested, the best performance was exhibited by an aqueous citrate—phosphate buffer mixed with ethanol in various ratios. The dependence of the logarithms of the capacity ratios on the ethanol content in the mobile phase were approximately linear within a range of 10—20 vol% (Fig. 9). As can be seen, the optimal composition of the mobile phase is 0.1 M aqueous citrate—phosphate buffer containing 10 vol% ethanol. An example of the separation under these conditions is given in Fig. 10.

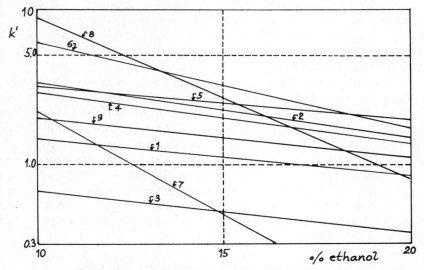

Figure 9. Dependence of capacity factors on the ethanol content in the mobile phase. 1. tryptophan; 2. tryptamine; 3. 5-hydroxytryptamine; 4. 5-methoxytryptamine; 5. *N*-methyltryptamine; 6. *N,N*-dimethyltryptamine; 7. 5-hydroxy-*N,N*-dimethyltryptamine; 8. 5-methoxy-*N,N*-dimethyltryptamine; 9. 5-hydroxy-*N*-acetyltryptamine. From Kysilka *et al.* (1985).

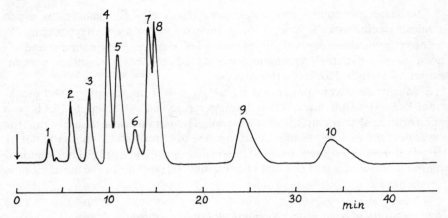

Figure 10. A chromatogram of a mixture of indole amines. 1. methanol; 2. 5-hydroxy-tryptamine; 3. tryptophan; 4. 5-hydroxy-*N*-acetyltryptamine; 5. 5-hydroxy-*N,N*-dimethyl-tryptamine; 6. 5-methoxytryptamine; 7. *N*-methyltryptamine; 8. tryptamine; 9. *N,N*-dimethyl-tryptamine; 10. 5-methoxy-*N,N*-dimethyltryptamine. For the conditions see the text. From Kysilka *et al.* (1985).

The studied substances absorb in spectral regions around 220 and 260—290 nm. The optimal wavelength for the detection is 280 nm where absorption of the mobile phase is minimal. The substances can further be electrochemically oxidized on a carbon electrode with half-wave potentials around +0.5 V for hydroxy derivatives, the other substances have half-wave potentials from +0.9 to +1.0 V. Therefore, the potential +1.0 V was used for detection. Under the optimal conditions for HPLC separation and detection, the calibration curves and detection limits were measured and are given in Table 8. As follows from Table 8, electrochemical detection is more sensitive for all the studied compounds and an especially high sensitivity is obtained for the hydroxy derivatives.

The method was applied to the determination of these compounds in cerebrospinal fluid. The recovery was determined by adding known amounts of the standard substances (30 ng of each of the substances). The measurements were carried out in quintuplicate and the results are given in Table 9. It follows from the table that in the procedure employed about 10% of the test substances are lost. The relative standard deviation amounts to 3—6% which is satisfactory. A chromatogram of a CSF sample with the addition of 10 ng of five hallucinogenic indole amines is shown in Fig. 11. As can be seen, the electrochemical detector is sufficiently sensitive to reliably detect the test substances in these amounts. The UV photometric detector yields no response under these conditions.

Determination of hallucinogenic indole alkaloids in mushrooms
Under the same experimental conditions as described above, standard solutions of psilocin and psilocybin and samples of the extracts of the mushrooms Psilocybe Bohemica Šebek were analysed, peak heights and areas evaluated,

Table 8. Calibration data for the studied indole derivatives (for the conditions see the text; six parallel determinations, $\alpha = 0.05$)

Substance	UV photometric detection			Voltammetric detection		
	Correlation coefficient	Detection limit (ng)	Relative standard deviation (%)	Correlation coefficient	Detection limit (ng)	Relative standard deviation (%)
Tryptophan	0.9933	4.0	2.0	0.9980	3.0	2.4
Tryptamine	0.9991	9.0	2.7	0.9928	6.0	2.5
5-Hydroxytryptamine	0.9967	1.7	3.2	0.9986	0.2	3.2
5-Methoxytryptamine	0.9946	80.0	4.8	0.9996	7.0	2.7
N-Methyltryptamine	0.9917	50.0	4.3	0.9972	9.0	3.7
N,N-Dimethyl-tryptamine	0.9874	20.0	4.1	0.9897	8.0	3.5
5-Hydroxy-N,N-dimethyltryptamine	0.9989	5.5	3.1	0.9994	0.2	3.5
5-Methoxy-N,N-dimethyltryptamine	0.9762	70.0	4.5	0.9823	5.0	3.1
5-Hydroxy-N-acetyltryptamine	0.9993	4.4	2.9	0.9995	0.2	2.6
4-Hydroxy-N,N-dimethyltryptamine	0.9941	80.0	3.0	0.9941	2.0	2.8
4-Phosphoryloxy-N-methyltryptamine	0.9996	93.0	2.8	0.9964	12.0	3.5

Table 9. The reproducibility and recovery of the analysis of CSF with electrochemical detection (taking 30 ng of each substance; 5 parallel measurements; $\alpha = 0.05$)

Substance	Found (ng)	S.D. (%)	Recovery (%)
5-Hydroxytryptamine	28.4	3.0	94.7
N-Methyltryptamine	27.4	3.7	91.3
5-Hydroxy-N-acetyltryptamine	26.3	6.3	87.7
Tryptophan	28.0	5.0	93.3

and contents of these substances determined from both UV photometric and electrochemical detection. The average results and the standard deviations are given in Table 10. The chromatograms of the mushroom extract using the two detection techniques are given in Fig. 12. The detection limits were obtained as the amounts of substances causing the detector response to equal

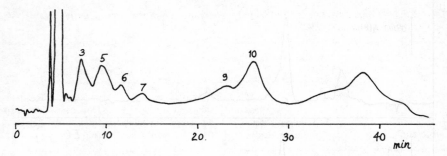

Figure 11. HPLC analysis of cerebrospinal fluid enriched by the following substances (10 ng each) (Electrochemical detection): 3. tryptophan; 5. 5-hydroxy-*N,N*-dimethyltryptamine; 6. 5-methoxytryptamine; 7. *N*-methyltryptamine; 9. *N,N*-dimethyltryptamine; 10. 5-methoxy-*N,N*-dimethyltryptamine No response was obtained from UV photometric detector; for the conditions see text. From Kysilka *et al.* (1985)

Table 10. Determination of psilocybin and psilocin in the extract of Psilocybe Bohemica Šebek using UV photometric and electro-chemical detection

	UV Photometric		Electrochemical	
Substance	$\bar{x}$ (%)	Relative S.D. (%)	$\bar{x}$ (%)	Relative S.D. (%)
Psilocybin	0.58	3.02	0.57	2.35
Psilocin	0.058	10.10	0.061	1.37

twice the absolute noise value and amounted to 93 and 12 ng for psilocybin with UV photometric and voltammetric detection, respectively, the corresponding values for psilocin being 80 and 2 ng.

It follows from these results that voltammetric detection is superior to UV photometric detection in determination of hallucinogenic substances in mushrooms. Psilocin is more readily oxidized than psilocybin, owing to the presence of an OH group in the molecule, and its determination with the voltammetric detector is much more sensitive than with the UV photometric detector. The chromatogram of the mushroom extract is simpler when voltammetric detection is used, as other components extracted from the mushrooms together with the test substances do not interfere with the determination.

Carcinogenic amines
Many aromatic amines are proven or suspect human carcinogens and therefore reliable methods must be found for the determination of traces of these compounds. Various methods have been recommended for the analysis of aromatic amines [for a review, see Egan (1981)]. Titration and spectrophoto-

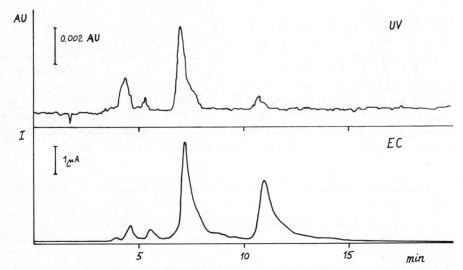

Figure 12. HPLC analysis of the extract of Psilocybe Bohemica Sébek. UV photometric and electrochemical detection; 10 μl sample; for experimental conditions see text. The retention times are 7.25 min for psilocybin and 11.0 min for psilocin. From Kysilka *et al.* (to be published).

metric methods are insufficiently selective for analyses of complex samples. Combination of spectrofluorimetry with TLC have given better results (Jakovljevic *et al.*, 1975). Gas chromatography, which has been widely used, often requires derivatization in view of the high polarity and low volatility of amines. In contrast to GC, HPLC enables direct analysis of aromatic amines. For their separation, non-polar chemically-bonded phases in combination with aqueous solutions of methanol or acetonitrile have been recommended. With a UV detector, a detection limit of 5×10^{-6} mol l^{-1} has been achieved (Castegnaro *et al.*, in press). To increase the sensitivity, electrochemical detection has been used (e.g. Armentrout and Cutie, 1980; Concialini *et al.*, 1983; Mefford *et al.*, 1977; Purnell and Warwick, 1980; Rice and Kissinger, 1979; Riggin and Howard, 1979). Aromatic amines can be easily oxidized at solid electrodes, monoamines at potentials close to $+1.0$ V and diamines around $+0.5$ V (SCE).

Several methods have been proposed for the destruction of aromatic amines in laboratory wastes (Castegnaro, 1985). Highly efficient is oxidation with permanganate in an acidic solution to non-mutagenic products (Castegnaro *et al.*, in press). This method, however, cannot be used in the presence of large amounts of oxidizable substances, e.g. ethanol. In this case, oxidation by hydrogen peroxide catalysed by horseradish peroxidase to yield a precipitate has been suggested. The solid residues which are, however, mutagenic can be further destroyed by the permanganate method.

In our study we have determined the conditions for HPLC separation and UV—electrochemical detection of selected aromatic amines. We have applied this method to the monitoring of traces of aromatic amines in wastes after their chemical degradation. The efficiencies of simple and combined per-manganate degradation methods were compared (Barek *et al.*, 1985).

The chemicals used were benzidine (Fluka, Switzerland), 3,3'-dimethoxy-benzidine, 3,3'-dichlorobenzidine (Merck, GFR), 1-naphthylamine, 2-naphthylamine, 4-aminobiphenyl, 4-nitrobiphenyl, *m*-toluenediamine (Sigma, USA), 3,3'-diaminobenzidine (Lachema, Czechoslovakia), 4,4'-methylene-bis(*o*-chloroaniline) (Serlabo, France) and horseradish peroxidase of specific activity 175 purpurogallin units per mg (Sigma, USA). All other chemicals used were of p.a. purity (Lachema, Czechoslovakia).

Procedures. Laboratory wastes were decontaminated in the following ways.

1. The permanganate method was carried out by dissolving *c.* 9 mg of a test amine in 10 ml 0.1 M HCl (or in 10 ml concentrated acetic acid for substances Nos. 6 and 11 in Table 11). Then 5 ml 0.2 M $KMnO_4$ and 5 ml 2 M H_2SO_4 were added and the solution allowed to react overnight, after which it was subjected to HPLC analysis.

2. The combined permanganate and enzymatic methods. A waste solution containing up to 100 mg aromatic amine per litre and up to 20% methanol or ethanol were adjusted to pH 5—7 with addition of NaOH or H_2SO_4. A 3% solution of hydrogen peroxide and 1000 units of horseradish peroxidase (*ca.* 6 mg) were added per litre of waste solution. After 3 h the precipitate formed was filtered off using an S4 frit. The frit with the precipitate was then immersed in the solution obtained by mixing 50 ml 0.2 M $KMnO_4$ with 50 ml 2 M sulphuric acid and allowed to react overnight while stirring magnetically. The precipitate was thus completely dissolved and the solution analysed by HPLC.

HPLC determination was carried out by gradually adding solid ascorbic acid to an aliquot of the solution after permanganate oxidation until the solution became colourless, the pH was adjusted to about 8 with 10 M NaOH and the solution was centrifuged. To 1 ml supernatant, 3 ml methanol were added, centrifuged and 20 µl supernatant injected onto the column. The standard solutions were prepared by dissolving the test substances in 0.1 M hydrochloric acid to obtain a 5×10^{-3} M concentration. Dissolution can be enhanced by sonication. The concentrations of the amines in the solutions after degradation were determined by the standard addition method. Stand-ard amine solutions were added to the sample immediately after the addition of ascorbic acid.

The column used (25 × 0.3 cm ID) contained the reverse phase Partisil ODS, 10 µl (Pye Unicam, England). The mobile phase consisted of 0.1 M aqueous ammonium acetate with various amounts of methanol; the pH was adjusted by aditions of perchloric acid and sodium hydroxide. The flow-rate was 1.0 ml/min. The chromatographic measurements were performed on a Pye Unicam LC—XP liquid chromatograph with an LC—UV photometric detector (Pye Unicam, England) operated at 280 nm and with a carbon-fibre voltammetric detector set at a potential of +0.9 or +0.6 V. The two detectors were connected in series by a short stainless-steel capillary (5 cm long, 0.2 mm ID). The chromatograms were recorded with a TZ 4200 dual-line recorder (Laboratorní Přístroje, Czechoslovakia). During measurement, the mobile phase was continuously deaerated by the passage of helium. All the measurements were carried out at laboratory temperature.

Separation of aromatic amines. The effects of methanol content and pH of the mobile phase on the separation of aromatic amines (see Table 11) were studied. The dependences of the logarithms of the capacity ratios on the methanol cencentration are given in Fig. 13. All the amines exhibit similar

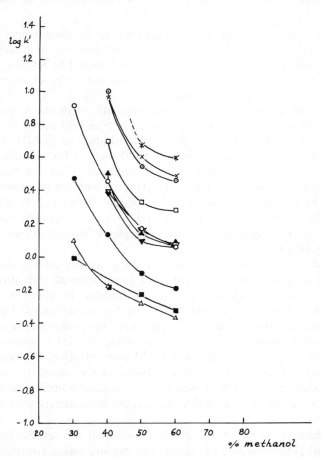

Figure 13. Dependence of log k' on the methanol content in the mobile phase for some carcinogenic amines: ○ — benzidine; ● — *o*-dianisidine; x — 3,3'-dichlorobenzidine; △ — 3,3'-diaminobenzidine; ▲ — *o*-tolidine; ▽ — 4-aminobiphenyl; ▼ — 3,3'-dimethoxy-benzidine; □ — 1-naphthylamine; ■ — 2-naphthylamine; ⊙ — 4,4'-methylene-bis(*o*-chloro-aniline). 0.1 M aqueous ammonium acetate +*x* vol% methanol, pH 7.1; flow-rate, 1 ml min⁻¹. From Barek *et al.* (1985).

Table 11. Characteristics of the studied substances and their detection limits

No.	Substance	pK_{a1}	pK_{a2}	λ_{max} (nm)	$E_{1/2}(E_p)$ (V)	Detection limit (pg)	
						UV	Electrochemical
1	Benzidine (4,4-diaminobiphenyl)	4.7	3.7	268	+0.36	3000	3
2	o-Dianisidine (3,3'-dimethoxybenzidine)	4.7	3.6	212; 303	+0.29 (+1.23)	4000	50
3	3,3'-Dichlorobenzidine			285	(+0.51)	16000	450
4	3,3'-Diaminobenzidine			211; 224; 278	+0.17	3000	50
5	o-Tolidine (3,3'-dimethylbenzidine)	4.7	3.7	282	+0.33	4000	30
6	4-Aminobiphenyl	4.2		278	(+0.58)	10000	1100
7	4-Nitrobiphenyl			222; 305	−0.75 (DME)	12000	–
8	1-Naphthylamine	4.0		243; 318; 328	(+0.51)	11400	1400
9	2-Naphthylamine	4.2		237; 259; 291; 337	(+0.58)	11400	9000
10	m-Toluenediamine	5.1	3.3	294	+0.45	12200	60
11	MOCA (4,4'-methylene-bis-(o-chlorodianiline))	3.5	3.0	247; 298	+0.29 (+0.63)	5000	4600

dependences and therefore separation cannot be improved by changing the methanol content. From the point of view of the speed of analysis (values of the capacity ratio of < 10.0), a content of 40% methanol is optimal.

The dependence of the log k' on the mobile phase pH was studied (Fig. 14). The greatest changes in the capacity ratios occur between pH 3 and 5 where pK_a values of most of the substances are found (see Table 11). The pK_a values were determined by potentiometric titration of free amine bases with a 0.01 M standard solution of HCl. At lower pH values the retention times decrease as the dissociation of the amines is enhanced, and thus their polarity is increased. The elution order of some of the substances is also changed (Substances Nos 1, 5—7). As follows from Fig. 14, the optimal pH of the mobile phase for separation is between 3.5 and 4.0.

Detection of aromatic amines. A UV photometric detector may be used with an optimal wavelength of *c.* 280 nm (see Table 11, where the appro-

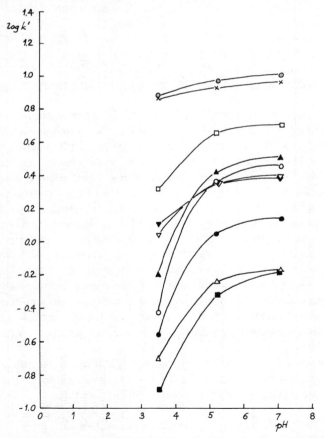

Figure 14. Dependence of log k' on the mobile phase pH for some carcinogenic amines. For the list of substances see Fig. 13. 0.1 M aqueous ammonium acetate +40 vol% methanol; flow-rate, 1 ml min⁻¹. From Barek *et al.* (1985).

priate absorption maxima are given). The detection limits given in Table 11 are of the order of tens or units of nanograms.

All the studied compounds can be oxidized at a glassy carbon electrode with the exception of 4-nitrobiphenyl. The half-wave potentials obtained in the given mobile phase at pH 7 are in Table 11. Substances Nos 3, 6, 8, 9, whose oxidation products are strongly adsorbed on the electrode surface, have the shape of peaks rather than normal waves, therefore the peak potentials are given in Table 11. It follows from Table 11 that a potential of +0.6 V is sufficient for all amines. The oxidation potentials depend on the pH; with decreasing pH the anodic waves shift to more positive potentials, about 80 mV per pH unit. At pH 3.5, the electrode working potential must thus be maintained at a value of +0.9 V. The detection limits for the voltammetric detector given in Table 11 are substantially lower for most of the amines (by two to three orders of magnitude) than those obtained with UV photometric detector (see substances Nos. 1, 2, 4, 5, 10). It follows from the results in Table 11 that the electrochemical detection is clearly preferable to UV photometric detection for these amines. The calibration curves exhibit good linearity (linear correlation coefficient varies from 0.9996 to 0.9978) and satisfactory precision (the relative standard deviations of the peak height for five parallel measurements, $\alpha = 0.05$, are a few per cent and do not exceed 10—12% even at the lowest measured concentrations).

Determination of aromatic amines in wastes after degradation. Both the degradation methods described previously have been tested by measuring the concentrations of the residual amines by HPLC with electrochemical detection. The UV photometric detector was insufficiently sensitive for this purpose. The results are summarized in Table 12. It can be seen that both degradation methods are highly efficient and lead to the destruction of more than 99.95% of most substances. A somewhat less favourable degradation efficiency found for substances Nos. 9 and 11 is caused by lower detection sensitivity. The enzymatic method could be used for all the studied amines, except for *m*-toluenediamine with which no precipitate is formed. It was demonstrated earlier that the solutions obtained after these degradation procedures have no mutagenic effects and thus are environmentally harmless. The combined enzymatic and permanganate methods are suitable for wastes containing small amounts of amines in the presence of oxidizable compounds where simple permanganate methods fails and are somewhat more effective (see Table 12).

The described HPLC method with electrochemical detection can be used not only for the monitoring of the effectiveness of degradation procedures, but also for trace analyses of these compounds in general.

HPLC of biologically important pyrimidine derivatives with UV photometric—voltammetric—polarographic detection

Great attention has been paid to the analysis of pyrimidine derivatives because of their importance for human organisms. Uracil, cytosine and

Table 12. The efficiency of the degradation of the studied substances as determined by HPLC with electrochemical detection (four parallel determinations; $\alpha = 0.05$)

No.	Substance	Initial concentration taken (M)	Final concentration, found (M)		Efficiency of degradation (%)	
			Procedure[a]	Procedure[b]	Procedure[a]	Procedure[b]
1	Benzidine	4.07×10^{-3}	$1.2 \pm 0.1 \times 10^{-7}$	$<4.4 \pm 0.7 \times 10^{-8}$	99.997	>99.999
2	o-Dianisidine	3.07×10^{-3}	$<4.1 \pm 0.8 \times 10^{-8}$	$<4.1 \pm 0.8 \times 10^{-8}$	>99.998	>99.998
3	3,3'-Dichlorobenzidine	2.30×10^{-3}	$<2.8 \pm 0.3 \times 10^{-7}$	$<2.8 \pm 0.2 \times 10^{-7}$	>99.988	>99.988
4	3,3'-Diaminobenzidine	2.10×10^{-3}	$5.1 \pm 0.4 \times 10^{-7}$	$<2.8 \pm 0.6 \times 10^{-8}$	99.976	>99.989
5	o-Tolidine	3.54×10^{-3}	$7.1 \pm 0.4 \times 10^{-7}$	$<8.0 \pm 0.7 \times 10^{-8}$	99.980	>99.998
6	4-Aminobiphenyl	4.44×10^{-3}	$1.3 \pm 0.1 \times 10^{-6}$	$4.0 \pm 0.2 \times 10^{-6}$	99.971	99.910
8	1-Naphthylamine	5.24×10^{-3}	$2.1 \pm 0.2 \times 10^{-6}$	$2.0 \pm 0.1 \times 10^{-6}$	99.960	99.962
9	2-Naphthylamine	5.24×10^{-3}	$<1.26 \pm 0.04 \times 10^{-5}$	$<1.30 \pm 0.02 \times 10^{-5}$	>99.718	>99.752
10	m-Toluenediamine	6.14×10^{-3}	$5.2 \pm 0.4 \times 10^{-8}$	c	99.999	c
11	MOCA	2.61×10^{-3}	$<3.60 \pm 0.08 \times 10^{-6}$	$<3.6 \pm 0.08 \times 10^{-6}$	>99.862	>99.862

[a] After simple permanganate method; [b] after oxidation of the residues formed by the enzymatic oxidation of the solution containing 10 mg of test amine per 100 ml; [c] no precipitate formed during the enzymatic oxidation.

Note: The concentration values given in Table 12 are recalculated to the original volume of the waste solution.

thymine, components of the nucleic acids, are among the most significant biological substances, many derivatives, e.g. fluoro- and aza-derivatives, are used as cytostatics, and some pyrimidine derivatives are important intermediates in organic syntheses. Gas chromatographic determination of these substances (e.g. Gehrke *et al.*, 1967, 1968; Hashizume and Sasaki, 1968; Miller *et al*; 1976) requires derivatization, mostly to the trimethylsilyl derivatives. Direct analysis can be performed by HPLC [reviewed by Zakaria and Brown, (1981)] and the determination is faster, more precise and sometimes also more sensitive than the GC determination [the HPLC determination of 5-fluorouracil in plasma is twenty times more sensitive than GC determination (Christophidis *et al.*, 1979)].

A great variety of stationary phases have been employed for the separation of pyrimidine derivatives. The presence of basic and acidic groups in these substances makes it possible to use ion exchangers. Both cation (Brown *et al.*, 1974; Falchuk and Hardy, 1978; van Haastert, 1981) and anion (Bakay *et al.*, 1978; Cohen and Brown, 1978; Eksteen *et al.*, 1978; Floridi *et al.*, 1977) exchangers have been employed, and the type and concentration of the counter-ion, the effect of the composition and the pH of the mobile phase have been studied. An adsorption mechanism was involved in the separation of pyrimidines on silica gel, using a mobile phase containing dichloromethane, methanol and an aqueous solution of a salt (Brugmann *et al.*, 1982; Evans *et al.*, 1979; Ryba and Beránek, 1981). A macroporous styrene— divinyl-benzene copolymer has also been used by Lee and Kindsvater (1980). The separation was greatly improved when using chemcially-bonded reverse phases. A comparison of various phases has shown that the best results can be obtained with the Spherisorb ODS-2 phase (Miller *et al.*, 1982). The separation is usually performed in the ion-pairing mode with organic acids (Knox and Jurand, 1981; Kraak *et al.*, 1981) or quaternary ammonium salts (Au *et al.*, 1982; Voelter *et al.*, 1980) as counter-ions. The effects of the temperature, pH, the mobile phase composition and the counter-ion concentration have been studied (e.g. Kraak *et al.*, 1981; Ryba, 1981; Voelter *et al.*, 1980). The method has been applied to serum and plasma (Hartwick *et al.*, 1979a, b; Krstulovic *et al.*, 1977; Taylor *et al.*, 1980).

For detection of pyrimidine bases, fluorescence has been employed in addition to UV photometry, or the two methods have been combined for identification purposes (Hartwick *et al.*, 1979a). Identification is also facilitated by measuring the absorbance at two wavelengths, 254 and 280 nm (Krstulovic *et al.*, 1977).

In our work (Štulík and Pacáková, 1983), we combined UV photometric detection with electrochemical detectors (both voltammetric and polarographic) to facilitate identification of pyrimidine derivatives in complex biological samples and to improve the resolution by utilizing the selectivity of the electrochemical detectors. Whereas most works published so far deal primarily with nucleobases and 5-fluorouacil, the present paper studies many more pyrimidine derivatives (see Table 13), which may be encountered in the production of pharmaceuticals and in the study of their chemical reactions and metabolism.

Table 13. The detector response and capacity ratios of various pyrimidine derivatives (average values of three determinations)

Compound	Detector response		Capacity ratio			
			pH			
	Voltam-metric	Polaro-graphic	2.5	3.5	4.5	6.0
2,4-Dihydroxypyrimidine (uracil)	−	−	0.99	1.13	1.25	1.26
2-Amino-4,6-dihydroxypyrimidine	+	−	0.29	0.65	0.60	0.34
4,5-Diamino-6-hydroxypyrimidine	+	−	0.22	0.58	1.15	1.24
2-Hydroxy-4-aminopyrimidine (cytosine)	−	−	0.43	0.66	0.79	1.04
2,4,5-Triamino-6-hydroxypyrimidine	+	−	0.13	0.28	0.30	0.65
2,4-Diamino-6-hydroxypyrimidine	+	−	0.88	1.72	1.50	1.39
2,4-Dihydroxy-6-aminopyrimidine	+	−	1.04	1.18	1.21	1.12
4,5,6-Triaminopyrimidine	+	−	0.87	1.58	1.43	2.12
2,4-Dihydroxy-5-methylpyrimidine (thymine)	−	−	3.36	3.73	4.35	4.60
5-Methylcytosine	−	−	1.02	1.78	2.28	3.49
5-Nitrouracil	−	+	1.12	1.50	1.73	1.75
2-Mercaptouracil	+	−	1.61	1.83	2.18	2.23
4-Mercaptouracil	+	−	2.73	3.12	3.72	3.85
2,4-Dimercaptouracil	+	−	1.60	1.81	2.17	2.20
2-Mercapto-4-hydroxy-6-aminopyrimidine	+	−	1.95	2.23	2.49	2.34
2-Mercapto-4,6-diaminopyrimidine	+	−	1.98	3.30	3.28	4.31
6-Azacytosine	−	+	0.58	0.70	0.73	0.75
6-Azauracil	−	+	0.75	0.88	0.95	0.95
6-Azathymine	−	+	2.30	2.77	3.17	3.65
2-Mercapto-6-azathymine	+	+	3.19	3.86	4.69	−
5-Fluorouracil	−	−	1.01	1.03	1.37	1.37
5-Bromouracil	−	−	3.25	3.47	4.53	4.31
5-Iodouracil	−	−	5.25	4.94	7.49	−
5-Bromocytosine	+	−	1.90	3.90	5.36	−
5,6-Dihydrouracil	−	−	0.87	1.04	0.54	1.15
5,6-Dihydrothymine	−	−	3.23	3.05	3.74	3.72

The chemicals used were 6-azathymine (Fluka, Buchs, Switzerland), 5-bromouracil, 5-iodouracil, 6-azauracil (Calbiochem, Los Angeles, USA), uracil, cytosine, 6-azacytosine, 5-bromocytosine, 5,6-dihydrothymine, 2-amino-4,6-dihydroxypyrimidine, 2,4-diamino-6-hydroxypyrimidine, 2,4,5-triamino-6-hydroxypyrimidine (Lachema, Brno, Czechoslovakia), 5-methyl-cytosine, 5,6-dihydrouracil (Sigma, St. Louis, USA), 5-fluorouracil (Hoff-mann-La Roche, Basle, Switzerland), thymine (Koch-Light, Colnbrook, Great Britain). Other pyrimidine derivatives were synthesized by Dr J. Černohorský of the Department of Clinical Biochemistry, Charles University, Prague, Czechoslovakia. All other chemicals were of p.a. purity, obtained from Lachema, Brno, Czechoslovakia.

Procedure. The standard solutions of the substances were prepared by dissolving about 3 mg of the preparation in 10 ml of the mobile phase; the solutions were appropriately diluted before use. The mobile phase was prepared by mixing 300 ml of 0.1 M citric acid (6.3 g in 300 ml) and 160 ml of 0.1 M $Na_2HPO_4 \cdot 12H_2O$ (5.73 g in 160 ml). The pH values were adjusted by addition of H_3PO_4 or NaOH. The mobile phase was deaerated by the passage of helium. The flow-rate used was 1 ml/min and all the measurements were performed at laboratory temperature. An LC—XP liquid chromatograph (Pye Unicam, Cambridge, England) with a Separon C-18 column (10 μm, 25 × 0.25 cm, Laboratorní Přístroje, Prague, Czechoslovakia) was used; the samples were injected through a 20 μl loop. The separated compounds were detected with three detectors connected in series: an LC—UV variable wavelength photometric detector (Pye Unicam, England), a voltammetric detector of our own construction operated in the wall-jet system with a polymeric carbon paste electrode, and a polarographic detector EDLC (Laboratorní Přístroje, Prague, Czechoslovakia). The UV photometric detector was operated at 254 nm, the voltammetric detector at +1.4 V and the polarographic detector at −1.0 V (vs Ag/AgCl reference electrode). The detectors were interconnected by a stainless steel capillary (0.2 mm ID).

HPLC separation of pyrimidine bases was carried out using a reversed-phase system with a citrate—phosphate buffer, selected on the basis of the good results obtained earlier. Citrate has also been recommended as an optimal counter-ion in ion-exchange separation of nucleobases and nucleotides (Eksteen *et al.*, 1978). The capacity ratios for 26 pyrimidine derivatives were measured at different pH values and are given in Table 13. Pyrimidines can exist in two tautomeric forms, lactam and lactim, as can be shown, for example, for uracil

lactam   lactim

The molecules of pyrimidines can be substituted simultaneously by acidic (−OH, −SH) and basic (−NH₂) groups and by other substituents (-F, -Br, -I and others), which can influence the retention behaviour of the substances. The pyrimidines with which the lactam form predominates in an aqueous mobile phase exhibit lower retention times. This behaviour can be explained by self-association (vertical stacking) of the lactim form (Brown and Grushka, 1980), leading to hydrophobization of the molecules, and thus to longer retention times. The presence of the methyl group more than doubles the retention time, as it enhances the formation of the lactim form (cf. the k' values for uracil and thymine, cytosine and 5-methylcytosine in Table 13).

On the other hand, substituents supporting the formation of the lactam form (e.g. -OH, -SH) cause a decrease in the retention times. The retention time of variously substituted pyrimidines increases in the series -OH < -H < -NH$_2$ < -CH$_3$, the greatest effect being exerted by the substituent in position 5. In the series of halogeno derivatives, the retention order is inversely proportional to the electronegativity of the halogen atoms (5-fluoro < 5-bromo < 5-iodo derivatives); 6-aza and 5,6-dihydrogeno substitution leads to only small changes in the retention times.

From the dependence of the capacity ratios on the pH of the mobile phase (Fig. 15) it can be seen that the optimal pH value is 3.5. At higher pH values, the elution peaks exhibited tailing and sometimes the double peaks were obtained. The direct correlation of the capacity ratios with the ionization constants of pyrimidine derivatives is impossible because of the simultaneous effect of various acidic and basic substituents. The effect of the position of

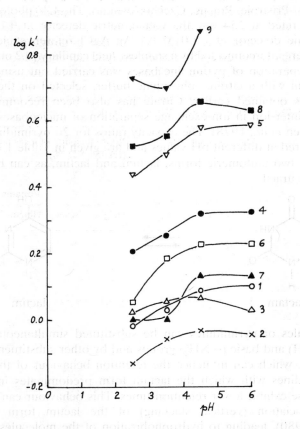

Figure 15. Dependence of log k' for some uracil derivatives on the pH of the mobile phase. 1. uracil; 2. azauracil; 3. 6-aminouracil; 4. 2-mercaptouracil; 5. 4-mercaptouracil; 6. 5-nitrouracil; 7. 5-fluorouracil; 8. 5-bromouracil; 9. 5-iodouracil. For the conditions see the text. From Štulík and Pacáková (1983).

the substituent on the heterocycle is definitely more pronounced (cf. the k' values of 2- and 4-mercaptouracil).

The effect of addition of dodecylsulphate to the mobile phase on the separation of pyrimidines was also tested, but the separation was not improved.

UV photometric—voltammetric—polarographic detection. Pyrimidine derivatives can be readily detected by the measurement of the absorbance at 254 nm. However, electrochemical detection offers a very useful alternative, especially for the identification of unresolved compounds. The half-wave potentials of the voltammetric waves vary widely, depending on the number and type of substituents (e.g. +0.45 V for 2,4,5-triamino-6-hydroxypyrimidine, +1.15 V for 2,4-diamino-6-hydroxypyrimidine and +1.2 V for 2-amino-4,6-dihydroxypyrimidine). Therefore, the selectivity can be varied to a certain extent by variation of the working electrode potential. To study the possibilities of electrochemical detection, the potentials of the working electrodes were set at limiting values of +1.4 V for the carbon paste voltammetric detector and −1.0 V for the polarographic detector, the potential limits being given by the electrolytic decomposition of the mobile phase. The results are given in Tables 13 and 14. As can be seen, the

Table 14. Typical detection limits and reproducibility of measurement (detection limits were determined as three times the peak-to-peak noise values)

Substance	Detection limit (ng)			Reproducibility of measurement (relative S.D. %)				
	UV	Voltam-metric	Polaro-graphic	UV		Voltammetric		Polaro-graphic
				Height	Area	Height	Area	Height
2-Amino-4,6-dihydroxypyrimidine	2	25	—					
2,4-Diamino-6-hydroxypyrimidine	2	10	—					
2,4,5-Triamino-6-hydroxypyrimidine	2	5	—	1.53[a]	1.24[a]	5.18[a]	5.92[a]	—
4,5,6-Triamino-pyrimidine	20	7	—					
2-Mercaptouracil	10	50	—					
5-Nitrouracil	2	—	1000	1.72[b]	—	—	—	2.25[b]
6-Azauracil	2	—	800					
6-Azacytosine	5	—	400					
2-Mercapto-6-azathymine	10	40	800					

[a] 1.32 µg, seven measurements; [b] 2.8 µg, four measurements.

voltammetric detector responds to amino and thio derivatives; the highest sensitivity has been attained for triamino derivatives. (The signal is proportional to the number of the electroactive groups in the molecule.) The polarographic detector responds to nitro- and aza-derivatives that can be reduced at a dropping mercury electrode (DME). The detection limits obtained with the electrochemical detectors are not lower than those obtained with the UV detector (except for 4,5,6-triaminopyrimidine — see Table 14) and those obained with the polarographic detector are even considerably higher. However, the combination of these three detectors can substantially improve the resolution (see Fig. 16) and identification of the components in complex mixtures, especially in biological fluids where UV-absorbing ballast components do not yield an electrochemical response. Fig. 17 shows the chromatogram of a mixture of six pyrimidine derivatives detected by UV photometric (a) and voltammetric detectors set at two different potentials: +1.4 V (b) where all the compounds with the exception of uracil (No. 4) are detected; and +0.8 V (c) where only 4,5,6-triamino-pyridmidine yields a signal.

The connection of the three detectors in series by the shortest possible stainless steel capillaries does not lead to extra peak broadening (i.e. to a decrease in column efficiency). The voltammetric detector is partially destructive, but the degree of electrochemical conversion is so low that the products

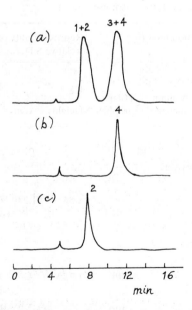

Figure 16. Chromatograms of: 1. cytosine (0.65 µg); 2. 6-azacytosine (1.2 µg); 3. uracil (0.56 µg); and 4. 6-aminouracil (0.60 µg). (a) UV photometric detection at 254 nm (sensitivity 0.32 AU/scale); (b) voltammetric detection, +1.4 V (sensitivity 0.2 µA/scale); (c) polarographic detection, −1.0 V (sensitivity 0.05 µA/scale). For the conditions see the text. From Štulík and Pacáková (1983).

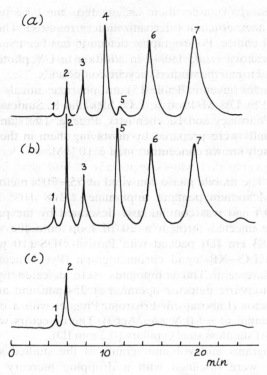

Figure 17. Separation of several pyrimidine derivatives. (a) UV photometric detection at 254 nm (sensitivity 0.32 AU/scale); (b) voltammetric detection, +1.4 V (sensitivity 0.2 μA/scale); (c) voltammetric detection, +0.8 V (sensitivity 0.2 μA/scale). 1. Hold-up time; 2. 2,4,5-triamino-6-hydroxypyrimidine, 1,12 μg; 3. 2-amino-4,6-dihydroxypyrimidine, 0.88 μg; 4. uracil, 0.56 μg; 5. 4,5,6-triaminopyrimidine, 0.64 μg; 6. 6-amino-4-hydroxy-2-mercaptopyrimidine, 0.40 μg; 7. 4-mercaptouracil, 1.50 μg. For the conditions see text. From Štulík and Pacáková (1983).

formed do not interfere in the response of the polarographic detector. The precision of the determination of pyrimidine derivatives with electrochemical detection is satisfactory for quantitative analysis (see Table 14). This chromatographic method with combined UV—voltammetric—polarographic detection can be utilized for the identification and determination of pyrimidine derivatives in complex matrices, e.g. in the study of the metabolism of pharmaceuticals in body fluids.

Determination of some tricyclic neuroleptics by HPLC with UV photometric—polarographic detection
Derivatives of {10-(4-methylpiperazino)-10,11-dihydrodibenzo[b, f]thiepin} that have been synthesized in the Research Institute for Pharmacy and Biochemistry in Prague (see e.g. Jílek *et al.*, 1965; Protiva, 1979) exhibit important neuroleptic properties. They can be determined by gas chromatography (Tomková, 1984; Tomková *et al.*, 1981), but because these com-

pounds are poorly volatile, their GC analysis must be performed at high temperatures and sometimes derivatization is required. Therefore, HPLC is the method of choice. Polarographic detection has been successfully used in our work (Pacáková *et al.*, 1984), in addition to UV photometric detection that is universal for all the studied tricyclic compounds.

The substances (given in Table 15) and pharmaceuticals used in this work were provided by Drs M. Protiva, J. O. Jílek and K. Šindelář of the Research Institute for Pharmacy and Biochemistry, Prague. The standard solutions of these compounds were prepared by dissolving them in the mobile phase to obtain a precisely known concentration of *c.* 10^{-3} M.

Procedure. The mobile phase consisted of 55—70% methanol, 1.8% acetic acid, 0.005 M sodium pentane sulphonate, 1.2×10^{-4} M EDTA, 0.5 g $Na_2SO_4 \cdot H_2O/l$ and was continuously deaerated by the passage of helium. Samples were injected through a 20 µl loop onto the column (stainless steel, 25 × 0.3 cm ID) packed with Partisil ODS (10 µm, Pye Unicam, England). An LC—XP liquid chromatograph (Pye Unicam, England) was used in measurement. The compounds were detected by a Pye Unicam UV—LC photometric detector operated at 254 nm and an EDLC electro-chemical detector (Laboratorní Přístroje, Prague) with a dropping mercury electrode potential of −1.0 V (Ag/AgCl). The detectors were connected in series by a short stainless steel capillary (0.2 mm ID).

The polarograms and voltammograms of the studied substances in the mobile phase were obtained with a dropping mercury electrode and a rotating disc glassy carbon electrode with a PA-3 polarographic analyser (Laboratorní Přístroje, Prague). The UV absorption spectra of the studied substances dissolved in the mobile phase were obtained using a Unicam SP-800 spectrophotometer in 1 cm quartz cuvettes.

Analysis of the pharmaceuticals involved finely pulverising 20 pills, suspending them in a 250 ml standard flask containing the mobile phase and stirring the suspension for about 15 min at room temperature. After settling, 20 µl of the supernatant were injected onto the HPLC column. Accurately weighed amounts of *c.* 6 mg of the standard substances were dissolved in the mobile phase in 10 ml standard flasks.

HPLC separation conditions for tricyclic compounds in a reversed-phase system with an ion-pairing agent (sodium pentane sulphonate) and an aqueous—methanolic solution were studied. EDTA was added to mask any metal ions present, and acetic acid to increase the dissociation of the studied weakly basic substances.

The capacity ratios of the compounds derived from dibenzo[*b,f*]thiepins, -oxepins, -selepins and others (see Table 16) were measured in dependence on the methanol content in the mobile phase. The results are given in Fig. 18. The optimal methanol content was found to be 70%.

An increase in the column temperature leads to a decrease in the retention times as does an increase in the mathanol content. A typical linear relation-ship between the logarithms of the capacity ratios of dibenzo[*b,f*]thiepin

Table 15. The compounds studied

Parent substances	Group	Substituents	Trivial name	Code
	I	$X = CH_2, R = CH_3$	—	I-1
	II	$X = S, R = CH_3$	Perathiepin	II-1
		8-F	—	II-2
		8—Cl	Octoclothepin	II-3
		8-Br	—	II-4
		8-CH_3	—	II-5
		8-C_2H_5	—	II-6
		8-OCH_3	—	II-7
		8-SCH_3	Methiothepin	II-8
		8-NO_2	—	II-9
		8-NH_2	—	II-10
		8-COOH	—	II-11
		8-CF_3	—	II-12
		8-OC_2H_5	—	II-13
		6-Cl	—	II-14
	III	$X = S, R = NH_2$	—	III-1
	IV	$X = S, R = CH_2CH_2OH$	—	IV-1
		3-F, 8-isopropyl	Isofloxythepin	IV-2
	V	$X = S, R = CH_2CH_2OH$,		
		8-SCH_3	Oxyprothepin	V-1
	VI	$X = S, R =$		
		$CH_2CH_2CH_2OCOCH_3$	—	VI-1
	VII	$X = S, R = CH_3$,		
		10,11-dehydro, 6,8-diCl	—	VII-1
	VIII	$X = O, R = CH_3$,	—	VIII-1
		8-CH_3	—	VIII-2
		8-Cl	—	VIII-3
		8-SCH_3	—	VIII-4
	IX	$X = Se, R = CH_3$	—	IX-1
	X	$X = Si(CH_3)_2, R = CH_3$	—	X-1
	XI	$R = OCH_3$	—	XI-1
		$R = Cl$	—	XI-2
		$R = CF_3$	—	XI-3
	XII	$R = Cl$	—	XII-1
		$R = CF_3$	—	XII-2

Table 16. UV spectrophotometric, electrochemical and chromatographic characteristics of the studied compounds

Code	λ_{max} (nm)	$E_{1/2}$(DME) (V)	k' 60% methanol	70% methanol	Detection limit (ng) UV	Polaro-graphic
Thioxanthene	235,250,265[a]	—	—	1.30	29	—
I-1	235	−0.95	—	1.62	95	19
II-1	233	−0.96	4.01	2.24	18	9
II-2	233	−1.05	3.46	1.86	15	13
II-3	236,265[a]	−0.98	—	2.62	15	19
II-4	233,258−268[b]	−1.01	5.33	2.60	42	15
II-5	233	−0.96	5.39	2.71	32	19
II-6	233−253[a]	df	6.60	3.11	21	14
II-7	d	d	4.40	2.09	—	—
II-8	233,275[a]	−1.03	5.77	3.03	19	14
II-9	233,340[a]	−0.40	3.14	1.66	26	18
II-10	233,260[a]	ef	2.04	1.79	8	21
II-11	233,257−267[b]	−0.93	3.23	1.54	19	15
II-12	233,273[c]	−1.07	4.83	2.11	20	30
II-13	233,253	f	5.35	2.85	20	12
II-14	233,257−267[b]	−0.95	—	3.08	52	9
III-1	233,253[b]	−0.93	1.22	0.58	16	17
IV-1	233,257−267[b]	−0.95	3.14	1.50	11	7
IV-2	235,252,275	f	—	2.65	10	2
V-1	236,277[a]	f	—	1.66	9	4
VI-1	233	−0.98	2.19	1.02	54	10
VII-1	233,272[a]	f	—	4.20	32	16
VIII-1	233,270[b]	−0.95	2.13	1.45	11	9
VIII-2	233,257[c]	−0.86	3.63	1.80	31	7
VIII-3	233,272[c]	−0.97	3.37	1.87	29	18
VIII-4	233,257[a]	−0.94	3.58	2.04	7	5
IX-1	235,276[b]	−0.93	—	2.71	25	8
X-1	235	−0.97	—	2.58	28	15
XI-1	233[a],278[a]	−1.00	3.92	2.31	14	25
XI-2	233,254,273	—	4.27	2.15	11	—
XI-3	233,253[b],282[b]	f	3.92	1.82	16	50
XII-1	233,254	−1.03	2.85	1.47	8	129
XII-2	253	—	2.97	1.37	12	—

[a] High peak; [b] flat peak; [c] low peak; [d] pure substance was not available; [e] non-reducible, but anodic wave at +1.2 V (glassy carbon); [f] shift in H_2 evolution.

derivatives and the reciprocal absolute temperature is shown in Fig. 19. Although temperature exerts no substantial influence on the separation itself, an increase in the experimental temperature is beneficial because of an increase in the speed of analysis.

Detection. The absorption curves were measured for the compounds given in Table 15, and the λ_{max} values are given in Table 16. Most of the substances exhibit a sharper absorption maximum at $\lambda = 233-236$ nm and a flatter one at $\lambda = 272-278$ nm. For HPLC detection both these maxima

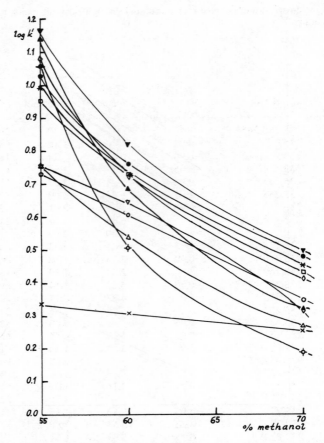

Figure 18. Dependence of log k' on the methanol content in the mobile phase for some tricyclic neuroleptics. Compounds: ▼ — II-6; ▲ — II-12; ◇ — II-4; ⋄ — II-11; ● — II-8; x — II-3; ∗ — II-13; □ — II-5; △ — II-2; ▽ — II-7; ○ — II-1. For the substance codes see Table 15. For other conditions see the text. From Pacáková *et al.* (1984).

can be used, but a maximum close to 254 nm is more advantageous as the absorbance in less dependent on the wavelength.

Electrochemical oxidation and reduction of the compounds given in Table 15 was studied at a rotating disc glassy carbon electrode from −0.4 to +1.5 V and at a dropping mercury electrode from +0.1 to −1.5 V (Ag/AgCl). Only two substances containing an amino group (II-10 and III-1) can be oxidized with half-wave potentials around +1.2 V; however, most of the compounds can be reduced, with the corresponding half-wave potentials close to −1.0 V. The measured half-wave potential values are given in Table 16. The heights of the polarographic waves are proportional to the substance concentration; the $\log[I/(I_1 - I)]$ vs E plots have a reciprocal slope of about 0.08 V. Therefore, the limiting current is controlled by diffusion and the charge-transfer reaction is irreversible.

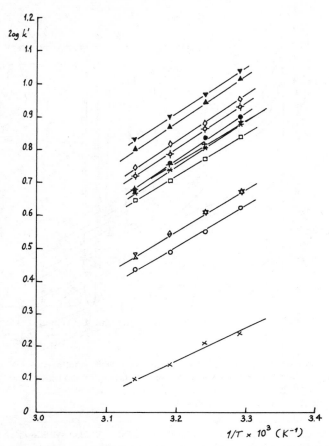

Figure 19. Dependence of log k' on the reciprocal column temperature for some tricyclic neuroleptics. For the compounds see Fig. 18 and for the conditions the text. From Pacáková *et al.* (1984).

Compounds XI-2, XII-2 and thioxanthene cannot be polarographically detected. The substances with two heteroatoms (O, S) can be polarographically detected, but the sensitivity is low. The other compounds can be detected with a sufficient sensitivity, even if some of them (II-6, II-10, II-13, IV-2, VII-1, XI-3) do not yield a polarographic wave, but they shift the potential of hydrogen evolution to more positive values.

The detection limits for UV and polarographic detection were determined and are given in Table 16. For most of the substances, polarographic detection is more sensitive than UV detection. An example of the separation of a mixture of the studied substances using UV and polarographic detection is shown in Fig. 20.

Determination in pharmaceuticals. The procedure described above has been applied to the determination of isofloxythepin and oxyprothepin in pills.

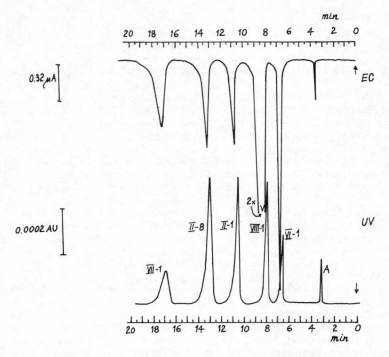

Figure 20. Chromatogram of some tricyclic neuroleptics with simultaneous UV photometric and polarographic detection. Stationary phase: Partisil ODS; for the other conditions see the text. Compounds: VI-1, 0.48 μg; VIII-1, 0.37 μg; II-1, 0.30 μg; II-8, 0.25 μg; VII-1, 0.23 μg. A — hold-up time. From Pacáková *et al.* (1984).

The absolute calibration method was used. The average values of 5 measurements, obtained with UV photometric and polarograpic detection, are given in Table 17. It can be seen that the agreement of the results is good for both detectors. Because of the low detection limits for isofloxythepin and oxyprothepin (2 and 4 ng, respectively), the precision of polarographic detection is better than that of UV detection. The high sensitivity and selectivity of polarographic detection is very promising for the study of metabolic processes of these pharmaceuticals in the body fluids.

Table 17. Determination of isofloxythepin and oxyprothepin in pills

Substance	Theoretical amount per pill (ng)	Found UV		Polarographic	
		Amount per pill (ng)	S.D. (%)	Amount per pill (ng)	S.D. (%)
Isofloxythepin	4	4.43	4.7	4.40	0.8
Oxyprothepin	5	5.41	2.6	5.38	2.6

REFERENCES

Application Note No. 14 (1981). Bioanalytical Systems Inc., West Lafayette, USA.
Armentrout, D. N., and Cutie, J. (1980). *J. Chromatogr. 18*, 370.
Au, J. L.-S., Wientjes, M. G., Luccioni, C. M., and Rustum, Y. M. (1982). *J. Chromatogr. 228*, 245.
Bakay, B., Nissinen, E., and Sweetman, L. (1978). *Anal. Biochem. 88*, 65.
Barek, J., Pacáková, V., Štulík, K., and Zima, J. (1985). *Talanta 32*, 279.
Brown, P. R., Bobick, S., and Hanley, F. L. (1974). *J. Chromatogr. 99*, 587.
Brown, P. R., and Grushka, E. (1980). *Anal. Chem. 52*, 1210.
Brugman, W. J. Th., Heemstra, S., and Kraak, J. C. (1982). *Chromatographia 15*, 282.
Castegnaro, M. (ed.) (1985). *Laboratory Decontamination and Destruction of Aromatic Amines in Laboratory Wastes.* International Agency for Research on Cancer, Lyon.
Castegnaro, M., Malaveille, Ch., Brouet, I., and Barek, J. (1985). *Am. Ind. Hyg. Assoc. J. 46*, 187.
Christiansen, A. L., and Rasmussen, K. E. (1982). *J. Chromatogr. 244*, 357.
Christiansen, A. L., Rasmussen, K. E., and Tønnesen, X. (1981). *J. Chromatogr. 210*. 163.
Christophidis, N., Mihály, G., Vajda, F., and Lois, W. (1979). *Clin. Chem. 25*, 83.
Cohen, J. L., and Brown, R. E. (1978). *J. Chromatogr. 151*, 237.
Concialini, V., Chiavari, G., and Vitali, P. (1983). *J. Chromatogr. 258*, 244.
Drake, B. (1950). *Acta Chem. Scand. 4*, 554.
Egan, H. (ed.) (1981). *Environmental Carcinogens — Selected Methods of Analysis. Vol. 4 — Some Aromatic Amines and Azo Dyes in the General and Industrial Environment.* IARC Scientific Publication No. 40, Lyon.
Eksteen, R. E., Kraak, J. C., and Linssen, P. (1978). *J. Chromatogr. 148*, 413.
Evans, J. F., Tieckelmann, H., Naylor, E. W., and Guthrie, R. (1979). *J. Chromatogr. 163*, 29.
Falchuk, K. H., and Hardy, C. (1978). *Anal. Biochem. 89*, 385.
Floridi, A., Palmerini, C. A., and Fini, C. (1977). *J. Chromatogr. 138*, 203.
Gehrke, C. W., and Ruyle, C. D. (1968). *J. Chromatogr. 38*, 473.
Gehrke, C. W., Stalling, D. L., and Ruyle, C. D. (1967). *Biochem. Biophys. Res. Commun. 28*, 869.
Hanekamp, H. B., and van Nieuwkerk, H. J. (1980). *Anal. Chim. Acta. 121*, 13.
Hartwick, R. A., Assenza, S. P., and Brown, P. R. (1979a). *J. Chromatogr. 186*, 647.
Hartwick, R. A., Krstulovic, A. M., and Brown, P. R. (1979b), *J. Chromatogr. 186*, 659.
Hashizume, T., and Sasaki, Y. (1968). *Anal. Biochem. 24*, 232.
Heineman, W. R., and Kissinger, P. T. (1980). *Anal. Chem. 52*, 138R.
Jakovljevic, I. M., Zynger, J., and Bishara, R. H. (1975). *Anal. Chem. 47*, 2045.
Jílek, O., Seidlová, V., Svátek, E., and Protiva, M. (1965). *Monatsh. Chem. 96*, 182.
Jordan, J., Javick, R. A., and Ranz, W. E. (1958), *J. Amer. Chem. Soc. 80*, 3846.
Kemula, W. (1952). *Rocz. Chem. 26*, 281.
Kimla, A., and Štráfelda, F. (1964). *Coll. Czech. Chem. Commun. 29*, 2913.
Kissinger, P. T. (1977). *Anal. Chem. 49*, 447A.
Kissinger, P. T., Bruntlett, C. S., Davis, G. C., Felice, L. J., Riggin R. M., and Shoup, R. E. (1977). *Clin. Chem. 23*, 1449.
Kissinger, P. T., Bruntlett, C. S., and Shoup, R. E. (1981). *Life Sci. 28*, 455.
Knox, J. H., and Jurand, J. (1981). *J. Chromatogr. 203*, 85.
Kraak, J. C., Ahn, C. X., and Fraanje, J. (1981). *J. Chromatogr. 209*, 369.
Krstulovic, A. M. (1979). *Adv. Chromatogr. (N.Y.) 17*, 279.
Krstulovic, A. M., Brown, P. R., and Rosie, D. M. (1977). *Anal. Chem. 49*, 2237.
Kysilka, R., Wurst, M., Pacáková, V., and Štulík, K. (1985). *J. Chromatogr. 320*, 414.
Lee, D. P., and Kindsvater, J. H. (1980). *Anal. Chem. 52*, 2425.
Levich, V. G. (1942). *Acta Physicochem. URSS 17*, 257.
Levich, V. G. (1944a). *Acta Physicochem. URSS 19*, 117.
Levich, V. G. (1944b). *Acta Physicochem. URSS 19*, 133.
Levich, V. G. (1947). *Disc. Faraday Soc. 1*, 37.
Levich, V. G. (1962). *Physico-Chemical Hydrodynamics*, Prentice-Hall, Englewood Cliffs., NJ, USA.

Lown, J. A., Koile, R., and Johnson, D. C. (1980). *Anal. Chim. Acta. 116*, 33.
Majer, V., Veselý, J., and Štulík, K. (1973). *J. Electroanal. Chem. Interfacial Electrochem. 45*, 113.
Matsuda, H. (1967). *J. Electroanal. Chem. 15*, 109.
Mefford, I., Keller, R. W., Adams, R. N., Sternson, C. A., and Yelo, M. S. (1977). *Anal. Chem. 49*, 683.
Miller, A. A., Benvenuto, J. A., and Loo, T. L. (1982). *J. Chromatogr. 228*, 165.
Miller, V., Pacáková, V., and Smolková, E. (1976). *J. Chromatogr. 119*, 355.
Myers, D. J., Osteryoung, R. A., and Osteryoung, J. (1974). *Anal. Chem. 46*, 2089.
Okinaka, Y., and Kolthoff, I. M. (1957). *J. Amer. Chem. Soc. 79*, 3326.
Pacáková, V., Štulík, K., and Tomková, H. (1984). *J. Chromatogr. 298*, 309.
Parry, E. P., and Osteryoung, R. A. (1965). *Anal. Chem. 37*, 1634.
Perkal, M., Blackman, G. L., Ottrey, A. L., and Turner, L. K. (1980). *J. Chromatogr. 196*, 180.
Podolák, M., Štulík, K., and Pacáková, V. (1982). *Chem. Listy 76*, 1106.
Protiva, M. (1979). *Die Pharmazie 34*, 274.
Purnell, C. J., and Warwick, C. J. (1980). *Analyst 105*, 861.
Rice, J. R., and Kissinger, P. T. (1979). *J. Anal. Toxicol. 3*, 64.
Riggin, R. I., and Howard, C. C. (1979). *Anal. Chem. 51*, 210.
Ross, T. K., and Wragg, A. A. (1965). *Electrochim. Acta 10*, 1093.
Rucki, R. (1980). *Talanta 27*, 147.
Ryba, M. (1981). *J. Chromatogr. 219*, 245.
Ryba, M., and Beránek, J. (1981). *J. Chromatogr. 211*, 337.
Štráfelda, F., and Kimla, A. (1965). *Coll. Czech. Chem. Commun. 30*, 3606.
Štulík, K., and Hora, V. (1976). *J. Electroanal. Chem. Interfacial Electrochem. 70*, 253.
Štulík, K., and Pacáková, V. (1980). *J. Chromatogr. 192*, 135.
Štulík, K., and Pacáková, V. (1981a). *J. Electroanal. Chem. Interfacial Electrochem. 129*, 1.
Štulík, K., and Pacáková, V. (1981b). *J. Chromatogr. 208*, 269.
Štulík, K., and Pacáková, V. (1983). *J. Chromatogr. Biomed. Appl. 273*, 77.
Štulík, K., and Pacáková, V. (1984). *CRC Crit. Rev. Anal. Chem. 14*, 297.
Štulík, K., Pacáková, V., and Podolák, M. (1983). *J. Chromatogr. 262*, 85.
Štulík, K., Pacáková, V., and Podolák, M. (1984). *J. Chromatogr. 298*, 225.
Štulík, K., Pacáková, V., and Stárková, B. (1981). *J. Chromatogr. 213*, 41.
Taylor, C. A., Dady, P. J., and Harrap, K. R. (1980). *J. Chromatogr. 183*, 421.
Tomková, H. (1984). Ph.D Thesis, Charles University, Prague, Czechoslovakia.
Tomková, H., Pacáková, V., and Smolková, E. (1981). *J. Chromatogr. 207*, 403.
van Haastert, P. J. M. (1981). *J. Chromatogr. 210*, 241.
Voelter, W., Zech, K., Arnold, P., and Ludwig, G. (1980). *J. Chromatogr. 199*, 345.
Wagner, J., Vitali, P., Palfreyman, M. G., Zraika, M., and Huot, S. (1982). *J. Neurochem. 38*, 1241.
White, P. C. (1979). *J. Chromatogr. 169*, 453.
Wurst, M., Semerdžieva, M., and Vokoun, J. (1984). *J. Chromatogr. 286*, 229.
Yamada, J., and Matsuda, H. (1973). *J. Electroanal. Chem. 44*, 189.
Yamaguchi, T., Yokota, K., and Uematsu, F. (1982). *J. Chromatogr. 231*, 166.
Young, S. N. (1982). *J. Chromatogr. 228*, 155.
Zakaria, M., and Brown, P. R. (1981). *J. Chromatogr. 226*, 267.

HPLC—EC APPLICATIONS IN MEDICINE AND CHEMISTRY

Progress in HPLC, Vol. 2, pp. 181—191
Parvez *et al.* (Eds)
© 1987 VNU Science Press

Analysis of choline and acetylcholine in tissue by HPLC with electrochemical detection

N. H. NEFF, J. L. MEEK, M. HADJICONSTANTINOU, and
H. E. LAIRD II
Laboratory of Preclinical Pharmacology, National Institute of Mental Health,
Saint Elizabeths Hospital, Washington, DC 20032, USA

INTRODUCTION

Acetylcholine (ACh) is the first known neurotransmitter. There are chemical techniques for the analysis of ACh but many are technically complicated or require expensive reagents or equipment (see Hanin, 1982). Moreover, in some studies it is desirable to know the tissue content of choline (Ch) as well as the rate of formation of ACh rather than only the steady-state tissue content of ACh. The rate of formation of ACh in neurons is apparently related to neuronal impulse flow (Moroni *et al.*, 1978), and is therefore an important parameter to measure for understanding normal brain function and the action of drugs (Cheney and Costa, 1978). This parameter of cholinergic function is not possible to ascertain with many methods for ACh measurement. We describe a method that has many advantages over other methods: It is simple in concept; it is easy and fast; the equipment and reagents are relatively inexpensive; it is sensitive, specific and reproducible; and ACh turnover rates can be measured. Since we first described the ACh method (Potter *et al.*, 1983), several modifications have been made to improve it (Eva *et al.*, 1984; Potter *et al.*, 1984). We will describe the improved method.

Our method is based on the separation of ACh and Ch by reverse-phase HPLC, passing the column effluent through a short anion exchange column containing adsorbed acetylcholinesterase (AChE; EC 3.1.1.7) and Ch oxidase (EC 1.1.3.17), followed by an electrochemical detector with a platinum electrode (Fig. 1). The heart of the detector system is the enzyme-loaded column and the platinum electrode electrochemical detector. Within the enzyme column Ch oxidase converts Ch to betaine and two moles of hydrogen peroxide (Fig. 2) (Ikuta *et al.*, 1977). The generated hydrogen peroxide is qualitatively detected by the platinum electrode. For ACh, the AChE bound to the column hydrolyzes ACh to Ch and acetate, and Ch oxidase completes the reaction.

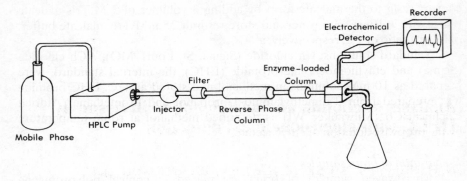

Figure 1. Design of HPLC equipment for ACh and Ch analysis.

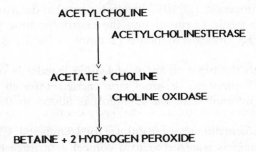

Figure 2. Acetylcholinesterase and choline oxidase convert each mole of acetylcholine and choline to two moles of hydrogen peroxide.

EXTRACTION OF ACh AND Ch FROM TISSUE

There are rapid post mortem changes of Ch and ACh if endogenous enzymatic activity is not stopped. We kill animals by focused microwave irradiation (Metabolic Vivostat, Cober Electronics, Stamford, CT). For samples of brain the instrument is set to 2.5 s, 5 kW, 2.5 GHz (Guidotti *et al.*, 1974). Ch and ACh are extracted from samples by precipitation with Reinecke salt (Moroni *et al.*, 1978). The reineckate is removed and the resulting solution injected directly into the HPLC system.

Reagents and solutions
The reagents and solutions used are listed below.

1. Perchloric acid: 0.4 M.
2. Potassium acetate: 7.5 M.
3. Tetraethylammonium: 5 mM.
4. Ammonium reineckate: 56 mM. Prepared fresh daily; let stand for 1 h on ice and filter before use.
5. Anion exchange resin, AG 1 × 2, mesh 50—100 (Bio Rad, Richmond, VA). Convert resin to the hydroxide form with 1 M sodium hydroxide. Then

convert resin to the maleate form by adding 3 volumes of 3 M Tris-maleate buffer, pH 7. Rinse with water and store resin in 25 mM Tris-maleate buffer, pH 7 in a ratio of 1:2, respectively, at 4° C.

6. Standard solutions: Ch chloride (Sigma, St. Louis, MO), ACh chloride (Sigma) and ethylhomocholine bromide (EHC), the internal standard, were prepared as 10 mM solutions in methanol and stored at 4° C. EHC bromide was prepared from 3-dimethylamino-1-propanol and bromethane (Aldrich Chemical Co., Milwaukee WI) as described methanol at room temperature by the dropwise addition of ethyl acetate.

Preparation of brain samples

1. Microwaved samples of brain are placed in conical polypropylene centrifuge tubes (1.5 ml) and 0.4 M perchloric acid (300 μl) added. The tissue is homogenized directly in the centrifuge tube with a glass pestle. A portion of the homogenate (20 μl) is taken for protein determination (Lowry *et al.*, 1951). The pestle is rinsed into the centrifuge tube with 400 μl of perchloric acid. EHC internal standard (2 nmol in 50 μl) is added to each sample.

2. In addition to the brain samples, at least 6 standards of Ch and ACh ranging from 125 pmol to 1 nmol are carried through the extraction procedure. EHC internal standard (2 nmol) is added to these samples as well.

3. The tubes are centrifuged at $10\,000 \times g$ for 15 min at 4° C.

4. The supernatant is transferred to a conical centrifuge tube, potassium acetate reagent added (100 μl), and the samples mixed and centrifuged as for sample 3.

5. The supernantant is transferred to another conical centrifuge tube and tetraethylammonium reagent added (20 μl), followed by ammonium reineckate reagent (500 μl), the tubes are mixed and allowed to stand on ice for 1 h.

6. The tubes are centrifuged as for sample 3 and the supernatant discarded. Samples may be stored at −20° C at this step or carried to completion.

7. To complete the extraction, the pellet is washed once with Tris-maleate buffer (200 μl), The biogenic amines are freed from the reineckate by adding the resin AG 1 × 2 Tris-maleate buffer solution (200 μl). The tubes are mixed and allowed to stand for 5 min. The mixing is repeated once again and the tubes centrifuged as for sample 3.

THE HPLC SYSTEM

The apparatus (see Fig. 1) is composed of an Altex 110 A pump with pulse dampener (Beckman, Palo Alto, CA); a Rheodyne 7125 injector with a 100 μl sample loop (Ranin, Woburn, MA); an in-line filter cartridge with 0.5 μm filter (Ranin); an analytical column—Altex Ultrasphere ODS 5 μm C-18, 150 × 4.6 mm (Beckman); an enzyme-loaded column, Brownlee AX-300, 30 × 2.1 mm in an appropriate column holder (Ranin); an electrochemical detector LC-4B (BAS, W. Lafayette, IN) with a platinum electrode set to a potential of +0.5 V vs an Ag/AgCl reference electrode; and a chart recorder. The mobile phase is pumped through the system at a rate of 1 ml/min.

Preparation of mobile phase

1. Tris-maleate: 0.2 M stock solution. Prepare with Tris-base, 24.2 g plus maleic acid 23.2 g dissolved in water to make 1 l.

2. Sodium hydroxide stock solution: 0.2 M.

3. Combine Tris-maleate stock solution (125 ml) plus sodium hydroxide stock solution (120 ml), and bring to 1 l with water. The resulting pH should be 7.0.

4. To prepare the mobile phase add tetramethylammonium (TMA) (150 mg/l), and sodium octylsulfate (10 mg/l). Filter through 0.45 μm Millipore filter (Bedford, MA) and de-gas for about 2 h by stirring under vacuum.

Loading the enzyme column

A Brownlee MPLC microbore cartridge, AX-300 (10 μm particle size, 30 × 2.1 mm) is placed in an MPLC guard holder (Ranin). Teflon tubing (0.012 in ID) of about 4 cm length is connected to the guard column holder with appropriate fittings. First AChE [EC 3.1.1.7; electric eel type III, 125 units (C-2629, Sigma)] and then Ch oxidase [EC 1.1.3.17; 75 units (C-5896, Sigma)], dissolved in water are injected into the inflow port of the guard column via the Teflon tubing using a small syringe with needle. The enzyme column is placed after the analytical column (Fig. 1) with high pressure fittings and about 10 ml of mobile phase passed through the system before attaching the detector. The enzymes are probably held to the anionic resin by ionic attraction (Messing, 1976).

ESTABLISHING THE SPECIFICITY, LINEARITY AND REPRODUCIBILITY OF THE METHOD

Figure 3 shows a chromatogram for Ch, EHC and ACh. There is adequate separation of all three compounds in standard solutions as well as for brain samples and the chromatographic analysis is completed in less than ten minutes. Moreover, there does not appear to be interfering material in samples of brain after the solvent front. The chromatographic characteristics can be changed by varying the concentration of sodium octylsulfate and TMA. Increasing sodium octylsulfate concentrations increases the retention time of the biogenic amines. The presence of TMA improves the peak shape and reduces adsorption of ACh to the column. TMA can inhibit Ch oxidase activity and therefore the concentration should not be increased much above 1.2 mM (Potter *et al.*, 1983). The pH of the mobile phase is a compromise between the optimal pH for Ch oxidase activity (pH 8) (Ikuto *et al.*, 1977) and the breakdown pH of the reverse phase column (> pH 7.5).

When evaluating, the specificity of the detector system of the reverse-phase column was removed from the chromatographic system (Fig. 1). Then

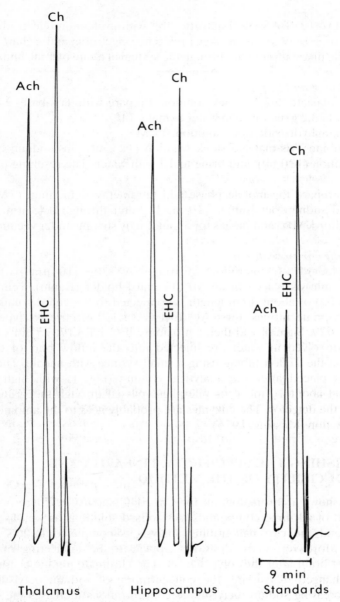

Figure 3. Chromatogram of standards and samples from rat hippocampus and thalamus. Samples were extracted and EHC added as detailed in the text.

a variety of Ch-related compounds were injected and the response of the detector was compared with Ch as the standard. As shown in Table 1, a number of compounds can be readily detected. For example, ethylcholine, homocholine and EHC are all detected, and any of them might be used as an internal standard. We have used EHC. It should be noted that no response

Table 1. Specificity of detection system

Compound	Response Relative to Ch
Choline	1
Acetylcholine	0.96
Butyrylcholine	0.07
Valerylcholine	0.01
Acetyl-β-methylcholine	0.01
Phosphorylcholine	0.01
Diethylethanolamine	0.01
Diethyl-3-amino-1-propanol	0.01
Diethylcholine	0.05
Diethylhomocholine	0.05
Dimethyl-3-amino-1-propanol	0.1
Ethylcholine	0.95
Ethylhomocholine (EHC)	0.95
Homocholine	0.95

Standards were dissolved in mobile phase and injected with the reverse phase column removed from the system. Two hundred pmol/20 µl injection of each compound was evaluated (Potter et al., 1983).

will be observed for any of the compounds in Table 1 if the enzymes are not included on the anionic exchange resin. When samples of Ch (250 pmol), ACh (50 pmol) and EHC (500 pmol) were injected repeatedly into the system we obtained detector responses, with sensitivity setting at 20 nA full scale, of (158 ± 5), (129 ± 3) and (129 ± 6) mm ± S.E.M., $n = 6$, respectively.

There is a linear relationship between the quantity of ACh and Ch extracted and injected into the system and the response of the detector (Fig. 4). With standard solutions the minimal quantities of ACh and Ch that produce a signal twice that of background noise is 1 and 0.5 pmol for ACh and Ch, respectively. The platinum electrode should be polished every several weeks with a polishing kit for a gold electrode (BAS) to maintain maximum sensitivity. To compensate for possible changes in the detector system sensitivity during long periods of operation and to obtain values for recovery, internal standards of EHC are added to tissue homogenates. Calculations, therefore, are based on the ratio of response for increasing concentrations of Ch or ACh to EHC vs concentrations of Ch and ACh.

The sensitivity as well as linearity depend on the conversion of ACh to Ch and the conversion of Ch to hydrogen peroxide. Two moles of hydrogen peroxide should be formed from one mole of Ch injected (Fig. 2). The enzyme column can be evaluated most conveniently by removing the reverse-phase column and injecting standards of hydrogen peroxide, ACh and Ch. In

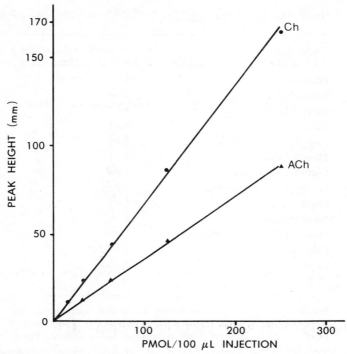

Figure 4. Linearity of detector system for standards of ACh and Ch.

theory, ACh and Ch should give responses twice that of hydrogen peroxide if there is no tailing of the bases due to adsorption. As we have outlined the system, the conversion of ACh and Ch to hydrogen peroxide is 95% or greater. The enzymes on the column retain their activity for at least 6 weeks at room temperature. We found that at the end of 6 weeks about 30% of the original enzyme activity was still present. This was evaluated by injecting standards periodically. When necessary the column can be recharged with new enzyme. There is excessive tailing and double peaks when the enzyme column deteriorates. We usually replace the enzyme column after two months of daily usage. If the system is to be left idle for several days, remove the analytical and enzyme column and store them at 4° C, and flush the tubing with water. With continual use of the HPLC system we occasionally observed a rise of pressure. This can be corrected by washing the reverse-phase column with Tris-maleate buffer without the TMA or sodium octyl-sulfate.

Brain contains many materials that oxidize at a detector potention of +0.5 V, such as the catecholamines. The presence of such substances can be evaluated by removing the enzyme column from the system and determining if a response is still observed with a sample. Moreover, a sample can be subjected to basic hydrolysis before injection into the system. After this treatment, ACh should be lost and there should be a quantitative increase of the Ch peak.

Table 2 shows the distribution of Ch and ACh in selected regions of rat brain.

Table 2. Distribution of Ch and ACh content in various areas of rat brain

Area	Ch (pmol/mg protein ± S.E.M.)	ACh (pmol/mg protein ± S.E.M.)
Striatum	179 ± 10	650 ± 50
Thalamus	246 ± 30	491 ± 30
Midbrain	174 ± 20	386 ± 60
Hippocampus	191 ± 20	311 ± 10
Frontal cortex	184 ± 20	311 ± 30
Parietal cortex	128 ± 10	247 ± 20
Occipital cortex	186 ± 20	282 ± 20
Cerebellum	110 ± 6	074 ± 5

Animals were killed by microwave irradiation. Ch and ACh were estimated by HPLC with electrochemical detection. $n = 9-12$.

MEASURING THE TURNOVER RATE OF BRAIN ACh

The measurement of the turnover rate of ACh is rather simple once the analytical method is in operation. High specific radioactivity methyl[^{3}H]Ch is administered to an animal. The animal is killed by microwave irradiation, and brain samples processed and assayed as usual except that the effluent fractions emerging from the electrochemical detector that correspond to Ch and ACh are collected separately and counted for radioactivity. Thus, the specific radioactivity of endogenous Ch and ACh can be obtained from the same sample. The results generated by this analytical procedure can be applied to any mathematical model for kinetic analysis and turnover rate estimations (Potter *et al.*, 1984).

To perform a turnover study, we have injected mice intravenously with methyl[^{3}H]Ch chloride, 80 Ci/mmol, 4 µCi/mouse, which corresponds to 50 pmol Ch. They were killed by whole-body microwave irradiation, 5 kW, 1.8 s at several time intervals over a 5 min period, see Fig. 5. The brain parts were processed as already described. The detector cell was modified for collecting the effluent. A short length of tubing was added to the bottom half of the cell so that the effluent could be collected directly into scintillation counting vials. The outlet of the reference electrode was pinched off so the flow did not pass through the reference. For turnover studies three fractions were collected, one while observing the detector response for Ch (about 1 ml), one while observing the response for ACh (about 1.5 ml) and one between Ch and ACh (about 1.5 ml) which served for background measurement. Ten milliliters of ACS (Amersham, Arlington Heights, IL) were added to the vials and radioactivity counted in a liquid scintillation spectrometer. Samples were

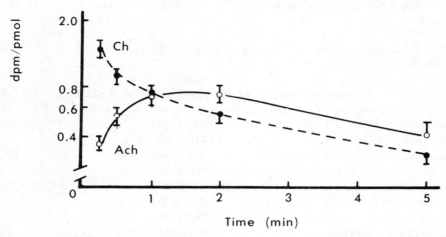

Figure 5. Change of the specific radioactivity of Ch and ACh with tissue in the mouse cerebral cortex after an intravenous injection of methyl[^{3}H]Ch, 4 μCi. Data used to estimate the turnover rate of ACh shown on Table 3. Data from Potter *et al.* (1984).

corrected to 100% counting efficiency by adding internal standards labelled with tritium. Table 3 shows the calculated turnover rate for ACh in three regions of mouse brain.

We have determined that the radioactivity corresponding to the ACh and Ch fractions are indeed pure for the 5 min period we studied (Fig. 4) after the injection of radioactive Ch into mice (Potter *et al.*, 1984). For example, total radioactivity injected into the HPLC was recovered in the Ch and ACh fractions. Moreover, high-voltage paper electrophoresis (Salens *et al.*, 1970) of the detector effluents yielded only radioactive ACh and Ch when the enzyme-loaded column was removed from the HPLC system. When the enzyme-loaded column was present only radioactive betaine was present in each fraction. ACh, Ch and betaine were detected on the electrophoretograms with Dragendorff's reagent (Krebs *et al.*, 1969).

Table 3. Estimated turnover rate of ACh in regions of mouse brain

Brain region	ACh content (pmol/mg prot ± S.E.M.)	Fractional rate constant (min^{-1} ± S.E.M.)	Turnover rate (pmol/mg prot/min)
Cortex	160 ± 10	0.44 ± 0.06	70
Striatum	881 ± 20	0.53 ± 0.02	466
Hippocampus	121 ± 4	0.33 ± 0.05	39

Methyl[^{3}H]Ch (80 Ci/mmol), 4 μCi/mouse i.v. Mice were killed by microwave irradiation at the time intervals shown on Fig. 5. Analyses were performed as described in the text. A finite difference equation was used to estimate the fractional rate constant for ACh (Neff *et al.*, 1971). The turnover rate was found by multiplying the tissue content of ACh by the fractional rate constant. Data from Potter *et al.* (1984).

EVOLUTION OF THE ACh HPLC METHOD

The idea for the HPLC method originated from two reports. In 1981, Israel and Lesbats described a procedure and equipment which they designed for the analysis of ACh in a sample based on the same general enzymatic reaction we have used. In their detector the generated hydrogen peroxide was used to oxidize luminol to emit light. Clark et al. (1981) reported on the design of a platinum electrode for an aqueous system that could be used for the amperometric determination of hydrogen peroxide. We combined the elements from both reports to develop the HPLC system.

In our original paper (Potter et al., 1983) we added the enzymes to the reverse-phase column effluent with a second pump which was followed by a 30 m enzyme reaction coil. This, of course, was costly and wasteful because the enzymes were continuously pumped even when there was no Ch or ACh in the system. The enzyme-loaded column eliminates the need for a second pump, the reaction coil and the continuous waste of enzyme (Eva et al., 1984). Chromatography is now faster with better efficiency and the analysis is cheaper. The Reinicke salt preparation of amines has also been simplified by eliminating an evaporation step to speed analysis time. The cromatography will eventually be improved as better analytical HPLC columns become available.

SUMMARY

In this report we summarize our studies of HPLC method for Ch and ACh. The method is relatively simple, rapid and sensitive, and can be used to study the turnover of ACh in brain. The specificity for Ch and ACh is provided by the HPLC column, two specific enzymatic reactions and the electrochemical detection of hydrogen peroxide. The method should be useful for studying the function of cholinergic neurons. It should also be recognized that the enzyme-loaded post column reactor system might be used to assay many other endogenous constituents if the product of the enzymatic reaction can be either oxidized or reduced by the electrochemical detector.

REFERENCES

Cheney, D. L., and Costa, E. (1978). Biochemical pharmacology of cholinergic neurons. In: *Psychopharmacology: A Generation of Progress*, M. A. Lipton, A. D. DiMascio, and K. F. Killam, (eds.), Raven Press, New York, pp. 283—291.

Clark, L. C. Jr., Duggan, C. A., Grooms, T. A., Hart, L. M., and Moore, M. E. (1981). One-minute electrochemical enzymic assay for cholesterol in biological materials. *Clin. Chem. 27*, 1978—1982.

Eva, C., Hadjiconstantinou, M., Neff, N. H., and Meek, J. L. (1984). Acetylcholine measurement by HPLC using an enzyme loaded post-column reactor. *Anal. Biochem.*, (in press).

Guidotti, A., Cheney, D. L., Tabucchi, M., Doteuchi, M., and Wang, C. T. (1974). Focussed microwave radiation: A technique to minimize *post mortem* changes of cyclic nucleotides, DOPA and choline and to preserve brain morphology. *Neuropharmacol. 13*, 1115—1122.

Hanin, I. (1982). Methods for the analysis and measurement of acetylcholine: An overview. In: *Modern Methods in Pharmacology*, S. Spector, and N. Back, (eds.), Alan R. Liss, New York, pp. 29—38.

Ikuta, S., Imamura, S., Misabi, H., and Horiuti, Y. (1977). Purification and characterization of choline oxidase from arthrobacter globiformis. *J. Biochem. 82*, 1714—1749.

Israel, M., and Lesbats, B. (1981). Chemiluminescent determination of acetylcholine, and continuous detection of its release from torpedo electric organ synapses and synaptosomes. *Neurochem. Int. 3*, 81—90.

Krebs, K. G., Heusser, D., and Wimmer, H. (1969). *Spray Reagents, in Thin-Layer Chromatography. A Laboratory Handbook*, E. Stahl, (ed.), Springer-Verlag, New York, pp. 854—911.

Lowry, O. H., Rosebrough, N. J., Farr, A. L., and Randall, R. J. (1951). Protein measurement with the Folin Phenol reagent. *J. Biol. Chem. 193*, 265—275.

Messing, R. A. (1976). Adsorption and inorganic bridge formation. In: *Methods in Enzymology XLIV: Immobilized Enzymes*, H. Mosbach, (ed.), Academic Press, New York, pp. 148—169.

Moroni, F., Malthe-Sorenssen, D., Cheney, D. L., and Costa, E. (1978). Modulation of ACh turnover in the septal-hippocampal pathway by electrical stimulation and lesioning. *Brain Res. 150*, 333—341.

Neff, N. H., Spano, P. F., Groppetti, A., Wang, C. T., and Costa, E. (1977). A simple procedure for calculating the synthesis rate of brain norepinephrine, dopamine and serotonin after a pulse injection of radioactive tyrosine and tryptophan. *J. Pharmacol. Exp. Ther. 176*, 701—710.

Potter, P. E., Meek, J. L., and Neff, N. H. (1983). Acetylcholine and choline in neuronal tissue measured by HPLC with electrochemical detection. *J. Neurochem. 41*, 188—194.

Potter, P. E., Hadjiconstantinou, M., Meek, J. L., and Neff, N. H. (1984). Measurement of acetylcholine turnover rate in brain: An adjunct to a simple HPLC method for choline and acetylcholine. *J. Neurochem. 43*, 288—290.

Saelens, J. K., Allen, M. P., and Simke, J. P. (1970). Determination of acetylcholine and choline by an enzymatic assay. *Arch. Int. Pharmacodyn. 186*, 179—286.

Progress in HPLC, Vol. 2, pp. 193—217
Parvez *et al.* (Eds)

THE MEASUREMENT OF ACETAMINOPHEN AND ITS MAJOR AND MINOR METABOLITES IN BIOLOGICAL FLUIDS VIA ELECTRO-CHEMICAL DETECTION

JOHN WILSON

Department of Pharmacy and Clinical Pharmacy, University of Massachusetts Medical Center, Worcester, MA 01605, USA

INTRODUCTION

The introduction of the electrochemical detector has provided analytical chemists and biological scientists a uniquely qualified tool with which to achieve their goals. Its high sensitivity for a variety of important bio-chemicals, its adaptability to advances in liquid chromatography and its flexibility in meeting the diverse challenges presented by complex materials combined with its relative low cost and ease of use have reserved for this approach a substantial place in the future of analytical and biological chemistry.

Certainly one area of research in which the utility of electrochemical detection has gone untapped is that of drug metabolism. A recent computer search of published procedures involving liquid chromatography and electro-chemical detection from 1982 to 1984 located 240 separate references of which only four involved the measurement of a drug metabolite. It is inevitable that this current dearth of literature representation will not long endure. With the current emphasis of governments on accountability and the realization within the pharmacological community of the importance of pharmacologically active, toxic, carcinogenic, mutagenic and teratogenic drug products an increased emphasis will naturally be placed on the detection, identification and quantitation of drug metabolites.

The practicality of amperometric detection is reserved for select chemicals capable of oxidation or reduction within the electrochemical cell. For many parent drugs this capacity is retained on metabolism. For example, many drugs contain carbon—nitrogen systems which act as sites of metabolism (Low and Castagnoli, 1979). *N*-Oxidation, *N*-hydroxylation and *N*-dealkylation are common in biological systems. Retention of the carbon—nitrogen system generally means retention of electrochemical activity due to the relative ease of conversion of nitrogen between its multiple oxidation states. Loss of the nitrogen such as by oxidative deamination, can produce an electrochemically inert metabolite though the amine, produced as a result of metabolism, could be monitored.

In a number of very important instances metabolism can produce a metabolite with a greater capacity for detection than the parent. A number of these are listed in Table 1. A common biological oxidation is aryl hydroxylation producing a phenolic species which is readily measured by oxidative electrochemistry. In some instances, the result of the oxidation process may produce a species that can best be measured in the reducing mode. An example is the conversion of the amine in amphetamine to a nitro group through an oxime. A number of bioreductions catalysed by liver enzymes or

TABLE 1

BIOCHEMICAL TRANSFORMATIONS LEADING TO ELECTROCHEMICALLY ACTIVE METABOLITES

1. *Aryl Hydroxylation*

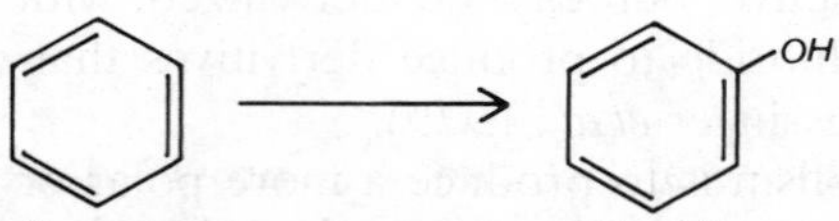
OH
Benzene
Phenol

2. *Nitro Reduction*

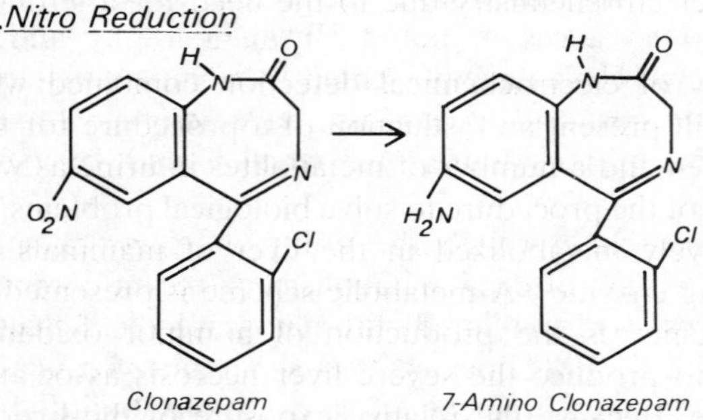
H
N
O
N
O_2N
Cl
H_2N
Clonazepam
7-Amino Clonazepam

3. *Azo Reduction*

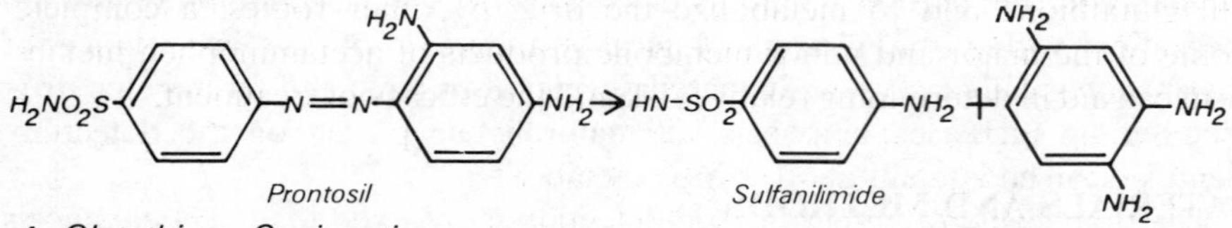
H_2NO_2S
N=N
H_2N
NH_2
$HN-SO_2$
NH_2
NH_2
NH_2
NH_2
Prontosil
Sulfanilimide

4. *Glutathione Conjugation*

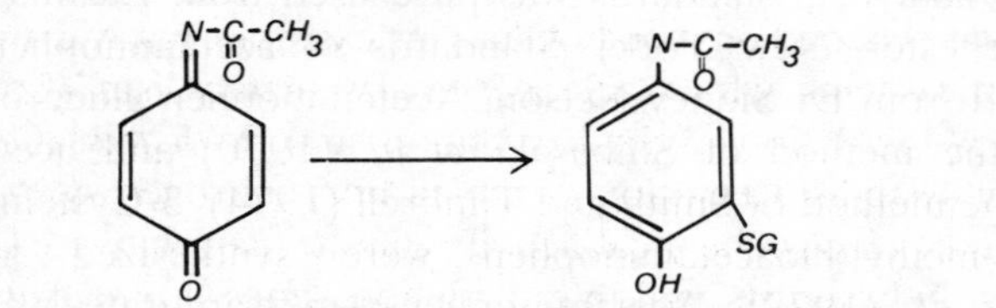
$N-C-CH_3$
O
O
$N-C-CH_3$
O
SG
OH
N- Acetyl
Parabenzoquinone Imine
3-Glutathionyl Acetaminophen

micro-organisms within the GI tract can produce species measured more easily by oxidative than reductive electrochemistry. For example, the reduction of nitro groups and aromatic azo compounds produces a primary amine readily measurable in the oxidative mode.

In general conjugation reactions such as glucuronidation and sulfation reduce the electrochemical activity of a parent drug, however, these conjugates are often readily hydrolysed either enzymatically or by acidification and heating to restore the parent active drugs.

Gluthathione conjugation can sometimes produce highly electrochemically active metabolites as with the nucleophilic addition of glutathione to arene oxides, or quinones produced by oxidative metabolism with e.g. acetaminophen or 6-hydroxydopamine.

There are additional instances where metabolites, though electrochemically inactive, can undergo pre- or post-column derivitization to produce electrochemically active species. For example, aliphatic alcohols produced by oxidative hydoxylation are not active but can be derivitized with active reagents such as dinitrobenzoyl chloride to produce derivitives that can be measured in the reductive mode (Kissinger *et al.*, 1979).

In general, the thrust of metabolism is to produce a more polar or water-soluble species to further its elimination by the kidneys. The increased solubility is an advantage in electrochemistry due to the need for a strongly conducting mobile phase.

To illustrate the capability of electrochemical detection combined with liquid chromatography, we will present an evaluation of a procedure for the measurement of acetaminophen and a number of metabolites in urine as well as examples of the application of the procedure to solve biological problems.

Acetaminophen is extensively metabolized in the liver of mammals by both oxidative and conjugating enzymes. A metabolic scheme is presented in Fig. 1. Of particular importance is the production of a minor oxidative metabolite which is thought to produce the severe liver necrosis associated with acetaminophen overdose. Because the relative exposure of the liver is dependent on the capacity of the body to detoxify the reactive metabolite with glutathione and to metabolize the drug by other routes, a complete profile of the major and minor metabolic products of acetaminophen metabolism can aid in determining relative risk and the efficacy of treatment.

MATERIALS AND METHODS

Acetaminophen used for analytical standards was purchased from Eastman Organic Chemicals (Rochester, New York). Standards of acetaminophen metabolites were obtained from Dr Sidney Nelson. Acetaminophen glucuronide was prepared by the method of Shibasaki *et al.* (1970) and acetaminophen sulphate by the method of Smith and Timbrell (1974). 3-Cysteinacetaminophen and 3-methylthioacetaminophen were synthesized as described by Focella *et al.* (1972). Acetaminophen-3-mercapturate was synthesized by a phase reaction involving *N*-acetyl-*p*-benzoquinone imine

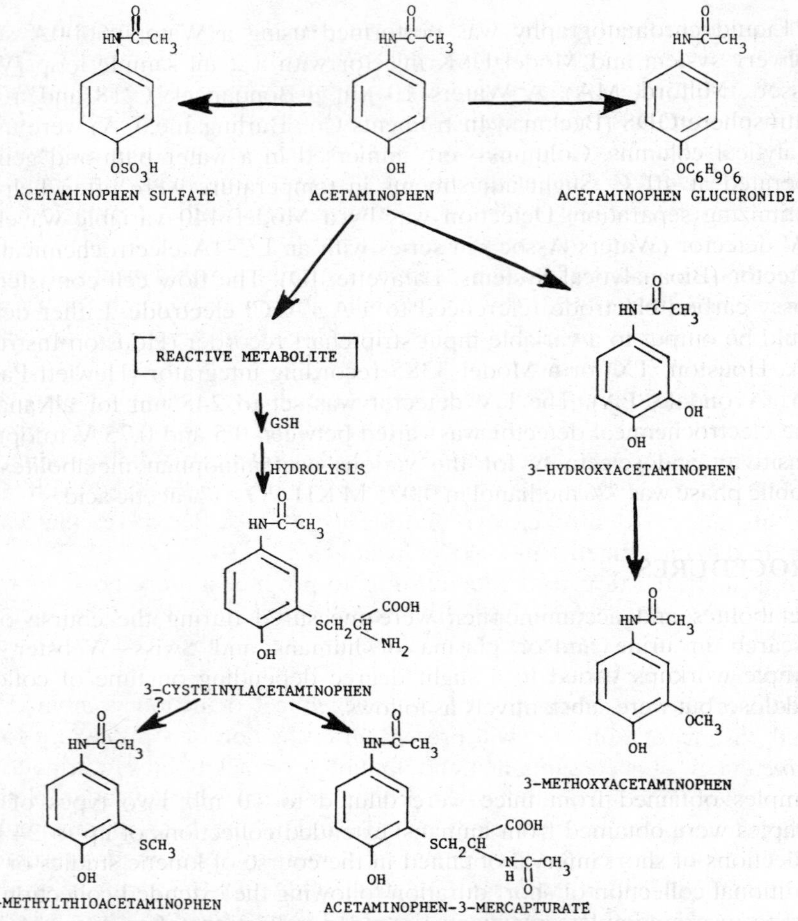

Figure 1. Metabolic scheme for the biotransformation of acetaminophen.

and *N*-acetylcysteine. This standard was subsequently found to be contaminated and unsuitable for quantitative purposes. However, it was useful in the preparation of sample chromatograms, the verification of retention times and for conducting linearity studies. Quantitation of the mercapturate metabolite was based on peak-area comparisons of known cysteine metabolite concentrations. This calculation was confirmed with the use of ring-labelled, ^{14}C-acetaminophen. Data for this experiment will be presented later in the text. 3-Hydroxyacetaminophen was synthesized according to the method of Szent-Gyorgi *et al.* (1976). 3-Methoxyacetaminophen was prepared in a similar reductive acetylation reaction starting with nitroguaiacol as described by Forte (1981).

All chemicals were reagent grade and solvents were LC grade and obtained from local suppliers. β-Glucuronidase was obtained from the Sigma Chemical Co. (St Louis, MO). 2-Hydroxyacetanilide was purchased from the Aldrich Chemical Co. (Milwaukie, WI) and used as an internal standard.

Liquid chromatography was performed using a Waters 6000A solvent delivery system and Model U6K injector with a 2 ml sample loop (Waters Assoc., Milford, MA). A Waters 10 μm μ-Bondapack C-18 and a 5 μm Ultrasphere ODS (Beckman Instruments Co., Burlingame, CA) were used as analytical columns. Columns were immersed in a water bath and generally operated at 40° C. Slight adjustments in temperature were often helpful in optimizing separations.Detection was by a Model 440 variable wavelength UV detector (Waters Assoc.) in series with an LC-4A electrochemical (EC) detector (Bioanalytical Systems, Lafayette, ID). The flow cell consisted of a glassy carbon electrode referenced to a Ag/AgCl electrode. Either detector could be output to a variable-input strip chart recorder (Houston Instrument Co., Houston, TX) or a Model 3385 recording integrator (Hewlett-Packard Co., Avondale, PA). The UV detector was set to 248 nm for all analyses. The electrochemical detector was varied between 0.5 and 0.75 V to optimize sensitivity and selectivity for the various acetaminophen metabolites. The mobile phase was 7% methanol in 0.075 M KH_2PO_4/1% acetic acid.

PROCEDURES

Metabolites and acetaminophen were measured during the course of this research in urine and/or plasma of humans and Swiss—Webster mice. Sample workups varied to a slight degree depending on time of collection and dose, but were substantively as follows.

Urine
Samples obtained from mice were diluted to 10 ml. Two types of urine samples were obtained from humans: extended collections of up to 24 h and collections of short interval obtained in the course of kinetic studies or as an additional collection of short duration following the extended collection. This was to insure complete recovery. Extended collections after doses of 1 g or less were diluted to 2 l with distilled water. Collections after doses greater than 1 g were diluted to 2.5 l. Aliquots were taken from dilutions and stored frozen until analysis. Volumes were measured for urines of short duration and they were not diluted prior to analysis.

Determination of metabolite concentrations in urine samples or aliquots is a two-step procedure: the first, a dilution on the order of 1:50 in a solution containing internal standard (2-hydroxyacetanilide, 20 mg/l). This dilution was analysed for acetaminophen glucuronide, acetaminophen sulfate and acetaminophen. Since in the mouse, 3-cysteineacetaminophen is a major metabolite, it is also measured in the mouse urine dilution. The second step involves hydrolysis with a combined β-glucuronidase/sulfatase solution for measurement of the oxidative metabolites. A urine aliquot, generally 50 μl, is diluted 1 to 4 with an internal standard solution containing β-glucuronidase/sulphatase and ascorbic acid. Ascorbic acid is added to inhibit the degradation of labile oxidative metabolites. This solution is transferred to a 250 μl polyethylene microcentrifuge tube, capped and placed in 10 ml of water in a

20 ml scintillation vial. Vials were incubated overnight for a minimum of 10 h at 37° C. To avoid injection of suspended material from the incubation mixture onto the analytical column, the microcentrifuge tube was spun at 12 000 × g for 5 min in an Eppendorf microcentrifuge (Brinkmann Instruments Inc., Westbury, NY). 2—20 μl of the supernate was injected for analysis. Hydrolysed samples could be measured for 3-hydroxyacetaminophen, 3-methoxyacetaminophen, 3-cysteineacetaminophen, acetaminophen-3-mercapturate and 3-methylthioacetaminophen.

Standards were prepared in aqueous solution and handled in the same fashion as samples. A standard curve was prepared using either peak heights or area ratios to the internal standard, 2-hydroxyacetanilide. These could be generated using either the UV signal or that from the electrochemical detector. 2-Hydoxyacetanilide was chosen as internal standard because it could be separated not only from endogenous urine components but also other acetaminophen metabolites within a reasonable time. In addition, it possessed good electrochemical and UV response.

Plasma

To precipitate proteins, plasma samples, generally 50 μl, were diluted 1 to 3 with methanol containing the internal standard. Samples were agitated and allowed to sit for several minutes to insure complete precipitation, then transferred to a glass microhematocrit tube, capped with clay, and centrifuged for 5 min in a hematocrit centrifuge (International Equipment Co., Needham Heights, MA). Ten μl of the supernate could be withdrawn directly from the hematocrit tube for injection. Plasma samples also can be hydrolysed for measurement of oxidative metabolites.

Chromatograms obtained from the diluted and hydrolysed urine of human subjects are presented in Figs 2 and 3. Both the UV and EC responses are depicted. Chromatograms obtained from diluted and hydrolysed mouse urine are presented in Figs 4 and 5. Plasma chromatograms tended to contain fewer endogenous components and are not depicted here.

RESULTS

Performance of the electrochemical detector

Since each acetaminophen metabolite evokes a slightly different response from the electrochemical detector, it was important to characterize that response. One manner in which that can be done is to generate a hydrodynamic voltammogram for each compound. This was done by injecting 5 μl of a 200 μg/ml solution of each metabolite and measuring the peak area at a series of increasing electrode potentials. The response was plotted to form smooth curve for each metabolite and used to optimize electrode conditions. These curves are presented in Figs 6 and 7. In the figures, the curves have been normalized to 100% response to allow easy comparison.

It was decided that in those circumstances when measurement of oxidative metabolites was important the electrode could be set to 0.57 V to maximize

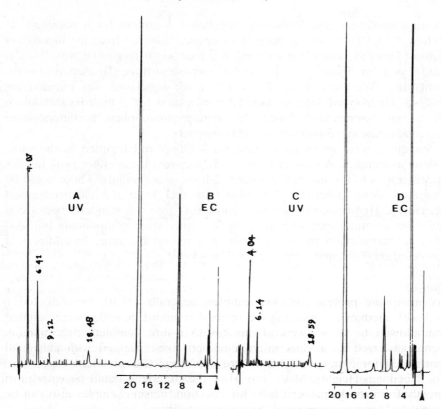

Figure 2. LC chromatograms demonstrating the UV, and EC response obtained from (A) and (B) aqueous standards and (C) and (D) human urine diluted 1 to 50. Identities of peaks are the glucuronide conjugate (4.07), the sulfate conjugate (6.4) acetaminophen (9.12) and the internal standard (18.48).

the catechol, methoxy and methylthio response, and to minimize that of acetaminophen and endogenous interferences. Further evidence for the differential response provided by the EC detector can be found in Figs 2—5. The major metabolites of acetaminophen, the glucuronide and the sulphate conjugates do not elicit an electrochemical response over the range tested.

Linearity
Standard curves were constructed for each metabolite for each study. Examples are shown in Figs 8—12. It was apparent that curves were linear over a wide range encompassing the range of metabolite concentrations found after the dilutions listed in the standard assay conditions.

Precision
Overall precision was evaluated in a number of ways. Within-run precision was assessed by making multiple injections of a standard solution. A UV precision and an EC precision study were conducted separetly. The UV

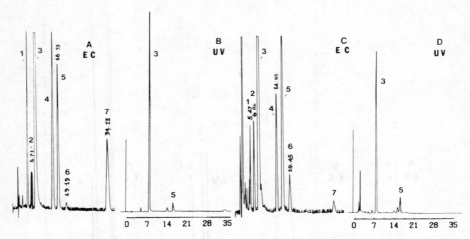

Figure 3. LC chromatograms demonstrating the (B) and (D) UV and (A) and (C) EC response obtained from (A) and (B) aqueous standards and (C) and (D) human urine diluted 1 to 4 and hydrolysed enzymatically. Identities of peaks are: 1. 3-hydroxyacetaminophen; 2. 3-cysteinyl-acetaminophen; 3. acetaminophen; 4. 3-methoxyacetaminophen; 5. internal standard; 6. acetaminophen-3-mercapturate; 7. 3-methylthioacetaminophen.

study involved making twenty 10 µl injections at the low end of the standard curve. On this day the cysteine conjugate was not separated from acetamino-phen to baseline and precision was higher than obtained on later dates. Since the molar response of the EC detector is higher for acetaminophen than for its cysteine conjugate, this slight overlap would affect the area measurement of the EC measurement more than the peak height measurement of the UV response. On the basis of these precision estimates, it was concluded that adequate precision could be obtained via external standardization; in addi-tion, an internal standard was included and we subsequently evaluated within-run precision by analysing study samples blind and in duplicate. The average coefficient of variation was 6.1% for acetaminophen, 2.8% for acetaminophen glucuronide and 4.0% for acetaminophen sulfate.

Detection limits
Detection capabilities were compared for UV and electrochemical detection and the data are presented in Table 2. It is obvious for the compounds studied that electrochemical detection provides greater sensitivity by an order of magnitude. Since the primary interest is in quantitation of oxidative metabolites in urine, the detection limits were also examined in a urine dilution to simulate sample workup.

Recovery of the catechol metabolites from urine was also investigated because these metabolites are present in very small quantities and there was concern over potential interferences from early eluting urine compo-nents. Urine was diluted and spiked with known concentrations of the catechol metabolites. Peak heights were measured and compared to aqueous standards. Urinary recovery for 3-hydroxyacetaminophen and 3-methoxy-

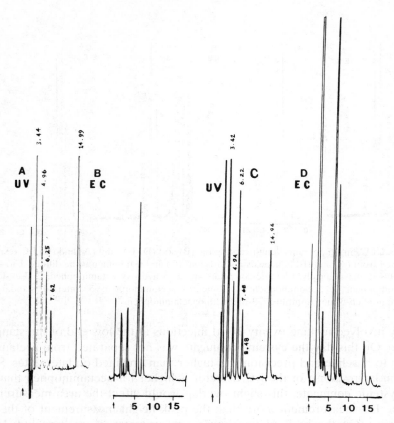

Figure 4. LC chromaograms demonstrating the (A) and (C) UV and (B) and (D) EC response obtained from (A) and (B) aqueous standards and (C) and (D) mouse urine diluted 1 to 50. Identities of peaks are acetaminophen glucuronide (3.44), acetaminophen sulphate (4.96), 3-cysteinylacetaminophen (6.75), acetaminophen (7.62) and internal standard (14.99).

acetaminophen is presented in Table 3. Recovery was complete and reproducible for both metabolites. Analysis was carried out on a μ-Bondapak C-18 analytical column (Waters Assoc.). Subsequent studies employed a 5 μm spherical packing material with much greater efficiency and endogeneous urine components were not a factor.

Effect of different columns
It is a well-observed phenomena that LC columns, even though they may be of the same type and manufacturer, will show differences in selectivity and retention due to variable particle sizes and unequivalent surface coverage (Goldberg, 1982). Table 4 lists elution times for the major and minor metabolites of acetaminophen as well as a number of structural analogues and other compounds commonly found in biological fluids. Times were produced on a 10 μm C-18 column produced by Waters Assoc. Columns prepared by other manufacturers may produce different results.

202   *J. Wilson*

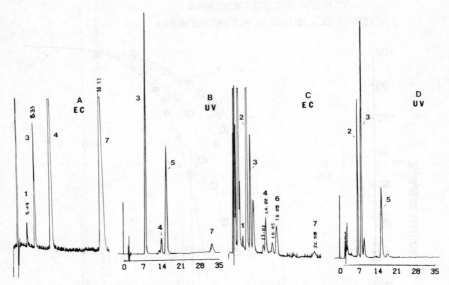

Figure 5. LC chromatograms demonstrating the (B) and (D) UV and (A) and (C) EC response of (A) and (B) aqueous standards and (C) and (D) mouse urine diluted 1:4, and hydrolysed enzymatically. Identities of peaks are: 1. 3-hydroxyacetaminophen; 2. 3-cysteinylacetaminophen; 3. acetaminophen; 4. 3-methoxyacetaminophen; 5. internal standard; 6. acetaminophen-3-mercapturate; 7. 3-methylthioacetaminophen.

Since it was planned to conduct a number of studies and process many samples, it was important to guarantee that separations should be reproduced from day-to-day and year-to-year. As a result, a number of ODS columns from a number of different manufacturers were tested to determine candidates for continued use. We also wanted to explore the possibility of shortening chromatographic times which initially, with a Waters μ-Bondapak 10 µm pellicular packing, was of the order of 35 min, and to optimize early peak separation for the quantitation of 3-hydroxyacetaminophen. Four different ODS columns from four manufacturers were evaluated using the same test mixture and mobile phase. Flow rate was optimized for each column. The temperature was maintained at 40° C. The difference can be seen in Figs 13 and 14. The Waters product separates 11 components into 8 fractions, while the Alltech 10 µm column separates all 11, though not to baseline. Going to a 5 µm spherical packing, such as the Altex 25 cm column, provides greater efficiency, which can be seen in the narrower peaks at the slower flow rate, but resolution is not as good due to different selectivity. The most impressive performer was a 3 µm spherical 10 cm column manufacturer by the Rainin Instrument Co. (Woburn, MA). Peaks are completely resolved and the overall time of analysis reduced to 22 min.

Evaluation of the method with ^{14}C-acetaminophen
A final check of analytical recovery was made by administering 9.1 mCi of ring-labelled ^{14}C-acetaminophen to two mice. The urine collected from the

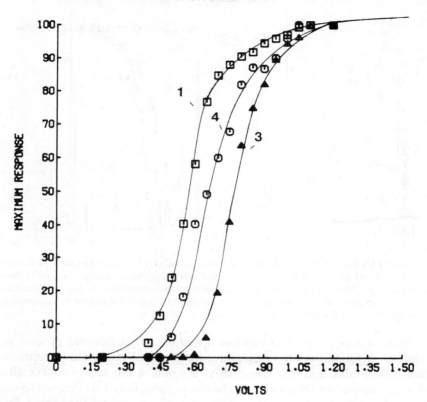

Figure 6. Normalized hydrodynamic voltammograms for: 1. 3-hydroxyacetaminophen; 3. acetaminophen; 4. 3-methoxyacetaminophen.

mice was taken through the sample workup; except that the nonhydrolysed urine dilution was 1 to 4 rather than 1 to 50 and larger quantities were injected on-column. Fractions were collected every 15 s on an LKB rotary fraction collector for 16 min and every 30 s thereafter. Figure 15 shows the UV and radiotracing for unhydrolysed urine. No unidentified peaks were noted, and when the fraction counts were totalled, the % of counts under each metabolite roughly corresponded to that determined by quantitation against authentic standard (Table 5). Feces were also collected, homogenized and tested for beta decay. The amount of radioactivity in the fecal samples was negligible.

APPLICATION OF THE ANALYTICAL METHOD TO PHARMACOKINETIC STUDIES

Eight normal, healthy subjects, seven men and one woman received a 1 g oral dose of Tylenol extra-strength elixir (McNeilab, Inc., Fort Washington,

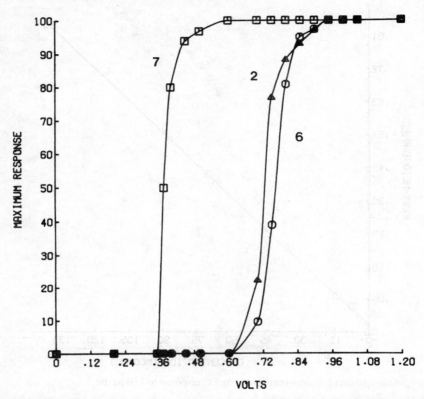

Figure 7. Normalized hydrodynamic voltammograms for: 2. 3-cysteineacetaminophen; 6. acetaminophen-3-mercapturate; 7. 3-methylthioacetaminophen.

PA). The amount of the elixir was premeasured using a graduated cylinder to insure equivalent dosing. Each subject was maintained drug-free for 1 week prior to starting the study and through the course of sample collection. Study participants were encouraged to load water to the extent of 1 l prior to taking the dose. Timed urine collections were made at 0, 45 and 90 min, and 2, 3, 4, 6, 8, 12, 18 and 24 h *post* dose. Urine volumes were measured and an aliquot retained for analysis. Samples were stored frozen and analysed for acetaminophen and metabolites on the day after completion of collection.

Two subjects in the study also took a 2 g dose of the elixir as a preliminary study to examine the effect of a change in dose. Urine was collected as before with the exception of an additional sample at 30 h.

Urine was analysed for individual metabolite concentrations and multiplied by urine volume to obtain an estimate of the amount of each metabolite in the sample. Excretion rate was obtained by dividing the amount by the collection interval. Excretion rate plots were prepared by plotting the log of

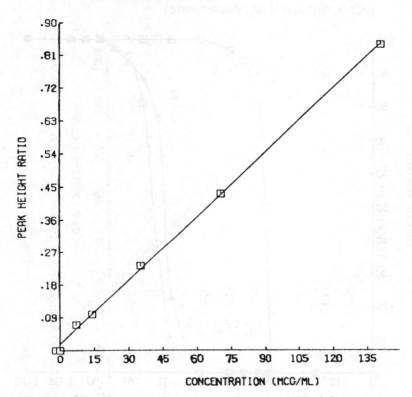

Figure 8. Standard curve demonstrating a linear EC response to 150 µg/ml.

the excretion rates against the midpoint of each collection period. Elimination rate constants for the metabolites were obtained by performing a linear regression on the log data points in the terminal portion of each curve and obtaining the slope of the regression line. Terminal half-lives were calculated from the elimination rate constant using the following equation:

$$t_{1/2} = 0.693/K \tag{1}$$

where K represents the elimination rate constant determined above. An estimate of the time to peak was also obtained for each metabolite by noting the maximum point on the curve and the corresponding point on the time axis. The fraction of the dose excreted at each metabolite was calculated by dividing the total amount (in moles) of each metabolite by the molar dose or by the total amount of acetaminophen and metabolites recovered in urine.

A representative excretion rate plot is presented in Fig. 16. From the data presented in this plot, calculations of pharmacokinetic parameters can be made. In addition, the relative amounts of each metabolite in the urine demonstrates the relative importance of the individual metabolic pathway.

J. Wilson

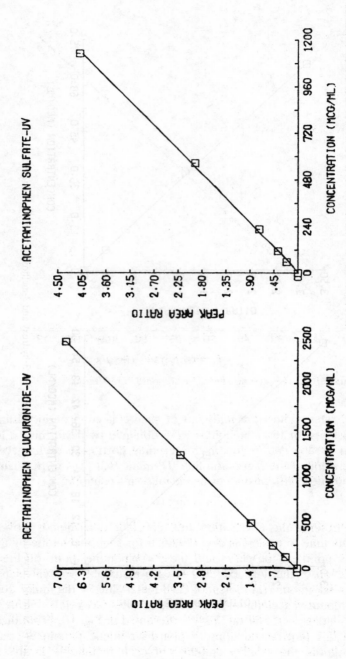

Figure 9. Standard curves demonstrating a linear UV response of acetaminophen glucuronide and sulfate conjugates to 2500 and 1200 μg/ml respectively.

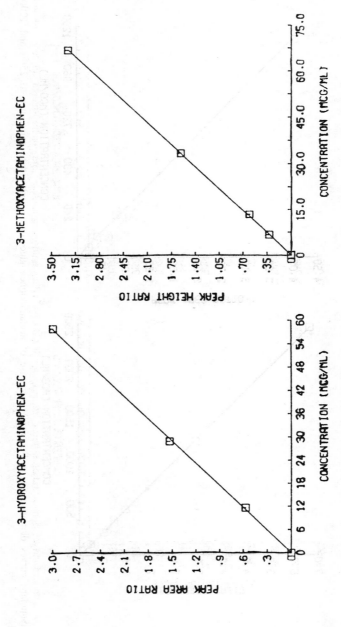

Figure 10. Standard curves demonstrating a linear EC response for 3-hydroxy and 3-methoxyacetaminophen to 60 μg/ml.

J. Wilson

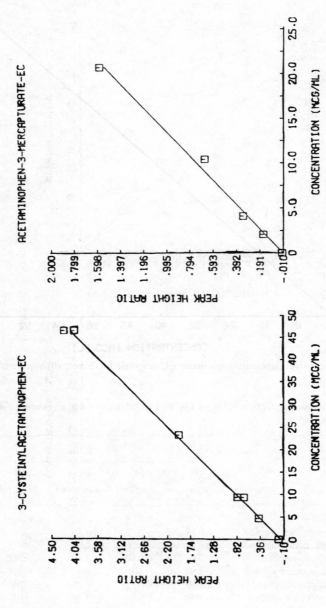

Figure 11. Standard curves demonstrating a linear EC response for 3-cysteinylacetaminophen and acetaminophen-3-mercapturate to 50 and 20 μg/ml respectively.

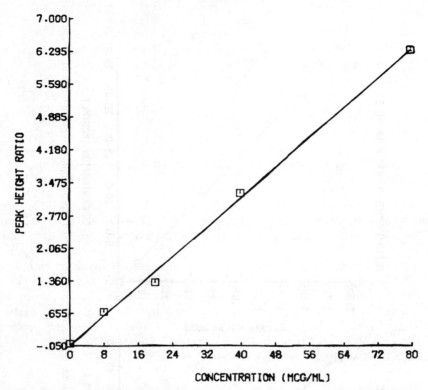

Figure 12. Standard curve demonstrating a linear EC response for 3-methylthioacetaminophen to 80 μg/ml.

Table 2. Minimum detection limits of UV and EC detection for acetaminophen and metabolites*

Substance	Aqueous		Urine	
	UV (μg/ml)	EC (μg/ml)	UV (μg/ml)	EC (μg/ml)
Acetaminophen	0.070	0.015	0.50	0.06
Acetaminophen glucuronide	0.035	—	0.50	—
Acetaminophen sulphate	0.045	—	1.00	—
3-Cysteinacetaminophen	0.250	0.050	2.40	0.30
3-Methylthioacetaminophen	0.250	0.100	2.80	0.30
3-Hydroxyacetaminophen	0.150	0.015	1.20	0.50
3-Methoxyacetaminophen	0.100	0.010	0.20	0.02

* Estimation of detection limits was based on a signal twice the detector noise level. Detection limits in urine were obtained by spiking urine diluted 1 to 4 to simulate normal assay conditions.

Table 3. Estimation of recovery of acetaminophen catechol metabolites in urine

	Spiked conc. (μg/ml)	N	Measured conc. (μg/ml)	% Recovery (% CV)
3-Hydroxyacetaminophen	24.89	4	28.20	113 (2.8)
	18.67	5	19.57	105 (4.1)
	12.44	4	12.46	100 (7.1)
3-Methoxyacetaminophen	14.64	4	14.50	99 (2.4)
	10.98	5	11.21	102 (1.4)
	7.32	4	6.97	95 (6.1)

Recovery was based on multiple analysis of spiked urine compared to an aqueous standard.

Table 6 presents the relative abundance of each metabolite in the urine and reinforces earlier reports (Prescott, 1980; Forrest *et al.*, 1982) that conjugation is the major metabolic route with oxidation representing minor amounts.

Table 7 presents the mean kinetic parameters obtained at two doses of acetaminophen. Important observations include the short half-lives of acetaminophen and the glucuronide and sulphate metabolites indicating that major metabolites are eliminated rapidly; the longer half-lives of the cysteine and mercapturate metabolites are as might be expected from the products of sequential metabolism; and the late time to peak for the methylthio metabolite possibly as a result of enterohepatic recirculation.

DISCUSSION

The development of this analytical method was aided by advances in two technologies. Liquid chromatography or HPLC, as it is often termed, underwent remarkable growth in sales and improvement in hardware from 1975 to 1980. Since that time there have been equally important advances in column technology. Efficiencies of reverse-phase columns have escalated from in the vicinity of 3500 theoretical plates per column to more than 15000. Without this improvement the separations presented in Fig. 15 would not have been possible. Of further importance was the introduction of the electrochemical detector, (Kissinger, 1977). By utilizing an extremely small flow cell, the electrochemical detector is ideally suited to take advantage of column innovations. More importantly, there is, with this detector, a new form of selectivity embodied in the half-cell potential of each molecule, which enables the user to desensitize the system to interfering elements while using the amplification capabilities of modern microelectronics to produce sub-nanogram detection limits.

These developments made possible this simplified approach to the measurement of acetaminophen metabolites. By limiting sample workup to a single dilution step, reproducibility of measurements has been ensured. The

Table 4. Relative retention times for acetaminophen and metabolites and related compounds

Metabolite/analog	Retention time (min)	Relative retention*
Acetaminophen glucuronide	3.6	0.45
3-Hydroxyacetaminophen	5.1	0.64
Acetaminophen sulfate	5.9	0.74
3-Cysteineacetaminophen	7.0	0.88
Acetaminophen	7.9	1.00
3-Methoxyacetaminophen	13.4	1.69
Acetaminophen-3-mercapturate	17.7	2.24
3-Methylthioacetaminophen	30.9	4.08
p-Hydroxypropionic acid	2.7	0.36
Hydroquinone	3.2	0.54
Phloroglucinol	3.5	0.47
Homogentisic acid	4.7	0.59
4-Acetamido-2-hydroxyphenylacetate	5.1	0.65
3-Hydroxyacetanilide	10.6	1.44
2-Hydroxyacetanilide (IS)	11.2	1.48
4-Flurophenol	12.9	2.23
Theobromine	15.1	2.20
Theophylline	15.1	2.20
3-Hydroxy-4-methoxyacetaminophen	19.4	2.39
Salicylic acid	19.6	2.47
Hydroxyethylphenacetin	21.1	2.71
Acetanilide	22.0	2.88
N-Methylacetaminophen	25.5	3.27
Caffeine	27.7	4.02
N-Butyrylacetaminophen	35.0	4.73
Vanillin	45.5	5.61
2-Hydroxyphenacetin	60.2	7.56
2,5-Dichloroacetanilide	65.2	8.21
p-Nitroacetanilide	> 30	
Methyl paraben	> 30	
Ethyl paraben	> 30	
Propyl paraben	> 30	

* Relative retention is reference to acetaminophen. The analytical column used in this study was a Waters μ-Bondapak 10 μm C18. Times may be approximate because samples were run on different days. Other reverse phase columns may give different results.

use of high efficiency columns has allowed the quantitation of many of the metabolites in a single injection and avoidance of the use of gradient techniques or multiple mobile phases. Unfortunately, synthetic standards of the glucuronide and sulphate conjugates of the catechol, methoxy and methylthio metabolites were not available and it was necessary to include a hydrolysis step and a second sample injection. Although the hydrolysis is over a 10 h period at 37° C, evaporation losses were avoided by maintaining

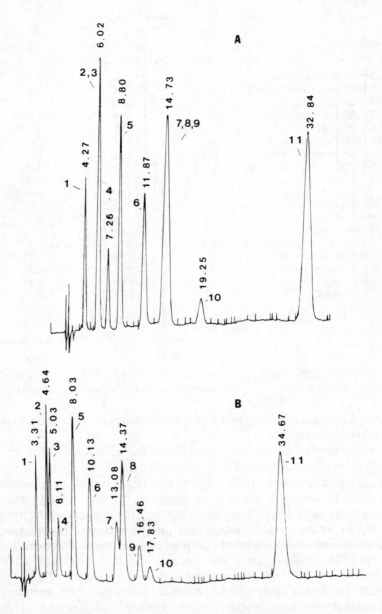

Figure 13. LC chromatograms demonstrating the resolution obtained from (A) A Waters U-Bondapak 10 μm C-18, and (B) an Alltech 10 μm C18. The standard mixture contains: 1. acetaminophen glucuronide; 2. 3-hydroxyacetaminophen; 3. acetaminophen sulfate; 4. 3-cysteineacetaminophen; 5. acetaminophen; 6. 3-glutathionylacetaminophen; 7. acetaminophen-3-sulfate; 8. 3-methoxyacetaminophen; 9. internal standard; 10. acetaminophen-3-mercapture; 11. 3-methylthioacetaminophen.

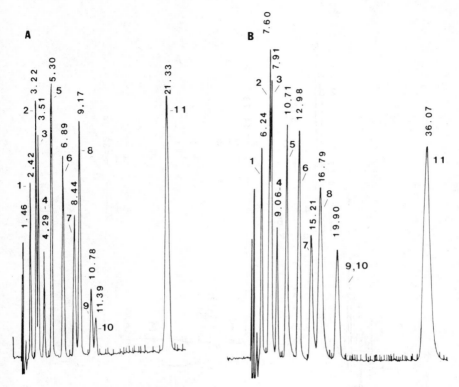

Figure 14. LC chromatograms demonstrating the resolution obtained from (A) a Rainin 3 μm C18 and (B) a 5 μm Altex 25 cm C18. Contents of the mixture are as enumerated in Fig. 13.

hydrolysates in microvessels that had a minimal surface area and which could be kept tightly sealed.

The addition of an internal standard added additional flexibility. Since metabolite concentrations and detector responses varied greatly, it was often impossible to measure the peak heights of all peaks in a single injection. Peak areas could be obtained, but in some instances, due to the high sensitivity of the electrochemical detector, would overpower the capacity of the integration system. The internal standard provided the possibility of injecting an occasional non-standard volume and compensating by measuring peak height or area ratio. In addition, since the internal standard produced an electrochemical response as well as absorbing in the UV range, it could be used as a reference for either detector, or in some instances, when an interference made quantitation impossible in one system, it could serve as an internal reference for the complementary chromatogram in the other system.

It was found that the quality of our results was highly dependent on column efficiency. A minimum of 8000 plates are necessary to achieve satisfactory performance. Experience with the 5 μm Altex column was very good. Though retention of the methylthio metabolite was in excess of 40 min

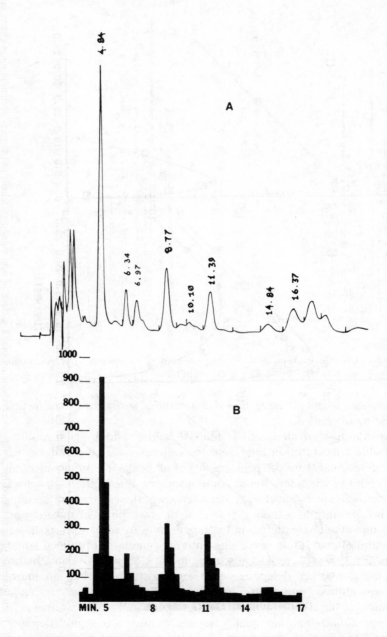

Figure 15. (A) LC and (B) radiotracing of mouse urine containing ^{14}C-labelled acetaminophen and metabolites. Major peak identities are acetaminophen glucuronide (4.84), acetaminophen sulfate (6.34), 3-cysteineacetaminophen (8.77) and acetaminophen (11.39).

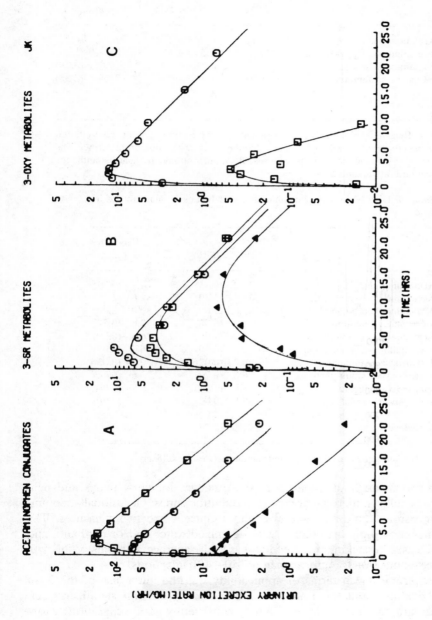

Figure 16. Excretion rate plots obtained for a subject receiving an oral 1 g dose of acetaminophen elixir. (A) depicts acetaminophen glucuronide (□), acetaminophen sulfate (○), and acetaminophen (▲). (B) depicts 3-cysteinylacetaminophen (□), acetaminophen-3-mercapturate (○) and 3-methyl-thioacetaminophen (▲). (C) depicts 3-hydroxyacetaminophen (□) and 3-methoxyacetaminophen (○).

Table 5. Metabolite recovery after administration of ^{14}C-acetaminophen to mice

	DPM	% Recovery*	UV reference mean (S.D.)	
Acetaminophen	702	20.9	11	(6.22)
Acetaminophen glucuronide	1745	51.9	45	(1.83)
Acetaminophen sulfate	292	8.7	7.5	(3.79)
3-Cysteineacetaminophen	660	19.6	23.5	(2.38)
% Recovered		80%	87%	

* % Recovered obtained for radioactivity by summation of radioactivity under chromatographic peaks compared to the total radioactivity in urine. The recovery of radioactivity is contrasted to measurements against authentic standards at the same dose.

Table 6. Distribution of acetaminophen metabolites in man after a 1 g dose*

	% Dose (S.E.)		% Recovered dose (S.E.)	
Acetaminophen	5.2	(0.86)	4.8	(0.75)
Acetaminophen glucuronide	55.6	(2.30)	52.0	(2.30)
Acetaminophen sulfate	32.6	(2.60)	30.0	(2.00)
3-Cysteineacetaminophen	2.8	(0.36)	2.6	(0.34)
Acetaminophen-3-mercapturate	7.2	(1.06)	6.8	(0.93)
3-Methylthioacetaminophen	0.3	(0.10)	0.3	(0.09)
3-Hydroxyacetaminophen	0.04	(0.026)	0.04	(0.026)
3-Methoxyacetaminophen	3.8	(0.76)	3.6	(0.74)
Sum GSH conjugates	10.4	(1.11)	9.7	(1.04)
Sum 3-oxymetabolites	3.8	(0.79)	3.6	(0.76)
% Recovery	107%	(2.22)		

* The data represents the mean and standard error in 8 subjects.

and limited throughput, separation was excellent for early peaks and peak shape was still symmetrical at very late retention times. Occasionally, column back pressure would be excessive and require corrective measures. This often meant washing the column with deionized water and/or methanol and, in some cases, repacking the precolumn. On the average, columns of this type could be counted on for approximately 250—400 injections.

More promising in terms of applicability was the initial use of the 3 μm 10 cm column from the Rainin Co. The shorter column significantly cut chromatography time while maintaining consistently good separations. However, the life of the column is significantly less than the others and it was unreasonable to expect more than 100 injections. In addition, care must be taken in the choice of precolumns. Precolumns with large internal volumes

Table 7. Pharmacokinetic parameters of acetaminophen and metabolites in humans

	1 g Dose*		2 g Dose[†]	
	Time to peak (h)	Half-life (h)	Time to Peak (h)	Half-life (h)
Acetaminophen	1.3 (0.34)	3.1 (0.78)	1.5 #[‡]	4.4 #[‡]
Acetaminophen glucuronide	2.9 (0.73)	4.1 (1.03)	2.9 (0.53)	4.1 (0.40)
Acetaminophen	1.7 (0.41)	4.1 (0.57)	2.6 (0.88)	5.4 (0.73)
3-Cysteineacetaminophen	3.8 (1.61)	5.7 (1.63)	3.9 (0.88)	9.7 (0.05)
Acetaminophen3-mercapturate	3.7 (1.41)	5.6 (1.32)	3.9 (0.88)	9.8 (1.74)
3-Methylthioacetaminophen	13.5 (3.87)	ND[‡]	12.3	ND[‡]
3-Hydroxyacetaminophen	ND[‡]	ND[‡]	ND[‡]	ND[‡]
3-Methoxyacetaminophen	2.0 (0.83)	5.2 (1.02)	1.6 (0.53)	6.0 (0.62)

* Mean and standard error of 8 subjects.
[†] Mean and standard error of 2 subjects.
[‡] Insufficient data points to allow calculation.
ND = not detectable.

effectively destroy the added efficiency of these small columns. It was found that the column could be protected somewhat and retain a high percentage of the efficiency by using a minimal volume precolumn produced by CM Laboratory (Nutley, NJ).

REFERENCES

Focella, A., Heslin, P., and Teitel, S. (1972). The synthesis of two phenacetin metabolites. *Can. J. Chem. 50*, 2025—2030.

Forrest, J. A. H., Clements, J. A., and Prescott, L. F. (1982). Clinical pharmacokinetics of paracetamol. *Clin. Pharmacokin. 7*, 93—107.

Forte, A. J. (1981). Doctoral dissertation.

Goldberg, A. P. (1982). Comparison of columns for reversed-phase liquid chromatography. *Anal. Chem. 54*, 342—345.

Kissinger, P. T. (1977). Amperometric and coulometric detectors for high-performance liquid chromatography. *Anal. Chem. 49*(4), 448A—456A.

Kissinger, P. T., Bratin, K., Davis, G. C., and Dachla, L. A. (1979). The potential utility of pre- and post-column chemical reactions with electrochemical detection in liquid chromatography, *J. Chromatogr. Sci. 17*, 137—146.

Low, L. K. and Castagnoli, N. (1979). *Drug Biotransformations, Burger's Medicinal Chemistry*, 4th edn., X. Wolf, (ed.), John Wiley and Sons, pp. 107—204.

Prescott, L. F. (1980). Kinetics and metabolism of paracetamol and phenacetin. *Br. J. Clin. Pharmac. 10*, 291S—298S.

Shibasaki, J., Sadokane, E., Konishi, R., and Koizumi, T. (1970). Drug absorption, metabolism, and excretion. VI.1. Preparation of ether-type glucuronide of 4-hydroxyacetanilide. *Chem. Pharm. Bull. 18*(11), 2340—2343.

Smith, R. L. and Timbrell, J. A. (1974). Factors affecting the metabolism of phenacetin. I. Influence of dose, chronic dosage, route of administration and species on the metabolism of (1-14 C-acetyl) phenacetin. *Xenobiotica 4*, 489—501.

Szent-Gyorgi, I., Chung, R. H., Boyajian, M. J., Tishler, M., Arison, B. H., Schoenwaldt, E. F., and Wittick, J. J. (1976). Agaridoxin, a mushroom metabolite. Isolation, structure, and synthesis. *J. Org. Chem. 41*(9), 1603—1606.

Progress in HPLC, Vol. 2, pp. 219—233
Parvez *et al.* (Eds)
© 1987 VNU Science Press

Pre- and post-column enzymatic hydrolysis and amperometric detection of glycosides. Applications to trimethoprim metabolites and cyanogenic glycosides

LARS DALGAARD

Royal Danish School of Pharmacy, Department of Chemistry BC, 2, Universitetsparken, DK-2100 Copenhagen, Denmark.

INTRODUCTION

Determination of compounds in biological material often requires extensive sample preparation before the actual measurement can be performed. Generally, the more universal a detection method is, the more extensive the work up has to be. Group selective detection is valuable because the necessity of tedious sample preparations is reduced. The cost is, however, that a more sophisticated equipment has to be used. One possibility is to combine the selectivity provided by two or more methods, e.g. pre- or post-column derivatization or series detection in HPLC. The amperometric detector has proved its value in combination with reversed-phase HPLC, not only because of the high sensitivity, but also because of the selectivity observed in some cases. Most of the literature has been concerned with applications using the glassy carbon or carbon paste electrodes, but in recent years the area has extended its growth due to new applications using metal electrodes for the detections of amino acids (Kok *et al.*, 1983), carbohydrates (Rocklin and Pohl, 1983) and sulphur compounds (Allison *et al.*, 1983).

Post-column reactions followed by electrochemical detection in HPLC have further extended the field to the determination of bile acids (Kamada *et al.*, 1982), reducing sugars (Watanabe and Inoue, 1983), glycosides of phenols including glucuronides of drugs (Dalgaard *et al.*, 1983), cyanogenic glycosides (Brimer and Dalgaard, 1984; Dalgaard and Brimer, 1984), sulpho-conjugates (Elchisak, 1983; Elchisak and Carlson, 1982) and choline (Eva *et al.*, 1984).

Pre-column reactions have been used for derivatization of amines and amino acids with nitroaromatic reagents (Jacobs and Kissinger, 1982) and hydrolysis of conjugates (e.g. Nordholm and Dalgaard, 1984).

Dual electrochemical detection may be used both in series or in parallel (Roston *et al.*, 1982; McClintock and Purdy, 1983). When the detectors are placed in series, compounds which cannot be easily oxidized, but on the other hand can be reduced electrochemically, might give reactive inter-mediates by reduction at the upstream electrode followed by oxidation of the intermediates at the second electrode. In the parallel mode two different potentials are applied to the electrodes in order to improve the selectivity of detection.

The use of catalysts, e.g. enzymes, in conjunction with HPLC is intriguing due to the high selectivity towards certain types of substrates inherent in enzymes. This property may be used to advantage for the group selective detection of compounds, provided that the detector can differentiate between the substrates and the products of the enzymatic reaction. For a number of reasons, it may, however, seem a fruitless task to use enzymes. Firstly, enzymes are often considered to be labile and easily inactivated. Secondly, optimum activity requires specific conditions with regard to temperature and pH, and sometimes co-factors have to be present. Thirdly, the linearity of standard curves may be questioned if the enzymatic reaction is not complete before the measurement. Fourthly, enzymes are usually costly. Luckily, all these hurdles can easily be circumvented in the applications investigated so far. Hydrolase enzymes can be immobilized in packed-bed reactors (Table 1) and used for post-column reactions followed by amperometric detection in HPLC. The method has been applied to the determination of drug conjugates, i.e. glucuronides of phenolic metabolites (Dalgaard *et al.*, 1983), and to natural products, i.e. cyanogenic glycosides (Brimer and Dalgaard, 1984; Dalgaard and Brimer, 1984). In the following, these applications will be described.

Table 1. Enzyme activity of reactors used for post-column cleavage of glycosides

Enzyme source/matrix	Enzyme activity (IU)*
Helix pomatia (crude)/CPG	3
Helix pomatia (purified)/CPG	10
Cassava/CPG	21
E. coli/CPG	40

* With *p*-nitrophenyl-β-*D*-glucoside used as substrate, except in the case of *E. coli* where the corresponding glucuronide is used.

DETERMINATION OF TRIMETHOPRIM AND METABOLITES IN URINE AND PLASMA

Trimethoprim [2,4-diamino-5-(3,4,5-trimethoxybenzyl)pyrimidine (TMP)] is a synthetic antibacterial, which interferes with folic acid metabolism by inhibiting dihydrofolate reductase (EC 1.5.1.3). The drug is administered either alone or in combination with a sulphonamide.

Various analytical methods for quantitative determination of the drug in biological material have been used, including microbiological assay, spectrofluorimetry, autoradiography, differential pulse polarography, thin-layer chromatography with densitometry and gas chromatography.

Several HPLC methods for determination of the drug have appeared. Normal phase as well as reversed-phase and ion-pair systems have been used. Sample preparation procedures vary from simple precipitation of proteins to extraction with chloroform, dichloromethane or ethyl acetate.

However, none of these methods are suitable for determination of TMP in small sample volumes at the 0.1 ppm level as required for the present assay, which is developed for pharmacokinetic studies of newborn pigs.

TMP has been reported to be metabolized to six different compounds in various species (Meshi and Sato, 1972; Schwartz *et al.*, 1970; Sigel *et al.*, 1973). Metabolites M1 and M4 (Fig. 1) are partly conjugated with glucuronic acid (M1—gln and M4—gln). In goats and pigs a sulpho-conjugate of M4 (M4—sulph) is formed along with the glucuronide (Nielsen and Dalgaard, 1978). In pigs, TMP is metabolized to M1, M4, M4—sulph, M4—gln and M1—gln. M1 and M4 can be synthesized (Nordholm and Dalgaard, 1983), while the conjugates can be isolated from urine (Nordholm and Dalgaard,

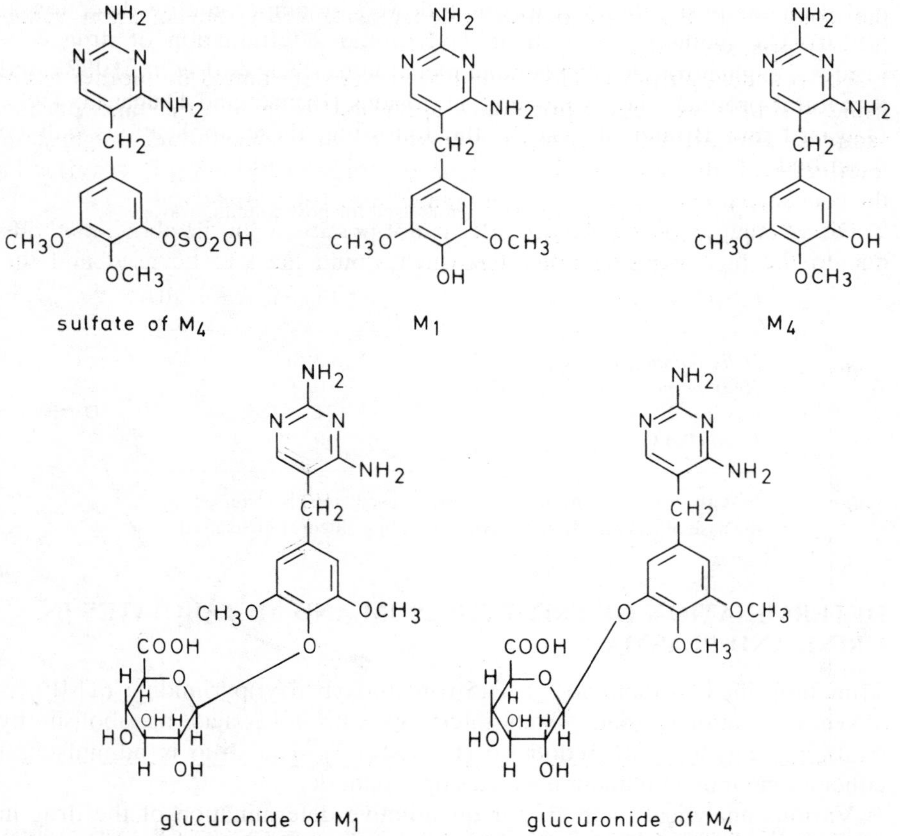

Figure 1. Metabolites of trimethoprim, found in pigs. M1 and M4 are phase I metabolites formed by oxidative demethylation. Both M1 and M4 are conjugated with glucuronic acid, while M4 is also conjugated with sulphate.

1984). Alternatively M1—gln and M4—gln are synthesized enzymatically by use of immobilized rabbit liver enzymes and the cofactor UDPGA, reacting with M1 and M4 (Dalgaard and Køppen, in preparation). In Fig. 2 the anodic current in response to the applied potential between the working glassy carbon and the Ag/AgCl reference electrode is shown for the metabolites present in pigs' urine. TMP and the conjugated metabolites are measurable above 900 mV. The metabolites M1 and M4 can be detected above 200 and 400 mV, respectively, and give equal response factors at 700 mV.

Quantitative determination of TMP and all its metabolites, except M1—gln and M4—gln, in urine can be performed with diluted urine using reversed phase ion-pair chromatography and combined UV and amperometric detection. M1—gln and M4—gln can be determined quantitatively by enzymatic hydrolysis to M1 and M4 prior to analysis. Subtraction of the content of M1 and M4 measured in the sample which has not been treated with enzyme, from the contents of the sample which has been treated with enzyme, yields the content of the glucuronides. Recoveries in the concentration range 0.3—10 ppm are 100%.

For practical purposes TMP might be better determined in another assay using a simple extraction procedure applicable both to urine and plasma samples. The extraction ensures removal of substances which might else interfere with the assay at the working potential (1200 mV). Recoveries in the concentration range 0.1—25 ppm range from 100% to 84%.

Direct analysis of the glucuronides is not possible with the above methods due to the high working potential which would have to be used and the

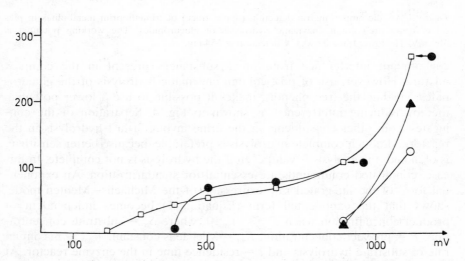

Figure 2. Amperometric response to the applied anodic working potential, using a glassy carbon electrode relative to the Ag/AgCl/KCl reference. Oxidation of M1—gln and M4—gln follow the same pattern as M4—sulph (▲) and TMP (o), which are only detectable above 900 m.V, while M1 (□) and M4 (•) are detectable above 200 mV and 400 mV respectively.

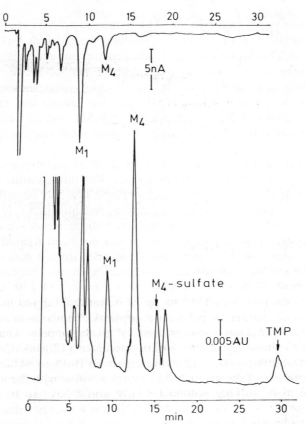

Figure 3. Anodic amperometric detection (upper trace) of trimethoprim metabolites in pigs urine, using pre-column enzymatic hydrolysis of glucuronides. The working potential is 500 mV. The lower trace shows UV detection at 254 nm.

concomitant interference from other substances present in the complex mixture. However, use of post-column enzymatic hydrolysis of the glucuronides, yielding the free phenols, makes it possible to use a lower potential, thereby reducing interference as shown in Fig. 4. Separation of the conjugates from other constituents in the urine favours a fast hydrolysis in the reactor. Although complete hydrolysis is preferable, because better sensitivity is obtained, the analysis is valid even if the hydrolysis is not complete. In this case, conjugated compounds are essential for standardization. An exponential form of the integrated rate expression of the Michaelis—Menten model shows that the exponential term (Equation 1) becomes independent of product concentration when $P \ll V_{max} t$, where S_o = substrate concentration, P = product concentration, K_m = Michaelis constant, V_{max} = maximum rate of substrate hydrolysis and t = residence time in the enzyme reactor. At low concentration the measured product concentration is therefore proportional to the substrate concentration.

$$S_o = P[1 - \exp(P/K_m - V_{max}/K_m)]^{-1} \qquad (1)$$

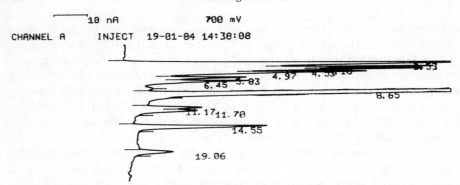

Figure 4. Anodic amperometric detection of trimethoprim metabolites using post-column enzymatic hydrolysis of glucuronides of M1 and M4. The working potential is 700 mV. M1—gln and M4—gln are eluted after 11.17 min and 14.55 min, respectively.

EXPERIMENTAL DETAILS

Preparation of enzyme reactors
Controlled pore glass (Pierce, USA) with a pore size of 550 A was refluxed with 3-aminopropyl triethoxysilane (EGA-Chemie, FGR) in toluene (Weetal and Filbert, 1974). The dried aminopropyl—CPG was poured into a steel column (50 × 3 mm ID) and packed by gentle tapping. The column was fitted with Swagelock end fitting and sintered filters (3 µm). The column was connected to a pump via a Rheodyne valve fitted with a 1 ml loop, and 2.5% glutaraldehyde was injected in 5 × 1 ml portions using a flow of 1 ml/min, followed by 3 × 100 µl enzyme (10—20 mg) delivered at a flow of 0.1 ml/min using a 100 µl loop. The mobile phase was either distilled water or 0.05 M phosphate buffer pH 7.0.

Assay of trimethoprim
The chromatographic system consisted of a Waters Model 6000 solvent delivery system connected to a Waters Model 440 UV detector (280 nm) and a Model 656/641 amperometric detector (Metrohm, Switzerland). The working and auxillary electrodes were glassy carbon and the reference electrode consisted of Ag/AgCl/KCl. The amperometric detector was connected downstream to the UV detector. The Rheodyne Model 7125 injection port contained a 100 µl loop. Chromatograms were traced on an Omniscribe 5111-5 recorder. The HPLC column (Knauer, 120 × 4.6 mm ID) was packed with Nucleosil C18 (5 µm) particles (Macherey-Nagel, GFR). The mobile phase consisted of 0.07 M potassium phosphate (pH 4.75) and methanol (3:1 by volume). The flow rate was 1.5 ml/min (200 bars).

Standards of TMP and the internal standard sulphamethoxazole were prepared from stock solutions in 10% methanol by dilution with the mobile phase.

Sample preparation. A 250 µl sample together with 50 µl of 0.1 M sodium hydroxide and 150 µl of ethyl acetate was pipetted into a 3 ml polypropylene tube. The tube was stoppered, mixed for 15 s on a vortex mixer and centrifuged for 1 min at 18 000 g. Then 1000 µl of the organic phase were transferred to another polypropylene tube and evaporated under a stream of nitrogen at 40° C. The residue was redissolved in 250 µl of mobile phase containing internal standard. After mixing for 30 s on a vortex mixer and centrifugation for 1 min at 18 000 g, 10—100 µl were injected into the HPLC system. Urine samples were treated in the same way after diluting ten times with distilled water.

Concentrations of TMP in plasma and urine were measured by comparing peak height ratios of TMP/internal standard with peak height ratios of extracts from blank plasma/urine samples with known amounts of TMP added. For recovery studies, blank plasma/urine samples with known amounts of TMP added, were compared with TMP standard solutions.

Assay of trimethoprim metabolites

The chromatographic system consisted of a Micromeretics solvent delivery system (Model 750), an autoinjector (Model 725) and a variable wavelength detector (Model 786) operated at 254 nm. The amperometric detector described in the assay of TMP was connected downstream. The HPLC column was 250 × 4.6 mm ID and the mobile phase consisted of 0.1 M potassium dihydrogen phosphate adjusted to pH 7.5 and acetonitrile (85:15 by volume). Tetrabutylammonium hydrogensulphate was added to the mobile phase giving a concentration of 0.7 mM. The flow rate was 1.0 ml/min (250 bars).

Sample preparation was carried out as follows for the determination of M1, M4, M4—sulph and TMP.

Urine was diluted ten times with distilled water and 10 µl injected into the HPLC system.

Quantitative determination of M1—gln and M4—gln by pre-column enzymatic hydrolysis was performed by incubating 1 ml of diluted urine with 100 µl glucuronidase (*E. coli*, 100 sigma units) at 38° C overnight. The hydrolysate (10 µl) was injected into the HPLC.

Post-column hydrolysis of M1—gln and M4—gln was performed on a chromatographic system consisting of a LDC Constametric model III solvent delivery system, equipped with a Rheodyne 7125 valve, a Spherisorb S5-C8 analytical column (250 × 4.6 mm ID), an enzyme reactor containing *E. coli* glucuronidase (Boehringer, FRG) and the amperometric detector described in the assay of TMP. The anodic working potential was 700 mV.

CYANOGENIC GLYCOSIDES AND CYANHYDRINS IN PLANT TISSUES

Gyanogenic glycosides, cyanogenic lipids and cyanhydrins (Fig. 5) are widely distributed in the plant kingdom, and the occurrence of such compounds in food is a problem in some parts of the world. Methods for the separation and quantitative analysis of cyanogenic compounds include thin-layer, gas

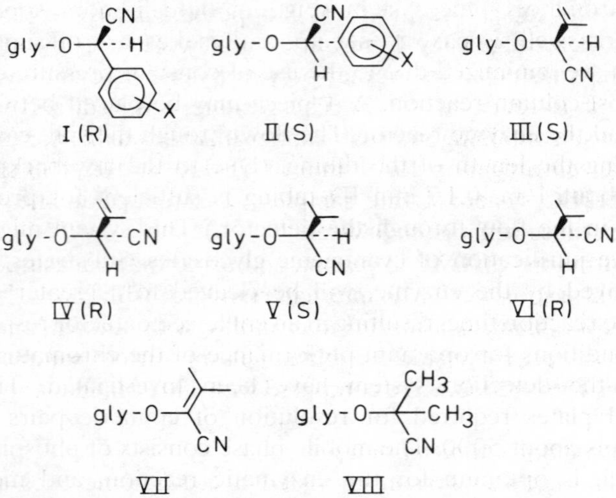

Figure 5. Structures of some cyanogenic glycosides.

and liquid chromatography. Selective detection requires a mass spectrometric measurement of the trimethylsilyl derivatives or hydrolysis of the glycoside to an aldehyde/ketone and hydrogen cyanide, which in turn can be measured by a number of methods. HPLC reversed phase techniques are useful in the separation of the genuine cyanogenic glycosides, including epimeric pairs. However, until recently (Brimer and Dalgaard, 1984; Dalgaard and Brimer, 1984), selective detection of this group of compounds, has not been available.

Enzymatic cleavage (Equation 2), covering a broad range of cyanogenic glycosides, is now possible using partly purified digestive juice of *Helix pomatia* (the vineyard snail). The intermediate cyanohydrin formed by the enzymatic reaction can easily be hydrolysed further to yield hydrogen cyanide and an aldehyde or a ketone (Equation 3). The enzyme possesses excellent stability and can easily be used in an immobilized state for several month. In post-column cleavage of cyanogenic glycosides, the immobilized enzymes have proved valuable in connection with amperometric detection of cyanide at a silver electrode (Equation 4).

$$
\begin{array}{c}
\text{H} \\
| \\
R-\text{C}-\text{CN} \\
| \\
\text{O}-\text{Gly}
\end{array}
\xrightarrow{\text{Glycosidase}}
\begin{array}{c}
\text{H} \\
| \\
R-\text{C}-\text{CN} \\
| \\
\text{OH}
\end{array}
+ \text{GlyOH} \qquad (2)
$$

$$
\begin{array}{c}
\text{H} \\
| \\
R-\text{C}-\text{CN} \\
| \\
\text{OH}
\end{array}
\xrightarrow{\text{OH}^-}
\text{RCHO} + \text{CN}^- + \text{H}_2\text{O} \qquad (3)
$$

$$
\text{Ag} + \text{CN}^- \longrightarrow \text{Ag(CN)}_2^- + e \qquad (4)
$$

The chromatographic system (Fig. 6), though more sophisticated than usual, works well, is easy to set up, and makes use of simple equipment. Pulsations are minimized due to the use of constant pressure delivery system for the post-column reaction. A T-piece may be placed between the HPLC column and the enzyme reactor. The flow through the split can be controlled by adjusting the length of the tubing. (Due to the low backpressure in the reactor, about 1 m, 0.15 mm ID tubing is sufficient for producing a 75% reduction in the flow through the detector.) This system might be used for preparative purification of cyanogenic glycosides. Substrates, which are not easily cleaved by the enzyme, will be cleaved to a greater extent by pro-longing the reaction time, resulting in an enhanced detector response (Fig. 7).

The conditions for optimum performance of the chromatographic system, including the detection system have been investigated. The number of theoretical plates required for resolution of epimeric pairs of cyanogenic glycosides is about 5000. The mobile phase consists of phosphate buffer (pH 5.0), which is optimum for the enzymatic reaction, and methanol (either 7.5% or 15% by volume). In order to resolve the more polar epimeric pairs, 7.5% methanol is used, but in most cases 15% methanol provides sufficient selectivity and a shorter time of analysis.

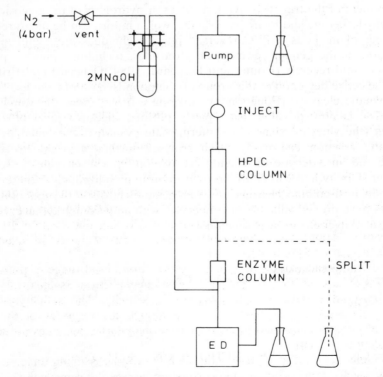

Figure 6. The chromatographic system used for post-column cleavage of cyanogenic glycosides and amperometric detection of cyanide at a silver electrode. The working potential is 0.0 V relative to the Ag/AgCl/KCl reference electrode.

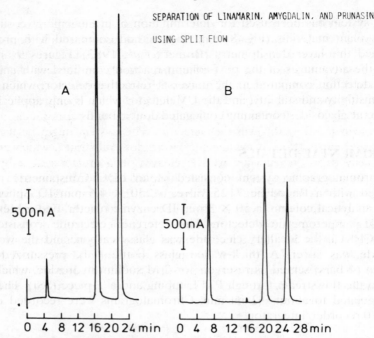

Figure 7. Separation of linamarin (VIII), amygdalin (I, X = H), and prunasin (I, X = H) (cf. Fig. 5) eluted in named order. Figure 7B shows the enhanced signal from linamarin using a split, which is not present in Fig. 7A.

Using purified enzymes (see Table 1) complete conversion of the glycosides occurs at 37° C and a flow of 1 ml/min. Under these conditions, all cyanogenic glycosides have the same response factor, because the final measurement at the silver electrode is based on cyanide, which is the product common for all substrates. A single internal standard (a cyanogenic glycoside) is therefore sufficient. In contrast, quantitative determination of substrates which are not completely cleaved by the enzyme, requires the individual cyanogenic glycosides for determination of the response factors.

Hydrolysis of the intermediate cyanhydrin is ensured by passing strong alkali (2 M NaOH) into the flow stream and allowing the reaction to take place within about 1.25 s, which corresponds to approximately 25 times the half life of cyanohydrins at pH 13.

The working electrode responds to cyanide in a broad range of potential, but 0.0 V relative to the Ag/AgCl/KCl reference electrode, seems to be the best compromise between sensitivity and selectivity. The stability of the working electrode is excellent, and even though the reservoir of the salt bridge in the reference electrode becomes strongly basic, this does not affect the detection.

A simple way to test whether complete hydrolysis is taking place in the enzyme reactor is to compare two chromatograms of the same sample using different flow rates. A change in peak height will indicate, that the conversion is not complete and the necessary precautions have to be taken.

The system has been used for determination of cyanogenic glycosides in various plant materials (Fig. 8), and the method compared with previous published thin-layer densitometry (Brimer *et al.*, 1983). Figures 9 and 10 shows the advantages of the post-column reaction combined with amperometric detection compared to the universal refractive detector, which lacks both sensitivity and stability, and the UV detector which is only applicable to cyanogenic glycosides containing conjugated double bonds.

EXPERIMENTAL DETAILS

The chromatographic system consisted of an LDC Constametric pump, equipped with a Rheodyne 7125 valve, a 250 × 4.6 mm ID Spherisorb S5-C8 analytical column, a 50 × 3 mm ID enzyme reactor, and a Metrohm 656/641 amperometric detector. The reference electrode consisted of Ag/AgCl/KCl, the auxillary electrode was glassy carbon, and the working electrode was silver. A thick-walled glass bottle (1 l), pressurized with nitrogen (4 bars) served as reservoir for 2 M sodium hydroxide, which was fed into the flowstream through PTFE tubing and a T-piece (SSI). The flow was regulated to about 0.2 ml/min. Chromatograms were recorded on an SP-4270 recorder/integrator.

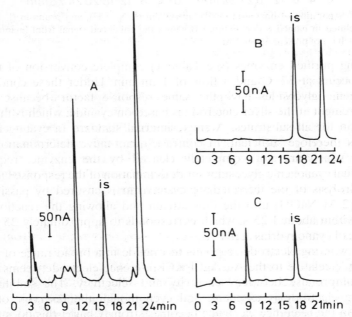

Figure 8. Amperometric detection. Chromatograms of plant extracts from (A) *S. nigra* (d.f. = 50); (B) *P. lauracerasus L*; and (C) *T. baccata* (d.f. = 50). Amygdalin is added as internal standard (d.f. = Dilution factor). *S. nigra* contains holocalin (I, X = *o*-OH, glucosyl), $V_r = 12$ min, prunasin (I, X = H, glucosyl), $V_r = 21$ min, and sambunigrin (II, X = H, glucosyl), $V_r = 23$ min. *P. lauracerasus* contains prunasin and *T. baccata* contains taxiphyllin (I, X = *p*-OH, glucosyl).

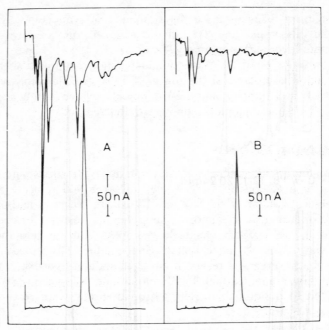

Figure 9. Series dual detection of extracts from leaves of *P. lauracerasus L.* (A) Crude extract diluted eleven times with mobile phase and (B) concentrated extract diluted 500 times with mobile phase. The upper trace is obtained with UV detection at 254 nm and the lower trace is obtained with amperometric detection.

Purification of glycosidases

Crude *H. pomatia* juice (Sigma, USA) (1 ml) was applied to a Sephadex G-200 column (40 × 2.5 cm ID) and eluted with 0.05 M phosphate buffer (pH 5.0). Fractions (10 ml) were collected. Fractions 10—24 were pooled, dialysed overnight and finally lyophilized. In some cases, linamarase (BDH, UK) from *Cassava*, was used for detection of linamarin.

Enzyme reactors were prepared as described in the assay of trimethoprim earlier in this chapter, using the appropriate enzymes.

Sample preparation

Extracts were prepared from leaves of *Sambucus nigra L.*, from leaves of *Prunus lauracerasus L.* and from needles of *Taxus baccata L.*

Plant materials were air-dried for 48 h at 40° C, and 1 part by weight of material was soaked with 20 parts of 70% boiling methanol. The mixture was treated in a Waring blender at 21 000 rpm. The mixture was decanted, the procedure repeated twice, and the extracts were combined and made up to 60 parts by volume. The crude extract was concentrated *in vacuo* at a bath temperature of 40° C to 10 parts by volume, and left overnight for precipitation at 5° C. The supernatant was evaporated *in vacuo* and dissolved in 15% methanol to make 2.5 parts by volume. The concentrated extract was diluted 50—500 times with the mobile phase and 20 µl injected on the HPLC column.

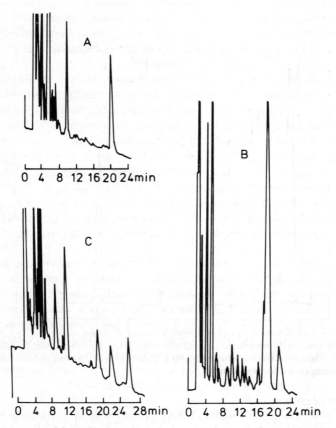

Figure 10. Refractive index detection of the same plant extracts (but undiluted) as in Fig. 8.

DISCUSSION

The use of amperometric detection in the study of trimethoprim and its metabolites has proved to be valuable in the study of the metabolism in newborn pigs, involving analysis of a large number of samples. The experience with the use of the glassy carbon electrode shows that it is only seldom passivated, and if so, it is very easy to polish the electrode surface. Stabilization of the output current can be time consuming if the detector has been shut off or the working potential is changed. The detector should be constantly switched on and if the potential has to be changed, e.g. because a current/voltage correlation of the analyte is going to be made, it is advisable to begin with the highest potential and then step towards lower potentials. Pulse damping devices and/or use of double reciprocating pumps improve the performance greatly.

Post-column reaction detection, generally result in decreased resolution, but usually this draw back is compensated for, by the increased selectivity of detection.

The immobilized enzymes from *Helix pomatia* are remarkably stable, using mobile phases in the pH range 4—7 and methanol contents of less than about 30%. With this in mind, the enzymes are sufficiently active for several months, and the cost of preparation is negligible.

The *E. coli* glucuronidase used in the present work was less stable than the glycosidases from *Helix pomatia*. However, recently (Boppana *et al.*, 1985) a stable preparation has been obtained and used for the determination of Fenoldopam glucuronide conjugates. Fenoldopam (6-chloro-2,3,4,5-tetrahydro-1-4-hydroxyphenyl)-1H-3-benzazepine-7,8-diol) is a new anti-hypertensive agent containing an easily oxidized catechol functional group.

It is believed that further applications using post-column reactions with enzymes and amperometric detections will appear. Preliminary experiments in my laboratory with other types of compounds, i.e. sulphate esters, carboxylic esters of phenols, and thioglycosides (glucosinolates) using suitable enzymes show that the method might be successfully applied to these types of compounds.

The success of the method is very much dependent on the stability of the enzyme. The denaturation of enzymes caused by organic solvents will generally limit the method to the analysis of polar compounds which elutes from the reversed phase HPLC column at low organic eluent concentration. However, one should be aware that enzymes from different sources can show great differences with regard to stability. This should be brought in mind before abandoning the method due to instability of the enzyme. Probably the industrial interest in developing stable and even artificial enzymes will be beneficial for the analytical chemistry in the present context.

ACKNOWLEDGMENT

The author is grateful to Lundbeckfonden for a donation to buy integration/plotter equipment.

REFERENCES

Allison, L. A., Keddington, J., and Shoup, R. E. (1983). Liquid chromatographic behaviour of biological thiols and the corresponding disulphides. *J. Liq. Chomatogr. 6*, 1785—1798.

Boppana, V. K., Fong, K.-L. L., Ziemniak, J. A., and Lynn, R. K. (1985). Use of post-column immobilized β-glucuronidase enzyme reactor for the determination of diastereomeric glucuronides of fenoldopam in plasma and urine by HPLC—ED. *9th International symposium on column liquid chromatography. Edinburgh 1985.*

Brimer, L., Christensen, S. B., Mølgaard, P., and Nartey, F. (1983). Determination of cyanogenic compounds by thin-layer chromatography. 1. A densitometric method for quantification of cyanogenic glycosides, employing enzyme preparations (β-glucuronidase) from *Helix pomatia* and picrate-impregnated ion-exchange sheets. *J. Agr. Food Chem. 31*, 789—793.

Brimer, L., and Dalgaard, L. (1984). Cyanogenic glycosides and cyanohydrins in plant tissues. Qualitative and quantitative determination by enzymatic post-column cleavage and electrochemical detection, after separation by high-performance liquid chromatography. *J. Chromatogr. 303*, 77—88.

Dalgaard, L., and Brimer, L. (1984). Electrochemical detection of cyanogenic glycosides after enzymatic post-column cleavage. *J. Chromatogr. 303*, 67—76.

Dalgaard, L., and Køppen, B., in preparation.

Dalgaard, L., Nordholm, L., and Brimer, L. (1983). Enzymatic post-column cleavage and electrochemical detection of glycosides separated by high-performance liquid chromatography. *J. Chromatogr. 265*, 183—192.

Elchisak, M. A. (1983). Determination of conjugated compounds by liquid electrochemical detection using post-column hydrolysis. Application to dopamine sulphate isomers. *J. Chromatogr. 255*, 475.

Elchisak, M. A., and Carlson, J. H. (1982). Assay of free and conjugated catecholamines by high-performance liquid chromatography with electrochemical detection. *J. Chromatogr. 233*, 79.

Eva, C., Hadjiconstantinou, M., Neff, N. H., and Meek, J. L. (1984). Acetylcholine measurement by high-performance liquid chromatography using an enzyme-loaded postcolumn reactor. *Anal. Biochemistry 143*, 320—324.

Jacobs, W. A., and Kissinger, P. T. (1982). Nitroaromatic reagents for determination of amines and amino acids by liquid chromatography/electrochemistry. *J. Liq. Chromatogr. 5*, 881—895.

Kamada, S., Maeda, M., Tsuji, A., Umezawa, Y., and Kurahashi, T. (1982). Separation and determination of bile acids by high-performance liquid chromatography using immobilized 3α-hydroxy-steroid dehydrogenase and an electrochemical detector. *J. Chromatogr. 239*, 773—783.

Kok, W. Th., Brinkman, U. A. Th., and Frei, R. W. (1983). Amperometric detection of amino acids in high-performance liquid chromatography with a copper electrode. *J. Chromatogr. 256*, 17—26.

McClintock, S. A., and Purdy, W. C. (1983). Dual working-electrode electrochemical detector for liquid chromatography. *Anal. Chim. Acta 148*, 127—133.

Meshi, T., and Sato, Y. (1972). Sulfamethoxazole/trimethoprim. Absorption, distribution, excretion, and metabolism of trimethoprim in rat. *Chem. Pharm. Bull. 20*, 2079—90.

Nielsen, P., and Dalgaard, L. (1978). A sulphate metabolite of trimethoprim in goats and pigs. *Xenobiotica 8*, 657—664.

Nordholm, L., and Dalgaard, L. (1982). Assay of trimethoprim in plasma and urine by high-performance liquid chromatography using electrochemical detection. *J. Chromatogr. 233*, 427—431.

Nordholm, L., and Dalgaard, L. (1983). Studies on the preparation of trimethoprim metabolites. *Arch. Pharm. Chem. Sci. Ed. 11*, 1—6.

Nordholm, L., and Dalgaard, L. (1984). Determination of trimethoprim metabolites including conjugates in urine using high-performance liquid chromatography with combined ultraviolet and electrochemical detection. *J. Chromatogr. 305*, 391—399.

Rocklin, R. D., and Pohl, C. A. (1983). Determination of carbohydrates by anion exchange chromatography with pulsed amperometric detection. *J. Liq. Chromatogr. 6*, 1577—1590.

Roston, D. A., and Kissinger, P. T. (1982). Series dual-electrode detector for liquid chromatography/electrochemistry. *Anal. Chem. 54*, 429—434.

Roston, D. A., Shoup, R. E., and Kissinger, P. T. (1982). Liquid chromatography/electrochemistry: thin-layer multiple electrode detection. *Anal. Chem. 54*, 1417A—1434A.

Schwartz, D. E., Vetter, W., and Englert, G. (1970). Trimethoprim metabolites in rat, dog, and man, qualitative and quantitative studies. *20*, 1867—71.

Sigel, C. W., Grace, M. E., and Nichol, C. A. (1973). Metabolism of trimethoprim in man and measurement of a new metabolite. New fluorescent assay. *128*, S580—S583.

Watanabe, N., and Inoue, M. (1983). Amperometric detection of reducing carbohydrates in liquid chromatography. *Anal. Chem. 55*, 1016—1019.

Weetall, H. H., and Filbert, A. M. (1974). *Methods Enzymol. 34*, 59.

Progress in HPLC, Vol. 2, pp. 235–260
Parvez *et al.* (Eds)
© 1987 VNU Science Press

Analysis of estrogens by high-performance liquid chromatotraphy with electrochemical detection

KAZUTAKE SHIMADA and TOSHIO NAMBARA

Pharmaceutical Institute, Tohoku University, Aobayama, Sendai 980, Japan

INTRODUCTION

Determination of estrogens in peripheral blood is remarkably important for understanding the physiological significance of female hormones. Estimation of estrogen level in biological fluids is also useful for monitoring fetal well-being. Common methods for quantification of estrogens involve colorimetry (Kober, 1931), fluorometry (Ittrich,1960; Manner *et al.*, 1963; Slaunwhite and Sandberg, 1959) and gas—liquid chromatography (GC) (Fishman and Brown, 1962; Luukkainen *et al.*, 1961; Wotiz and Martin, 1962). These methods, however, are not necessarily satisfactory with respect to sensitivity and reliability because of their complicated and tedious procedure. Numerous approaches have been advanced to overcome the difficulties and in consequence, excellent methods have recently been developed. One of these is an immunological method such as radioimmunoassay (Crosignani *et al.*, 1975) and non-isotopic immunoassay (Riad-Fahmy *et al.*, 1981). Various chromatographic techniques, GC, GC—mass spectrometry (MS) and high-performance liquid chromatography (HPLC), have also been used for the analysis of estrogens. Among these methods, HPLC developed by Kirkland (1969), appears to be of great use for steroid hormones and their metabolites in biological samples. Numerous efforts have been directed to development of the specific and sensitive detection system in HPLC. The electrochemical detector (EC), developed by Kissinger (1976), appears to be most promising for the trace analysis of electrochemically active compounds. A sensitive and selective EC is widely used for phenolic compounds, in particular catecholamines in biological materials (Fehshauge *et al.*, 1974; Hansson *et al.*, 1979; Hashimoto and Maruyama, 1978; Kissinger *et al.*, 1974; Pachla and Kissinger, 1976). The use of EC for the detection of phenolic steroids, estrogens, has also been undertaken. In this chapter applications of HPLC/ EC to the analysis of classical and catechol estrogens in biological fluids are described.

ANALYSIS OF CLASSICAL ESTROGENS BY HPLC/EC

EC is *c.* 100 times more sensitive than an ultraviolet (UV) detector and is highly specific for classical estrogens, i.e. estrone, estradiol and estriol (Fig. 1) (Shimada *et al.*, 1979). The estrogens are electrochemically oxidized on the surface of a glassy carbon electrode. A high potential ($> +1.1$ V) must be applied to reach the limiting current plateau. At the high applied potential the mobile phase is oxidized, resulting in considerably high background currents. The optimum operating potential for any substrate is a trade-off between minimizing the background current and maximizing the limiting current. For the phenolic group in estrogens, optimum applied potential seems to be from $+0.9$ to $+1.0$ V (Fig. 2). HPLC with fluorescence or UV detection has been widely used for estrogens. In HPLC with UV detection, a liquid—liquid extraction technique is usually employed for the separation of estrogen fraction. Because of the insufficient sensitivity a urine sample needs a large volume of extracting solvent. Hiroshima *et al.* (1980) and Inderstrodt (1981) reported the use of HPLC/EC for the assay of urinary estriol after prior hydrolysis of the conjugates. This method is more sensitive (detection limit, 10 ng) and selective for urinary estriol than the known methods. Sagara *et al.* (1981) also developed a method for simultaneous quantification of unconjugated estrone, estradiol, estriol and estetrol in serum and amniotic fluid by HPLC/EC. The ethereal extract was purified by chromatography on a microcolumn of Sephadex LH-20 using benzene:methanol (85:15) as an eluent. The desired steroid fraction was collected and then applied to HPLC/EC where the detection limit was 50 pg.

These methods, however, have an inevitable disadvantage, that is the loss of information about the conjugated form. In this respect, we developed a new method for the simultaneous determination of estriol 16- and 17-glucuronides in biological fluids by HPLC/EC (Shimada *et al.*, 1982a, b). The separation of estrogen glucuronides by HPLC has previously been reported by several groups. In spite of considerable efforts the resolution of estriol 16- and 17-glucuronide still remains unsatisfactory (Hermansson, 1980; Musey *et al.*, 1978; Van der Wal and Huber, 1974). Among the typical columns suitable for the polar compounds, TSKgel ODS-120A (LS-410 ODS-SIL) was chosen for the separartion of these two compounds.

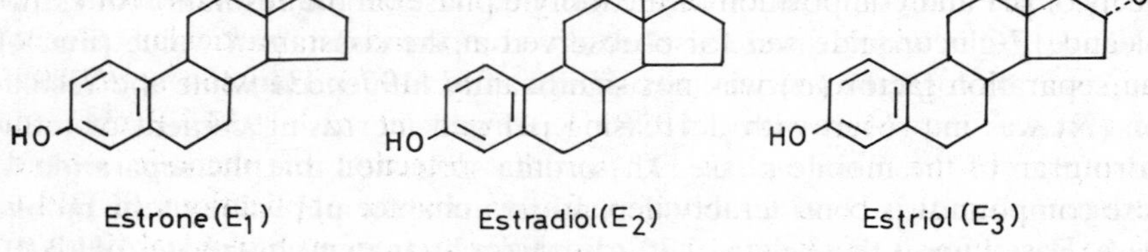

Estrone (E_1)   Estradiol(E_2)   Estriol (E_3)

Figure 1. Structures of classical estrogens.

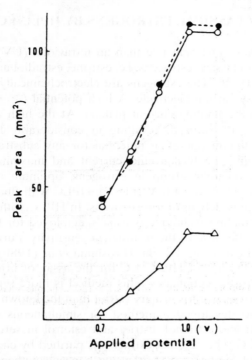

Figure 2. Relationship between applied potential and sensitivity in electrochemical detection. From Shimada *et al.* (1979). $\Delta - E_3$ (30.5 ng); $\bullet - 2\text{-OHE}_1$ (31 ng); $\circ - 2\text{-OMeE}_1$ (30 ng). Conditions: column, μBondapak C_{18}; mobile phase, methanol/0.5% $NH_4H_2PO_4$ (pH 3.0) (6:5), 1 ml/min; detection, Yanagimoto VMD-101 ECD, 16 nA full scale. OMe: methyl ether.

The effect of pH of a mobile phase on the capacity ratio (k') was investigated with the 0.7% disodium hydrogen phosphate/tetrahydrofuran system. The k' values of typical estrogen glucuronides relative to estriol 17-glucuronide were plotted against pH of the mobile phase (Fig. 3). A close similarity in the chromatographic behaviour was observed between estriol 16-glucuronide and 17-glucuronide in the pH range 5.0—7.0. The relative k' value of estriol 16-glucuronide increases with the decreasing pH value from 5.0 to 3.0. There was also seen a similar relation between 16-epiestriol 16-glucuronide and 17-glucuronide. This phenomenon can be explained in terms of dissociation of steroid conjugates having a glucuronic acid moiety (pK 3.20) in acidic medium where undissociated species are dominant. The effects of pH and composition of the mobile phase on the resolution of estriol 16- and 17-glucuronide were also observed at the constant k' value (Fig. 4). The separation factor (α) was not significantly influenced while the resolution (R) was improved with decreasing pH and increasing content of tetrahydrofuran of the mobile phase. These data suggested that the separation of these compounds is considerably dependent upon the pH value of the mobile phase. Based upon these data, 0.7% disodium hydrogen phosphate (pH 3.0)/tetrahydrofuran (6:1) was chosen as a suitable mobile phase. A synthetic

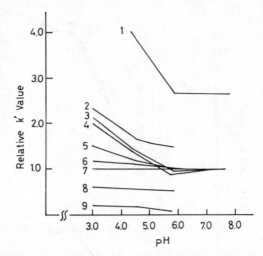

Figure 3. Effect of pH on relative k' value. From Shimada *et al.* (1982b). 1. E_2 17-G; 2. 16-epiE$_3$ 16-G; 3. E_2 3-G; 4. E_1 3-G; 5. 16-epiE$_3$ 17-G; 6. E_3 16-G; 7. E_3 17-G; 8, 16-epiE$_3$ 3-G; 9. E_3 3-G. Conditions: column, TSKgel ODS-120A; mobile phase, 0.7% Na$_2$HPO$_4$/ tetrahydrofuran (6:1), 2 ml/min; detection, UV 280 nm. G: glucuronide.

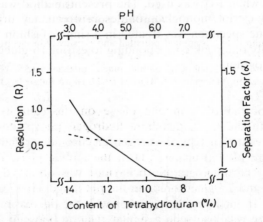

Figure 4. Effect of pH and composition of mobile phase on the separation of estriol 16- and 17-glucuronides. From Shimada *et al.* (1982b). ——: Resolution; - - -: separation factor. Conditions: mobile phase, 0.7% Na$_2$HPO$_4$/tetrahydrofuran. Other conditions were the same as in Fig. 3.

mixture of 3-, 16- and 17-glucuronides of both estriol and 16-epiestriol was efficiently separated as shown in Fig. 5.

Next effort was directed to the determination of estriol 16- and 17-glucuronides in rat bile by using a UV detector. Numerous interfering peaks appeared on the chromatogram even when several clean-up steps were performed (Fig. 6b). In order to overcome this problem the use of EC was undertaken. One-tenth or one-twentieth milliliter of rat bile was subjected to chromatography on Amberlite XAD-2 resin, followed by ion-exchange

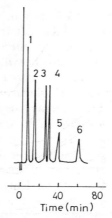

Figure 5. Separation of a synthetic mixture of estrogen monoglucuronides. From Shimada *et al.* (1982b). 1. E_3 3-G; 2. 16-epiE$_3$ 3-G; 3. E_3 17-G; 4. E_3 16-G; 5. 16-epiE$_3$ 17-G; 6. 16-epiE$_3$ 16-G. Conditions: mobile phase, 0.7% Na_2HPO_4 (pH 3.0)/tetrahydrofuran (6:1), 1.5 ml/min. Other conditions were the same as in Fig. 3.

chromatography. As illustrated in Fig. 6a, two isomeric estriol monoglucuronides were excreted in rat bile and no interfering peaks were observed on the chromatogram when EC was used. The present method was then applied to the analysis of estriol monoglucuronides in pregnancy urine. A one-tenth milliliter of urine specimen was treated in a similar fashion to that described for rat bile. Only one peak corresponding to estriol 16-glucuronide appeared on the chromatogram.

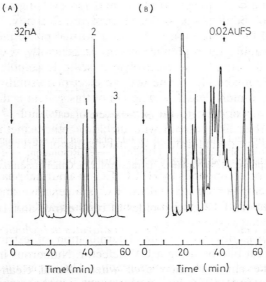

Figure 6. Separation of estriol monoglucuronides in rat bile by HPLC. From Shimada *et al.* (1982a). 1. E_3 17-G; 2. E_3 16-G; 3. 16-epiE$_3$ 17-G (IS). Conditions: flow rate, 1 ml/min; detection, (A) EC, (B) UV detector at 280 nm. Other conditions were the same as in Fig. 5.

It has been demonstrated that EC is much superior in selectivity and sensitivity to the UV detector for the determination of estrogen mono-glucuronides in biological fluids. The excretion of estriol 17-glucuronide in pregnancy urine has been ambiguous since the reports by Carpenter and Kellie (1962) and Hashimoto and Neeman (1963). The present method revealed that both estriol 16- and 17-glucuronide are excreted in rat bile while only the former is present in human pregnancy urine. The multiplicity of glucuronyltransferase which catalyzes the formation of estrogen glucuro-nides, appears to be of interest.

These methods also have disadvantages in that only conjugates having a phenolic group can be detected, but A-ring conjugates such as estrone glucuronide are not responsive to EC. Dalgaard and Nordholm (1983) reported enzymic hydrolysis followed by the electrochemical detection of deconjugated estrogen. β-Glucuronidase from beef liver was immobilized on Agarose beads and used as a post-column reactor in HPLC of phenolic glucuronides. Electrochemical oxidation on a glassy carbon electrode was used for the detection of the phenolic products formed by the enzymic reaction. It has proved to be useful for the unequivocal characterization and sensitive detection of estrone glucuronide. The detection limit of the glucuro-nide was 13 pmol at a signal-to-noise ratio of 3. This technique is similarly used for the determination of dopamine 3-O-sulfate and 4-O-sulfate where post-column deconjugation with perchloric acid was employed (Elchisak, 1983).

Reversed-phase and ion-exchange chromatography is of disadvantage in that the separation of geometric and steric isomers is difficult. In con-trast, normal phase chromatography is favorable for these purposes. HPLC/ EC has been limited to reversed-phase and ion-exchange chromatography because of the solubility of supporting electrolytes. Recently, Hiroshima et al. (1981) extended the application of EC to a normal phase mode. In adsorp-tion chromatography, hydrophobic solvent is generally used as a mobile phase in which the supporting electrolyte is scarcely soluble. This problem was overcome by mixing the eluate with an electrolyte solution in alcohol at the outlet of the column. The flow diagram of this system is illustrated in Fig. 7. The method was applied to the separation of estradiol-17α and -17β by HPLC on a Zorbax SIL column with dichloromethane:methanol (98.5:1.5) using 0.1 M sodium perchlorate in methanol:ethanol (9:1) as an electrolyte. The newly developed detection system should be widely applicable to the analysis of isomeric compounds by HPLC/EC on a normal phase column.

It should be emphasized that HPLC/EC is a sensitive and selective tool for the determination of classical estrogens in biological fluids.

Procedure for the determination of estriol conjugates in biological fluids (Shimada et al., 1982a)

High-performance liquid chromatography. The apparatus used for this work was a Toyo Soda 803A high-performance liquid chromatograph (Toyo Soda Co., Tokyo) equipped with a Yanagimoto VMD-101 electrochemical

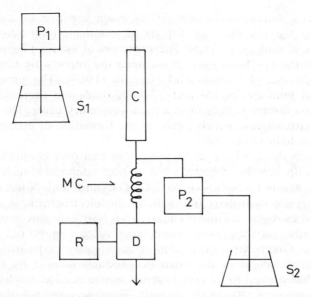

Figure 7. Flow diagram for high-performance liquid chromatography. From Hiroshima *et al.* (1981). C: column, D: detector, R: recorder, P_1, P_2: pump, MC: mixing coil, S_1: mobile phase, S_2: electrolyte solution.

detector (Yanagimoto Co., Kyoto). The potential of the detector was set at +1.0 V vs Ag/AgCl reference electrode. A TSKgel ODS-120A (5 µm) column (30 × 0.4 cm ID) (Toyo Sodo Co.) was employed under ambient conditions. The pH of the mobile phase was adjusted with H_3PO_4.

Bile samples from rats. Male Wistar rats weighing *c.* 200 g were anesthesized with ether, cannulated to the bile duct with polyethylene tube (PE 10) (Clay Adams, Parsippany, NJ) by surgical operation and housed in a Bollman cage for collection of bile. All animals were starved overnight prior to administration of estriol. A suspension of estriol (50 mg) in dimethylsulfoxide (0.1 ml) with saline (0.7 ml) and Tween 80 (0.2 ml) was given orally to each rat and bile was collected every 2 h over a period of 26 h following administration.

Urine samples from human pregnancy. Twenty-four-hour urine samples were collected without preservative from pregnant women (30, 34 and 38 weeks of gestation). Each aliquot was taken and stored at −20°C until analysis.

Procedure. To an aliquot of bile or urine specimen was added 16-epiestriol; 17-glucuronide (5 µg) as an internal standard (IS), and the sample was percolated through a column (5 × 0.6 cm ID) packed with Amberlite XAD-2 resin. After thorough washing with distilled water (5 ml), the conjugate fraction was eluted with 5 ml of methanol (pH 10.0 adjusted with

NH₄OH). Evaporation of the solvent under reduced pressure gave a residue
which in turn, was redissolved in methanol and subjected to a piperidino-
hydroxypropyl Sephadex LH-20 (PHP-LH-20) column (2 × 0.65 cm ID)
(Goto *et al.*, 1978). The column was washed with 90% methanol (5 ml) and
the estrogen conjugate fraction was then eluted with 5 ml of 2.5% ammonium
carbonate in 70% methanol. After evaporation of the solvent under reduced
pressure below 45°C, the residue was dissolved in methanol and an aliquot of
the solution was subjected to HPLC.

ANALYSIS OF CATECHOL ESTROGENS BY HPLC/EC

Aryl hydroxylation of estradiol and estrone yielding catechol estrogens is
recognized as a major metabolic pathway of estrogens in human and other
species (Fig. 8). Several excellent reviews on catechol estrogens have been
published (Ball and Knuppen, 1980; Ball *et al.*, 1982; Fishman, 1983;
Gelbke *et al.*, 1977; MacLusky *et al.*, 1981; Parvizi and Ellendorff, 1980,
1983). Recent studies disclosed that catechol estrogens are not necessarily
final biotransformation products but active metabolites having biological and
endocrine potencies. These compounds may possibly participate in the
mechanism of estrogenic action on the brain and pituitary tissues. By virtue
of their catechol structure, catechol estrogens are capable of interacting with
sites of catecholamine synthesis, metabolism and binding as well as classical
estrogen receptor sites. The biological activities of catechol estrogens are
summarized in Table 1.

In recent years, various methods for the determination of catechol estrogens in biological fluids have been presented. For example, 2-hydroxyestrone
is converted into a stable phenazine derivative (Gelbke and Knuppen, 1973a,
b), which was then subjected to quantitation by means of UV spectrometry.

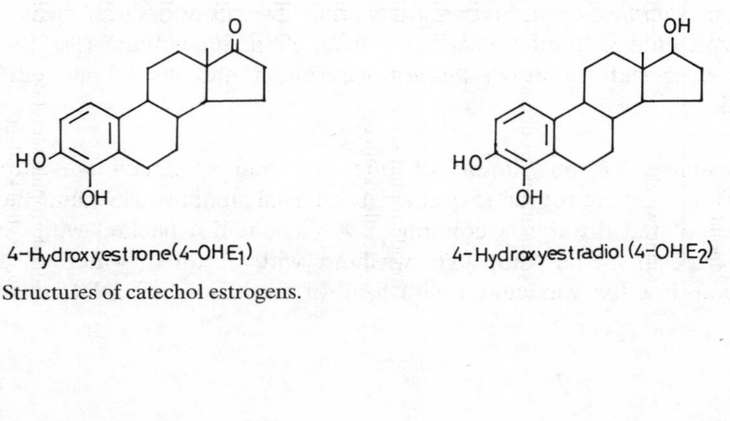

2-Hydroxyestrone(2-OHE₁)

2-Hydroxyestradiol(2-OHE₂)

4-Hydroxyestrone(4-OHE₁)

4-Hydroxyestradiol(4-OHE₂)

Figure 8. Structures of catechol estrogens.

Table 1. Biological activities of catechol estrogens (Parvizi and Ellendorff, 1983)

Endpoint	Steroid	Effect	Potency relative to primary estrogens	References
Uterine weight	$2\text{-}OHE_2$	Increased	Catechol estrogens < primary estrogens	(Martucci and Fishman, 1977, 1979)
	$2\text{-}OHE_3$	Increased	Catechol estrogens < primary estrogens	(Jellinck et al., 1981)
Prostaglandin production	$2\text{-}OHE_2$	PGE_2, PGF_2 increased PGF_1 decreased	$2\text{-}OHE_2$ more effective than E_2	(Kelly and Abel, 1980)
	$2\text{-}OHE_1$, $2\text{-}OHE_2$ $4\text{-}OHE_1$, $4\text{-}OHE_2$	PGE, PGF increased $6\text{-}oxoPGE_1$ decreased		(Kelly and Abel, 1981)
Gonadotrophin secretion +ve feedback effects	$2\text{-}OHE_1$	LH levels increased	Effect opposite of that produced by primary estrogens	(Morishita et al., 1975; Naftolin et al., 1975)
	$2\text{-}OHE_1$	Potentiation of E_2-induced LH surge		(Gethman and Knuppen, 1976)
	$2\text{-}OHE_2$, $4\text{-}OHE_2$	Advancement of PMS induced ovulation	$2\text{-}OHE_2$ ineffective, $4\text{-}OHE_2$ and E_2 equally ineffective	(Franks et al., 1980)
	$2\text{-}OHE_2$	Sensitivity to LHRH increased	$2\text{-}OHE_2 < 4\text{-}OHE_2 < E_2$	(Hsueh et al., 1979; Franks, 1980)
	$2\text{-}OHE_2$	LH levels decreased and then increased FSH no effect		(Miyabo et al., 1984)
	$4\text{-}OHE_2$	LH, FSH levels decreased and then increased		(Miyabo et al., 1984)
Gonadotrophin secretion −ve feedback effect	$2\text{-}OHE_2$	LH levels decreased		(Parvizi and Ellendorf, 1975)
	$2\text{-}OHE_1$	LH levels decreased after $4\text{-}OHE_2$; unaffected by $2\text{-}OHE_1$	$2\text{-}OHE_1$, $2\text{-}OHE_2 \ll 4\text{-}OHE_2$, E_2	(Franks et al., 1981b)

Table 1 (continued)

Endpoint	Steroid	Effect	Potency relative to primary estrogens	References
Prolactin secretion	2-OHE$_1$, 2-OHE$_2$	Prolactin levels reduced	Effect opposite of that of primary estrogens	(Barbieri et al., 1980)
	2-OHE$_1$	Prolactin levels reduced	Effect opposite of that of primary estrogens	(Fishman and Tulchinsky, 1980; Schinfeld et al., 1980)
	2-OHE$_1$	Prolactin levels unchanged		(Franks et al., 1981a; Merriam et al., 1980, 1982)
	2-OHE$_2$	Prolactin levels reduced		(Linton et al., 1981)
	2-OHE$_1$	Prolactin levels reduced		(Katayama and Fishman, 1982)
	2-OHE$_2$	No effect		(Miyabo et al., 1984)
	4-OHE$_2$	Prolactin levels reduced		(Miyabo et al., 1984)
Female sex behavior	2-OHE$_2$	Lordosis behavior potentiated	2-OHE$_2$ ≪ E$_2$	(Luttge and Jasper, 1977; Marrone et al., 1977)
	2-OHE$_2$, 4-OHE$_2$	Lordosis behavior potentiated	2-OHE$_2$ ≪ 4-OHE$_2$, E$_2$	(Jellinck et al., 1981)
Differentiation of the CNS	2-OHE$_1$, 2-OHE$_2$	'Defeminization' of patterns of gonadotrophin secretion	2-OHE$_1$, 2-OHE$_2$ < E$_2$	(Parvizi and Naftolin, 1977)
	4-OHE$_2$	'Defeminization' of patterns of gonadotrophin secretion	2-OHE$_2$ < E$_2$ = 4-OHE$_2$	(Naftolin et al., 1981)

2-Hydroxyestradiol and 2-hydroxyestriol are determined as trimethylsilylated derivatives by GC (Gelbke and Knuppen, 1974; Gelbke *et al.*, 1975). Cohen *et al.* (1978) described the analysis of catechol estrogens in pregnancy urine by mass fragmentography. These methods, however, have inevitable disadvantages, i.e. a need for troublesome and tedious pretreatments involving derivatization and clean-up of the biological sample. The double isotope technique, radioenzymic assay and radioimmunoassay methods require highly specific radioactive compounds (Ball *et al.*, 1978a; Chatoraj *et al.*, 1978; Yoshizawa and Fishman, 1971). Catechol estrogens undergo facile oxidative decomposition, especially in alkaline media. Accordingly, the development of a more sensitive and reliable method for the analysis of catechol estrogens is desirable. The separation and determination of catechol estrogens and related compounds by HPLC/EC has been undertaken (Shimada *et al.*, 1979).

Catechol estrogens are electrochemically oxidized on a glassy carbon electrode at the same applied potential as that for classical estrogens, but the sensitivity is more than three times high as that of classical estrogens (see Fig. 2). The quantitation limit of catechol estrogens was determined to be 1 ng per injection (signal-to-noise = 8 at 4 nA full scale) (Shimada *et al.*, 1979). The value obtained with an UV detector (280 nm) was 500 ng per injection (signal-to-noise = 10 at 0.04 AUFS). Separation of the four isomeric catechol estrogens (2-hydroxyestrone, 4-hydroxyestrone, 2-hydroxyestradiol and 4-hydroxyestradiol) (see Fig. 8) has been attained by HPLC/EC on a μBondapak C_{18} column using acetonitrile/0.5% $NH_4H_2PO_4$ (pH 3.0) (1:2.1) as a mobile phase (Shimada *et al.*, 1981a).

The separation and determination of catechol estrogens in human pregnancy urine was undertaken employing this system. The urine specimen containing labile catechol estrogens was successfully collected and preserved, oxidative decomposition being prevented by the addition of ascorbic acid. The deconjugated estrogen fraction was obtained by hot acid hydrolysis followed by extraction with ethyl acetate. The clean-up procedure for urine catecholamines appeared to be similarly effective for urine catechol estrogens. The catecholic compounds were selectively adsorbed on acid-washed alumina and then recovered by eluting with hydrochloric acid at a rate of > 74%. An aliquot of the eluate was subjected to HPLC/EC according to the procedure previously established. As shown in Fig. 9, the four isomeric catechol estrogens in pregnancy urine were distinctly separated and identified on the chromatogram by comparison with the authentic samples. The structures of these four peaks were unequivocally characterized by means of GC—MS. Fotsis *et al.* (1980) also identified 4-hydroxyestriol in pregnancy urine by GC—MS. The present method was then applied to quantitation of the four catechol estrogens excreted in pregnancy urine. In almost all cases 2-hydroxyestrone showed the highest level of these four (Table 2).

The excretion of catechol estrogens in urine has also been investigated using both chemical and radioimmunoassay methods. The ring-A hydroxylated products of estrone rather than estradiol or estriol ring-A hydroxylation, are of the greatest quantitative significance. The urine levels of

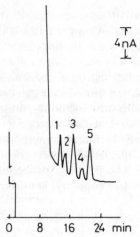

Figure 9. Separation of catechol estrogens in pregnancy urine by HPLC/EC. From Shimada *et al.* (1981a). 1. 2-OHE$_1$ (3.4 ng); 2. 4-OHE$_1$ (<1 ng); 3. 2-OHE$_2$ (1.2 ng); 4. 4-OHE$_2$ (1.0 ng); 5. 4-hydroxy-16-oxoE$_2$ 17-acetate (IS) (5.0 ng). Conditions: column, μBondapak C$_{18}$; mobile phase, acetonitrile/0.5% NH$_4$H$_2$PO$_4$ (pH 3.0) (1:2.1), 1 ml/min; detection, Yanagimoto VMD-101 ECD. Interference due to saturation of the detector with polar substances was overcome by taking off the connector between the column and detector over a period of 8 min after injection of a sample.

Table 2. Urine levels of catechol estrogens in normal pregnancy (Shimada *et al.*, 1981b)

Subject	Determined (ng/ml)			
	2-OHE$_1$	2-OHE$_2$	4-OHE$_1$	4-OHE$_2$
A (16 w)	125	103	51	44
B (16 w)	138	36	n.d.*	21
C (25 w)	128	138	n.d.	58
D (29 w)	276	59	n.d.	104
E (34 w)	145	305	n.d.	95
F (37 w)	382	130	36	39
G (38 w)	203	385	116	n.d.
H (38 w)	264	104	22	136
I (38 w)	749	109	73	72
J (39 w)	109	168	65	n.d.

* n.d. = not detectble.

2-hydroxyestrone and 4-hydroxyestrone vary during the menstrual cycle and become ten times as high in pregnancy (Ball and Knuppen, 1980). Paul and Axelrod (1977a) demonstrated the presence of catechol estrogens (4—30 ng/g tissue) in rat tissues, i.e. hypothalamus, pituitary and liver using the catechol *O*-methyltransferase (COMT) method. On the other hand, Fishman and Martucci (1979), using radioimmunoassay and radioimmunoassay combined with enzymic *O*-methylation, estimated the amount of 2-hydroxy-estrone in rat brain to be less than 6 pg/g tissue. Further study is necessary to resolve the disparity between these findings and establish reliable estimates of tissue catechol estrogens.

The mechanism of catechol estrogen formation has been the subject of investigation. It has been demonstrated that catechol estrogens are formed in the normal and neoplastic tissues including brain and pituitary (Table 3). Owing to extreme lability of catechol estrogens, the measurement of aryl hydroxylase activities presents formidable methodological problems. The aryl hydroxylase is assayed by means of paper chromatography, radioenzymic and radiometric methods. These assay procedures, however, have some disadvantages with respects to simplicity, sensitivity and versatility, and are

Table 3. Tissues in which catechol estrogen formation has been demonstrated (MacLusky *et al.*, 1981)

Tissue	Species	References
Liver	Rat, human fetus[*ox]	(Ball and Knuppen, 1978; Ball *et al.*, 1978b; Barbieri *et al.*, 1978; Fishman *et al.*, 1976; Hoffman *et al.*, 1979a)
Brain	Rat, human fetus[*ox]	(Ball and Knuppen, 1978; Ball *et al.*, 1978b; Barbieri *et al.*, 1978; Fishman *et al.*, 1976; Hoffman *et al.*, 1979a; Paul *et al.*, 1977b)
Pituitary	Rat, human fetus[*ox]	(Ball and Knuppen, 1978; Ball *et al.*, 1978b; Barbieri *et al.*, 1978; Fishman *et al.*, 1976; Hoffman *et al.*, 1979a)
Kidney	Rat[x]	(Ball and Knuppen, 1978; Hoffman *et al.*, 1979a)
Testis	Rat[x]	(Ball and Knuppen, 1978; Hoffman *et al.*, 1979a)
Ovary	Rat[x]	(Hoffman *et al.*, 1979a)
Adrenal	Rat[x]	(Ball and Knuppen, 1978)
Lung	Rat[x]	(Ball and Knuppen, 1978)
Heart	Rat[x]	(Ball and Knuppen, 1978)
Placenta	Baboon, man[o]	(Milewich and Axelrod, 1971; Smith and Axelrod, 1969)
Prostate (hypertrophic and carcinomatous)	Man[o]	(Acevedo and Goldzieher, 1965b)
Mammary tumor	Man[x]	(Hoffman *et al.*, 1979b)
Pheochromoocytoma	Man[o]	(Acevedo and Beering, 1965a)

Assay techniques utilized: [*] tritium release, [o] direct product isolation, [x] COMT coupled product isolation.

applicable solely to estrogen 2-hydroxylase. The properties of 2-hydroxylase have been characterized whereas those of 4-hydroxylase still remain unclear owing to the lack of a satisfactory assay method. A new method has been developed for the simultaneous assay of estrogen 2- and 4-hydroxylase activities (Shimada *et al.*, 1981b). It consists of determination of catechol estrogens enzymatically produced from the phenolic substrate by means of HPLC/EC. Estradiol was incubated with the enzyme preparation in the presence of NADPH. 2-Hydroxyestrone, 2-hydroxyestradiol, 4-hydroxy-estrone and 4-hydroxyestradiol formed were extracted with a boric acid solution and then separated by reversed-phase HPLC on a Hitachi 3053 column. The peaks on a chromatogram were characterized by GC—MS and microchemical reactions involving borohydride reduction of the 17-ketone and alkali decomposition of the catechols. The amounts of the four catechol estrogens could be determined using 2-hydroxy-16-oxoestradiol 17-acetate as internal standard with satisfactory accuracy and precision. The assay method is superior in simplicity, sensitivity and versatility. A typical chro-matogram obtained by the standard assay procedure for 2- and 4-hydroxy-lases in the rat liver homogenate is shown in Fig. 10. These enzymes are localized in the microsomal fraction of the homogenate. Besides two main peaks of 2-hydroxyestrone and 2-hydroxyestradiol, two small ones assigned to 4-hydroxyestrone and 4-hydroxyestradiol were recognized. The chromato-gram indicates enzymic activities (nmol/mg protein/min) as follows: 2-hydroxylase — 0.123 for estrone and 0.745 for estradiol; 4-hydroxylase — 0.002 for estrone and 0.014 for estradiol.

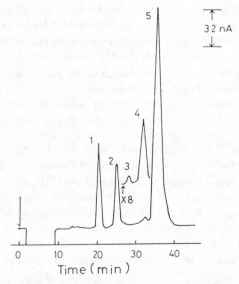

Figure 10. A chromatogram of catechol estrogens formed from estradiol with rat liver homogenate by the assay procedure. From Shimada *et al.* (1981b). 1. 2-hydroxy-16-oxoE$_2$ 17-acetate (IS); 2. 2-OHE$_1$; 3. 4-OHE$_1$; 4. 4-OHE$_2$; 5. 2-OHE$_2$. Interference due to saturation of the detector with polar substances was overcome by the same procedure as described in Fig. 9.

Watanabe and Yoshizawa (1983a, b, c) also used the HPLC/EC for the assay of 2- and 4-hydroxylase activities employing conjugated estradiol as a substrate. When estradiol 17-sulfate was incubated with rat liver microsomes fortified with a NADPH-generating system, a small amount of 4-hydroxy-estradiol 17-sulfate was formed together with the major product, 2-hydroxy-esradiol 17-sulfate. It is to be noted that aryl hydroxylation did take place with retention of the intact conjugated form. The ratio of 2- to 4-hydroxylase activities was approximately 60:1. The availability of a simple and sensitive method for the determination of catechol estrogens forming enzyme activities may serve to clarify the physiological function and metabolic significance of catechol estrogens in living animals.

The metabolism of catechol estrogens proceeds in two major pathways. One is *O*-methylation by COMT resulting in the formation of the mono-methyl ethers (Ball *et al.*, 1972a, b; Breuer *et al.*, 1962) and the other is conjugation. Catechol estrogens and their monomethyl ethers serve as substrates for sulfation and glucuronidation. *In vitro* metabolic conjugation of catechol estrogens, 2-hydroxyestrone and 4-hydroxyestrone has also been investigated by HPLC/EC (Shimada *et al.*, 1982c, 1984b). Separation of guaiacol estrogens, monosulfates and monoglucuronides of catechol estro-gens was attained by HPLC/EC using TSKgel ODS-120A or -120T (LS-410 or -420 ODS-SIL) as a reversed-phase column. Isomeric guaiacol estrogens were distinctly separated when the acetonitrile/0.5% ammonium dihydrogen phosphate system was used as a mobile phase. The pH of the mobile phase is an important factor for the separation of positionally isomeric sulfates and glucuronides (Shimada *et al.*, 1984a). Three mobile phase systems, 5% ammonium dihydrogen phosphate (pH 3.0)/tetrahydrofuran/acetonitrile (16:3:2, 16:2:3 and 40:9:3), were chosen for 2-hydroxyestrogen sulfates, 4-hydroxyestrogen sulfates and 4-hydroxyestrogen glucuronides, respectively. For 2-hydroxyestrogen glucuronides 0.5% ammonium dihydrogen phosphate (pH 3.5)/tetrahydrofuran/acetonitrile (20:3:3) was found to be the most suitable mobile phase. Each group of isomeric monosulfates and mono-glucuronides of 2- and 4-hydroxyestrogens was efficiently resolved with the above solvent systems. The k' values of these catechol estrogen conjugates are listed in Table 4. A typical chromatogram of 2-hydroxyestrogen glucuronides is illustrated in Fig. 11. Guaiacol estrogens and D-ring conjugates showed almost the same sensitivity as that of catechol estrogens. The detection limits (signal-to-noise = 2 at 2 nA full scale) were 1 and 5 ng for A-ring sulfates and glucuronides, respectively. For the determination of guaiacol estrogens formed from catechol estrogens and *S*-adenosyl-L-methionine (SAM), the incubation mixture was deproteinized with heat, extracted with ethyl acetate, and then subjected to HPLC/EC. This procedure is simple and completed within 1 h and therefore, is suitable for the routine assay. The proposed method is also satisfactory for accuracy and precision (coefficients of variation < 5%). 2-Hydroxyestrone was transformed *in vivo* mainly into the 2-methyl ether while *in vitro* into the 2- and 3-methyl ethers in an equal amount. On the other hand, in the case of 4-hydroxyestrone the yielded amount of the 4-methyl ether was 12 times more than that of the 3-methyl ether (Table 5).

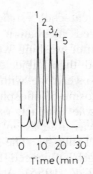

Figure 11. HPLC separation of a synthetic mixture of 2-hydroxyestrogen monoglucuronides. From Shimada *et al.* (1984a). 1. 2-OHE$_2$ 3-G; 2. 2-OHE$_1$ 3-G; 3. 2-OHE$_2$ 17-G; 4. 2-OHE$_1$ 2-G; 5. 2-OHE$_2$ 2-G. Conditions: column, TSKgel ODS-120T; mobile phase, 0.5% NH$_4$H$_2$PO$_4$ (pH 3.5)/tetrahydrofuran/acetonitrile (20:3:3), 1 ml/min; detection, EC at +0.9 V.

Table 4. k' values of catechol estrogen conjugates (Shimada *et al.*, 1984a)

Compound	k'^{*}	Compound	k'^{**}
2-OHE$_2$ 17-S	2.9	2-OHE$_2$ 3-G	0.7
2-OHE$_2$ 3-S	3.7	2-OHE$_1$ 3-G	1.4
2-OHE$_1$ 3-S	5.4	2-OHE$_2$ 17-G	2.2
2-OHE$_1$ 2-S	6.8	2-OHE$_1$ 2-G	2.7
2-OHE$_2$ 2-S	7.3	2-OHE$_2$ 2-G	3.4
4-OHE$_2$ 17-S	1.7	4-OHE$_2$ 3-G	3.0
4-OHE$_2$ 4-S	4.6	4-OHE$_2$ 4-G	3.4
4-OHE$_2$ 3-S	5.1	4-OHE$_2$ 17-G	3.8
4-OHE$_1$ 3-S	7.6	4-OHE$_1$ 3-G	4.6
4-OHE$_1$ 4-S	8.7	4-OHE$_1$ 4-G	6.2

* t_0 = 4.4 min; ** t_0 = 4.8 min.

Table 5. In vitro formation of catechol estrogen conjugates from 2- and 4-hydroxyestrones (Shimada *et al.*, 1984b)

Source	Co-factor		
	PAPS	UDPGA	SAM
Human		2-G	
		4-G	
Rat	2-S	2-G	2-OMe/3-OMe (1:1)
	4-S	4-G	4-OMe/3-OMe (12:1)
Guinea pig	2-S, 3-S	2-G, 3-G	2-OMe/3-OMe (3:1)
	3-S, 4-S	3-G, 4-G	4-OMe/3-OMe (7:1)

Substrate; upper line: 2-OHE$_1$, lower line: 4-OHE$_1$

For the determination of catechol estrogen monosulfates and mono-glucuronides, the deproteinized incubation mixture was applied to an Amberlite XAD-4 column. After washing with water, the desired fraction was eluted with methanol and then subjected to HPLC/EC. Sulfation of 2-hydroxyestrone and 4-hydroxyestrone with the rat liver cytosol fraction fortified with 3'-phosphoadenosine-5'-phosphosulfate (PAPS) provided the 2- and 4-monosulfates as a sole product, rspecitvely (Fig. 12). When the guinea pig liver homogenate was used as an enzyme source, both phenolic groups underwent sulfation (Table 5). The conjugated position of these catechol estrogens was strongly influenced by the pH of the incubation medium (Fig. 13) (Shimada *et al.*, 1985). A similar phenomenon was also observed on the *O*-methylation of catechol estrogens. The variability of conjugation of catechol estrogens (Ball *et al.*, 1972a; Nakagawa *et al.*, 1979) under different experimental conditions may be explained as follows: changes in the pK values of both hydroxy groups of catechol estrogens, alterations of the tertiary structure of the enzyme molecule, and existence of isoenzymes. Further studies including enzyme purification are necessary to arrive at the conclusion.

Glucuronidation of the two catechols with the rat and human liver homo-genates in the presence of uridine-5'-diphosphoglucuronic acid (UDPGA) gave the 2- and 4-glucuronides, respectively. In contrast, incubation with the

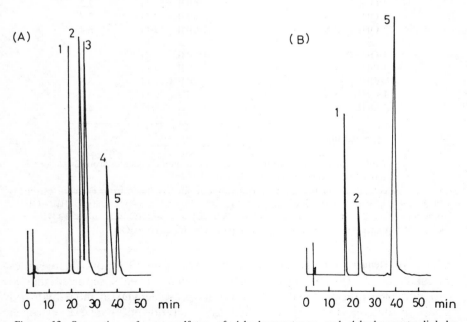

Figure 12. Separation of monosulfates of 4-hydroxyestrone and 4-hydroxyestradiol by HPLC/EC. From Shimada *et al.* (1984b). (A) A synthetic mixture of standard samples, (B) 4-hydroxyestrogen monosulfates formed from 4-hydroxyestrone with rat liver. 1. *p*-dimethyl-aminobenzoic acid (IS); 2. 4-OHE$_2$ 4-S; 3. 4-OHE$_2$ 3-S; 4. 4-OHE$_1$ 3-S; 5. 4-OHE$_1$ 4-S. S = sulfate.

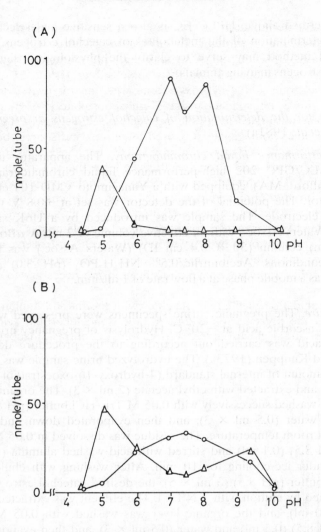

Figure 13. Effect of pH on the formation of isomeric monosulfates from catechol estrogens with guinea pig liver. From Shimada *et al.* (1985). Substrate: (A) 2-OHE$_1$, (B) 4-OHE$_1$. Product: (A) o-o 2-OHE$_1$ 2-S, △-△ 2-OHE$_1$ 3-S; (B) o-o 4-OHE$_1$ 4-S, △-△ 4-OHE$_1$ 3-S. Each point represents the mean value obtained by duplicate experiments.

guinea pig liver homogenate yielded both isomeric monoglucuronides (Shimada *et al.*, 1984b). Yoshizawa and Fishman (1969) reported that *in vivo* glucuronidatin of 2-hydroxyestrone occurs selectively at the 3-hydroxyl group in human urine. In contrast, the excretion of catechol estrogen 2-glucuronide in rat urine and its formation by the guinea pig liver 1500 g supernatant were observed (Yoshizawa *et al.*, 1971). No plausible explanation is at present available for these inconsistent results obtained in the *in vivo* and *in vitro* studies.

As described above, HPLC/EC is also a sensitive and selective method for the determination of the metabolites of catechol estrogens. The newly developed method may serve to clarify the physiological significance of catechol estrogens in living animals.

Procedure for the determination of catechol estrogens in pregnancy urine (Shimada *et al.*, 1981a)

High-performance liquid chromatography. The apparatus used was a Waters ALC/GPC 202 high-performance liquid chromatograph (Waters Assoc., Milford, MA), equipped with a Yanagimoto VMD-101 electrochemical detector. The potential of the detector was set at +0.8 V vs Ag/AgCl reference electrode. The sample was introduced by a U6K sample loop injector (Waters Assoc.) with an effective volume of 2 ml. A μBondapak C_{18} (8—10 μm) column (30 × 0.4 cm ID) (Waters Assoc.) was used under ambient conditions. Acetonitrile/0.5% $NH_4H_2PO_4$ (pH 3.0) (1:2.1) was employed as a mobile phase at a flow rate of 1 ml/min.

Procedure. The pregnancy urine specimens were preserved with a small amount of ascorbic acid at −20° C. Hydrolysis of pregnancy urine (0.2 ml) with hot acid was carried out according to the procedure described by Gelbke and Knuppen (1973a). The hydrolyzed urine sample was mixed with a known amount of internal standard (4-hydroxy-16-oxoestradiol 17-acetate; *c.* 100 ng) and extracted with ethyl acetate (2 ml × 3). The organic layer was combined, washed successively with 0.05 M Tris-HCl buffer (pH 8.5) (0.5 ml × 3) and water (0.5 ml × 3), and then evaporated down under reduced pressure at room temperature. The residue was dissolved in 0.05 M Tris-HCl buffer (pH 8.5) (0.1 ml) and stirred with acid-washed alumina (500 mg) at pH 8.5 under ice-cooling for 10 min. After washing with chilled 0.05 M Tris-HCl buffer (pH 8.5) (3 ml × 5), the desired catechol estrogen fraction was obtained by elution with 1 N HCl. The effluent was extracted with ethyl acetate (10 ml), and the organic layer was washed with 0.05 M Tris-HCl buffer (pH 8.5) (0.5 ml) and water (0.5 ml × 3), and then evaporated down under reduced pressure. The residue was dissolved in methanol (0.1 ml) and an aliquot of the solution was injected into HPLC.

Procedure for the determination of estrogen 2- and 4-hydroxylase activities (Shimada *et al.*, 1981b)

High-performance liquid chromatography. The apparatus used was the same as mentioned above. The stainless steel column (15 × 0.4 cm ID) was packed by the balanced-density technique with Hitachi gel 3053 (Hitachi Co., Tokyo), that is octadecylsilanized silica gel (particle diameter 4—6 μm), and used under ambient condition. A mobile phase 0.5% $NH_4H_2PH_4$ (pH 2.5)/methanol (15:11) was employed at a flow rate of 1 ml/min.

Procedure. The rat liver homogenate (2 mg protein), NADPH (2 mg), estradiol (200 nmole) in 50% methanol (0.1 ml) and sufficient 0.05 M Tris-HCl buffer (pH 7.4) were used to make the total volume 1.2 ml. Incubation was carried out at 37° C for 20 min under aerobic conditions. After addition of 5% HCl (1 ml) to terminate the reaction, the incubation mixture was immediately cooled in an ice-bath and added with internal standard (2-hydroxy-16-oxoestradiol 17-acetate) and ascorbic acid (2 mg). The following procedure was carried out at 0–4° C. The reaction mixture was centrifuged at 3000 *g* for 30 min to give a sediment which in turn was extracted with ethyl acetate (1 ml). The supernatant was saturated with NaCl (2 g) and extracted with ethyl acetate (2 ml × 3). The organic layer was combined, washed with H_2O (2 ml × 2), and extracted with borate buffer (pH 10.7) (3 ml × 3). The aqueous layer was acidified with 15% HCl (1 ml), saturated with NaCl, and then extracted with ethyl acetate (3 ml × 3). The organic layer was washed successively with ascorbic acid buffer (pH 10.2) (1 ml), 5% HCl (1 ml) and H_2O (1 ml × 3), and then evaporated to dryness under a nitrogen gas stream below 50° C. The residue was dissolved in methanol (0.1 ml) and an aliquot of the solution was applied to HPLC.

Characterization of peaks in HPLC. The peaks corresponding to 2-hydroxyestrone and 4-hydroxyestrone on the chromatogram in HPLC were shifted to the 2-hydroxyestradiol and 4-hydroxyestradiol peaks respectively, when the sample was reduced with $NaBH_3CN$. Upon exposure to alkali (pH 11.0) at 60° C for 15 min followed by acidification with 5% HCl, the four catechol estrogens disappeared on the chromatogram.

Procedure for the determination of in vitro conjugation of catechol estrogens (Shimada *et al.,* 1984b)

High-performance liquid chromatography. The apparatus used was a Toyo Soda 803A liquid chromatograph equipped with an EC 8 electrochemical detector (Toyo Soda Co.). The applied potential was set at +0.9 V vs Ag/AgCl reference electrode. A TSKgel ODS-120A column (25 × 0.4 cm ID) was employed under ambient conditions unless otherwise stated. The pH of the mobile phase was adjusted with H_3PO_4. The flow rate was set at 1 ml/min.

Enzyme preparation. Male Wistar rats weighing 200–300 g were used. Fresh liver was homogenized in ice-cold 0.25 M sucrose solution to bring a final concentration to 20% (w/v). The homogenate was centrifuged for 15 min at 1500 *g*, and the supernatant was further centrifuged for 60 min at 105000 *g*. Fresh liver homogenates (20% in 0.25 M sucrose) of Hartley strain male guinea pigs (400–700 g) and human (male 50 years old, female 22 years old: liver ruptura by traffic accident) were also used for enzymic glucuronidation experiments.

Assay procedure for enzymic sulfation. The assay medium (16 ml) contained the substrate (700 nmol in 0.1 ml of methanol), PAPS (700 nmol in 5 ml of water) (Singer, 1979), enzyme preparation (105 000 g supernatant: 20 mg protein), and 0.1 M Tris-HCl buffer (pH 7.5) containing 0.1 mM dithiothreitol. The mixture was incubated in air at 37° C for 120 min. After addition of *p*-dimethylaminobenzoic acid as internal standard, the incubation mixture was deproteinized with heat and centrifuged. The supernatant was separated and percolated through an Amberlite XAD-4 column (15 × 1 cm ID). After washing with water (10 ml), the desired fraction was eluted with methanol (5 ml), evaporated down *in vacuo* and then subjected to HPLC/EC using 0.5% $NH_4H_2PO_4$ (pH 3.0)/tetrahydrofuran/methanol/acetonitrile (35:1:5:10) and 0.5% $NH_4H_2PO_4$ (pH 3.0)/tetrahydrofuran/acetonitrile (16:4:1) for 4- and 2-hydroxyestrogens as a mobile phase, respectively.

Assay procedure for enzymic glucuronidation. The assay medium (5 ml) contained the substrate (350 nmol in 0.1 ml of methanol), UDPGA (1 mol), D-glucaro-1,4-lactone (400 nmol), enzyme preparation (1500 g supernatant: 30 mg protein) and 0.1 M phosphate buffer (pH 7.4). The mixture was incubated in air at 37° C for 120 min. After addition of internal standard (2-hydroxyestradiol 3-glucuronide for 4-hydroxyestrogens; 4-hydroxyestrone 4-glucuronide for 2-hydroxyestrogens), the incubation mixture was deproteinized with heat and centrifuged. The supernatant was separated and percolated through an Amberlite XAD-4 column (15 × 1 cm ID). After washing with water (10 ml), the desired fraction was eluted with methanol (5 ml), evaporated *in vacuo* and then subjected to HPLC/EC on a TSKgel ODS-120T (5 μm) column (25 × 0.4 cm ID) (Toyo Soda Co.) using 0.5% $NH_4H_2PO_4$ (pH 3.0)/tetrahydrofuran/acetonitrile (20:5:1) as a mobile phase.

Assay procedure for enzymic O-methylation. The assay medium (2 ml) contained the substrate (200 nmol in 0.1 ml of methanol), SAM (250 nmol), $MgCl_2$ (5 μmol), enzyme preparation (1500 g supernatant: 10 mg protein) and 0.07 M phosphate buffer (pH 7.6). The mixture was incubated under anaerobic condition at 37° C for 90 min. After addition of internal standard (4-nitroestradiol for 4-hydroxyestrogens; 4-nitroestrone for 2-hydroxyestrogens), the incubation mixture was deproteinized with heat and centrifuged. The supernatant was separated and extracted with ethyl acetate. A portion of the extract was subjected to HPLC/EC using 0.5% $NH_4H_2PO_4$/acetonitrile (3:2) as a mobile phase for the determination of guaiacol estrogens.

Characterization of peaks in HPLC. The amount ratio of guaiacol estrogens formed was not changed with another solvent system, methanol/0.5% $NH_4H_2PO_4$ (3:2). On the chromatogram the 17-ketone was quantitatively transformed into the corresponding 17β-hydroxyl compound when reduced with $NaBH_4$ under ice-cooling. Upon solvolysis or enzymic hydrolysis, catechol estrogen monosulfates and monoglucuronides disappeared on the chromatogram.

Enzymic hydrolysis of catechol estrogen conjugates. The eluate corresponding to the peak on the chromatogram was dissolved in 0.1 M acetate buffer (pH 5.4) and incubated with the acetone powder of snail digestive juice at 37° C overnight.

REFERENCES

Acevedo, H. F., and Beering, S. C. (1965a). The mechanism of 4-^{14}C estradiol-17β by pheochromocytoma tissue. *Steroids 6*, 531—541.

Acevedo, H. F., and Goldzieher, J. W. (1965b). The metabolism of [^{14}C] estrone by hypertrophic and carcinomatous human prostate tissue. *Biochim. Biophys. Acta 97*, 571—578.

Ball, P., and Knuppen, R. (1978). Formation of 2- and 4- hydroxyestrogens by brain, pituitary and liver of the human fetus. *J. Clin. Endocr. Metab. 47*, 732—737.

Ball, P., and Knuppen, R. (1980). Catechol estrogens. *Acta Endocr. Suppl. 232*, 1—127.

Ball, P., Emons, G., and Knuppen, R. (1982). Importance of catecholestrogens in the regulation of the ovarian cycle. *Arch. Gynecol. 231*, 315—320.

Ball, P., Emons, G., Haupt, O., Hoppen, H. O., and Knuppen, R. (1978a). Radioimmunoassay of 2-hydroxyestrone. *Steroids 31*, 249—258.

Ball, P., Haupt, M., and Knuppen, R. (1978b). Comparative studies on the metabolism of oestradiol in the brain, the pituitary and liver of the rat. *Acta Endocr. 87*, 1—11.

Ball, P., Knuppen, R., and Breuer, H. (1972a). Kinetic properties of a soluble catechol *O*-methyltransferase of human liver. *Eur. J. Biochem. 26*, 560—569.

Ball, P., Knuppen, R., Haupt, M., and Breuer, H. (1972b). Interactions between estrogens and catechol amines III. Studies on the methylation of catechol estrogens, catechol amines and other catechols by the catechol-*O*-methyltransferase of human liver. *J. Clin. Endocr. Metab. 34*, 736—746.

Barbieri, R. L., Canick, J. A., and Ryan, K. J. (1978). Estrogen 2-hydroxylase activity in rat tissues. *Steroids 32*, 529—538.

Barbieri, R. L., Todd, R., Morishita, H., Ryan, K. J., and Fishman, J. (1980). Response of serum prolactin to catechol estrogen in the immature rat. *Fertil. Steril. 34*, 391—393.

Breuer, H., Vogel, W., and Knuppen, R. (1962). Enzymatische Methylierung von 2-Hydroxy-östradiol-(17β) durch eine *S*-Adenosylmethionin: Acceptor-*O*-methyltransferase der Rattenleber. *Hopper-Seyler's Z. Physiol. Chem. 327*, 217—224.

Carpenter, J. G. D., and Kellie, A. E. (1962). The structure of urinary oestriol mono-glucuronide. *Biochem. J. 84*, 303—307.

Chatoraj, S. C., Fanous, A. S., Cecchini, D., and Low, E. W. (1978). A radioimmunoassay method for urinary catechol estrogens. *Steroids 31*, 375—391.

Cohen, S. L., Ho, P., Suzuki, Y., and Alspector, F. E. (1978). The preparation of pregnancy urine for an estrogen profile. *Steroids 32*, 279—293.

Crosignani, P. G., Trojsi, L., Attanasio, A., Lombroso Finzi, G. C., and Malvano, R. (1975). Estradiol and estriol direct radioimmunoassay in pregnancy-procedure, validation and normal values. In: *Radioimmunoassay of steroid hormones*, D. Gupta (ed.), Verlag Chemie, Weinheim, pp. 105—116.

Dalgaard, L., and Nordholm, L. (1983). Enzymatic post-column cleavage and electrochemical detection of glycosides separated by high-performance liquid chromatography. *J. Chromatogr. 265*, 183—192.

Elchisak, M. A. (1983). Determination of conjugated compounds by liquid chromatography with electrochemical detection using post-column hydrolysis. *J. Chromatogr. 255*, 475—482.

Fehshauge, C., Kissinger, P. T., Drelling, R., Band, L., Freeman, R., and Adams, R. N. (1974). New high performance liquid chromatographic analysis of brain catecholamine. *Life Sci. 14*, 311—322.

Fishman, J. (1983). Aromatic hydroxylation of estrogens. *Ann. Rev. Physiol. 45*, 61—72.

Fishman, J., and Brown, J. B. (1962). Quantitation of urinary estrogens by gas chromatography. *J. Chromatogr. 8*, 21—24.

Fishman, J., and Martucci, S. (1979). Absence of measurable 2-hydroxyestrone in the rat brain: evidence for rapid turnover. *J. Clin. Endocr. Metab. 49*, 940—942.

Fishman, J., Naftolin, F., Davies, I. J., Ryan, K. J., and Petro, Z. (1976). Catechol estrogen formation by the human fetal brain and pituitary. *J. Clin. Endocr. Metab. 42*, 177—180.

Fishman, J., and Tulchinsky, D. (1980). Suppression of prolactin secretion in normal young women by 2-hydroxyestrone. *Science 210*, 73—74.

Fotsis, T., Järvenpää, P., and Adlercreutz, H. (1980). Identification of 4-hydroxyestriol in pregnancy urine. *J. Clin. Endocr. Metab. 51*, 148—151.

Franks, S. (1980). The role of oestradiol and catechol oestrogens in the control of gonadotrophin and prolactin secretion. In: *Endocrinology of human infertility; New aspects*, P. G. Crosignani (ed.), Academic Press, London, pp. 27—37.

Franks, S., Ball, P., Naftolin, F., and Ruf, K. (1980). Effect of catechol estrogens on induced ovulation in the immature rat. *J. Endocr. 86*, 263—268.

Franks, S., Lightman, S. L., MacLusky, N. J., Naftolin, F., Lynsh, S. S., Butt, W. R., and Jacobs, H. S. (1981a). Failure of 2-hydroxyoestrone to lower prolactin concentrations in hyperprolactinaemic women. *Clin. Endocr. 15*, 385—389.

Franks, S., MacLusky, N. J., Naish, S. J., and Naftolin, F. (1981b). Actions of catechol estrogens on concentrations of serum luteinizing hormone in the adult castrated rat: various effects of 4-hydroxyestradiol and 2-hydroxyestradiol. *J. Endocr. 89*, 289—295.

Gelbke, H. P., and Knuppen, R. (1973a). A chemical method for the determination of 2-hydroxyestrone in human urine. *Acta Endocr. Suppl. 173*, 110.

Gelbke, H. P., and Knuppen, R. (1973b). Synthesis of specific phenazine derivatives of 2-hydroxyestrogens. *Steroids 21*, 689—702.

Gelbke, H. P., and Knuppen, R. (1974). Identification and quantitative determination of 2-hydroxyoestriol in human late-pregnancy urine. *J. Steroid Biochem. 5*, 1—7.

Gelbke, H. P., Ball, P., and Knuppen, R. (1977). 2-Hydroxyestrogens: chemistry, biogenesis, metabolism and physiological significance. In: *Advances in Steroid Biochemistry and Pharmacology, Vol. 6*, M. H. Briggs, and G. A. Christie (eds.), Academic Press, London, pp. 81—154.

Gelbke, H. P., Hoogen, H., and Knuppen, R. (1975). Identification of 2-hydroxyestradiol and the pattern of catechol oestrogens in human pregnancy urine. *J. Steroid Biochem. 6*, 1187—1191.

Gethman, U., and Knuppen, R. (1976). Effect of 2-hydroxyestrone on lutropin (LH) and follitropin (FSH) secretion in the ovariectomized primed rat. *Hoppe-Seyler's Z. Physiol. Chem. 357*, 1011—1013.

Goto, J., Hasegawa, M., Kato, H., and Nambara, T. (1978). A new method for simultaneous determination of bile acids in human bile without hydrolysis.*Clin. Chim. Acta 87*, 141—147.

Hansson, C., Agrup, G., Rorsman, H., Rosengren, A. M., and Rosengren, E. (1979). Chromatographic separation of catecholic amino acids and catecholamines on immobilised phenylboric acid. *J. Chromatogr. 161*, 352—355.

Hashimoto, H., and Maruyama, Y. (1978). Development of an electrochemical detector for high-performance liquid chromatographic assay of brain catecholamines. *J. Chromatogr. 152*, 387—393.

Hashimoto, Y., and Neeman, M. (1963). Isolation and characterization of estriol 16α-glucosiduronic acid from human pregnancy urine. *J. Biol. Chem. 238*, 1273—1286.

Hermansson, J. (1980). Separation of steroid glucuronides by reversed-phase liquid column chromatography. *J. Chromatogr. 194*, 80—84.

Hiroshima, O., Ikenoya, S., Ohmae, M., and Kawabe, K. (1980). Electrochemical detector for high-performance liquid chromatography III. Determination of estriol in human urine during pregnancy. *Chem. Pharm. Bull. 28*, 2512—2514.

Hiroshima, O., Ikenoya, S., Ohmae, M. and Kawabe, K. (1981). Electrochemical detector for high-performance liquid chromatography V. Application to adsorption chromatography. *Chem. Pharm. Bull. 29*, 451—455.

Hoffman, A. R., Paul, S. M., and Axelrod, J. (1979a). Estrogen 2-hydroxylase in the rat: distribution and response to hormonal manipulation. *Biochem. Pharmacol. 29*, 83—87.

Hoffman, A. R., Paul, S. M., and Axelrod, J. (1979b). Catechol estrogens: synthesis and metabolism by human breast tumors *in vitro. Cancer Res. 39*, 4584—4587.

Hsueh, A. J. W., Erickson, G. F., and Yen, S. S. C. (1979). The sensitizing effect of estrogens and catechol estrogen on cultured pituitary cells to luteinizing hormone-releasing hormone: its antagonism by progestins. *Endocrinology 104*, 807—813.

Inderstrodt, J. (1981). Assay for urinary estriol by LCEC. *Current Sep. 3*, 1—2.

Ittrich, G. (1960). Eine Methode zur chemischen Bestimmung von östrogenen Hormonen in Blut, Milch und Colostrum. *Hoppe-Seyler's Z. Physiol. Chem. 320*, 103—110.

Jellinck, P. H., Krey, L., Davis, P. G., Kamel, F., Luine, V., Parsons, B. P., Roy, E. J., and McEwen, B. S. (1981). Central and peripheral actions of estradiol and catechol estrogens administered at low concentrations at constant infusion. *Endocrinology 108*, 1848—1854.

Katayama, S., and Fishman, J. (1982). 2-Hydroxyestrone suppresses and 2-methoxyestrone augments the preovulatory prolactin surge in the cycling rat. *Endocrinology 110*, 1448—1450.

Kelly, R. W., and Abel, M. H. (1980). Catechol estrogens stimulate and direct prostaglandin synthesis. *Prostaglandins 20*, 613—626.

Kelly, R. W., and Abel, M. H. (1981). A comparison of the effects of 4-catechol oestrogens and 2-pyrogallol oestrogens on prostaglandin synthesis by the rat and human uterus. *J. Steroid Biochem. 14*, 787—791.

Kirkland, J. J. (1969). High-speed liquid chromatography with controlled-surface-porosity supports. *J. Chromatogr. Sci. 7*, 7—12.

Kissinger, P. T. (1976). Analytical electrochemistry: Methodology and applications of dynamic techniques. *Anal. Chem. 48*, 17R—23R.

Kissinger, P. T., Felice, L. J., Riggin, R. M., Pachla, L. A., and Wenk, D. C. (1974). Electrochemical detection of selected organic components in the eluate from high-performance liquid chromatography. *Clin. Chem. 20*, 992—997.

Kober, S. (1931). Eine kolorimetrische Bestimmung des Brunsthormons (Menformon). *Biochem. Z. 239*, 209—223.

Linton, E. A., White, N., Tineo, O. L., and Jeffcoate, S. L. (1981). 2-Hydroxyoestradiol inhibits prolactin release from the superfused rat pituitary gland. *J. Endocrin. 90*, 315—322.

Luttge, W. G., and Jasper, T. W. (1977). Studies on the possible role of 2-OH estradiol in the control of sexual behavior in female rats. *Life Sci. 20*, 419—426.

Luukkainen, T., VandenHeuvel, W. J. A., Haahti, E. O. A., and Horning, E. C. (1961). Gas-chromatographic behavior of trimethylsilyl ethers of steroids. *Biochem. Biophys. Acta 52*, 599—601.

MacLusky, N. J., Naftolin, F., Krey, L. C., and Franks, S. (1981). The catechol estrogens. *J. Steroid Biochem. 15*, 111—124.

Manner, F. D., Saffan, E. D., Wiggins, R. A., Thompson, J. D., and Preedy, J. R. K. (1963). Interrelationship of estrogen concentrations in the maternal circulation, fetal circulation and maternal urine in late pregnancy. *J. Clin. Endocrinol. Metab. 23*, 445—458.

Marrone, B. L., Rodriguez-Sierra, J. F., and Feder, H. H. (1977). Role of catechol estrogens in activation of lordosis in female rats and guinea pigs. *Pharmacol. Biochem. Behav. 7*, 13—17.

Martucci, C., and Fishman, J. (1977). Direction of estradiol metabolism as a control of its hormonal action-uterotrophic activity of estradiol metabolites. *Endocrinology 101*, 1709—1715.

Martucci, C., and Fishman, J. (1979). Impact of continuously administered catechol estrogens on uterine growth and LH secretion. *Endocrinology 105*, 1288—1292.

Merriam, G. R., Kono, S., Brandon, D. D., Loriaux, D. L., and Lipsett, M. B. (1982). Does 2-hydroxyestrone suppress prolactin in women? *J. Clin. Endocr. Metab. 54*, 753—756.

Merriam, G. R., Kono, S., Keiser, H. R., and Lipsett, M. B. (1980). Effects of catechol estrogen infusions upon gonadotrophin and prolactin levels in men. *Proc. Endocr. Soc. 62nd Annual Meeting.* Abstract No. 723.

Milewich, L., and Axelrod, L. R. (1971). Metabolism of [4-^{14}C]-testosterone by lypophilized baboon placental microsomes. *Endocrinology 88*, 589—595.

Miyabo, S., Kishida, S., Sugiyama, T., Hashimoto, T., Fujii, Y., and Teranishi, M. (1984). Effects of 2-hydroxyoestradiol and 4-hydroxyoestradiol on gonadotrophin and prolactin secretion in women. *Acta Endocr. 105*, 1—5.

Morishita, H., Adachi, N., Naftolin, F., Ryan, K. J., and Fishman, J. (1975). Elevation of serum gonadotrophins in male rats by catechol estrogens. *Acta Obstet. Gynaecol. Jpn (Engl. Ed.) 23*, 421—426.

Musey, P. I., Collins, D. C., and Preedy, J. R. K. (1978). Separation of estrogen conjugates by high pressure liquid chromatography. *Steroids 31*, 583—592.

Naftolin, F., MacLusky, N. J., Riskalla, M., and Krey, L. C. (1981). Estrogen actions on developing brain: sexual differentiation and nuclear receptors after exogenous estradiol (E$_2$), 4-hydroxyestradiol (4-OHE$_2$) and 2-hydroxyestradiol (2-OHE$_2$). In: *Proceedings of the Society of Gynecologic Investigation, 28th Annual Meeting.* Abstract No. 69.

Naftolin, F., Morishita, H., Davies, I. J., Todd, R., Ryan, K. J., and Fishman, J. (1975). 2-Hydroxyestrone induced rise in serum luteinizing hormone in the immature male rat. *Biochem. Biophys. Res. Commun. 64*, 905—910.

Nakagawa, A., Oh'uchi, R., and Yoshizawa, I. (1979). The effect of hydrogen ion concentration on enzymatic *O*-methylation of catechol estrogens. *J. Pharm. Dyn. 2*, 365—373.

Pachla, L. A., and Kissinger, P. T. (1976). Precolumn for high performance liquid chromatography. *Anal. Chem. 48*, 237.

Parvizi, N., and Ellendorff, F. (1975). 2-Hydroxyestradiol-17β as a possible link in steroid-brain interaction. *Nature 256*, 59—60.

Parvizi, N., and Ellendorff, F. (1980). Recent views on endocrine effects of catecholestrogens. *J. Steroid Biochem. 12*, 331—335.

Parvizi, N., and Ellendorff, F. (1983). Catechol estrogens in the brain: neuroendocrine integration. *J. Steroid Biochem. 19*, 615—618.

Parvizi, N. and Naftolin, F. (1977). Effects of catechol estrogens on sexual differentiation in neonatal female rats. *Psychoneuroendocrinology 2*, 409—411.

Paul, S. M., and Axelrod, J. (1977a). Catechol estrogens: presence in brain and endocrine tissues. *Science 167*, 657—659.

Paul, S. M., Axelrod, J., and Diliberto, E. J. (1977b). Catechol estrogen-forming enzyme of brain: demonstration of a cytochrome p 450 monooxygenase. *Endocrinology 101*, 1604—1610.

Riad-Fahmy, D., Read, G. F., Joyce, B. G., and Walker, R. F. (1981). Steroid immunoassays in endocrinology. In: *Immunoassays for the 80s*, A. Voller, A. Bartlett and D. Bidwell (eds.), MTP Press, Lancaster, pp. 205—261.

Sagara, Y., Okatani, Y., Takeda, Y., and Kambegawa, A. (1981). The determination of unconjugated estrone, estradiol, estriol and estetrol in serum or amniotic fluid by high performance liquid chromatography with an amperometric detector. *Folia endocrinol. Jp 57*, 963—973.

Schinfeld, J. S., Tulchinsky, D., Schiff, I., and Fishman, J. (1980). Suppression of prolactin and gonadotrophin secretion in post-menopausal women by 2-hydroxyestrone. *J. Clin. Endocr. Metab. 50*, 408—410.

Shimada, K., Tanaka, T., and Nambara, T. (1979). Studies on steroids CL. Separation of catechol estrogens by high-performance liquid chromatography with electrochemical detection. *J. Chromatogr. 178*, 350—354.

Shimada, K., Tanaka, T., and Nambara, T. (1981a). Studies on steroids CLXV. Determination of isomeric catechol estrogens in pregnancy urine by high-performance liquid chromatography with electrochemical detection. *J. Chromatogr. 223*, 33—39.

Shimada, K., Kumai, M., Shinkai, H., and Nambara, T. (1981b). The simultaneous determination of the products of estrogen 2- and 4-hydroxylase action; the use of high-performance liquid chromatography with electrochemical detection. *Anal. Biochem. 116*, 287—291.

Shimada, K., Xie, F., and Nambara, T. (1982a). Studies on steroids CLXXIX. Determination of estriol 16- and 17-glucuronide in biological fluids by high-performance liquid chromatography with electrochemical detection. *J. Chromatogr. 232*, 13—18.

Shimada, K., Kaji, M., Xie, F., and Nambara, T. (1982b). Studies on steroids CLXXVIII. Separation of estrogen glucuronides by high-performance liquid chromatography. *J. Liquid Chromatogr. 5*, 1763—1770.

Shimada, K., Yumura, Y., and Nambara, T. (1982c). *In vitro* O-methylation of 4-hydroxy-estrone monosulfates. *J. Pharm. Dyn. 5*, 448—450.

Shimada, K., Xie, F., Terashima, E., and Nambara, T. (1984a). Studies on steroids CLXXXXIV. Separation of catechol estrogen monoglucuronides and monosulfates by high-performance liquid chromatography with electrochemical detection. *J. Liquid Chromatogr. 7*, 925—934.

Shimada, K., Yumura, Y., Terashima, E., and Nambara, T. (1984b). *In vitro* metabolic conjugation of catechol estrogens. *J. Steroid Biochem. 20*, 1163—1167.

Shimada, K., Hara, M., and Nambara, T. (1985). Studies on high-performance liquid chromatography with electrochemical detection. The pH-dependency of enzymic sulfation of catechol estrogens. *Chem. Pharm. Bull. 33*, 685—689.

Singer, S. S. (1979). Enzymatic sulfation of steroids VI. A simple, rapid method for routine enzymatic preparation of 3′-phosphoadenosine-5′-phosphosulfate. *Anal. Biochem. 96*, 34—38.

Slaunwhite, W. R., and Sandberg, A. A. (1959). Phenolic steroids in human subjects III. Estrogens in plasma of pregnant women. *Proc. Soc. Exptl. Biol. Med. 101*, 544—546.

Smith, S. W., and Axelrod, L. R. (1969). Studies on the metabolism of steroid hormones and their precursors by the human placenta at various stages of gestation I. *In vitro* metabolism of 1,3,5-(10)-estratriene-3,17β-diol. *J. Clin. Endocr. Metab. 29*, 85—91.

Van der Wal, Sj., and Huber, J. F. K. (1974). High-pressure liquid chromatography with ion-exchange celluloses and its application to the separation of estrogen glucuronides. *J. Chromatogr. 102*, 353—374.

Watanabe, K., and Yoshizawa, I. (1983a). Evidence of 2-hydroxylation of estradiol-17β 17-glucuronide by male rat liver microsomes. *Steroids 42*, 163—172.

Watanabe, K., and Yoshizawa, I. (1983b). Assay of estradiol 17-sulfate 2-hydroxylase activity by high-performance liquid chromatography. *J. Chromatogr. 277*, 71—77.

Watanabe, K., and Yoshizawa, I. (1983c). Induction of liver microsomal 2-hydroxylation of estradiol 17-sulfate by phenobarbital in male rat. *J. Pharm. Dyn. 6*, 438—440.

Wotiz, H. H., and Martin, H. F. (1962). Studies in steroid metabolism XI. Gas chromatographic determination of estrogens in human pregnancy urine. *Anal. Biochem. 3*, 97—108.

Yoshizawa, I., and Fishman, J. (1969). Conjugation of 2-hydroxyestrone in man. *J. Clin. Endocr. Metab. 29*, 1123—1125.

Yoshizawa, I., and Fishman, J. (1971). Radioimmunoassay of 2-hydroxyestrone in human plasma. *J. Clin. Endocr. Metab. 32*, 3—6.

Yoshizawa, I., Fujimori, K., and Kimura, M. (1971). Glucuronidation of 2-hydroxyestrone with guinea pig liver homogenate. *Chem. Pharm. Bull. 19*, 2431—2432.

Progress in HPLC, Vol. 2, pp. 261—290
Parvez *et al.* (Eds)
© 1987 VNU Science Press

Separation of steroids by HPLC: tentative applications of electrochemical detection for high sensitivity yield

MICHELLE BASTART-MALSOT, HASAN PARVEZ*,
SIMONE PARVEZ and GEORGES CARPENTIER

Unité de Neuropharmacologie, Université de Paris XI, Centre d'Orsay-Bât, 440-91405 Orsay, France

TOSHIHARU NAGATSU

Laboratory of Cell Physiology, Tokyo Institute of Technology, 4259 Nagatsuta, Yokohama, Japan

INTRODUCTION

For several years HPLC has been successfully employed for the analysis of steroids in biological samples as well as in drug preparations. This methodology is a new addition to the techniques already existing such as gas chromatography (GC) and radioimmunoassays (RIA). First of all HPLC has the advantage of requiring little or no sample pretreatment, and secondly it allows simultaneous separation and quantification of several substances by a single injection. This is an easy method of analysis and permits full automation for routine clinical and chemical determinations. HPLC can be applied to the analysis of different classes of steroids such as sexual steroids, glucocorticoids and mineralocorticoids regardless of their natural or synthetic origin. Detailed metabolic pathways of steroids often needed in pharmacokinetic studies, during pregnancy, drug therapy and hormonal deficiency can also be monitored by HPLC. These are some of the useful applications of HPLC which are widely documented in published literature with direct access through computer assisted by bibliography service. The steroid terminology designates an identical chemical structure possessing the same nucleus at four cycles. However, a small chemical modification of the structure leads to modified pattern of analysis relative to functional polarity of the specific member steroid. The most hydrophilic products are those which are eliminated in the form of conjugated derivative in urine such as glucuroconjugates.

---

* To whom correspondence should be addressed.

Published data also provide a great choice of methodology such as normal silca gel, bonded silica gel, ion-exchange resin and ion-pair chromatography.

Since many of the estrogens have high molecular weights and low vapor pressure, are relatively polar and, in some instances, chemically unstable, the general utility of GC techniques would appear to be compromised. Recent advances in column technology have resulted in column efficiencies up to 200 000 theoretical plates by availability of small particle supports (2.5—5 µm). This efficiency now rivals that obtained by GC. Now HPLC is used more frequently in biomedical research because heavy molecules can be analyzed at ambient temperature without derivatization to volatile compounds.

The developments and accomplishments in stationary and mobile phase composition now allow separation of several groups of steroids at very high resolution. The two decades of HPLC history is a direct reflexion of such achievements and demonstrated in the experimental part of this chapter where a comparison of separations with the use of stationary phase of different granulometry is shown. Now the HPLC technology has reached an optimal level as far as selectivity and specificity are concerned since the chemical separation fully discriminates between two classes of compounds or analogs. This is not yet the case with classical RIA which lack specific discrimination in cross-reactivity such as in the separation of enkephalin.

Despite the highlight of optimal separation of steroids by HPLC, the interface of this method of chemical analysis lacked sensitivity barrier due to relatively poor detection limits. Ultraviolet detection is the most applicable media of quantification since it can be directly applied to molecules possessing a chromophore before or after derivatization. However, the limit of detection with the use of classical columns is often lower than the required levels in biological samples. The availability of microbore columns and microflow cell enhances the sensitivity limits to the range of biological concentrations. The combination of fluorescence or radioisotopic methodology with HPLC becomes the most powerful and economical tool for rapid and sensitive analysis of steroids.

Electrochemical detection coupled with HPLC provided a new impulsion to break the sensitivity barrier in biomedical research. The analysis of neurotransmitters, vitamins and several other groups of compounds is a direct proof of this methodological development. The application of electrochemical detection in steroid analysis is very new but it seems possible that steroid molecules possess all the characteristics of oxidative electrochemistry. Though the theme of this book is directly related to HPLC—EC methodology, we shall analyze all the current methods of steroid analysis in use; the data published on steroid quantification by HPLC—EC have been obtained by different authors or by our laboratory.

SEPARATION

Normal-phase chromatography
Tables 1 and 2 contain the key experimental details of a total of ten HPLC

Table 1. Normal-phase chromatography: separations of synthetic steroids

Compounds	Column	Elution (components in parts by volume)	Detection	References
Methylprednisolone cortisol MP hemisuccinate	Zorbax SIL 5—6 μm 250 × 4.6 mm precolumn	Hexane: 26 Dichloromethane: 69 Ethanol: 3.4 Acetic acid: 1 Flow-rate: 2 ml/min	UV 254 nm	Ebling *et al.* (1984)
Prednisone prednisolone 20-OH-metabolites	Zorbax SIL 5—6 μm 250 × 4.6 mm	Heptane: 60.8 Dichloromethane: 29.5 Ethanol: 8.1 Acetic acid: 1.6 Flow-rate: 1.4 ml/min	UV 254 nm	Rocci and Josko (1981)
Methylprednisolone MP acetate	Silica gel 10 μm 100 × 8 mm radial compression system	Dichloromethane: 96.8 Methanol: 2.4 Acetic acid: 0.8 8 min Flow-rate: 1.4 ml/min	UV 254 nm	Alvinerie and Toutain (1984)
Prednisolone	Silica gel 5 μm 250 × 3.2 mm	*n*-Hexane: 63.8 Dichloromethane: 30 Ethanol: 6 Acetic acid: 0.2 Flow-rate: 2 ml/mm	UV 254 nm	Loo *et al.* (1977)
Prednisone prednisolone dexamethasone hydrocortisone	Silica gel 5 μm 250 × 3.2 mm	*n*-Hexane: 65.8 Dichloromethane: 30 Ethanol: 4 Acetic acid: 0.2 Flow-rate: 2.5 ml/mn	UV 254 nm	Loo and Jordan (1977)

separations for steroids using different columns operated in normal-phase chromatographic modes. Table 1 summarizes the experimental details of five separations of synthetic glucocorticoids used in human or veterinary medicine. Several methods have been described to measure selectively the concentration of methylprednisolone (MP) in plasma by use of HPLC (Garg *et al.*, 1977; McGinley *et al.*, 1982). Ebling *et al.* (1984) developed an assay which can simultaneously measure endogenous cortisol in addition to serum MP. Incubation of a sample with carboxylesterase allows the determination of the hemisuccinate ester of MP by the difference of results with and without the use of enzyme. Samples are extracted by methylene chloride, washed first with aqueous sodium hydroxide and then water, and chromatographed. Using a silica gel column and a quaternary solvent system the MP elutes in approximately 12 min and the cortisol in 0 min. Dexamethasone is the internal standard. Sensitivity is greater than 10 ng/ml and the intra-day coefficient of variation is less than 5% for both steroids. The mean assay

Table 2. Normal-plase chromatography: separations of natural steroids

Compounds	Column	Elution (components in parts by volume)	Detection	References
17-oxosteroids	Silica gel 5 μm 250 × 4.6 mm 35° C	Dichloromethane: 400 Ethanol: 1 Water: 2	Fluorophotometer excitation: 365 nm emission: 505 nm	Kawasaki *et al.* (1981)
17-OH corticosterolds	Silica gel 5 μm 250 × 4.6 mm	Dichloromethane: 900 Ethanol: 60 Water: 40	Fluorophotometer excitation: 365 nm emission: 505 nm	Kawasaki *et al.* (1982a)
Pregnenolone 17-OH pregnenolone	Diol 5 μm 250 × 4.6 mm 40° C	1.3 ml/min gradient mode A = *n*-hexane B = *n*-hexane: 85 Isopropanol: 15 from 15% to 100%B within 41 min	UV RIA	Schoeneshoefer *et al.* (1981b)
15 Steroid hormones	Diol 5 μm 250 × 4.6 mm 40° C	1.3 ml/min two gradient modes *HPLC system 1:* A = *n*-hexane B = *n*-hexane: 75 Isopropanol: 25 from 20% to 100%B within 40 min *HPLC system 2:* A = *n*-hexane B = *n*-hexane: 85 Isopropanol: 15 within 30 min	UV 254 nm	Eibs and Schoeneshoefer (1984)
Cortexolone cortisol	Silica gel 5—6 μm 250 × 4.6 mm	*n*-Hexane: 26 Dichloromethane: 69 Ethanol: 34 Acetic acid: 1 2 ml/min	UV	Carson and Jusko (1984)

recovery of MP, cortisol and dexamethasone was about 62% and was independent of concentration. According to the authors the method is selective and clearly separate other synthetic and endogenous glucosteroids.

Alvinerie and Toutain (1984) developed a sensitive and specific procedure for the concomitant determination of MP and MPA (methylprednisolone acetate) in synovial fluid. Flumethasone is the internal standard. Samples are extracted by dichloromethane in the presence of sodium hydroxide. The organic layer is evaporated and reconstituted with eluent before being injected in the column. They use a silica gel column placed in a radial compression system. The eluent is a ternary solvent system. A good separation was obtained in 8 min. The recovery of all steroids from synovial fluid was 80—85%

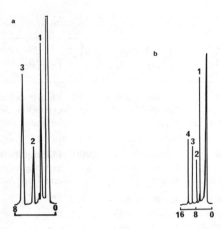

Figure 1. Separation of steroid hormones by normal-phase chromatography. (a) Separation of methylprednisolone acetate (I), flumethasone (2), methylprednisolone (3) extracted from synovial fluid on a 10 μm silica gel column. (b) Separation of methylprednisone (I), dexamethasone (2), cortisol (3) and methylprednisolone (4) extracted from plasma on a 5—6 μm silica gel column.

and coefficients of variation were 5.8% (MP) and 4.9% (MPA), respectively. Rocci and Jusko (1981) described the HPLC technique for the determination of prednisone, prednisolone and their 20β-hydroxylated metabolites in perfusate and urine samples. Modification of the chromatographic conditions permits quantification of their compounds in plasma. The column is the same as in the paper by Ebling *et al.* (1984). The mobile phase consists of a heptane:methylene chloride:ethanol:acetic acid (60.8:29.5:8.1:1.6 by volume) mixture and is pumped through the column at a solvent speed of 3 ml/mn. The mean assay recovery ranges from 54.1% for prednisone to 63.2% for 20β-hydroxyprednisone. The approximate sensitivity limits of the assay range from 4—10 ng/ml. The coefficients of variation are less than 10% for each steroid examined. The separation is achieved in 12 min. The method also detects other glucocorticoids and their metabolites. Through the use of prednisone, prednisolone or their 20β-hydroxylated metabolites as internal standards, 6β-hydroxycortisol determinations are possible. The excretion rate of this last compound has been used to reflect enzyme induction or inhibition in man.

Loo *et al.* (1977) present a method sufficiently sensitive and specific for the determination of plasma samples containing 25 ng/ml of prednisolone. An aliquot of plasma sample is extracted by ether:methylene chloride (60:40 by volume), and washed by 0.1 N aqueous hydrochloric acid. Metabolites and endogenous hydrocortisone do not interfere with prednisolone. The elution is achieved in less than 12 min. The coefficient of variation for prednisolone is 1.5% and the extraction recovery is about 75%. Loo and Jordan (1977) reported the adaptation of the above procedure for the analysis of dexa-methasone, as well as the simultaneous determination of prednisone and

prednisolone in plasma. For this simultaneous determination, the internal
standard was β-methasone. The mobile phase was the same as before with
6% of ethanol instead of 4% (Table 1). Loo *et al.* (1977) noted that the
percentage of ethanol used influences the resolution of the chromatogram.
The optimal percentage of ethanol necessary for a particular column was
determined empirically.

Table 2 summarizes the experimental details of four separations of natural
steroids.

In some clinical conditions such as carcinoma of the adrenal cortex,
gonadal disorders etc., it is desirable to obtain information about the urinary
excretion of individual 17-oxosteroids. Kawasaki *et al.* (1981) developed
a highly sensitive fluorescence — HPLC method for the determination
of 17-oxosteroids in the biological samples. The 17-oxosteroids in urine
samples are extracted with dichloromethane after enzymatic hydrolysis, and
dihydroepiandrosterone sulfate, in serum samples, is solvolyzed with sulfuric
acid in ethyl acetate. The 17-oxosteroids are labeled with dansyl hydrazine.
The eluent is the organic layer separated from the mixture of dichloro-
methane:ethanol:water (400:1:2 by volume) and the column temperature is
35° C. The detection limit of the 17-oxosteroids is about 60 pg from the
working curves. Using 0.1 ml of serum or 1.0 ml of urine as sample in routine
assay, the detection limits are 0.5 or 0.7 µg/dl, respectively. Cortisol is
excreted as glucosiduronates of tetrahydrocortisol and tetrahydrocortisone, so
called 17-hydroxycorticosteroids. Measurement of these steroids has been
used for screening the abnormalities of adrenocortical function. Kawasaki *et
al.* (1982a) described a method useful for the routine analysis of urinary
17-hydroxycorticosteroids using fluorescence HPLC. The 17-hydroxycortico-
steroids were extracted after enzymatic hydrolysis and chromatographied
on a microparticulate silica gel column; the mobile phase was dichloro-
methane:ethanol:water (900:60:40 by volume). Linearity of the fluorescence
intensities (peak heights) of various 17-hydroxycorticosteroids were obtained
between 60 and 20 ng. The assay seems sensitive, precise and accurate.

Pregnenolone and 17-hydroxypregnenolone are precursors of gonadal and
adrenal steroid hormones. Their estimation in serum is of special importance
for the chemical diagnosis of adrenal disorders. Schoeneshoefer *et al.* (1981b)
presented an automatic HPLC procedure with subsequent quantitation of the
purified steroid fractions by radioimmunoassay (RIA). The extraction procedure has been previously described by the authors (Schoeneshoeffer and
Fenner 1981). *n*-Hexane is used as the mobile phase in gradient mode elution
(Schoeneshoefer and Dulce, 1979). The detection limit amounted 0.69 nmol/l
for pregnenolone and 0.21 nmol/l for 17-OH—pregnenolone.

The precise biochemical diagnosis of inborn errors of steroid biosynthesis
requires the estimation of the steroidal substrates and products of all the
enzymes involved in steroid biosynthesis. Eibs and Schoeneshoefer (1984)
describe the evaluation of a routine-suited method, allowing the simultaneous
estimation of fifteen steroid hormones from a single, small-volume sample.
These steroid hormones are listed in Table 3 with their abbreviations.

Table 3. Fifteen steroid hormones analysed by HPLC

Progesterone   P
Androstenedione   AD
Pregnenolone   PL
5-Dihydrotestosterone   DHT
Dihydroepiandrosterone   DHEA
Testosterone   T
11-Deoxycorticosterone   DOC
17-Hydroxyprogesterone   17-OHP
17-Hydroxypregnenolone   17-PL
11-Deoxicortisol   S
18-Hydroxy,11-deoxycorticosterone   18-OH-DOC
Corticosterone   B
Cortisol   F
18-Hydroxycorticosterone   18-OH-B
Aldosterone   Aldo

The procedure involves separation by systems HPLC followed by radioim-munological quantification. Two gradient modes with *n*-hexane—isopropanol are used for elution (Table 2). The polar phase column is heated at 40° C. Steroids were simply extracted from serum samples by diethylether. The HPLC system 1 permits the efficient separation of eleven steroid hormones in a single run in about 35 min. It is evident that the more unpolar steroids (PL, DHT, DHEA, T and DOC) are not sufficiently resolved from each other. The unresolved non-polar steroids are cumulatively collected and subjected to a second system which provides sufficient resolution of these steroids (Table 2). The authors assure that the analytical characteristics (recovery, sensitivity, specificity, precision, accuracy and practicability) are comparable to those obtained in the eleven assay (Schoeneshoefer *et al.*, 1981a). Evidently, the recovery of the non-polar steroids is lower than those of the other steroids.

Carson and Jusko (1984) presented a simultaneous analysis of cortexolone and cortisol in plasma. This analysis has become important to appreciate the integrity of hypothalamic pituitary—adrenal axis in various diseases. Many of the existing methods lacked specificity in separation to distinguish various endogenous steroids. This may be particularly important in the metyrapone test. The extraction procedure is simple. The two steroids and the internal standard are eluted in 10 min with a quaternary solvent. The mean assay recoveries are 59% for cortexolone and 50% for cortisol. The minimum quantitation limit for both cortexolone and cortisol is 5 ng/ml. These results are comparable to those of Ebling *et al.* (1984) who gave 10 ng/ml for cortisol.

The five HPLC methods summarized in Table 1 and 2 appear to be efficient, sensitive, selective and rapid. Extraction procedures are very short. Samples are extracted with organic solvent, washed with aqueous solution and chromatographed. The eluents are ternary or quaternary solvent mixtures. The use of solvent systems with small amounts of water in the mobile phase increases column efficiency. It appears that with many steroids, peak tailing is reduced by addition of a small amount of water when silica gel columns are used. Authors (Kawasaki *et al.*, 1981, 1982a; Schoeneshoefer *et al.*, 1981b, Carson and Jusko, 1984) compare the results of HPLC methods to radio-immunoassay (RIA). Kawasaki *et al.* (1981) obtained good correlation ($r =$ 0.964) between the serum values of DHEA—sulfate determined by the HPLC method proposed and by RIA. Some investigators (Kawasaki *et al.*, 1982a) obtained a coefficient of correlation of 0.932 for tetrahydrocortisol and 0.930 for tetrahydrocortisone. Eibs and Schoeneshoefer (1984) did not compare their results directly with RIA but they conclude that the serum concentrations of the steroids measured by their HPLC method are comparable with the results of previous publications. Carson and Jusko (1984) analyzed the same serum samples by both HPLC and RIA methods. Two assays showed excellent correlation of cortexolone with a regression line slope of 1.05. The RIA method yielded similar results as the HPLC method for cortisol in the absence of metyrapone test. However, the RIA yielded higher values than the HPLC method in the presence of metyrapone and elevated cortexolone concentration. The RIA values were 4—66% higher than the HPLC values.

Reversed-phase chromatography
Reversed-phase chromatography has been very popular for HPLC of steroids to overcome many limitations of earlier TLC and GC methods such as formulation excipient interference, low time analysis, thermal instability or poor resolution. This technique appears to be the most interesting for separation of highly polar, non-volatile and thermally labile compounds. Burgess (1978) and Lafosse *et al.* (1976) discussed the optimization of the conditions for the application of reversed-phase HPLC to the analysis of steroids products. Burgess (1978) described the influence of temperature and flow-rate on column efficiency. For a given column geometry, the optimal flow-rate can be doubled without loss of efficiency by operating at 60° C, hence halving the analysis time. On-column injection techniques are discussed and data presented, showing that for corticosteroids on-column loop injection can be efficient as septum injection using the techniques described. Darney *et al.* (1983) developed a method for the simultaneous measurement of testosterone, androstenedione, 17-hydroxyprogesterone and progesterone in venous effluent from *in vitro* perfused rat testes. The assay uses isocratic HPLC and UV absorbance detection at 240 nm. Two isocratic HPLC systems are described: tetrahydrofuran:methanol:water (16:28:56 by volume) and methanol:acetonitrile:water (9:36:55 by volume). Separation and measurement of all four steroids are resolved in both HPLC systems presented with the same μBondapak C18 column. Separations are achieved in 20 and in 25 min with HPLC system 1 and HPLC system 2, respectively. The elution order and the retention times are different. The resolution of androstenedione and 17-hydroxyprogesterone is better with the HPLC system 1. The main influence of eluent is shown by this example.

The separation of four free 17-ketosteroids (androsterone, epiandrosterone, etiocholanolone and dihydroepiandrosterone), their ester sulfates and glucuronides was described by Lafosse *et al.* (1976). In order to improve the separation, several parameters have been studied. They observed the influence of the temperature on column efficiency. Retention data allow comparison of the hydrophilic behaviour of free and conjugated steroids. Generally, the increasing of lipophilic behaviour is the same with free and conjugated steroids. However, these authors could not separate different types of conjugated, glucuronides and sulfates, and their determination range was at the microgram level.

In the first part of this revue we reported a paper of Kawasaki *et al.* (1981) which described a sensitive HPLC method with fluorescence detection for the determination of urinary 17-oxosteroids after enzymatic hydrolysis and extraction of the liberated steroids. In a second paper (Kawasaki *et al.*, 1982b) the same authors report a highly sensitive fluorescence HPLC method for the direct determination of conjugated 17-oxosteroids in urine and serum samples without hydrolysis. After dilution, urine or serum samples are applied onto a Sep-Pak C18 cartridge and washed with aqueous solution. Conjugated 17-oxosteroids are eluted with methanol. The experimental details of chromatography are summarized in Table 4. Many solvent systems were examined to obtain the complete separation of 17-oxosteroids and conjugates. Sodium acetate concentration appears to be important for the separation of these products. Therefore, 0.01 M sodium acetate in methanol:water:acetic acid (65:35:1 by volume) was used as eluent because glucuronides and sulfates are

Table 4. Reversed-phase chromatography: separations of various steroids from biological fluids

Compounds	Column	Elution (components in parts by volume)	Detection	References
17-oxosteroids and their conjugated	C-18 10 μm 300 × 3.9 mm	0.01 M Sodium acetate Methanol: 65 Water: 35 Acetic acid: 1 1 ml/min	Fluorophotometer excitation: 365 nm emission: 520 nm	Kawasaki *et al.* (1982b)
18-OH corticosterone	C-18 5 μm 250 × 4 mm 37° C precolumn	0.005 M Hydrochloric acid: 35 Methanol: 65 1 ml/min	UV 245 nm RIA	Imaizumi *et al.* (1984)
Estrogens	C-18 10 μm 250 × 4.6 mm	Methanol: 55 0.1% Ammonium carbonate: 45 2 ml/min	UV 280 nm	Dolphin and Pergande (1977)
Glucuronides of: estrone estradiol estriol	C-8 5 μm 150 × 4.5 mm 25° C	Phosphate buffer at pH 6.5 1.25 at 2.5% 1-pentanol	UV 280 nm	Hermansson (1980)

separated completely from the peak of dansyl hydrazine excess, the reactive
part of the fluorescence reaction. 17-oxosteroids are recovered in the range
94.9—105.2%. The reliability of this HPLC method is assessed by comparing
the results with those obtained by the previous method (Kawasaki *et al.*, 1981)
involving enzymatic hydrolysis of urine sample. The regression times obtained
are summarized in Table 5.

Table 5. 17-Oxosteroids in urine: comparison between the direct
method and the method with hydrolysis

Steroid	Regression line (*r*)	*n*
Androsterone glucuronide	0.894	20
Etiocholanolone glucuronide	0.948	20
Androsterone sulfate	0.972	16
Etiocholanolone sulfate	0.987	16
Dehydroepiandrosterone sulfate	0.914	18

The latter authors assure that with a very short assay time, this method is
very sensitive and gives information on the rate of conjugation with glucuronic
or sulfuric acid.

Measurement of plasmatic or urinary 18-hydroxycorticosterone (18-OHB)
together with aldosterone has been proved very useful in the differential
diagnosis of hypertensive disorders. Imaizumi *et al.* (1984) proposed a
method for the separation of 18-OHB from cross-reacting steroids by
reversed-phase HPLC prior to its RIA. The experimental details are summarized in Table 4. After adding sodium hydroxide, plasma samples were
extracted with dichloromethane, washed by water and chromatographed.
After separation by isocratic mode, the 18-OHB-containing fraction is
collected and subjected to RIA. The mean recoveries of 18-OHB before and
after HPLC are 89.7 ± 5.3% and 69.9 ± 5.1%, respectively. The recovery of
this method after HPLC is slightly higher than that (55.3 ± 4.4%) reported by
Schoeneshoefer *et al.* (1981a) who used a normal-phase chromatographic
system. A chromatogram with UV detection of the steroid standards shows
that the separation of the 18-OHB from other steroids is satisfactory. The
18-OHB is eluted at a relatively early retention time, 6.53 min, compared with
that using of normal-phase HPLC system. The chromatographic isolation of
plasma 18-OHB makes its RIA highly appropriate for clinial use.

During pregnancy, large quantities of estrogens, particularly estriol, are
produced and excreted in the urine, mainly in the form of sulfate and
glucuronide metabolites. Trefz *et al.* (1975) have analyzed human plasma for
cortisol and related compounds. Synthetic mixtures of estrogens have been
separated in an isocratic system by Butterfield *et al.* (1973) or by the way of
gradient elution by Majors and Hopper (1974). However, the determination
of estrogens in urine samples was normally performed using other techniques
(TLC—spectrodensitometry; LC—GC—MS). Dolphin and Pergande (1977)
described analysis of estrogens in pregnancy by both normal-phase and

reversed-phase chromatography. The experimental details are summarized in Table 4. Urine samples are hydrolyzed. Two identical extracts are prepared. One extract is analyzed on silica gel with isocratic eluent. A second extract is processed by reversed-phase HPLC. Samples were injected by means of 75 μl loop valve to give improved sensitivity with little degradation of column performances. The results obtained from the two chromatographic systems show good coherence (Fig. 2). The combined use of normal-phase and reversed-phase HPLC has led to reduced uncertainty in peak identification and given more reliable quantitation.

During pregnancy, a decrease in the excretion level of conjugated estriol indicates a possible malfunction of the placenta. These metabolites are measured routinely after hydrolysis and extraction of the liberated aglucone. Hermansson (1980) describes separation of glucuronides of estrone, estradiol and estriol from untreated pregnancy urine by reversed-phase HPLC (Table 4) with LiChosorb RP-8 as solid phase and phosophate buffer with ionic strength of 0.1, pH 6.5, containing 1.25—2.5% of 1-pentanol as mobile phase. Increase of the pentanol content in the mobile phase gives rise to changes in the separation selectivity and, in some case, even changes in retention order. The separation efficiency is good in the system containing 1.25 and 1.9% pentanol. Increase of the pentanol content to 2.5% strongly reduces the separation efficiency. Since estriol—16-glucuronide is clearly separated from

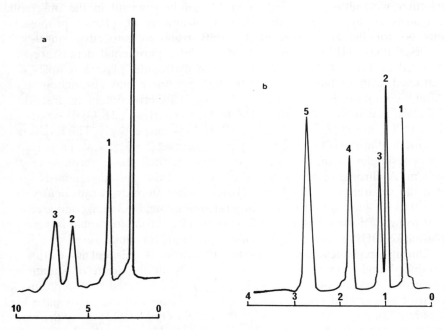

Figure 2. Separation of steroid hormones by reversed-phase chromatography. (a) Separation of estriol (1), estrone (2) and estradiol (3) standards on a 10 μm C18 silica gel column. (b) Rapid separation of estrogen conjugates: E3—3-G (I) E3—3-S (2), E3—17-G (3), E2—3-G (4), E1—3-G (5) on a 5 μm C18 silica gel column.

other compounds, the chromatographic system could be used for its isolation from untreated human urine. Further investigations (Ager and Oliver, 1984) showed that this separation, in terms of elution time, is critically dependent on column temperature and is of limited use for studying pregnancy urine because of low sensitivity.

Some authors have developed multiple-techniques to analyze complex mixtures of natural steroids. In order to compare the functional activity of normal and neoplasic human adrenal or testis cells in monolayer culture, it is necessary to identify and measure a wide range of steroids. RIA is not suitable for this task because it does not recognize unusual or unexpected products that may occur in pathological tissues. O'Hare *et al.* (1976) used reversed-phase HPLC with three related systems of gradient elution on Zorbax—ODS columns to separate, identify and measure the steroids. The levels of most steroids secreted by the cultures were within the capabilities of the techniques previously described. They needed reversed-phase chromatography with gradient elution because of the wide range of polarities of the steroids to separate the samples. Empirical studies showed that the best solution appeared to be a concave exponential gradient of methanol—water to separate adrenal steroids an acetonitrile—water gradient to separate testis steroids, and a dioxane—water gradient to separate polar steroids such as aldosterone. Details of these systems are summarized in Table 6. UV absorbing compounds were detected at 240 or 254 nm and non-UV-absorbing radioactive

Table 6. Reversed-phase chromatography: chromatographic conditions for the separation of adrenal, testis and polar steroids

Compounds	Column	Elution (components in parts by volume)	Detection	References
	C-18 250 × 2.1 mm 45° C	Gradient mode $y = x3$ Flow start: 0.38 ml/min	UV and liquid scintillation counting	O'Hare *et al.* (1976)
Adrenal steroids		*Solvent start*: Methanol: 40 Water: 60 *Solvent finish*: Methanol 100%	UV 240 nm	
Testis steroids		*Solvent start*: Acetonitrile: 32 Water: 68 *Solvent finish*: Acetonitrile: 100%	UV 240 nm	
Polar steroids		*Solvent start*: Dioxane: 20 Water: 80 *Solvent finish*: Dioxane: 100%	UV 254 nm	

steroids by liquid scintillation counting of eluated fractions. Quantification of steroids was routinely based on peak heights. The minimum detectable quantity of adrenal steroid in the least sensitive region of the gradient was 2 ng. These three systems together permit the resolution of at least 43 naturally occurring steroids, plus four synthetic steroids with adrenocortical activity, with overall total elution time of 1 h or less for each system. Authors gave retention data for these steroids.

In view of the fact that specific estrogen conjugates are associated with specific physiological states or toxicities, methods providing a complete estrogen metabolic profile are needed. Slikker *et al.* (1981) described a series of HPLC systems to resolve conjugated and non-conjugated estrogens. Experimental details are summarized in Table 7. Five different reversed-phase columns and one normal-phase column are used. Two gradient elution and three isocratic systems are developed. Twenty-eight estrogens and estrogen conjugates were separated and detected by a UV detector at 280 nm. Radioactive fractions were collected and counted by a liquid scintillation system. An HPLC-purified dose of radiolabelled estradiol-17β is administered via the renal vein to a pregnant rhesus monkey of 130 days gestational age. Maternal blood and urine, fetal blood and tissue samples are collected. The process for tissue extraction has been published previously by Newport *et al.* (1980). The LiChrosorb RP-18 gradient system A2 separated the compounds into six Groups I—VI, from highly polar diconjugates (Group I) to least polar compounds consisting mainly of non-conjugated metabolites (Group VI). The recovery of standards on the reversed-phase system A2 ranged between 87 and 111% for all compounds with the exception of the catechol estrogens 2-hydroxyestrone (74%), 2-hydroxyestradiol (65%) and 2-hydroxyestriol (71%). The sensitivity of the present radioisotopic method is greater than 0.5 pg per 0.5 ml biological fluid. The polar conjugates were separated with the use of more polar solvent mixture, i.e. larger water—methanol ratio. The normal-phase system (E) adequately resolved the 14 non-conjugated and one glycosid conjugates. This normal phase packing has already been used for the separation of ethynylated estrogens. The coefficients of variation ranged from 11 to 17% for the most abundant and least polar metabolites and up to 47% for the more polar metabolites. The initial LiChrosorb RP-18 system (A1 or A2) can resolve 14 metabolites both free and conjugated. The plasma conjugates of estradiol in new born are relatively few in number and may be adequately defined by this single chromatographic system. The main disadvantage of this technique is the use of radioisotopes for quantification of the various estrogen metabolites. This use is generally restricted to animal studies. We believe that such a complex analytical system is not suitable for routine clinical analysis but only for research purposes.

In 1984 Ager and Oliver published a comparative study of several methods to resolve separation of estrogen conjugates in urine and synthetic mixtures by HPLC. They searched for an efficient method to separate and quantitate complex aqueous mixtures in short periods of time, with little or no pretreatment of sample and easy automation for routine use. They summarized a

Table 7. Separation of estrogens and estrogen metabolites by various HPLC systems

Extract	*System B*	
System A1 or A2	2 columns in series	Estriol 3-glucuronide
C-18 5 µm	C-18 (10 µm)	Estradiol 3-sulfate,
250 × 10 mm (A1)	300 × 3.9 mm	17-glucuronide
250 × 9 mm (A2)	Methanol: 45	
50 min convex gradient	0.01 M Ammonium acetate	
Solvent start:	buffer: 55	
Methanol 10	pH 3.97, 1.0 ml/min	
0.01 M Ammonium acetate		
buffer: 90	*System C*	
pH 6.9	Similar to B except:	Estradiol 3,17-disulfate
Solvent finish:	Methanol: 35	Estriol 3-sulfate
Methanol 100%	0.01 M Ammonium acetate	
2 ml/min	buffer: 65	
	pH 7.74	
	System B	
		Estriol 16-glucuronide
		Estriol 17-glucuronide
		Estriol 17-sulfate
	System B	
		Estrone 3-glucuronide
		Estradiol 3-glucuronide
		Estradiol 17-D-glucuronide
	System D	
	C-2 10 µm	Estrone 3-sulfate
	250 × 3.2 mm	Estradiol 3-sulfate
	Methanol: 25	15-Hydroxyestriol
	0.01 M Ammonium acetate	
	buffer: 75	
	pH 7.56, 1.0 ml/min	
	System E	
	Silica gel Diol 10 µm	Estrone
	300 × 4.6 mm	Estradiol 17
	10 mm linear gradient	Estriol
	Solvent start:	Catechols
	Hexane: 100%	non-conjugated
	Solvent finish:	
	Hexane: 80	
	Isopropanol: 20	
	1.5 ml/min	

total of eighteen HPLC separations of one or more of the four following estrogen conjugates: estriol—16-glucuronide (E3—16-G), estriol—3-glucuronide (E3—3-G), estriol—3-sulfate—16-glucuronide (E3—3-S—16-G) and estriol—3-sulfate (E—3-S). The separations are divided into three groups according to the apparent mechanism of separation: anion-exchange, ion-pair and reversed-phase chromatography. Anion-exchange and ion-pair chromatography will be examined later. In the reversed-phase system, Ager and Oliver (1984) included the methods described by Hermansson (1980) and Slikker *et al.* (1981) already discussed. Furthermore, they summarized three other papers. Keravis *et al.* (1977) have separated a synthetic mixture on a μBondapak C18 column (10 μm) with water—methanol (9.34—0.66 by weight) as eluent at 0.89 ml/min. Detection was operated by a refractive index detector and the total elution time was 34 or 54 min.

Shimada *et al.* (1982a, b) reported methods for analysis of estrogen conjugates in biological fluids, using TSK-gel column. The first paper (Shimada *et al.*, 1982a) describes the simultaneous determination of E3—16-G and E3—17-G in rat bile and human pregnancy urine. The authors adopted a very complex treatment procedure. An aliquot of bile or urine sample is adsorbed onto a column of amberlite XAD-2 resin which is washed with water and then eluted with methanol. After evaporation the residue is subjected to a Sephadex LH 20 column. After washing, the estrogen conjugates are eluted with 2.5% ammonium carbonate in 70% methanol. The residue obtained after evaporation is subjected to HPLC. A TSK-gel LS-410 ODS-SIL (5 μm) column (300 × 4 mm) is employed under ambient conditions. The elution is carried out by 0.05 M sodium phosphate (pH 3.0)—tetrahydrofuran (6:1 by volume). Electrochemical detection is used. The predominant estriol conjugate, E3—16-G, is separated after an elution time of 45 min. The two compounds added to rat bile are recovered at a rate of more than 78%. A linear response to each glucuronide is observed in the range 0—10 μg. The detection limit of these compounds is 5 ng per injection. Authors have concluded that this method was satisfactory in accuracy and precision. The complex pretreatments procedure together with its long time analysis are not consistent with a routine use.

In a second paper, Shimada *et al.* (1982b) have applied the same method for the separation of a synthetic mixture of 6 estrogen conjugates with UV detection at 280 nm. The elution time is much longer: 61 min. The resolution of the TSK-gel column with respect to the separation of these isomeric estriol glucuronides progressively decreased as the pH is raised.

Van der Wal and Huber (1978) described a method using LiChrosorb RP-18 column (5 μm) (150 × 3 mm) in isocratic mode with 0.05 M phosphate pH 8.0—acetonitrile (8:2 by volume) as a mobile phase. It is a high-speed separation of three estriol conjugates from a synthetic mixture achieved in approproximately 1 min with a total elution of 4 min.

Keravis and Durand (1980) published a short note where they proposed an analysis of diluted pregnancy urine on a short Spherisorb ODS column (5 μm) by a ternary eluent water:acetonitrile:acetic acid (26.5:6:2 by weight). The eluate is subjected to a one-line Kober fluorimetric procedure which leads to a completely resolved chromatogram containing only three peaks attributed to E3—16-G, E3—3-S and E3—3-G. Because of the brevity of the paper, the exact chromatographic conditions are somewhat ambiguous.

The complete separation of complex natural mixtures of steroid hormones pose many problems due to the wide range of polarities. Their separation using complex analysis has been described (Sliker *et al.*, 1981). Standardization of bonded-phase technology (Engelhardt *et al.*, 1982) has increasingly limited the opportunity to exploit mixed mode chromatography for difficult separations. In an alternative strategy D'Agostino *et al.* (1984) have used computer-aided optimization of mobile phase. The general methods, described by Glajch *et al.* (1981, 1982) to optimize complex separations, have been modified to take into account the behavior of the steroids encountered in biological samples. Separations are carried out isocratically on 150 × 5 mm or 250 × 5 mm ODS—Hypersil C18 column. Steroids are eluted at a solvent flow rate of 1 ml/mn at 45° C and detected with spectrophotometer at 240 nm. Four binary mobile phases compatible with UV detection at 240 nm are examined. They consisted respectively of 35% methanol, 20% dioxane, 20% acetonitrile and 12% tetrahydrofuran in water. These four organic mixtures have approximately equal solvent strength. Under isocratic conditions this solvent strength appeared to be the best practical compromise between resolution and analysis time. Incomplete resolutions are observed in each case. A method based on a seven-step procedure for calculation of the Chromato-graphic Optimization Function (COF) has been used. In this method, peaks cross-over, overlaps, maximum acceptable analysis time factor, are incor-porated for priority separations.

The optimum calculated mobile phase under these conditions was methanol:tetrahydrofuran:water (22.4:4.3:73.3 by volume). When this system is used to separate the steroid mixture the predicted retention order is fulfilled, and observed retention times are closer to predicted values. This isocratic reversed-phase gave the separation of ten polar adrenocortical steroids: isoaldosterone, 18-hydroxy-11-dehydrocorticosterone, aldosterone, 20-dihydrocortisone, 20-dihydrocortisol, 20β-dihydrocortisone, 18-hydroxy-corticosterone, cortisone, 20β-dihydrocortisol and cortisol.

In a paper by Wortman *et al.* (1973) the reversed-phase was used for quantitation of corticosteroids from plasma. Corticosteroids were separated on a SIL XRP column (500 × 3 mm) using methanol:water (40:60 by volume). The steroids were separated in groups: cortisol, cortisone and aldosterone showed almost the same retention time and were separated from other corticosteroids. With the increase of column efficiency and optimization of eluent it became possible to separate these three steroids.

The calculated mobile phase is not necessarily the best obtainable but simply the best in relation to the original choice of solvents and column characteristics. Although flow-programming or gradient elution can be used to facilitate the elution of individual strongly retained compounds, they present certain practical problems due to difficulties in ensuring reproducibility and re-equilibration. The seven-step method described required a total calculation time of approximately 1 h to search all combinations of mobile phase with ten compounds.

Ion-pair analysis of steroids of hormones
Van der Wal and Huber (1978) evaluated the merits of column liquid chromatography on mixed-bed anion-exchange cellulose—diatomite columns

for the separation of estrogens conjugates relative to other phase systems suitable for HPLC.

The choice of adsorption and ion-pair chromatography with polar stationary phases was made in order to avoid the complicated and time-consuming extraction from urine. The dependence of retention, selectivity and efficiency on the type of stationary phase, composition, pH and viscosity of the mobile phase, and the temperature are investigated. The choice of chromatographic system is strongly dependent on the specific composition of the sample. For example, the cellulose and polystyrene anion-exchangers show a large selectivity for the site of conjugation; the use of other phase systems consisting of a polar chemically bonded-phase and an aqueous eluent was found to be suitable for the separation of individual estrogen conjugates.

Hermansson (1978) presented methods for the separation of steroid glucuronides as acids and ion-pairs by HPLC. When the glucuronides are separated as acids, a phosphate buffer is employed as the mobile phase and 1-pentanol as stationary phase. When glucuronides are separated as ion-pairs, 1-pentanol is used as the stationary phase with tetrapropylammonium as counter-ion in the mobile phase. The glucuronides showed good chromatographic behavior in this latter system. The experimental details are summarized in Table 8. The concentration and also the nature of the quaternary ammonium ion affect the capacity ratio, and the possibility to regulate the retention is considerably better when the glucuronides are retained as ion-pairs. The separation efficiency is better. Dixon *et al.* (1979) described a method using Hypersil ODS column with a buffer phosphate as mobile phase and cetyltrimethylammonium bromide as the ion-pair forming surfactant. The two major estriol conjugates, E3—16-G and E3—3-G, have been rapidly separated. This system gives much shorter retention time than that required by

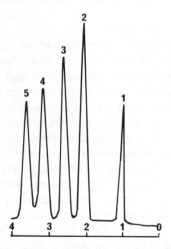

Figure 3. Rapid separation of estrogen conjugates by ion-pair chromatography. HPLC of E3—3S (1), E2—17S (2), E2—3S (3), 17 Eq—3S (4) and E1—3S (5) on a C18 column (5 μm) by 70% methanol + 30% mixture (0.02 M phosphate pH 5.0 and 0.1% CTMABr).

Table 8. Ion-pair chromatography: separation of estrogen conjugates

Compounds	Column	Elution	Detection	References
Estrogen glucuronide synthetic	C-18 coated with n-pentanol 10 μm 150 × 3.2 mm 25° C	0.27 ml/min Sodium phosphate buffer pH 6.4 saturated with n-pentanol and 0.041 M TPA Br Analysis time: 20 min	UV 254 nm	Hermansson (1978)
Estrogen glucuronide pregnancy urine	C-18 5 μm 125 × 45 mm	1 ml/min 0.05 M phosphate CTMA Br—Methanol (5:5 by volume) Analysis time: 13 min	UV 220 nm UV 278 nm ELCD	Dixon et al. (1979)

methods using modified cellulose. The concentration levels of the former conjugates are quantified using a UV detector (220 and 278 nm) and also an electrochemical monitor. In this system, the estriol glucuronides are eluted before the estriol sulfates, estriol ring-A conjugates are eluted before the ring-D conjugated isomers, and the elution order is determined by the nature of the estrogen moiety, and is estriol, estradiol, estrone. The extraction procedure is simplified. The urine specimens are adsorbed on a XAD-2 column and extracted with a 60% methanol aqueous mixture before injection onto a Hypersil ODS column and subsequent ion-pair chromatography. This method is not markedly affected by temperature variations. It is rapid and convenient for routine assay of at least the estriol monoglucuronides.

Anion-exchange chromatography

Ion-exchange chromatography has been utilized for the analysis of steroids for a long time. For example, Henry et al. (1961) showed the separation of dexamethasone disodium phosphate from the other ingredients in a tablet formulation. After extraction, the steroid is chromatographed on anion-exchange on Zipax with 1% amino-substituted polyamide. The authors observed that HPLC is particularly suited to routine tablet analysis because experimental conditions can be adjusted so that other formulation ingredients do not interfere with the determination of the active compound. Ion-exchange chromatography, taking advantage of the ionic character of the estrogen glucuronides, makes it possible to separate the steroids conjugates. More recently, Van der Wal and Huber have published many papers on ion-exchange HPLC and its application to the separation of estrogen glucuronides (Van der Wal and Huber, 1974, 1977a, b). The experimental details from four publications (Musey et al., 1978; Van der Wal and Huber, 1974, 1977a, b) are summarized in Table 9. In the first paper, Van der Wal and Huber (1974) have investigated the selectivity of 6 different types of anion-exchangers of

Table 9. Anion exchange chromatography: separation of estrogen conjugates

Compounds	Column	Elution	Detection	References
Estrogen glucuronides synthetic	ECTEOLA—cellulose 13 μm 250 × 3 mm 25° C	0.125 M Sodium chloride and 0.05 M Sodium acetate pH 5.0 Analysis time: 14 min	UV 220 nm	Van der Wal and Huber (1974)
Estrogen glucuronide synthetic	ECTEOLA—cellulose— Diatomite 5:1 by volume 7 μm 100 × mm	0.025 M perchlorate + 0.01 M phosphate pH 6.8 Analysis time: 28 min	UV 220 nm	Van der Wal and Huber (1977a)
Estrogen glucuronide synthetic	ECTEOLA—cellulose 13 μm 250 × 3 mm 25° C	0.025 M perchlorate + 0.01 M phosphate pH 7.0 Analysis time: 25 min	UV 220 nm	Van der Wal and Huber (1977b)
Estrogen glucuronide synthetic	ECTEOLA—cellulose 11 μm 250 × 3 mm 70° C	0.25 M perchlorate + 0.01 M phosphate pH 8.5 Analysis time: 31 min	UV 220 nm	
Estrogen glucuronide synthetic	ECTEOLA—cellulose 11 μm 250 × 3 mm 70° C	0.025 M perchlorate + 0.01 M phosphate pH 6.8 Analysis time: 75 min	UV 220 nm Kober- fluorimeter	
Estrogen glucuronide synthetic	Strong Anion Exchange Silica 10 μm two columns 250 × 4.6 mm	0.8 ml/min 0.1 M Sodium chloride pH 4.8	UV 254 nm	Musey *et al.* (1978)

estrogen glucuronides. The anion-exchanger materials tested are: polydextran gel Sephadex G-15; dextran-based anion-exchanger DEAE—Sephadex A-25; polyalkylene amino-based anion-exchanger Bio-Rex; aminoethylcellulose Cellex AE and ECTEOLA—cellulose Cellex E. The liquid chromatograph is assembled with a thermostated eluent reservoir and a UV detector. The best system for the separation of all 6 estrogen glucuronides is the first system although the separation of E3—16-G and E3—17-G is difficult and takes 130 min. Excluding E3—17-G, the separation can be performed in 17 min. Finally, the better method uses ECTEOLA—cellulose support and 0.125 M sodium chloride + 0.05 M sodium acetate at pH 5 as eluent. The total elution time is 14 min.

In a second paper, Van der Wal and Huber (1977a) have compared columns containing ion-exchange cellulose of small particle size with mixed-bed columns containing ion-exchange cellulose and diatomite. ECTEOLA—cellulose and cellulose phosphate are used together with and without diatomite as column packings. In column chromatography with ion-exchangers having a cellulose matrix, the flow velocity cannot be increased because of the

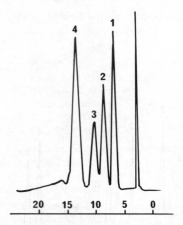

Figure 4. Separation of estrogen monophosphates by anion-exchange chromatography. E3—3P (1), E1—3P (2), E2—3P (2), E2—3P (3) and E2—17P (4) on a ECTEOLA—cellulose (11 μm); 0.25 M perchlorate and 0.01 M phosphate pH 8.5.

non-rigid nature of the support and this leads to a generally long elution time. The mixing of diatomite powder with the cellulose allows the highest elution pressure employed in this column. The mixed-columns containing ion-exchange cellulose and diatomite, with a 5:1 ratio of ion-exchange cellulose and diatomite, were found to be superior in almost every aspect. The experimental details of the separation of a test mixture of seven steroid conjugates are given in Table 9.

In continuation of previous work, Van der Wal and Huber (1977b) studied the selectivity of anion-exchange cellulose systems with respect to thirty estrogen conjugates. Five types of anion-exchange celluloses are used as column packings: polyethyleneiminecellulose Cellex PEI; aminoethylcellulose Cellex AE; triethylaminoethylcellulose Cellex T; guanidinoethylcellulose Cellex GE; ECTEOLA—cellulose cellex E. Polyethyleneiminecellulose gives a large selectivity in the separation of estriol—3-conjugates and estrone—3-conjugates. The influence of the type of counter anion, pH of the eluent and temperature were investigated. The results of the optimization are visualized by three examples with the analysis of the estrogen conjugates in synthetic mixture or in human pregnancy urine. The experimental details of these analysis are summarized in Table 9. The commercially available monophosphates can be separated within 15 min. Half an hour was needed for the separation of seven sulfates. The urine extract, after adsorption on Amberlite XAD-2 and elution with methanol, is too impure for direct UV-detection of the effluent. Authors try to improve the specificity of the detection by using a fluorimetric detector. Such an estrogen pattern can be obtained within 2 h. The detection limit is less than 20 nmol of estriol—16-glucuronides. From cellulose anion-exchangers it is found that the elution order of estrogen glucuronides is determined primarily by the site of conjugation. Ring-A conjugates are eluted before ring-D conjugates. The elution order according to

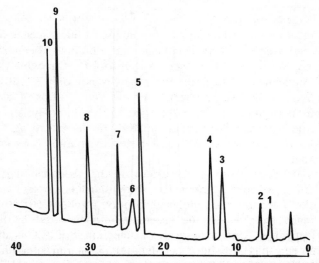

Figure 5. Chromatogram of a mixture of steroid standards by gradient mode chromatography. Separation of progesterone (1), androstenedione (2), testosterone (3), 17-hydroxyprogesterone (4), 11-deoxycortisol (5), 18-hydroxydesoxycorticosterone (6), corticosterone (7), aldosterone (8), cortisol (9) and 18-hydroxycorticosterone (10) on normal-phase column (5 μm); gradient mode from 20 to 100% hexane—isopropanol (75:25 by volume) in hexane within 40 min.

the type of steroid is estriol, estrone and estradiol. Estrogen conjugates with sulfate groups are eluted after the corresponding glucuronide conjugates.

A second type of anion-exchange chromatographic system is abstracted in Table 9. Musey *et al.* (1978) described a single separation of a synthetic mixture of estrogen conjugates by the use of two columns of Partisil SAX in series. Analysis is performed in 23 min. In this case, it is showed that the specific nature of the estrogen moiety is the major determinant of the elution order: estriol, estradiol, estrone. But in all of the anion-exchange systems it is found that glucuronides are eluted before the corresponding sulfates.

DETECTION

The developments in stationary phases up to 3 μm granulometry has led to nearly perfect separation of different biological and chemical substances. However, this accomplishment still lacks a complete takeover of other analytical methods by HPLC. The sensitivity of detection remains the main cause for current utilization of RIA, laser scattering, diode array, biolumine-scence and fluorescence enzyme assays in trace analysis. The present decade has been mainly devoted to the better interface of highly sensitive detection devices with HPLC systems. It is evident that UV still remains the classical means of detection when sensitivity is not required. At present, other modes of detection such as the index of refraction, fluorescence, radioactivity flow-throw monitoring, electrochemical detection and even bioluminescence have been successfully interfaced with HPLC. Fluorescence and amperometry provide a new dimension for sensitive analysis of biomedical samples such as aminoacids, neuropeptides, hormones and transmitters. However, many investigators still remain reluctant for routine use of fluorescence due to post-column derivatization. Electrochemical detection has been successfully developed and constitutes the most powerful tool for trace analysis of drugs and natural substances. The latter technique is not only sensitive but also highly selective for the characterization of different groups of biochemicals. The applications of amperometric detection to steroid analysis is rather new but opens a new horizon for clinical analysis in the future. A summary of different types of detection employed for steroid analysis is provided in the following paragraphs.

The detection mode used is dependent on the chemical structure of the steroid and of its concentration level. Many steroids possess the chromophoric Δ^4-3-keto-group with a molar absorption of about 10000 at 254 nm. Therefore, many steroids can be successfully quantitated by HPLC—UV without derivatization. Universal mode of detection in HPLC is still inappropriate. Monitoring has also been achieved by RIA, refractometer, spectrofluorimeter or electrochemical detector.

UV Detection
A UV detector is the most commonly used detector in HPLC. The glucocorticoids are Δ^4-keto-steroids and possess a strong absorption at 254 nm. They are directly detected with a UV detector at 254 nm. Table 10 summarized detection details given in 8 publications (Alvinerie and Toutain, 1984; Carson and Jusko, 1984; Darnay *et al.*, 1983; Ebling *et al.*, 1984; Loo *et al.*, 1977; Loo and Jordan, 1977; Rocci and Jusko, 1981; Wortman *et al.*, 1973). Response of the UV detector is linear in the range of 0 to 1500 ng/ml. Detection of glucocorticoids by UV is sensitive (commonly less than 10 ng) and precise with a coefficient of variation of less than 6%, except for

Table 10. Characteristics of some UV detections in steroid analysis

Products		Linearity	Precision	Sensitivity	References
Prednisone Prednisolone	254	0—1500 ng/ml	≤10%	4—10 ng/ml	Rocci and Jusko (1981)
Cortisol Methylprednisolone	254	10—1000 ng/ml	≤5%	≤10 ng/ml	Ebling *et al.* (1984)
Methylprednisolone	254	0.02—1 µg/ml	5.8%		Alvinerie and Toutain (1984)
Prednisolone	254		3.9%		Loo *et al.* (1977)
Cortexolone Cortisol		5—1000 ng/ml		5 mg/ml	Carson and Jusko (1984)
Δ^4-3-ketosteroids	240	0—1600 ng/ml	3.3%	10 ng T 25 ng P	Darnay *et al.* (1983)

prednisone and prednisolone (Rocci and Jusko, 1981) (Table 10). These characteristics are reliable with good detection of glucuronides in biological samples. The molar absorption of estrogen glucuronides is about 10^{-3} at 275 nm and higher at 220 nm. Testosterone glucuronide and testosterone sulfate show a maximum absorption at 248 nm. In the majority of publications studying the separation of estrogen conjugates, the detection is monitored at 280 nm (Hermansson, 1980) and 220 nm (Van der Wal and Huber, 1974, 1978). Hermansson (1978) used a UV detector at 254 nm to quantitate estrogen glucuronides.

When steroid compounds do not possess a strong absorbant moiety it becomes difficult to detect them with sufficient sensitivity. Fitzpatrick and Siggia (1973) have recommended the use of benzoesters as derivatives for the detection of hydroxysteroids which do not possess an ultraviolet chromophore. The derivative is formed by reacting the steroid with benzylchloride before chromatography. The authors also showed that p-nitrobenzoate esters are useful. When these techniques were applied to the determination of steroids in urinary extracts, considerable interference from other hydroxylated compounds was observed and purification was required.

Radioisotopic detection
To identify and measure a wide range of steroids it is necessary to use different modes of detection. O'Hare *et al.* (1976) analyzed the steroids secreted by both human adrenal and testis cells. UV-absorbing steroids were detected at 240 nm (254 nm with dioxane gradient). While 5-en-3-one steroids can be measured at 200 nm (more than 150 ng) and estrogens at 280 nm (more than 50 ng), the reduced sensitivity of direct spectophotometric detection renders it, in general, unsuitable for the measurement of these steroids in extracts from culture cells. Radioimmunoassay is used in conjunction with the resolving power of HPLC to measure the non-UV-absorbing steroids. It is necessary to add $3H$-pregnenolone as an exogenous precursor in human adrenocortical cultures and human testis cultures.

Slikker *et al.* (1981) developed a method with UV detection at 280 nm to resolve estradiol-17β and its various metabolites in biological fluids and tissues. Approximately 25 conjugated and non-conjugated standards could be resolved by the sequential use of 6 systems. They administered radiolabelled E2 to a rhesus monkey and samples of urine, blood and tissue were collected, extracted and chromatographed with appropriate estrogen standards. The advantage of radioisotope study is that the sensitivity of the method is determined by the specific activity of the radioisotope in question. In the case of an isotope of estradiol-17β which has a specific activity of over 100 Ci/nmol, based on the ability to quantify 200 dpm of radiolabelled estrogen, the sensitivity of the method is greater than 0.5 pg per 0.5 ml biological fluid. The radioactive steroids were collected and detected by liquid scintillation counting of eluted fraction. This detection method is very sensitive but, generally, the use of radioisotopes is restricted to animal or cell culture studies and can rarely be applied to human experiments.

Imaizumi *et al.* (1984) apply a HPLC method for the separation of 18-hydroxycorticosterone (18-OHB) from cross-reacting steroids prior to its radioimmunoassay. The 18-OHB-containing fraction of plasma extract eluted by HPLC was collected in a tube, neutralized with ammonium hydroxide and evaporated at 37%. The resulting residue was subjected to RIA. The detection limit of the RIA of 18-OHB was 0.0016 pmol. The coefficient of variation ranged up to 15%. The interference of the steroids eluted around 18-OHB was tested by adding 0.28 μmol of aldosterone, cortisol or corticosterone to 0.28 nmol of 18-OHB. No interference of these steroids was observed.

Schoeneshoefer *et al.* (1981b) described a HPLC separation of pregneno-lone and 17-OH pregnenolone in human serum and subsequent quantitation of the purified steroid fractions by RIA. Quantitation with photometric detection is scarcely feasible because these steroids lack UV-absorbing configuration in the molecule. After HPLC, the specific fractions containing pregnenolone and 17-OH pregnenolone were automatically collected. The organic fractions were evaporated and an aliquot is subjected to RIA. The detection limit of the total assay is dictated by the affinity of the antisera used. They amounted to 0.69 nmol/l for pregnenolone. The chromatographic purification eliminates unspecific pregnenolone and 17-OH pregnenolone immunoreactivities arising in the ether extracts of normal serum samples. Eibs and Schoeneshoefer (1984) developed two successive HPLC methods with gradient mode to separate fifteen steroid hormones. The individual organic fractions were collected and quantitated by RIA and computer evaluation (Schoeneshoefer *et al.*, 1981a). These radioimmunological quantitations give good sensitivity but require pretreatment of sample (O'Hare *et al.*, 1976; Garg *et al.*, 1977) or collection of the fractions for RIA processing.

Spectrofluorimetric detection
Van der Wal and Huber (1977b) studied the separation of estrogen con-jugates on an ion-exchanger cellulose by HPLC with UV detection. But, with an extract too impure for direct UV detection of the effluent, instead of, or in addition to, improving the specificity of the extraction, the authors used an off-line Kober—Ittricht analysis of the chromatogram obtained with a seg-mented-flow apparatus. The first stage is a Kober reaction at 127° C for approximately 10 mn, followed by enhancement of fluorescence with acetic acid. An estrogen pattern can be obtained within 2 h with little manual work; the detection limit is less than 20 nmol of estriol—16 glucuronides.

Kawasaki *et al.* developed various methods for determination of 17-oxosteroids (Kawasaki *et al.*, 1981, 1982b) and 17-hydroxycorticosteroids (Kawasaki *et al.*, 1982a). These steroids have no strong UV-absorbing groups in their structures. After enzymatic hydrolysis (Kawasaki *et al.*, 1981) 17-oxosteroids are labelled with dansyl hydrazine in trichloroacetic acid—benzene solution and then chromatographed. The elution is monitored by a fluorophotometer at 365 nm (excitation) and 505 nm (emission). The reaction conditions of the labelling were examined and the optimal conditions were selected. The chromatographic conditions were also selected to give a complete separation between dansyl hydrazones of 17-oxosteroids and the fluorescent co-existing substances in serum or urine samples. Linearity of the fluorescence intensities with the amounts of various 17-oxosteroids were obtained between 60 and 1000 pg. The assay proved satisfactory with respect to sensitivity precision and accuracy. Authors obtained good correlation between results obtained by HPLC—fluorescence and RIA ($r = 0.964$, $n = 81$) for serum dihydroepiandrosterone sulfate. The derivatization conditions of conjugated 17-oxosteroids are examined (Kawasaki *et al.*, 1982b) and the optimal conditions were similar to those of free 17-oxosteroids (Kawasaki *et al.*, 1981). The eluate is monitored by a fluorophotometer at 365 nm (excitation) and 520 nm (emission). Linearity is obtained between 10 and 1000 pmol. The sensitivity of this method is superior to other HPLC methods using a UV detector.

Kawasaki *et al.* (1982a) described a similar method for the determination of 17-hydroxycorticosteroids in urine using dansylhydrazine as a pre-column labelling reagent. The eluate is monitored on a fluorophotometer at 365 nm (excitation) and 505 nm (emission). Linearity of the fluorescence intensities are obtained between 60 pg and 20 ng of various 17-hydroxycorticosteroids. Comparison with the results obtained by RIA give correlation coefficients of 0.932 for tetrahydrocortisol and 0.930 for the tetrahydrocortisone.

Electrochemical detection
The electrochemical detector is more selective because it discriminates selectively for electrochemically active material.

A sensitive and selective ELCD is widely used for HPLC of phenolic compounds (catecholamines, etc.) in biological materials. For the first time Shimada *et al.* (1979) reported the utilization of ELCD for HPLC of phenolic

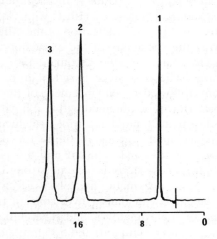

Figure 6. HPLC of steroid hormones with electrochemical detection. Separation of E3 (1), E2 (2) and E1 (3) on a C18 (5 μm) column.

steroids. As a compromise between sensitivity and stability, the applied potential was set at +0.8—1.0 V vs the reference electrode. The response of ELCD for catechol estrogens and their methyl ethers are almost identical. The detection limits of 2-hydroxyestradiol and estriol were determined to be 1 and 5 ng per injection, respectively. These results imply that the HPLC—ELCD method should be capable of determining catechol estrogens in rat brain.

Later, Shimada *et al.* (1982a) tentatively estimate estriol—16-glucuronide and 17-glucuronide in rat bile using a UV' detector. Even when several clean-up procedures were carried out, numerous interfering peaks appeared on the chromatogram. They overcame this problem by the use of ELCD. The potential of the electrochemical detector was set at +1.0 V vs an Ag/AgCl reference electrode. A linear response to each glucuronide is observed in the range 0—10 μg. The detection limit of these compounds was 5 ng per injection. It was demonstrated the ELCD is much superior in selectivity and sensitivity than a UV detector for the determination of estrogen mono-glucuronides in biological fluids. Shimada *et al.* (1980) already described a procedure to determine 17-ketosteroids after solvolysis or hydrolysis in human serum using *p*-nitrophenylhydrazine as a derivatization reagent for HPLC—EC. Furthermore, they developed the method (Shimada *et al.*, 1984) for the direct determination of 17-ketosteroids sulfates without solvolysis.

After deproteinization of a serum sample with acetonitrile, 17-ketosteroids in the supernatant were derivatized with *p*-nitrophenylhydrazine in trichlor-acetic acid—benzene solution. The solution is chromatographied on a μ-Bondapak C18 (5 μm) column using methanol:0.5% $NH_4H_2PO_4$ (pH 3) (8:3 by volume) as a mobile phase at a flow-rate of 1 ml/mn. The applied potential was set at +0.8 V vs an Ag/AgCl reference electrode. A satisfactory linearity was observed in the range of 2.5 to 15 ng. The detection limit of dehydroepi-androsterone sulfate in 0.1 ml of serum sample was 80 ng/ml. It is comparable to that obtained by the fluorescence—HPLC method (Kawasaki *et al.*, 1982b). Authors find good correlation with direct RIA without hydrolysis ($r = 0.952$, $n = 14$).

Separation of estriol, estradiol and estrone were successfully accomplished on a C18 reversed-phase column in our laboratory in close collaboration with Dr Carpentier of Metrohm, France. The mobile phase utilized was acetone:water (1:1) containing 5 g/l each of acetic acid and lithium per-chlorate. The applied potential was +1 V and the limit of detection was in the range of 2—5 ng of standard mixture. The applied potential can be modified if the mixture of synthetic or natural steroids is pure. At 900 mV one can observe a good range of detection and this offers a better possibility of operating the potentiostat at higher sensitivity. Lithium perchlorate is essential for better electrolytic conductivity. This phenomenon is often observed when detection is made with the use of solvents such as acetonitrile or acetone. The time of separation with a 5 μm column can be rather long since the optimal flow rate for HPLC—EC was evaluated to be 0.7 ml/min. However, using rapid 3 μm short columns this process can be greatly

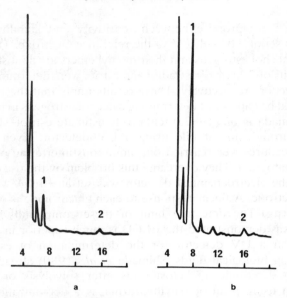

Figure 7. Separation and quantification of estradiol (1) and 17B-estradiol (2) by HPLC—EC using reversed phase 5 μm column and 60% methanol containing 0.5% acetic acid and sodium perchlorate as mobile phase. The applied potential was 1100 mV and the flow rate was 1 ml/min. (a) is a chromatogram before hydrolysis whereas (b) is after hydrolysis of urine.

shortened. Higher flow rate will lead to a drift of base line without loss of sensitivity.

Simultaneous separation and quantification of estriol and 17β-estradiol by HPLC—EC can be easily made using 60% methanol and 50 mM NaClO$_4$ as mobile phase and a regular reversed-phase column. The potential is again very high (+1150 mV) and the flow selected should be near 1 ml/min. Estradiol and 17β-estradiol elute, respectively, at 8 and 16 min under these conditions. Hydrolysis can greatly enhance this yield for clinical evaluation. Dynamic correlation with linear elution is the main criteria in this method of separation.

CONCLUSION

The present contribution reports different accomplishments in chemical analysis of steroid hormones in biological and chemical samples. All the methods of separation by HPLC have been discussed thoroughly. The importance of mobile phase granulometry, isocratic or gradient mode of elution have been provided in detail with direct access to new investigators. A review of other means of quantification such as fluorescence and RIA methodology has been compared with HPLC techniques. An attempt to couple different methods of detection is discussed from a practical point of view. Electrochemical detection, a very new method of detection, appears to be a possible replacement for laborious RIA methodology. The application of HPLC—EC in steroid analysis using microbore column and differential polarographic techniques is suggested to be the optimal tool of future steroid analysis of a high sensitivity of HPLC—EC. The authors hope that this contribution will serve a useful purpose in the selection of an appropriate method for steroid analysis in biomedical or clinical research.

ACKNOWLEDGMENTS

The authors express their sincere gratitude to Dr Georges Carpentier, Director, Metrohm, France, for his kind help, advice and participation in some of the experimental work. We also acknowledge material support by some chemical companies such as Roucaire, Prolabo and LKB, France. The participation of all colleagues at the Unit of Neuropharmacology such as Drs Casimiri and Eberle is gratefully acknowledged.

REFERENCES

Ager, R. P., and Oliver, R. W. A. (1984). Separation of estrogen conjugates in urine and synthetic mixtures by high-performance liquid chromatographic methods. *J. Chromatogr.* *309*, 1—15.

Alvinerie, M., and Toutain, P. L. (1984). Determination of methylprednisolone and methyl-prednisolone acetate in synovial fluid using high-performance liquid chromatography. *J. Chromatogr.* *309*, 385—390.

Burgess, C. (1978). Rapid reversed-phase high-performance liquid chromatographic analysis of steroid products. *J. Chromatogr.* *149*, 233—240.

Butterfield, A. G., Lodge, B. A., and Pound, N. J. (1973). High-speed liquid chromatographic separation of equine estrogens. *J. Chromatogr. Sci. 11*, 401—405.

Carson, S. W., and Jusko, W. J. (1984). Simultaneous analysis of cortexolone and cortisol by high-performance liquid chromatography for use in the metyrapone test. *J. Chromatogr.* *306*, 345—350.

D'Agostino, G., Mitchell, F., Castagnetta, L., and O'Hare, M. J. (1984). Solvent optimization of reversed-phase high-performance liquid chromatography for polar adrenal steroids using computer predicted retentions. *J. Chromatogr. 305*, 13—26.

Darnay K. J. Jr., Wing, T. Y., and Ewing, L. L. (1983). Simultaneous measurement of four testicular Δ^4-3-ketosteroids by isocratic high-performance liquid chromatography with on-line ultraviolet absorbance detection. *J. Chromatogr. 257*, 81—90.

Dixon, P. F., Lukha, P., and Scott, N. R. (1979). Clinical analysis of steroids by HPLC. *Proc. Anal. Div. Chem. Soc. 16*, 302—305.

Dolphin, R. J., and Pergande, P. J. (1977). Improved method for the analysis of estrogenic steroids in pregnancy urine by high-performance liquid chromatography. *J. Chromatogr. 143*, 267—274.

Ebling, W. J., Szefler, S. J., and Jusko, W. J. (1984). Analysis of cortisol, methylprednisolone and methylprednisolone hemisuccinate. Absence of effects of troleandomycin on ester hydrolysis. *J. Chromatogr. 305*, 271—280.

Eibs, G., and Schoeneshoefer, M. (1984). Simultaneous determination of fifteen steroid hormones from a single serum sample by high-performance liquid chromatography and radioimmunoassay. *J. Chromatogr. 310*, 388—389.

Engelhardt, H., Dreyer, B., and Schmidt, H. (1982). Properties and diversity of C-18 bonded phases. *Chromatographia 16*, 11—17.

Fitzpatrick, F. A., and Siggia, S. (1973). High resolution liquid chromatography of derivatized nonultraviolet absorbing hydroxysteroids. *Anal. Chem. 45*, 2310—2314.

Fransson, B., Wahlund, K. G., Johansson, I. M., and Schill, G. (1976). Ion-pair chromatography of acedic drug metabolites and endogenous compounds. *J. Chromatogr.* 125, 327—344.

Garg, D. C., Ayres, J. W., and Wagner, J. G. (1977). Determination of methylprednisolone and hydrocortisone in plasma using high-pressure liquid chromatography. *Res. Commun. Chem. Pathol. Pharmacol.* 18, 137—146.

Glajch, J. L., Kirkland, J. J., Squire, K. M., and Minor, J. M. (1980). Optimization of solvent strength and selectivity for reversed-phase liquid chromatography using an interactive mixture-design statistical technique. *J. Chromatogr.* 199, 57—79.

Glajch, J. L., Kirkland, J. J., and Snyder, L. R. (1982). Practical optimization of solvent selectivity in liquid solid chromatography using a mixture-design statistical technique. *J. Chromatogr.* 238, 269—280.

Henry, R. A., Schmit, J. A., and Dieckman, J. F. (1971). Analysis of steroids and derivatized steroids and derivatized steroids by high-speed liquid chromatography. *J. Chromatogr. Sci.* 9, 513—520.

Hermansson, J. (1978). Reversed-phase liquid chromatography of steroid glucuronides. *J. Chromatogr.* 152, 437—445.

Hermansson, J. (1980). Separation of steroid glucuronides by reversed-phase liquid chromatography. *J. Chromatogr.* 194, 80—84.

Imaizumi, N., Morimoto, S., Kigoshi, T., Uchida, K., Hosojima, H., and Yamamoto, I. (1984). Application of reversed-phase high-performance liquid chromatography for radioimmunoassay of plasma 18-hydroxicorticosterone. *J. Chromatogr.* 308, 295—300.

Kawasaki, T., Maeda, M., and Tsuji, A. (1981). Determination of 17-oxosteroids in serum and urine by fluorescence high-performance liquid chromatography using dansyl hydrazine as a pre-labeling reagent. *J. Chromatogr.* 226, 1—12.

Kawasaki, T., Maeda, M., and Tsuji, A. (1982a). Determination of 17-hydroxicorticosteroids in urine by fluorescence high-performance liquid chromatography using Dns-hydrazine as a pre-column labeling reagent. *J. Chromatogr.* 232, 1—11.

Kawasaki, T., Maeda, M., and Tsuji, A. (1982b). Determination of 17-oxosteroid glucuronides and sulfates in urine and serum by fluorescence high-performance liquid chromatography using dansyl hydrazine as a pre-labeling reagent. *J. Chromatogr.* 233, 61—68.

Keravis, G., and Durand, M. H. (1980). Determination of urinary estriol 16 glucosiduronate in pregnancy using high-performance liquid chromatography. *Pathology 28*, 283—294.

Keravis, G., Lafosse, M. and Durand, M. M. (1977). Reversed-phase high-performance liquid chromatography of some free and some conjugated hormonal steroids. *Chromatographia* 10, 678—681.

Lafosse, M., Keravis, G., and Durand, M. H. (1976). Etude par chromatographie liquide haute pression de 17-cetosteroides libres et conjugués. *J. Chromatogr.* 118, 283—294.

Loo, J. C. K., Butterfield, A. G., Moffatt, J., and Jordan, N. (1977). Analysis of prednisolone in plasma by high-performance liquid chromatography. *J. Chromatogr.* 143, 275—280.

Loo, J. C. K., and Jordan, N. (1977). High-performance liquid chromatographic analysis of synthetic corticosteroids in plasma. *J. Chromatogr.* 143, 314—316.

Majors, R. E., and Hopper, M. J. (1974). Siloxane phases bonded to silica gel for use in high-performance liquid chromatography. *J. Chromatogr. Sci.* 12, 767—778.

McGinley, P. A., Braughler, J. M., and Hall, E. D. (1982). Determination of methylprednisolone in central nervous tissue and plasma using normal-phase high-performance liquid chromatography. *J. Chromatogr.* 230, 29—35.

Musey, P. I., Collins, D. C., and Preedy, J. R. K. (1978). Separation of estrogen conjugates by high-pressure liquid chromatography. *Steroids 31*, 583—592.

Newport, G. D., Headley, S. K., Freeman, J. P., and Slikker W. Jr. (1980). Separation of diethylstilbestrol and derivatives in biological fluids and tissues by high-pressure liquid chromatography. *J. Liquid Chromatogr.* 3, 1053—1070.

O'Hare, M. J., Nice, E. C., Magee-Brown, R., and Bullman, H. (1976). High-pressure liquid chromatography of steroids secreted by human adrenal and testis cells in monolayer culture. *J. Chromatogr.* 125, 357—367.

Rocci M. L. Jr., and Jusko, W. J. (1981). Analysis of prednisone, prednisolone and their

20β-hydroxylated metabolites by high-performance liquid chromatography. *J. Chromatogr.* 224, 221—227.

Schoeneshoefer, M., and Dulce, H. J. (1979). Comparison of different high-performance liquid chromatographic systems for the purification of adrenal and gonadal steroids prior to immunoassay. *J. Chromatogr.* 164, 17—28.

Schoeneshoefer, M., and Fenner, A. (1981). A convenient and efficient method for the extraction and fractionation of steroid hormones from serum or urine. *J. Clin. Chem. Clin. Biochem.* 19, 71—74.

Schoeneshoefer, M., Fenner, A., and Dulce, H. J. (1981a). Assessment of elevan adrenal steroids from a serum sample by combination of automatic high-performance liquid chromatography and radioimmunoassay (HPLC—RIA). *J. Steroid Biochem.* 14, 377—386.

Schoeneshoefer, M., Maxeiner, J., and Fenner, A. (1981b). Quantitation of pregnenolone and 17-hydroxypregnenolone in human serum by automatic high-performance liquid chromatography and subsequent radioimmunoassay. *J. Chromatogr.* 224, 229—237.

Shimada, K., Tanaka, M., and Nambara, T. (1979). Studies on steroids. CL. Separation of catechol estrogens by high-performance liquid chromatography with electrochemical detection. *J. Chromatogr.* 178, 350—354.

Shimada, K., Tanaka, M., and Nambara, T. (1980). Derivatization of ketosteroids for high-performance liquid chromatography with electrochemical detection. *Anal. Lett.* 13(B13), 1129—1136.

Shimada, K., Xie, F., and Nambara, T. (1982a). Studies on steroids. CLXXIX. Determination of estriol 16- and 17-glucuronide in biological fluids by high-performance liquid chromatography with electrochemical detection. *J. Chromatogr.* 232, 13—18.

Shimada, K., Kaji, M., Xie, F., and Nambara, T. (1982b). Studies on steroids. CLXXVIII. Separation of estrogen glucuronides by high-performance liquid chromatography. *J. Liquid Chromatogr.* 5, 1763—1770.

Shimada, K., Tanaka, M., and Nambara, T. (1984). Studies on steroids. CC. Determination of 17-ketosteroid sulfates in serum by high-performance liquid chromatography with electrochemical detection using pre-column derivatization. *J. Chromatogr.* 307, 23—28.

Slikker, W. Jr., Lipe, G. W., and Newport, G. D. (1981). High-performance liquid chromatographic analysis of estradiol-17β and metabolites in biological media. *J. Chromatogr.* 224, 205—219.

Trefz, F. K., Byrd, D. J., and Kochen, W. (1975). Quantitative determination of cortisol in human plasma by high-pressure liquid chromatography. *J. Chromatogr.* 107, 181—189.

Van der Wal, Sj., and Huber, J. F. K. (1974). High-pressure liquid chromatography with ion-exchange celluloses and its application to the separation of estrogen glucuronides. *J. Chromatogr.* 102, 353—374.

Van der Wal, Sj., and Huber, J. F. K. (1977a). Improvement of the efficiency of ion-exchange cellulose columns and their application. *J. Chromatogr.* 135, 287—303.

Van der Wal, Sj., and Huber, J. F. K. (1977b). Separation of estrogen glucuronides, sulphates and phosphates on ion-exchange cellulose by high-pressure liquid chromatography. *J. Chromatogr.* 135, 305—321.

Van der Wal, Sj., and Huber, J. F. K. (1978). Comparative study of several phase systems for separation of estrogen conjugates by high-pressure liquid chromatography. *J. Chromatogr.* 149, 431—453.

Wortmann, W., Schnabel, C., and Touchstone, L. C. (1973). Quantitative determination of corticosteroids from plasma by high-pressure liquid chromatography. *J. Chromatogr.* 84, 396—401.

HPLC—EC: ADAPTATION FOR ANALYSIS OF BIOGENIC AMINES AND RELATED ENZYMES

Progress in HPLC, Vol. 2, pp. 293—306
Parvez *et al.* (Eds)
© 1987 VNU Science Press

Analysis of enzyme activities by electrochemical detection coupled with HPLC

TOSHIHARU NAGATSU* and KOHICHI KOJIMA

Department of Biochemistry, Nagoya University School of Medicine, Nagoya 466, and Laboratory of Cell Physiology, Department of life Chemistry, Graduate School at Nagatusta, Tokyo Institute of Technology, Yokohama 227, Japan.

SIMONE PARVEZ and HASAN PARVEZ

Unité de Neuropharmacologie, Université de Paris XI, Centre d'Orsay, Bat. 440, 91405 Orsay Cedex, France

GENERAL PRINCIPLES

Chromatographic—photometric measurements of a substrate or a product in an enzyme reaction have been used for many years (Dixon *et al.*, 1979), but the time-consuming chromatographic procedure and the relatively low sensitivity of photometric detection have limited the wide application of chromatrographic methods to the study of enzymes; however, the introduction of HPLC has reduced the time from several hours to several minutes. Thus, it is now possible to measure several tens of samples after enzyme reactions within several hours by the combined use of HPLC, an auto-sampling apparatus and an automatic data processor. HPLC-assay of enzyme activity is now applicable even for purification of enzymes.

HPLC combined with electrochemical detection (ECD) have made the chromatographic method not only rapid but also highly sensitive, since the sensitivity for HPLC—ECD is generally in the order of femtomol.

The activity of an enzyme can be measured either by the rate of disappearance of a substance or by the rate of appearance of a product formed by the enzymic reaction. Any HPLC—ECD method for the assay of enzyme activity is generally based on the rate of appearance of a product. For following enzyme reaction, the continuous methods are considered to be superior to the sampling or discontinuous method. However, HPLC—ECD assay in principle is discontinuous and a sample is withdrawn for the incubation mixture at various times and subjected to HPLC assay.

---

* To whom correspondence should be addressed.

Since HPLC—ECD is extremely sensitive, only a small amount of the enzyme material is required. Even microgram amounts of tissue samples, and microliter amounts of body fluids [blood, plasma, serum, cerebrospinal fluid (CSF), urine or amniotic fluid] may be sufficient for many kinds of enzymes. This high-sensitivity of HPLC—ECD assays of enzyme activity makes the method especially useful in clinical biochemistry.

In the incubation mixture of an enzyme assay, the concentrations of substrate and cofactors should be optimal. However, in some HPLC—ECD assays, the substrate concentration cannot be saturating, because a high peak of a substrate inteferes with the peak of the product to be assayed. In such cases, the substrate must be removed by a preliminary purification procedure of the product from the incubation mixture either by solvent extraction or by rapid column chromatography.

The incubation mixture should be kept as small as possible to get high sensitivity, so that a large portion of the samples after deproteinization can be injected directly into HPLC. The volume of incubation mixture frequently used is from 50 to 1000 µl.

Since the enzyme assay by HPLC—ECD is highly sensitive but is not completely specific, the choice of the blank (control) is important to detect any blank peak due to non-enzymatic reaction, to contamination of the product in the reagents, or to other compounds which are eluted together with the product. Incubation with a specific enzyme inhibitor is most preferable, if such a specific inhibitor is available and does not interfere with the assay by HPLC—ECD. If the enzyme can recognize the stereospecificity of the substrate, incubation with the inactive stereoisomer of the substrate is ideal. However, this is not always the case, and either incubation without the enzyme and its addition after incubation, or incubation with the boiled, inactivated enzyme is frequently used.

Aliquots of the incubation mixture are withdrawn at various times, and deproteinized by conventional reagents such as trichloroacetic acid or perchloric acid contating an appropriate amount of an internal standard. The internal standard should be electrochemically active and eluted near the product, and the amount of the internal standard added should be similar to that of the product to be measured.

The conditions of HPLC should be chosen to separate completely the product from the peaks of the substrate and any interfering substances, and preferably to elute the product prior to the substrate, so that the large peak of the substrate does not interfere with the assay of the product. The bonded-phase chromatography columns are most frequently used (Snyder and Kirkland, 1979).

Either pre-column or post-column chemical or enzymatic derivatization can be applied for the assay of electrochemically inactive product.

The product from enzymatically in the incubation mixture is usually calculated based on the peak heigth of the internal standard by the equation

$$\frac{R(E) - R(B)}{R(B+S) - R(B)} \times \text{[amount of the standard product added into the blank incubation]}$$

where R is the ratio of peak height of the product/peak height of the internal standard, $R(E)$ (E means experiment) being that from the enzyme incubation, $R(B)$ (B means blank) from the blank incubation (for example, no-enzyme, boiled-enzyme or inhibitor-enzyme incubation), and $R(B + S)$ from the blank incubation plus a known amount of the product added as standard.

HPLC—ECD assays have been successfully applied for the enzymes of catecholamine biosynthesis, since the products are electrochemically active and can be measured with a high sensitivity. The sensitivity of HPLC—ECD assays is, in some cases, higher than that by radioassay. The substrate concentration can be saturating in HPLC—ECD assays to get V_{max} values, while this is not frequently possible in the radioassays due to the dilution of specific radioactivity of the radio-labeled substrate by the non-labeled substrate.

Several examples of HPLC—ECD assays of enzyme activity are described on the catecholamine-synthesizing enzymes.

TYROSINE HYDROXYLASE (TH)

TH [tyrosine 3-monooxoygenase; EC 1.14.16.2; L-tyrosine, tetrahydropteridine: oxygen oxidoreductase (3-hydroxylating)], which is a pteridine-requiring monooxygenase, catalyzes the first step in the biosynthesis of catecholamines (Nagatsu, 1983; Nagatsu *et al.*, 1964a).

TH couples *in vivo* with dihydropteridine reductase. Based on analogy with the phenylalanie-hydroxylating system, the dihydropteridine formed during the tyrosine hydroxylation reaction may be a quinonoid 7,8-dihydropteridine, and dihydropteridine reductase regenerates tetrahydropeteridine with NADH (Breeneman and Kaufman, 1964).

$$L\text{-Tyrosine} + O_2 + \text{tetrahydropteridine}$$
$$\rightarrow 3,4\text{-dihydroxy-}L\text{-phenylalanine (}L\text{-DOPA)} +$$
$$+ \text{quinonoid dihydropteridine} + H_2O$$

Quinonoid dihydropteridine $+ \text{NADH} + H^+ \rightarrow$ tetrahydropteridine $+ \text{NAD}^+$

Either NADH or NADPH is effective as an electron donor *in vitro*, but dihydropteridine reductase *in vivo* is the NADH form (Togari *et al.*, 1983).

(6R)-L-*erythro*-5,6,7,8-tetraphydrobiopterin may be the natural pteridine cofactor for TH (Oka *et al.*, 1981).

Since the activity of TH is very low, TH has been measured only by radioisotopic methods (Nagatsu *et al.*, 1964a b; Waymire *et al.*, 1971). Blank and Pike (1976) first reported an assay for TH by HPLC—ECD. We developed a highly sensitive method based on HPLC—ECD (Nagatsu *et al.*, 1979a, b; Oka *et al.*, 1982).

Principle
We use L-tyrosine as the natural substrate and D-tyrosine for the blank, since TH is specific for L-tyrosine.

We combine the simple and specific isolation of enzymatically formed
L-DOPA by our double-column procedure (the top column of Amberlite
CG-50 and the bottom column of alumina) with the highly sensitive assay of
DOPA by HPLC—ECD. α-MethylDOPA as an internal standard is added
to each sample after TH incubation. Use of α-methylDOPA as an internal
standard makes the assay highly accurate. Both the double columns and
HPLC permit nearly complete isolation of DOPA, and thus the blank values
are very low. The only interfering substance is endogenous DOPA in crude
tissues and nonenzymatically formed DOPA from both L- and D-tyrosine,
and this blank value can be completely cancelled by the control incubation
with D-tyrosine plus 3-iodo-L-tyrosine (an enzyme inhibitor). TH activity in
less than 1 mg of a brain nucleus can be assayed by this method.

Materials
In the TH reaction mixture, the enzymatically formed quinonoid dihydropteridine is reduced back to tetrahydropteridine either by a coupled enzymatic reaction with dihydropteridine reductase and NAD(P)H (Brenneman
and Kaufman, 1964), or by chemical reduction with mercaptoethanol
(Nagatsu *et al.*, 1964a, b), dithiothreitol (Nagatsu *et al.*, 1972) or ascorbic
acid (Kato *et al.*, 1981; Lerner *et al.*, 1978). The order of TH activity in the
presence of these reducing reagents is ascorbic acid, mercaptoethanol and
dithiothreitol. Ascorbic acid gives a higher TH activity, but frequently
interferes with the DOPA peak in HPLC. Therefore, mercaptoethanol is
most frequently used for chemical reduction of quinonoid dihydropteridine
and also to protect the tetrahydropteridine cofactor in the TH incubaion
mixture.
    The tissues are homogenized in 4 volumes of 0.25 M sucrose in a glass
Potter homogenizer.
    Amberlite CG-50 (Type 1, 100—200 mesh) is washed by cycling through
acid and sodium forms with 2 M HCl and 2 M NaOH and finally with water,
equilibrated with 0.4 M potassium phosphate buffer, pH 6.1, and stored as a
suspension in the same buffer.
    Alumina (aluminium oxide) is washed with acid and heated before use for
activation. Alumina (200 g) is boiled in 1 l of 2M HCl for 30 min in a reflux
condenser and the resulting cloudly supernatant is poured off. The alumina is
stirred with 1 l of water, allowed to settle for 5 min, and the supernatant is
decanted. This washing and decanting procedure is repeated with water
10—15 times until the washed water becomes clear after 5 min of settling
and is pH 4—5. The alumina is collected in a large suction funnel, allowed to
dry overnight in an open pan at room temperature, and then heated in an
oven at 100° C for 2 h. The acid-washed alumina is passed through a 200
mesh sieve to remove fine particles and is kept tightly closed so as not to be
exposed to moisture.
    The natural pteridine cofactor is (6*R*)-L-*erythro*-tetrahydrobiopterin
(Matsuura *et al.*, 1981), which is not commericaly available. (6*RS*)-L-*erythro*-
tetrahydrobiopterin is commercially available, and is prepared by chemical

reduction of L-*erythro*-biopterin by catalytic hydrogenation in 1 M HCl over platinum oxide catalyst (Kaufman, 1967). The (6*RS*)-tetrahydrobiopterin is separated to the natural (6*R*)-form and the unnatural (6*S*)-form by HPLC (Bailey and Ayling, 1978). Use of the natural cofactor, (6*R*)-L-*erythro*-tetrahydrobiopterin, is most preferable, especially for kinetic studies, but the (6*RS*)-L-*erythro*-tetrahydrobiopterin gives similar kinetic results as the (6*R*) from (Oka *et al.*, 1981). It should be noted that O_2 at 20% (220 µM) in air is inhibitory with (6*R*) or (6*RS*)-tetrahydrobiopterin as a cofactor, but not with unnatural (6*S*)-tetrahydrobiopterin and with an artificial cofactor, 6-methyl- or 6,7-dimethyltetrahydropterin. For most routine assays of TH activity, either (6*RS*)-6-methyltetrahydropterin or (6*RS*)-6,7-dimethyltetradhydro-pterin, which is commerically available, is generally used. (6*RS*)-Methyl-tetrahydropterin is structurally similar to the natural tetrahydrobiopterin and gives higher activity than (6*RS*)-6,7-dimethyltetrahydropterin.

Tetrahydropterins are generally very unstable and easily oxidized in air especially in alkaline pH. Tetrahydropteridines are dissolved in 0.005—0.01 M HCl or 1.0 M 2-mercaptoethanol and stored frozen at −20° C protected from light and prepared once a week. The concentration in solution is estimated based on the extinction coefficient of $18\,500$ M^{-1} cm^{-1} at 264 nm in 2 M HCl (Nagatsu *et al.*, 1972). Tetrahydropterin concentrations can also be estimated by diminshed absorbance of 2,6-dichlorophenolindophenol solu-tion based on the extinction coefficient of $18\,500$ M^{-1} cm^{-1} at 603 nm at a final pH of 6.99, 30 s after the addition of tetrahydropterin in 0.1 M. HCl (Oka *et al.*, 1981).

Procedures

The standard incubation mixture consists of the following components in a total volume of 100 µl (final concentrations in parentheses); 20 µl of 1 M acetate buffer pH 6.0 (0.2 M), 20 µl of 2 m L-tyrosine in 0.01 M HCl (0.2 mM), 10 µl (6*RS*)-methyl-5,6,7,8,-tetrahydropterin (1 mM) in 1 M 2-mercaptoethanol (100 mM), 30 µl of 0.25 M sucrose (75 mM) containing enzyme, 10 µl of 1 mg/ml catalase (10 µg/100 µl) or 10 µl of 10 mM ferrous ammonium sulfate (1 mM), water. For the blank incubation, D-tyrosine is used as substrate instead of L-tyrosine and 50 pmol of DOPA are added to another blank incubation as an internal standard for DOPA.

Incubation is done at 37° C for 10 min, and the reaction is stopped with 600 µl 0.5 M perchloric acid containing 50 pmol α-methylDOPA as an internal standard in an ice-bath. After 10 min, 20 µl 0.2 M EDTA and 300 µl 1 M potassium carbonate are added to adjust the pH to 8.0—8.5, and the mixture is centrifuged at 1600 × g for 10 min at 4° C. The clear supernatant is passed though the double columns, the upper column, containing 200 µl Amberlite CG-50 (12.5 × 0.5 cm ID), and the bottom column, containing 100 mg aluminium oxide (12.5 × 0.4 cm ID), fitted together sequentially. The effluent through both columns is discarded. Both columns are washed once with 1.5 ml water, and the washings are discarded, DOPA and α-methyl-DOPA pass through the first Amberlite column and are absorbed on the second aluminum oxide column, which is separated and washed with 1.5 ml of water twice, and with 100 µl of 0.5 M HCl once. DOPA and α-methylDOPA are eluted with 200 µl of 0.5 M HCl.

A 100 µl aliquot of the eluate is injected into the high-performance liquid chromatograph with an electrochemical detector and a column (25 × 0.4 cm ID) packed with Yanapak ODS (particle size 5 µm) (Yanagimoto, Kyoto, Japan). The mobile phase is 0.1 M postassium phosphate buffer (pH 3.5) with a flow-rate of 0.6 ml/min; the detector potential is set at 0.8 V vs the Ag/AgCl electrode. Under these conditions the retention times are: solvent front, 1.8 min; DOPA, 3.8 min: and α-methylDOPA, 5.5 min. Limit of sensitivity for DOPA is 0.1 pmol.

The DOPA formed enzymatically by TH is calculated by the question

$$\frac{R(L) - R(D)}{R(D + S) - R(D)} \times 50 \text{ pmol}$$

where R is the ratio to peak heights (peak height of DOPA: peak height of α-methylDOPA), $R(L)$ being that from the L-tyrosine incubation, $R(D)$ from the D-tyrosine incubation, and $R(D + S)$ that of D-tyrosine plus DOPA (internal standard, 50 pmol).

The highest sensitivity can be obtained by the double column prodecure to reduce the interfering peaks in HPLC, but the first Amberlite CG-50 column can be omitted in regular TH assay, and only the alumina column can be used.

AROMATIC L-AMINO ACID DECARBOXYLASE (AADC)

AADC (4.1.1.28) catalyzes the decarboxylation of various aromatic L-amino acids including L-DOPA and L-5-hydroxytryptophan (5HTP) (Lovenberg *et al.*, 1962). Thus, the enzyme is responsible for the formation of both dopamine and serotonin. Among a wide range of naturally occurring and synthetic aromatic amino acids L-DOPA and 5-hydroxytryptophan are good substrates. V_{max} for L-DOPA is approximately 10 times higher than that for L-5-hydroxytryptophan.

Various methods are available for AADC assay (Culvenor and Lovenberg, 1983). The assay method most frequently used to measure the activity of AADC has been the radioisotopic method with L-[1-^{14}C]DOPA or L-[1-^{14}C]-5-hydroxytryptophan as substrate, in which [14]CO$_2$ formed in the enzymatic decarboxylation is measured.

We have developed a HPLC method using either L-DOPA (Nagatsu *et al.*, 1979a, b) or L-5-hydroxytryptophan (Rahman *et al.*, 1980) as substrate to measure enzymatically formed dopamine or serotnin.

HPLC method using L-DOPA as substrate (Nagatsu *et al.*, 1979a, b).

Principle. L-DOPA is used as substrate, and the product, dopamine, is measured by HPLC—ECD. A large amount of the substrate, L-DOPA, which interferes with the dopamine peak in HPLC, is removed by a preliminary column of Amberlite CG-50.

Experimental procedures. Tissues are homogenized in 9 volumes of 0.25 M sucrose in a Potter glass homogenizer.

The standard incubation mixture contains (total volume 400 µl): 30 mM sodium phosphate buffer (pH 7.2), 0.3 mM EDTA, 0.17 mM ascorbic acid, 0.01 mM pyridoxal phosphate, 1.0 mM L-DOPA (or D-DOPA for the blank), 0.1 mM pargyline HCl (a monoamine oxidase inhibitor), and the enzyme. Incubation is done at 37° C for 20 min, and the reaction is stopped by adding 80 µl of 3 M trichloroacetic acid. After 10 min, 1.87 ml water and 50 µl 0.01 M HCl containing 50 pmol dihydroxybenzylamine as an internal standard are added, and the mixture is centrifuged at 3000 rpm for 10 min. The supernatant is passed through a column (packed volume 0.5 ml) of Amberlite CG-50—Na$^+$ equilibrated with 0.1 M potassium phosphate buffer at pH 6.5. The resin is washed with 2.5. ml buffer and 200 µl 1 M HCl, and the dopamine adsorbed is eluted with 700 µl 1 M HCl. One hundred microliters of the eluate are injected into the high-performance liquid chromatograph with an electrohemical detector and a Yanapak ODS column (25 × 0.4 cm ID). The carrier buffer for the liquid chromatography is 0.1 M potassium phosphate buffer, pH 3.0, with a flow rate of 0.6 cm/min. The detector potential is set at 0.8 V vs the Ag/AgC1 electrode. The peak height of dopamine is measured and converted to picomoles from the peak height of dihydroxybenzylamine added as an internal standard. The retention times under these conditions are: dihydroxybenzylamine, 4.2 min; DOPA, 4.5 min; and dopamine, 5.5 min.

HPLC method using L-5-hydroxytryptophan as substrate (Rahman *et al.,* 1980)

Principle. L-5-Hydroxytrytophan is used as substrate, and the product serotonin, is measured by HPLC—ECD. Although the V_{max} of AADC with L-5-hydroxytryptophan as substrate is approximately 1/10 of that with L-DOPA as a substrate, AADC reaction proceeds linearly at 37° C for 150 min with L-5-hydroxytryptophan as a substrate, while only for 20 min with L-DOPA as a substrate. Therefore, this HPLC method is also highly sensitive with longer incubation.

Experimental procedures. The tissues are homogenized with 0.32 M sucrose solution (1 part tissue plus 9 parts 0.32 M sucrose solution) in a Potter glass homogenizer.

The standard incubation mixture contains (total volume 400 µl, final pH 8.3): 30 mM sodium phosphate buffer (pH 9.0), 0.01 mM pyridoxal phosphate, 1.0 mM L-5-hydroxytryptophan (or D-5-hydroxytryptophan for the blank), 0.1 mM pargyline HCl and the enzyme. Incubation is done at 37° C for 20—120 min, and the reaction is stopped by adding 80 µl 3 M trichloroacetic acid. After 10 min 1.82 ml water and 100 μl 0.01 M HCl containing 100—500 pmol of N-methyldopamine as an internal standard are added, and the mixture is centrifuged at $1600 \times g$ for 10 min. The supernatant is passed through a column (packed volume 0.5 ml) of Amberlite CG-50 (Na^+) eqilibrated with 0.1 M potassium phosphate buffer (pH 6.5). The resin is washed twice with 4.5 ml buffer and with 200 μl 1 M HCl. The serotonin adsorbed is eluted with 1.4 ml 1 M HCl. A 100 μl aliquot of the eluate is injected into the high-performance liquid chromatograph with an electrochemical detector and Yanapak ODS-T reversed-phase column (particle size 10 μm, 25×0.4 cm ID). The carrier buffer for the liquid chromatography is 0.1 M potassium phosphate buffer containing 10% methanol, pH 3.2, with a flow rate of 0.5 ml/min. The detector potential is set at 0.8 V against the Ag/AgCl electrode. The peak height of serotonin is measured and converted to pmol from the peak height of N-methyldopamine added as an internal standard. The retention times under these conditions are: N-methyldopamine 5.0 min; serotonin, 9.25 min; and 5-hydroxytryptophan, 10.0 min.

DOPAMINE β-HYDROXYLASE (DBH)

DBH [dopamine β-monooxygenase; EC 1.14.17.1; 3,4-dihydroxyphenylethylamine, ascorbate: oxygen oxidoreductase (β-hydroxylating)] (Friedman and Kaufman, 1965; Levine et al., 1960), which catalyzes the conversion of dopamine to noradrenaline, is an ascorbate-requiring copper-containing monooxygenase and is localized in peripheral and central noradrenergic neurons and chromaffin cells of the adrenal medulla. Besides the formation of noradrenaline from dopamine, DBH generally catalyzes the conversion of various phenylethylamines to the corresponding phenylethanolamines (Creveling et al., 1962). DBH is secreted into the blood from peripheral sympathetic nerve endings and adrenal medulla, and probably into CSF from the brain, together with catecholamines, and therefore, is present in serum or plasma and in CSF (Nagatsu, 1977). Since DBH is an intraneuronal enzyme specific for noradrenergic neurons, the assay of its enzyme activity in blood to estimate the neuronal function has gained much attention.

The assay of DBH activity in crude tissue preparations such as homogenate or blood is difficult, mainly due to the presence of endogenous inhibitors (Nagatsu et al., 1967) and its low activity. Endogenous inhibitors may be mostly sulfhydryl compounds and can be eliminated by sulfhydryl-blocking agents such as N-ethylmaleimide or Cu^{2+} (Nagatsu et al., 1967).

Numerous methods have been reported on the assay of the activity of DBH (Nagatsu, 1983), and the HPLC—ECD method is highly sensitive (Kissinger et al., Matsui et al., 1981; Sperk et al., 1980).

Experimental Procedures (Matsui et al., 1981). Dopamine is used as a substrate and is incubated under optimal conditions. Noradrenaline formed enzymatically from dopamine, is isolated by a double-column procedure, the first column of Dowex-50—H$^+$ and the second column of aluminum oxide. Noradrenaline is adsorbed on the second aluminium oxide column and then eluted with 0.5 M HCl and assayed by HPLC—ECD. Adrenaline is added to each incubation mixture as an internal standard. The peak height in HPLC is linear from 500 fmol to 100 pmol of noradrenaline. The lower limit of detection for noradrenaline formed enzymatically is about 30 pmol. This HPLC—ECD method is highly sensitive and is applicable for the DBH assay in human CSF.

The incubation mixture contains (total volume 1.0 ml, each final concentration in parentheses): 500 µl of an enzyme solution (in case of CSF, 500 µl); 100 µl 2 M sodium acetate buffer, pH 5.0 (0.2 M); 150 µl 0.2 M N-ethylmaleimide (30 mM); 50 µl 100 µM CuSO$_4$ (5 µM); 25 µl aqueous solution (20 mg/ml) of catalase (25000 U, 500 µg); 25 µl 40 mM pargyline HCl (1 mM); 50 µl 0.2 M ascorbic acid (10 mM); 50 µl 0.2 M sodium fumarate (10 mM); and 50 µl 0.4 M dopamine HCl (20 mM). There is included in the blank incubation 50 µl 2 mM fusaric acid (0.1 mM). The reaction mixture is preincubated at 37° C for 5 min, and the reaction is started by addition of dopamine. Acetate buffer + N-ethylmaleimide + CuSO$_4$ (cocktail A), and catalase + pargyline + ascorbic acid + fumarate (cocktail B), can be mixed together; 500 µl of an enzyme preparation, 300 µl of cocktail A and 150 µl of cocktail B, are mixed in this order, and finally the reaction is started by adding 50 µl of dopamine solution.

The reaction is carried out in air at 37° C for 45 min. The reaction is stopped by adding 1 ml of ethanol containing 1 mM fusaric acid and 500 pmol adrenaline (internal standard), after which 100 µl 10% Na$_2$S$_2$O$_5$ and 100 µl 0.2 M disodium EDTA are added. The mixture is left at 0° C for at least 30 min and centrifuged at 15000 rpm for 10 min. The supernatant is transferred to a column (4 × 55mm) of Dowex-50W-X4—H$^+$ (200—400 mesh). The column is washed with 2 ml water three times and 3 ml 1 M HCl to remove dopamine and ascorbate. Then, noradrenaline and adrenaline are eluted with 2 ml 1 M HCl.

The eluate containing noradrenaline and adrenaline is adjusted to pH 8.4 by addition of 8 ml 3 M Tris-HCl buffer, pH 8.6, and is transferred to an aluminium oxide column (0.4 cm ID) containing 100 mg aluminium oxide. The column is washed three times with 4 ml water and 100 µl 0.5 M HCl. Noradrenaline and adrenaline are eluted with 200 µl 0.5 M HCl. Mean recovery of noradrenaline from the Dowex column is 73% and that from the aluminium oxide is 63%.

A 10—20 µl aliquot of the eluate from the aluminium oxide column is injected into a high performance liquid chromatograph with an electrochemical detector and a Yanapak ODS reversed-phase column (particle size 10 µm, 25 × 0.4 cm ID). The mobile phase is 0.1 M potassium phosphate buffer, pH 3.0, containing pentanesulfonic acid (20 mg/100 ml buffer) with a flow rate of 0.6 ml/min. The detector potential is set at 0.8 V vs the Ag/AgCl electrode. The peak height of noradrenaline is measured and converted to picmoles from the peak height of adrenaline added as an internal standard. The retention time under these conditions is dopamine, 12 mn; noradrenaline, 4.4 min; and adrenaline 7.2 min.

The peak of noradrenaline in the blank incubation may be due mainly to its nonenzymatic formation.

PHENYLETHANOLAMINE N-METHYLTRANSFERASE (PNMT)

Phenylethanolamine N-methyltransferase (PNMT), also referred to as nor-adrenaline N-methyltransferase (EC 2.1.1.28), is the enzyme that catalyzes the formation of adrenaline from noradrenaline (Axelrod, 1962, 1966). The enzyme activity is high in the adrenal gland, and is also detected in specific brain regions of rats by a highly sensitive radioassay (Saavedra et al., 1974). Highly sensitive radioisotopic methods have been used for the assay of PNMT (Parvez, H. and Parvez, S, 1973; Saavedra et al., 1974). PNMT activity is also demonstrated in human brain regions using radioassays (Nagatsu et al., 1977; Vogel et al., 1976). All of the investigators used radioassays to measure PNMT activity in brain, but a recently developed HPLC—ECD method provides a rapid, sensitive and accurate technique for measuring PNMT activity, and permits the measurement of PNMT activity in all rat and human brain regions (Borchardt et al., 1977; Parvez et al., 1983; Trocewicz et al., 1982a).

Principle (Trocewicz et al., 1982a, b). PNMT activity is assayed with noradrenaline as a substrate based on the measurement of adrenaline by HPLC with electrochemical detection. Commercially available noradrenaline contains about 0.27% of a contaminant which behaved like adrenaline in HPLC and can be removed by recrystallization to reduce the blank value. Enzymatically formed adrenaline is adsorbed on an aluminium oxide column, eluted with 0.5 M hydrochloric acid, separated by high-performance reversed-phase paired-ion chromatography and measured with ECD. 3,4-Dihydroxy-benzylamine is added to the incubation mixture as an internal standard after the reaction. This assay is very sensitive and 0.5 pmol of epinehrine formed enzymatically can be detected.

The brain and adrenal tissues are homogenized in 5 volumes of 0.32 M sucrose in a glass Potter homogenizer.

The standard incubation mixture consists of the following components to a total volume of 250 µl (final concentrations in parentheses): 10 µl 0.01 M pargyline (a monamine-oxidase inhibitor) in 0.01 M HC1 (0.4 mM); 50 µl 0.5 M Tris-HCl buffer, pH 8.0 (0.1 M); 15 µl 0.3 mM S-adenosylmethionine (18 µM); 20 µl 0.2 mM noradrenaline (16 µM); 100 µl 0.32 M sucrose (128 mM) containing homogenized tissues as enzyme, and water. The blank reaction mixture is incubated either without enzyme or with boiled enzyme (90° C for 5 min). Adrenaline (15 pmol) is added to another no-enzyme blank incubation as a standard.

Incubation is carried out at 37° C for 60 min, and the reaction is stopped with 600 µl 0.42 M perchloric acid containing 1.55 mg disodium EDTA and

3.12 mg $Na_2S_2O_5$, and 15 pmol dihydroxybenzylamine as an internal standard, in an ice-bath. After stopping the reaction, 100 µl homogenate are added to the no-enzyme blank and no-enzyme standard tubes. After 10 min, 200 µl 0.8 M potassium carbonate are added to remove excess perchloric acid, and 1 ml 0.5 M Tris-HCl buffer (pH 8.5) is added to adjust the pH to 8.0—8.5. The mixture is centrifuged at $1600 \times g$ for 10 min at 4° C. The clear supernatant is passed through a column (0.4 cm ID) containing 100 mg of aluminium oxide. The column is washed with 2 ml 0.05 M Tris-HCl buffer (pH 8.5) and 5 ml water twice, then 100 µl 0.5 M HCl. All of these washing solutions are previously cooled in ice before passing through the column. Adsorbed noradrenaline, adrenaline and dihydroxybenzylamine are eluted with 200 µl 0.5 M HCl; 50 µl the eluate are injected into the high-performance liquid chromatograph equipped with an electrochemical detector and a column packed with Nucleosil 7 C_{18} (25×0.4 cm ID). The mobile phase is 0.1 M sodium phospahte buffer (pH 2.6) containing 5 mM sodium pentanesulfonate and 0.5% acetonitrile, at a flow-rate of 0.9 ml/min; the detector potential is set at 0.6 V vs an Ag/AgCl electrode. The chromatography is peformed at 21° C. Under these conditions, the retention times are: solvent front 2.2 min, noradrenaline 5.0 min, adrenaline 8.0 min and dihydroxybenzylamine 9.8 min. The adrenaline formed enzymatically by PNMT is calculated from the equation

$$\frac{R(E) - R(B)}{R(B + S) - R(B)} \times 15 \ (\text{pmol})$$

where R is the ratio of peak height (peak height of adrenaline/peak height to dihydroxybenzylamine), $R(E)$ being that from the enzyme incubation, $R(B)$ from the no-enzyme or boiled-enzyme incubation, and $R(B + S)$ from the no-enzyme or boiled-enzyme plus adrenaline (internal standard, 15 pmol) incubation.

Adrenaline, noradrenaline and dihydroxybenzylamine can be measured with very high sensitivity by this HPLC—ECD method. A linear response of the peak height of the electrochemical detector for the amount of adrenaline injected is observed from 0.1 pmol to 1 nmol. L-Noradrenaline bitartrate as substrate contains about 0.27% of an adrenaline-like compound. This contaminant gives a high blank value, thus decreasing the sensitivity of the PNMT assay. Purification of noradrenaline by recrystallization decreases the contaminating substance down to 0.012%.

An assay procedure for PNMT activity using HPLC—ECD was first reported by Borchardt *et al.* (1977). They injected the supernatant of the deproteinized reaction mixture directly into the HPLC system. Although their method was very simple, the sensitivity was not enough to measure rat brain PNMT activity except in the hypothalamus. We have tried to increase the sensitivity of PNMT assay by HPLC—ECD. First, we isolate adrenaline formed enzymatically with an alumina column. Secondly, a reversed-phase column is used instead of an ion-exchange column. This reversed-phase column has a high capacity, and good separation is obtained for adrenaline.

Thirdly, we use highly purified noradrenaline as a substrate. Therefore, the limit of the sensitivity in the present method depends only on endogenous adrenaline in the crude enzyme. The content of endogenous adrenaline is low in the brain and non-enzymatic reaction hardly occurs. Consequently, about 0.5 pmol of adrenaline formed enzymatically can be detected in the present method.

CONCLUSION

HPLC—ECD methods for the assay of enzyme activity are highly sensitive. HPLC—ECD methods are inexpensive as compared to radioassays. With proper sample pre-treatments, one column can be used for several hundred samples. However, care must be taken over the fact that electrochemical detection is not highly specific, and the use of a proper blank (control) is important. Another drawback of HPLC—ECD methods is the time required for one assay, but the combined use of an automatic sampler and a data processor makes it possible to measure as many as 50 samples within 24 h.

REFERENCES

Axelrod, J. (1962). Purification and properties of phenylethanolamine-N-methyltransferase. *J. Biol. Chem. 237*, 1657—1660.
Axelrod, J. (1966). Methylation reactions in the formation and metabolism of catecholamines and other biogenic amines. *Pharmacol. Rev. 18*, 95—113.
Bailey, S. W., and Ayling J. E. (1978). Separation and properties of the 6-diastereoisomers of 1-*erythro*-tetrahydrobiopterin and their reactions with phenylalanine hydroxylase. *J. Biol. Chem. 253*, 1598—1605.
Blank, C. L. and Pike R. (1976). A novel, inexpensive, and sensitive method for analysis of tyrosine hydroxylase activity in tissue samples. *Life Sci. 18*, 859—866.
Borchardt, R. T., Vincek, W. C., and Grunewald, G. L. (1977). A liquid chromatographic assay for phenylethanolamine-N-methyltransferase. *Anal. Biochem. 82*, 149—157.
Brenneman, A. R., and Kaufman, S. (1964). The role of tetrahydropteridines in the enzymatic conversion of tyrosine to 3,4-dihydroxyphenylalanine. *Biochem. Biophys. Res. Commun. 17*, 177—183.
Creveling, C. R., Daily, J. W., Witkop, B., and Udenfriend, S. (1962). Substrates and inhibitors of dopamine-β-hydroxylase. *Biochim. Biophys. Acta 64*, 125—134.
Culvenor, A. J., and Lovenberg, W. (1983). Analysis of aromatic L-amino acid decarboxylase. In: *Methods in Biogenic Amine Research*, S. Parvez, T. Nagatsu, I. Nagatsu, and H. Parvez, (eds.), Elsevier Science Publishers b.v., Amsterdam, pp. 375—384.
Dixon, M., Webb, E. C., Thorne, C. J. R., and Tipton, K. F. (1979). Enzyme techniques. In: *Enzymes*, 3rd edn, Longman, London, pp. 7—22.
Friedman, S., and Kaufman, S. (1965). 3,4-Dihydroxyphenylethylamine hydroxylase. Physical properties, copper content, and role of copper in the catalytic activity. *J. Biol. Chem. 240*, 4763—4773.
Hidaka, H., Nagatsu, T., Takeya, K., Takeuchi, T., Suda, H., Kojiri, K., Matsuzaki, M., and Umezawa, H. (1969). Fusaric acid, a hypotensive agent produced by fungi. *J. Antibiot. 22*, 228—230.
Kato, T., Horiuchi, S., Togari, A., and Nagatsu, T. (1981). A sensitive and inexpensive high-performance liquid chromatographic assay for tyrosine hydroxylase. *Experientia 37*, 809—810.
Kaufman, S. (1967). Metabolism of the phenylalanine hydroxylase cofactor. *J. Biol. Chem. 242*, 3934—3943.

Kissinger, P. T., Brunlett, C. S., Kavis, G. C., Felice, L. J., Riggin, R. M., and Shoup, R. E. (1977). Recent developments in the clinical assessment of the metabolism of aromatics by high perfromance, reverse-phase chromatography with amperometric detection. *Clin. Chem. 23*, 1449—1455.

Lerner, P., Nosé, P., Ames, M. M. and Lovenberg, W. (1978). Modification of the tyrosine hydroxylase assay. Increased enzyme activity in the presence of ascorbic acid. *Neurochem. Res. 3*, 640—651.

Levine, E. Y., Levenberg, B., and Kaufman, S. (1960). The enzymatic conversion of 3,4-dihydroxyphenylethylamine to norepinephrine. *J. Biol Chem. 235*, 2080—2086.

Lovenberg, W., Weissbach, H., and Udenfriend, S. (1962). Aromatic L-amino acid decarboxylase. *J. Biol. Chem. 237*, 89—93.

Matsui, H., Kato, T., Yamamoto, C., Fujita, K., and Nagatsu, T. (1981). Highly sensitive assay for dopamine-β-hydroxylase activity in human cerebrospinal fluid by high performance liquid chromatography—electrochemical detection: properties of the enzyme. *J. Neurochem. 37*, 289—296.

Matsuura, S., Sugimoto, T., Hasegawa, H., Imaizumi, S., and Ichiyama, A. (1980). Studies on biologically active pteridines III. The absolute configuration at the C-6 chiral center of tetrahydrobiopterin cofactor and related compounds. *J. Biochem. 87*, 951—957.

Nagatsu, T. (1977). Dopamine-β-hydroxylase in blood and cerebrospinal fluid. *Trends Biochem. Sci. 2*, 217—219.

Nagatsu, T. (1983). Analysis of monooxygenases in catecholamine biosynthesis; tyrosine hydroxylase and dopamine β-hydroxlase. In: *Methods in Biogenic Amine Research* (S. Parvez, T. Nagatsu, I. Nagatsu, and H. Parvez, eds.), Elsevier Science Publishers b.v., Amsterdam, pp. 329—357.

Nagatsu, T., Kato, T., Numata-Sudo, Y., Ikuta, K., Sano, M., Nagatsu, I., Kondo, Y., Inagaki, S., Iizuka, R., Hori, A., and Naabayashi, H. (1977). Phenylethanolamine-*N*-methyltransferase and other enzymes of catecholamine metabolism in human brain. *Clin. Chim. Acta 75*, 221—231.

Nagatsu, T., Kuzuya, H., and Hidaka, H. (1967). Inhibition of dopamine-β-hydroxylase by sulfhydryl compounds and the nature of the natural inhibitors. *Biochim. Biophys. Acta 139*, 319—327.

Nagatsu, T., Levitt, M., and Udenfriend, S. (1964a). Tyrosine hydroxylase. The initial step in norepinephrine biosynthesis. *J. Biol. Chem. 239*, 2910—2917.

Nagatsu, T., Levitt, M., and Udenfriend, S. (1964b). A rapid and simple radioassay for tyrosine hydroxylase activity. *Anal. Biochem. 9*, 122—126.

Nagatsu, T., Mizutani, K., Sudo, K., and Nagatsu, I. (1972). Tyrosine hydroxylase in human adrenal glands and human pheochromocytoma. *Clin. Chim. Acta 39*, 417—424.

Nagatsu, T., Oka, K., and Kato, T. (1979a). Highly sensitive assay for tyrosine hydroxylase activity by high performance liquid chromatography. *J. Chromatogr. 163*, 247—252.

Nagatsu, T., Yamamoto, T., and Kato, T. (1976). A new and highly sensitive voltammetric assay for aromatic L-amino acid decarboxylase activity by high-performance liquid chromatography. *Anal. Biochem. 100*, 160—165.

Oka, K., Kato, T., Sugimoto, T., Matsuura, S., and Nagatsu, T. (1981). Kinetic properties of tyrosine hydroxylase with natural tetrahydrobiopterin as cofactor. *Biochim. Biophys. Acta 661*, 45—53.

Oka, K., Ashiba, G. Sugimoto, T., Matsuura, S., and Nagatsu, T. (1982). Kinetic properties of tyrosine hydroxylase purified from bovine adrenal medulla and bovine caudate nucleus. *Biochim. Biophys. Acta 706*, 188—196.

Parvez, H., and Parvez, S. (1973). Micro-radioisotopic determination of enzymes, catechol-*O*-methyltransferase, phenylethanolamine *N*-methyltransferase and monoamine oxidase in a single concentration of tissue homogenate. *Clin. Chim. Acta 46*, 85—90.

Parvez, H., Parvez, S., and Nagatsu, T. (1983). The assay of phenylethanolamine *N*-methyltransferase. In: *Methods in Biogenic Amine Research*, S. Parvez, T. Nagatsu, I. Nagatsu, and H. Parvez, eds.), Elsevier Science Publishers b.v., Amsterdam, pp. 399—415.

Rahman, M. K., Nagatsu, T., and Kato, T. (1980). New and highly sensitive assay for L-5-hydroxytryptophan decarboxylase activity by high-performance liquid chromatography—

voltammetry. *J. Chromatogr. 221*, 65—270.

Saavedra, J. M., Palkovitz, M., Brownstein, M., and Axelrod, J. (1974). Localization of phenylethanolamine-*N*-methyltransferase in rat brain nuclei. *Nature 248*, 695—696.

Snyder, L. R., and Kirkland, J. J. (1979). In: *Introduction to Modern Liquid Chromatography*, John Wiley and Sons, New York, pp. 168—268.

Sperk, G., Galhaup, I., Scholögl, E., Hörtnagl, H., and Hornykiewicz, O. (1980). A sensitive and reliable assay for dopamine-*β*-hydroxylase in tissues. *J. Neurchem. 35*, 972—976.

Togari, A., Kano, H., Oka, K., and Nagatsu, T. (1983). Simultaneous simple purification to tyrosine hydroxylase and dihydropteridine reductase. *Anal. Biochem. 132*, 183—189.

Trocewicz, J., Oka, K., and Nagatsu, T. (1982a). Highly sensitive assay for phenylethanolamine *N*-methyltransferase activity in rat brain by high-performance liquid chromatography with electrochemical detection. *J. Chromatogr. 277*, 407—413.

Trocewicz, J., Oka, K., Nagatsu, T., Nagatsu, I., Iizuka, R., and Narabayashi, H. (1982b). Phenylethanolamine-*N*-methyltransferase activity in human brains. *Biochem. Med. 27*, 317—324.

Vogel, W. H., Lewis, L. E., and Boehma, D. H. (1976). Phenylethanolamine-*N*-methyltransferase activity in various areas of human brain, tissues, and fluids. *Brain Res. 115*, 357—359.

Waymire, J. C., Bjur, R., and Weiner, N. (1971). Assay of tyrosine hydroxylase by coupled decarboxylation of dopa formed L-^{14}C-tyrosine. *Anal. Biochem. 43*, 588—600.

Progress in HPLC, Vol. 2, pp. 307—328
Parvez *et al.* (Eds)
© 1987 VNU Science Press

Coulometric detection of brain monoamines and their metabolites: relevance in the study of functionality of central monoaminergic systems

FRANCA PONZIO*, GUIDO ACHILLI, and CARLO PEREGO

Instituto di Ricerche Farmacologiche, 'Mario Negri', Via Eritrea 62, 20157 Milano, Italy

INTRODUCTION

In the recent history of neurochemistry, few substances have had the impact of the monoamines, serotonin (5-HT), dopamine (DA), noradrenaline (NA) and adrenaline (A) on the understanding of central effects of drugs. In the last 25—30 years an enormous number of publications has appeared, offering new theories or clarifications of their neurobiological roles. Research in this field has been helped too by the development of increasingly sensitive-analytical methods.

The first neurotransmitter identified in the mammalian central nervous system (CNS) was NA, described by Von Euler in 1946; its regional distribution was investigated by Vogt in 1954, and in the same year Amin *et al.* (1954) identified 5-HT in brain tissue. More recently, Weil-Malherbe and Bone (1952) used spectrophotofluorimetry to detect DA in mammalian CNS, and Carlsson (1959), using the same method, determined it quantitatively in various brain areas, finding a high concentration in the basal ganglia. In 1962 Falch *et al.* studied the distribution of these monoamines in histological preparations by histochemical fluorescence microscopy and by the same method Dahlstrom and Fuxe (1964) 'mapped' the neuronal pathways of DA, NA and 5-HT. Adrenaline was 'mapped' in the CNS more recently (Hökfelt *et al.*, 1974; Palkovits, 1980).

Although determination of brain neurotransmitter content is useful in locating and mapping monoaminergic neural systems from the neuroanatomical point of view, it gives only partial information on CNS function, and is not generally very useful in clarifying the mechanism of action of drugs. Basal neurotransmitter levels remain constant as the system tends towards equilibrium, changes in synthesis corresponding to like changes in use. The function of monoaminergic neuronal systems can thus be assessed by analysing the rate of formation or disappearance of a neurotransmitter after inhibiting its

---

* Present address: ISF SpA via L de Vinci 1 20090 Trezzano S/N (Milano) Italy.

synthesis or catabolism (Costa and Neff, 1970; Glowinski, 1970). Another method is to determine the changes in levels of the neurotransmitter's metabolite(s) assuming that the rate of elimination from the brain does not change in relation to changes in the rate of formation (Di Giulio et al., 1978; Ponzio et al., 1981a, b; Westerink, 1979a, b; and Westerink and Korf, 1976; Westerink and Spaan, 1982a, b).

The first kinetic model is theoretically more correct, but it raises technical problems since DA and NA follow more than one metabolic pathway, as well as practical difficulties since correct kinetic assessment calls for many experimental groups. This approach is therefore not very easy. The second system, determination of changes in metabolite levels, was therefore proposed as an index of neuronal function (Di Giulio et al., 1978; Ponzio et al., 1981a, b; Westerink, 1979a, b; Westernik and Korf, 1976; Westerink and Spaan, 1982a, b) on the basis of the fact that an increase in synthesis corresponds to increased metabolism, hence increased metabolite formation. Diminished synthesis results in reduced metabolism with a consequence drop in metabolite levels.

Much has been written on the debate on which neurotransmitter metabolite is the best index of neuronal function. The problem is straight forward for 5-HT, as it follows a sole metabolic pathway, forming 5-hydroxyindole-acetic acid (5HIAA) through the action of monoamine oxidase (MAO) (Cooper et al., 1982). For NA and DA, however, there is more than one catabolic route (Cooper et al., 1982) and since the catabolism of these two transmitters depends both on MAO inside the neuron and on catechol-O-methyltransferase (COMT) outside (Ponzio et al., 1981a, b; Westerink and Spaan, 1982b), it has been suggested that all the metabolites formed represent changes in synthesis or release of the neurotransmitter (Ponzio et al., 1981b; Racagni et al., 1982; Westerink and Spaan, 1982b).

Therefore simultaneous determination of all the main neurotransmitter metabolites might provide a fuller picture of the neuron's functional status.

There are four basic methods for determining metabolite levels: spectrofluorimetry, radioenzymatic assay, GCMS and HPLC with electrochemical detection.

1. Spectrofluorimetry was among the first methods used, but as its sensitivity is limited, samples must either be conjugated with fluorescent substances or transformed to fluorophores in order to improve it (Hjemdahl, 1984).

2. Radioenzymatic assays are highly selective, and offer the same sensitivity as GCMS. However, they require several manipulations, enzymatic preparation and purification, and are not very reproducible (Hjemdahl, 1984; Hjemdahl et al., 1979; Mefford et al., 1981). Additionally, as they are based on the concept of O-methylation, they can only be used for non-O-methylated compounds.

3. Gas chromatography coupled with mass spectrometry (GCMS) calls for costly instruments and lengthy preparation, and the sample must be derivatized before assay.

4. HPLC with electrochemical detection is one of the most widely used

methods for determining precursor neurotransmitter monoamines and their metabolites. It is inexpensive, easy to apply and highly flexible (Hjemdahl, 1984; Hjemdahl *et al.*, 1979; Mefford *et al.*, 1981). Greater details of this method are given below.

In the sections below, sample preparation methods, the eluent mobile phases and chromatographic columns, and the electrochemical detectors themselves will be discussed.

SAMPLE PREPARATION

Samples of biological fluids, plasma, urine or tissue generally require a purification step to eliminate any interfering compounds and to protect the column from solid residues which could damage it. The sample can be purified on alumina (Al$_2$O$_3$), by extraction with organic solvents, on an ion-exchange column, on boric acid gel, by ion-pairs or with diphenylborate. Generally these purification steps are followed by concentration of the sample before column chromatography; this increase in the sample concentration indirectly enhances the sensitivity of the method. Not all workers, however, insist on pre-purification and many methods describe direct injection of the supernatant phase after centrifugation (Anderson *et al.*, 1980; Dayton *et al.*, 1979; Frattini *et al.*, 1982; Lackovic *et al.*, 1981; Langlais *et al.*, 1980; Nielsen and Johnston, 1982; Santagostino *et al.*, 1982; Tagari *et al.*, 1984; Ward *et al.*, 1983; Wightman *et al.*, 1977).

An appraisal will be made here of the advantages and drawbacks of the various extraction methods in relation to the type of sample, bearing in mind that most of these methods were developed before HPLC—ED was available, and were later micronized to adapt them to this technique.

Adsorption and extraction using alumina

Separation of neurotransmitter catecholamines by alumina adsorption is one of the most frequently used approaches (Fenn *et al.*, 1978; Freed and Asmus, 1979; Hallman *et al.*, 1978; Hjemdahl, 1984; Hjemdahl *et al.*, 1979; Jonsson *et al.*, 1980; Keller *et al.*, 1976; Maruyama *et al.*, 1980; Mefford *et al.*, 1981; Moyer *et al.*, 1979; Palermo *et al.*, 1979; Perry *et al.*, 1983; Refshauge *et al.*, 1974; Riggin *et al.*, 1976; Riggin and Kissinger, 1977; Wagner *et al.*, 1979; Ward *et al.*, 1983); the specificity of adsorption of the catechol groups in an alkaline medium ensures an extract with relatively high purity. The procedure is simple and rapid and final recovery is usually high. However, both the final recovery and the purity of the extract depend closely on how well the alumina was initially purified and activated, on the pH of adsorption of the sample, and on the presence of antioxidants. This explains the differences in recovery reported by different authors (Anton and Sayre, 1962).

It is also worth noting that Wollm aluminum oxide has been indicated as the most suitable for ensuring good adsorption of the sample and an extract containing relatively few impurities (Adams and Marsden, 1982; Anton and Sayre, 1962).

Preparation of the alumina involves boiling it for about an hour with 2N HCl (1:5 by volume) and aspiring off the acid phase; the Al_2O_3 is then repeatedly washed with distilled water until the final pH is higher than 3.5. The alumina is then dried on a Büchner funnel, washed with permanganate re-distilled water, dried again and activated in an oven for 2 h at 200—300°C. Thereafter must be stored in an anhydrous atmosphere. Boiling in HCl and the final drying temperature are essential to a high yield. In fact, the adsorption capacity of alumina activated at 100°C is much lower (Weil-Malerbe, 1971).

The optimal pH for catechol adsorption on Al_2O_3 is 8.5 but in a range between 8.0 and 8.6 the percentage recovery does not vary much; under pH 8, however, it drops rapidly (Anton and Sayre, 1962).

Samples are stored carefully on ice and are brought to alkaline pH by adding a preset amount of Tris 0.5 M buffer, (pH 8.6) containing 0.1% EDTA and 0.05% sodium metabisulphite. These two additions are necessary to protect the sample from oxidation by atmospheric oxygen, which can occur easily in this pH range (Anton and Sayre, 1962) especially in the presence of heavy metal traces. In the absence of antioxidant, recovery may be reduced by up to 90%.

After adsorption of the sample, the alumina must be washed at least three times with permanganate—re-distilled water containing 1% Tris-buffer to buffer any acidity of the water which would interfere with the stable retention of the sample on the alumina. Final elution is in perchloric or acetic acid, the concentrations varying with different authors from 0.05 to 0.6 M for $HClO_4$ and from 0.5 to 1 M for acetic acid. The low molarities of perchloric acid are generally employed when the sample is eluted in large volumes (100 µl or more) and in cases where the acid front could interfere with peak resolution. The high molarities are used for eluting the O-methylated metabolites or their precursor L-DOPA together with the catecholamines (Benfenati et al., 1982). We generally use from 0.25 M for determining catecholamines in plasma and, up to 0.4 M for low concentrations in tissues, which have to be eluted in small volumes.

Recovery is calculated for each individual sample in relation to the internal standard (DHBA, α-met DA, epinine) added to the sample during initial preparation.

Purification with organic solvents
Using organic solvents to extract neurotransmitter monoamines is definitely not a selective method as it is based on separation of acid compounds from basic ones. However, it is rapid and gives a very high final recovery. The extraction is based on the concept that both the monoamines and their acid metabolites can be extracted from an aqueous solution in the presence of a sufficiently polar organic solvent such as butanol, ethyl acetate or acetone. The sample can be sonicated in 0.4 N perchloric acid (Koch and Kissinger, 1980; Mefford et al., 1980; Moyer et al., 1979; Ponzio et al., 1981a, b; Ponzio and Jonsson, 1978; Reinhard et al., 1980) or directly in the organic

solvent (Cross and Joseph, 1981; Fleisher *et al.*, 1979; Loullis *et al.*, 1979; Ponzio and Jonsson, 1979; Sasa and Blank, 1977, 1979; Sasa *et al.*, 1978). In the former case the organic phase is added to the aqueous phase 1:2 (by volume), and the mixture is shaken thoroughly to form an emulsion (which an also be produced using a sonifer, 30 s/sample).

Samples are eluted adding a non-polar solvent such as heptane, cyclohexane or iso-octane (1:2, 1:3 by volume) which, in the presence of an acid (or basic) solution, facilitates the transfer of the amine [or acid metabolite(s)] from the organic phase, which has become non-polar, to the aqueous phase (perchloric acid or 0.35 M sodium acetate). If the sample is heavily lipid-contaminated, further purification with chloroform (1:0.25 by volume) is necessary (Ponzio and Jonsson, 1979).

Possibly the greatest drawback of this extraction method is that contaminants appear sporadically in different batches of solvent; these are extracted with the amines (or acids) and produce unknown interfering peaks. To overcome this problem organic solvents must be pre-washed with the same perchloric or hydrochloric acid used for final elution.

Another problem lies in the nature of the method itself: various types of amines or acidic compounds being extracted simultaneously result in the possibility of interference or — to get round the problem — in longer sample retention times.

Purification with ion-exchange microcolumns

Microcolumns packed with ion-exchange resins have been used for pre-purification of monoamines and their metabolites, as well as tissue and biological fluid extracts. In the latter case in particular, purification on a column takes longer but is preferable to extraction on alumina as it eliminates many of the impurities that are co-extracted with the sample (e.g. uric acid) (Bunyagidj and Girard, 1982). These impurities are most evident in reverse phase chromatography (Hjemdahl, 1984) and are almost absent when the sample is chromatographed on an ion-exchange column. Thus the method for preparing the sample depends mainly on the type of column to be used for chromatography.

Sample preparation for purification on microcolumns varies according to the type of resin used for packing. Cation exchange Bio Rad (Cat. No. 1892202) (Orsulak *et al.*, 1983; Riggin and Kissinger, 1977) and Bio-Rex 70, 50—100 mesh (Na$^+$) (Jackman *et al.*, 1984; Lyness *et al.*, 1980; Sandhu and Sponsor, 1968; Shoup and Kissinger, 1977) are those most widely used for purification of biogenic amines from plasma, urine and tissue extracts. Westerink and co-workers (Westerink and Mulder, 1981; Westerink and Spaan, 1982a, b; Westerink and Van Oene, 1980) 'micronized' their previous purification methods on Sephadex G-10 resin (Pharmacia) (Westerink and Korf, 1977) adapting them to liquid chromatography with electrochemical detectors. They used fractioned elution methods; in one fraction the amines (DA, 3-MT) were eluted and in the second the acids (HVA, DOPAC). To determine brain contents of MHPG—SO$_4$ samples were prepurified on

DEAE Sephadex A-25 (Pharmacia) columns (Hornspenger et al., 1984), then undergo enzymatic hydrolysis. The resulting free MHPG is extracted in ethyl acetate, brought to dryness and resuspended in 0.1 N PCA.

Some authors use only hydrolysis without prepurification of the sample on the DEAE column (Krstulovic, 1981; Warnhoff, 1984) but in our experience this preliminary step is important to eliminate interfering peaks, an essential condition for good resolution in relatively short times.

Columns containing Amberlite (G-50, 100—200 mesh, Sigma), as packing material, are not generally widely used (Chiu et al., 1981; Warsh et al., 1979) but are indicated for separating DA from 5HT and normetanephrine (NA metabolite).

Generally on-column purification calls for large volumes of eluent, so samples must sometimes be concentrated before final chromatography (Jackman et al., 1984; Orsulak et al., 1983; Riggin and Kissinger, 1977; Shoup and Kissinger, 1977; Warsh et al., 1979).

Purification on boric acid gel
This method is not widely used although it is sufficiently specific and speedy. The principle of extraction is similar to that of alumina, based on the ability of boric acid gel to form, in a neutral or slightly alkaline medium, a specific complex with the *cis*-diols of various compounds including the catecholamines, and to release these compounds in an acid medium (Higa et al., 1977). The technique has been applied to purify and concentrate urine samples after adsorption on alumina or after prepurification on an ion-exchange column (Moyer et al., 1979).

Ion-pairs extraction
The ion pairs most widely described in the literature are *bis*-diethylhexylphosphoric acid (DEHPA) and diphenylborate (DPBEA). Alkylamines and long-chain alkylphosphoric acids can be used as liquid ion exchangers, in which case the separation method is the same as for conventional ion-exchange chromatography (McCaman et al., 1972; Temple and Gillespie, 1966). The liquid ion exchanger DEHPA is dissolved in an organic solvent immiscible with water, generally chloroform, and emulsified with the sample containing the ions to be separated.

The critical point of this method is the pH at which the cations in the sample are adsorbed on the ion pair, and at which the compound under investigation is eluted from the ion pair. The pH values for adsorption and elution are characteristic for each compound, and the literature describes curves of pH recovery (McCaman et al., 1972; Ponzio et al., 1981a; Temple and Gillespie, 1966). However, this method is neither easy nor straightforward, and calls for experience and manual skill in order to achieve high yields (70—80%) and reproducible results.

The principle on which DPBEA extraction is based is similar to that for extraction on boric acid gel: in this case too a complex is believed to form between the borate and the catecholamine diol group. This reaction occurs in an alkaline medium, with formation of an ion pair (Smedes *et al.*, 1982). DPBEA, an inorganic compound, is dissolved in an aqueous solution, such as phosphate buffer (pH 8—9) and left to react with the diol groups. The resulting ion pair is adsorbed in an organic phase and eluted in an acid medium.

For this method the pH of adsorption and elution are not so critical, but optimum operating conditions have nevertheless to be established carefully to ensure good recovery and avoid co-extraction of interfering compounds (Smedes *et al.*, 1982).

Direct injection
Direct injection into the column, with no pre-purification, offers the advantage of shortening sample preparation times, but has the disadvantage of involving a sample containing impurities, i.e. compounds that can interfere with the resolution of the monoamines and their metabolites. Samples must be carefully deproteinized and filtered before injection into the column which is generally protected by a pre-column. The sample can be deproteinized with 0.1—0.2 N perchloric acid (Anderson *et al.*, 1980; Dayton *et al.*, 1979; Nielsen and Johnston, 1982), 0.1 M acetic acid (Wightman *et al.*, 1977), trichloroacetic acid (Tagari *et al.*, 1984) or $ZnSO_4/NaOH$ (Lackovic *et al.*, 1981). The choice of acid is important as regards the width of chromatographic front and as regards interfering substances (Lackovic *et al.*, 1981). A compromise must thus be sought between the molarity of the acid which must be high enough to deproteinize the sample but must not give too large an injection front which will interfere with sample resolution.

It is advisable to filter samples before injection using commercial filters in disposable test-tubes or on glass wool.

This method is usually applied to establish the content of monoamines in cerebrospinal fluid (CSF) (Frattini *et al.*, 1982; Langlais *et al.*, 1980; Nielsen and Johnston, 1982; Santagostino *et al.*, 1982; Wightman *et al.*, 1977), plasma (Tagari *et al.*, 1984), and synaptosomal fractions p1, p2 and p3 (Dayton *et al.*, 1979). It is rarely used to analyse the endogenous levels of monoamines in tissues (Anderson *et al.*, 1980; Lackovic *et al.*, 1981) as the time-saving achieved by eliminating the initial extraction step is usually cancelled out during analysis of the sample, when longer elution times are required.

In this type of analysis critical points are the selection of the appropriate elution time, the pH value, the mobile phase, the column and the chromatography temperature, since even small variations can result in superimposed peaks, e.g. DA and ascorbic acid (Dayton *et al.*, 1979).

MOBILE PHASE AND CHROMATOGRAPHIC COLUMNS

Numerous types of commercially available chromatographic columns are used for separation of monoamines and their metabolites, but there are only two main types of packing materials: ion-exchange resin (anionic and cationic) and reverse-phase resins.

Ion-exchange resins

The stationary phase in this separation technique is an ion exchanger bound to a solid support (Wolf, 1969); the ions in the resin can be exchanged with an equivalent number of other ions of the same sign provided by the surrounding solution. Ion-exchange chromatography is thus particularly suited for analysis of ionized or ionizable compounds. Ion exchangers consist of two components: the matrix or support (Wolf, 1969) and the functional groups (strong or weak acids, strong or weak bases).

For liquid chromatography cationic-exchange resins are used, with a strong acid as functional group such as sulphonates ($-SO_3^-$), or anionic-exchange resins with a strong base as the functional group, such as quaternary amines (trimethylammonium, dimethylhydroxymethyl ammonium). These matrices were first used in liquid chromatography coupled with electrochemical detection because, in view of the nature of the detector electrode, the mobile phases containing organic solvents, so useful in reverse-phase chromatography, could not be used.

On account of the resin granule size (35—40 µm) these columns are characterized by low efficiency, although with high selectivity. The use of ion-exchange columns packed with smaller granule size resins (10 µm) improves the efficiency and gives better resolution of NA from the front (Adams and Marsden, 1982; Hjemdahl, 1984; Hjemdahl *et al.*, 1979). These types of resin are often used for analysing plasma and urine as many of the interfering peaks (e.g. uric acid) are not retained by the column and elute in the front (Bunyagidj and Girard, 1982; Hjemdahl, 1984).

One major problem with these resins is that one batch is likely to differ widely in efficiency and resolution from another (Adams and Marsden, 1982; Jonsson *et al.*, 1980). This is why only a few columns prepacked with ion-exchange resins are available on the market.

The eluent phase most widely used is a citrate/acetate buffer of pH 5.1—5.8 for chromatography of basic compounds (amines) on cation-exchange resins, and of pH 4—4.7 for chromatography of acid compounds (acid metabolites) on anion-exchange resins (Anderson *et al.*, 1980; Dayton *et al.*, 1979; Felice and Kissinger, 1978; Fenn *et al.*, 1978; Goldman *et al.*, 1980; Hallman *et al.*, 1978; Jonsson *et al.*, 1980; Keller *et al.*, 1976; Ponzio *et al.*, 1981a, b; Ponzio and Jonsson, 1978, 1979; Sasa *et al.*, 1978; Warsh *et al.*, 1979; Wightman *et al.*, 1977). Raising the molarity of the buffer increases the rate of elution which, naturally, is reduced when the molarity is reduced. Small variations in pH do not substantially affect retention time, but adding a small amount of methanol reduces it. The percentage of methanol is critical for the carbon paste electrode life (not more than 0.5—3%).

Reverse-phase resins

Reverse-phase chromatography is unquestionably the best method of separating a mixture of biogenic substances containing compounds with acid and basic functional groups. The stationary phase in this separation method is made of apolar groups (C_2, C_3, C_6, C_8, C_{18}) covalently bound to a silica support, hence the hydrophobic nature of the resin. Separation is governed by the hydrophobic nature of the molecules present in the surrounding solution and delay in the elution of individual compounds depends on the extent to which they interact stably with the hydrophobic aliphatic carbon chain (Molnar and Horvath, 1976; Snyder and Kirkland, 1978). Retention time (T_R) and efficiency are directly proportional to the number and length of the hydrocarbon groups of the bound phase, therefore octadecylsilane (C_{18}-ODS) is the most widely used packing material in reverse phase chromatography. However, the longer the carbon atom chain, the easier it is to damage it, so C_8 columns are often used (Molnar and Horvath, 1976; Snyder and Kirkland, 1978).

The eluent mobile phase is usually a mixture of buffer and organic solvents such as methanol or acetonitrile. The most widely used buffers are phosphate or citrated phosphate (Chiu *et al.*, 1981; Hjemdahl, 1984; Kilts *et al.*, 1981; Maruyama *et al.*, 1980; Moyer *et al.*, 1979; Nielsen and Johnston, 1982; Orsulak *et al.*, 1983; Shoup and Kissinger, 1977; Wagner *et al.*, 1979; Westerink and Mulder, 1981; Westerink and Spaan, 1982a, b; Westerink and Van Oene, 1980) and citrate—acetate (Alonso *et al.*, 1981; Frattini *et al.*, 1982; Jackman *et al.*, 1984; Mefford *et al.*, 1981; Perry *et al.*, 1983; Santagostino *et al.*, 1982; Tagari *et al.*, 1984), in a pH range of 3.0—3.6 depending on the substances to be chromatographed, though some authors use buffers at pH 2.7 (Langlais *et al.*, 1980) or even over 4 (Frattini *et al.*, 1982; Jackman *et al.*, 1984; Moyer *et al.*, 1979; Tagari *et al.*, 1984).

In our experience a pH 3.0 citrate—acetate buffer gives the best result for resolution of L-DOPA from the front and pH 3.75 for the monoamines and their acid metabolites. Citrate—acetate buffer causes the less interference in the baseline of our working conditions and at our detector sensitivity.

It must however be borne in mind that reverse-phase resins on a silica support are not stable at pH values over 8 or below 2, beyond which limits the resin is irremediably damaged (Molner and Horvath, 1976).

The pH, the molarity of the eluting buffer and the percentage of organic phase are critical in obtaining reproducible chromatographic separations.

The effects of mobile phase pH on retention times of monoamines and their metabolites has been amply described in the literature (Kilts *et al.*, 1981; Moleman and Borstrak, 1982; Molner and Horvath, 1976; Nielsen and Johnston, 1982). As the mobile phase pH increase, acidic samples, HVA, DOPAC and 5HIAA, are deprotonated to a greater extent in relation to their constant of dissociation (pK_a) with a consequent decrease in their capacity factor and hence a reduction in retention time. Aminic compounds are less affected by modifications of the mobile phase pH in a range of 3—6 since their pK_a values are higher (Kilts *et al.*, 1981; Molner and Horvath,1976; Nielsen and Johnston, 1982; Snyder and Kirkland, 1978).

These phenomena are always observed in function of the pK of dissociation of the compounds; separation of substances of similar polarity can be achieved by selecting a buffer with a pH value that ensures complete non-ionization of one of the substances to be analysed; this is thus retained longer on the column than the other compounds which are partially or totally ionized.

Increasing the molarity of the buffer results in longer retention times, the opposite of what happens with ion-exchange resin. This shows there are no appreciable ionic interactions between solutes and stationary phase, the longer retention times being due to hydrophobic interactions (Molner and Horvath, 1976).

Temperature changes also affect T_R, a rise in temperature prolonging T_R but not generally enough to have a substantial effect on peak resolution (Molnar and Horvath, 1976).

Addition of an organic solvent such as methanol or acetonitrile to the elution buffer reduces its polarity, thus increasing its hydrophobicity; raising the percentage of organic phase in the mobile phase results in a corresponding decrease in sample retention time (Kilts *et al.*, 1981; Moleman and Borstrok, 1982; Nielsen and Johnston, 1982).

Reverse-phase columns present considerable variations in efficiency and resolution from one batch to another. Changing the pH of the mobile phase or the percentage of organic solvent may compensate these variations. Variations in T_R may also be noted during routine column use, and can be compensated by small modifications in the eluent phase (Freed and Asmus,1979; Kilts *et al.*, 1981; Riggin and Kissinger, 1977).

Reverse-phase resins with ion pairs added
One way of improving chromatographic separation on reverse-phase columns, especially in the presence of ionized substances which are poorly retained by the column, is to add to the mobile phase a low concentration of counter-ion (less than 1 mM). Counter-ions have an apolar part with high affinity for the column aliphatic groups, and a polar part that interacts with the mobile phase, acting as ion exchanger. The ion pair does not influence retention time of undissociated compounds, which are eluted following the laws normally governing elution on a reverse-phase column, but acts selectively on the T_R of dissociated compounds, which are eluted following the laws governing elution on ion-exchange resin (Kilts *et al.*, 1981; Nielsen and Johnston, 1982; Snyder and Kirkland, 1978; Wagner *et al.*, 1979). The chain length of the counter-ion controls the hydrophobicity of the final ion-pair product and thus the extent of T_R modification of the solute. In this working condition mobile phase pH is critical.

The most widely used counter-ions are sodium octadecylsulphate (ODS) (Chiu *et al.*, 1981; Mefford *et al.*, 1981; Nielsen and Johnston, 1982; Riggin and Kissinger, 1977; Wagner *et al.*, 1979) and sodium heptane sulphonate (Jackman *et al.*, 1984; Kilts *et al.*, 1981; Moyer *et al.*, 1979). The polar part of these compounds is acid, giving the reverse-phase column characteristics of a cation-exchange resin. There are other ion pairs characterized by basic groups,such as tetraethylammonium and tetramethylamine (Potter *et al.*, 1983; Voelter *et al.*, 1980).

ELECTROCHEMICAL DETECTORS

Neurotransmitter monoamines and their metabolites were the first substances analysed using electrochemical detection as they are easily oxidized, constituting a good substrate for amperometric determination (Adams and Marsden, 1982; Keller *et al.*, 1976). As analytical techniques improved, other compounds, such as GABA and acetylcholine (Caudill *et al.*, 1982; Eva and Meek, 1984; Potter *et al.*, 1983), were determined with these detectors, after appropriate treatment of the molecules.

Electrochemical detection is basically very straightforward. The detector consists of an amplifier potentiostat that can amplify and detect low level microampere currents and maintain a constant pre-set potential (Keller *et al.*, 1976). The success of the whole system depends on the design of the electrochemical cell through which the eluate flows. The active part of most amperometric detectors is carbon paste or glassy carbon inserted on a tubular electrode (MetroHm), a rotating electrode disc (Westerink and Mulder, 1981) or a thin-layer cell electrode (BAS, LKB). The latter is the most widely used, since the chamber is small and various electro-active materials can be used for the working electrode (platinum, gold, carbon paste and glassy-carbon).

Electrochemical detection is based on Faraday's law: $Q = nFN$, where the number of coulombs, Q, is directly proportional to the number of moles, N, converted, and the number of electrons, n, involved in the reaction (Bunyagidj and Girard, 1982; Kissinger, 1977). Considering the laminar structure of the working cell, the efficiency of conversion of the electrodes never reaches 100% and is generally low (around 10%) (Bunyagidj and Girard, 1982; Kissinger, 1977). As this is true for the sample and the mobile phase, the noise-to-signal ratio should not be affected by an electrode with greater conversion efficiency.

The possibility of pre-oxidizing (or reducing) the mobile phase reduces the background current to about 40 nA, with a consequent reduction in baseline disturbance, so the electrode can develop greater sensitivity. To achieve this result, a guard cell (ESA model 5020) must be fitted between the pump and the injector and a difference in potential is applied to it, 0.05 V higher than that applied to the working electrode. This eliminates the electro-active impurities in the mobile phase, which is thus electrochemically 'clean' when it reaches the detector.

Using this 'trick' one can employ coulometric electrodes, which have 100% conversion efficiency (Achilli *et al.*, 1985, Bunyagidj and Girard, 1982; Fenn *et al.*, 1978). With the coulometric detector absolute concen-

tration determinations can be made. This is also a way of assessing the purity
of a standard.

Amperometric detectors have been amply described (Adams, 1969;
Adams and Marsden, 1982; Kissinger, 1977) and this part of the chapter will
concentrate mainly on analysis of the sensitivity, stability, reproducibility and
selectivity of coulometric detectors in investigations of neurotransmitter
monoamines and their metabolites. The more technical details can be found
in another chapter in this book.

Detectors used for these analyses (Model 5100 A, 5010 ESA, MA) have
a working electrode with a flow-cell made of porous graphite, through which
the column eluate flows. The total contact between eluate and working
electrode permits 100% oxidation or reduction of the sample (Achilli *et al.*,
1985; Bunyagidj and Girard, 1982; Fenn *et al.*, 1978).

Further details are discussed below regarding selectivity, stability and
sensitivity.

Selectivity

The possibility of two electro-active surfaces (two working electrodes in one
cell) greatly improves the selectivity of analysis. Applying different potentials
to the first and second electrode permits complete resolution of interfering
peaks, which would otherwise have required much longer T_R (Figs 1, 2),
moreover irreversible oxidated peaks can be removed from the chromato-

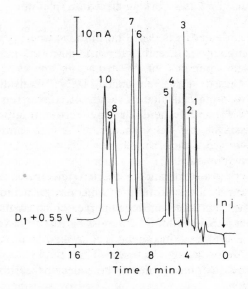

Figure 1. Chromatogram of 10 pmol of external standard: 1. L-DOPA; 2. NA; 3. MHPG; 4.
DOPAC; 5. DA; 6. 5HIAA; 7. TP; 8. HVA; 9. 3-MT; 10. 5-HT; samples were eluted with
the following mobile phase: 90:10 by volume mixture of 2 parts 0.04 M citric acid plus 1 part
of 0.04 M Na_2HPO_4 and methanol, with 2×10^{-3} M heptasulphonic acid sodium salt and 2×10^{-4} M EDTA, final pH 3.

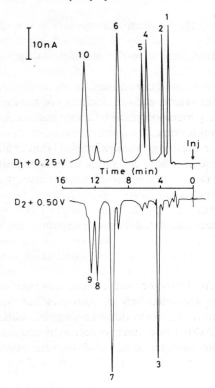

Figure 2. Chromatogram of samples reported in Fig. 1 detected by a dual electrode with an applied potential of +0.25 V at the first electrode (upper part of the chromatogram) and of +0.55 V at the second one (lower part).

gram. To get the best results from a dual-electrode cell, the lower potential should be applied to the first and the higher to the second. This permits separation of groups of compounds with a low oxidation potential (0—+0.25 V) (DA, 5TH, DOPAC, DOPA, 5HIAA, NA) and with high oxidation potential (+0.25—+0.5 V) (Fig. 3).

This method can only be used when the substances oxidized at low potential are completely oxidized at the first electrode, so as not to leave interfering residues at the second one. This can be achieved with working electrodes based on the coulometric principle, which ensure 100% oxidation (reduction) of the sample.

Stability
The average working life of an electrode, especially carbon paste, is erratic on account of various factors: contamination by substances non-specifically adsorbed on the electrode surface, brusque pressure changes, flow variations, microvortices created by scratches on the electrode or support, or by the difference in applied potential (Adams and Marsden, 1982). Glassy-carbon electrodes are less sensitive but more stable over time than carbon paste.

However, they need fairly frequent maintenance (every 20—30 days depending on the sensitivty required) to remove substances non-specifically
adsorbed on the active surface, which has a fairly small exchange area (0.1
cm^2).

The high sensitivity and reproducibility over time characteristic of the
porous graphite electrode are the result mainly of the larger exchange area of
the electrode (5 cm^2). Sensitivity is virtually unaffected in view of the high
ratio between electrode area and percentage contamination. This prolongs
the average working life and means there is no daily degradation. To
maintain high stability, however, 10—35% nitric acid must be flowed through
the electrode cell every 6—7 months (4000—5000 samples).

Sensitivity
The monoamines, their metabolites and precursors, can be easily oxidized or
reduced (Fig. 3). The sum of the intensity of current in oxidation and
reduction results in increased sensitivity. Combining two working electrodes
in a single cell means the sample can first be oxidized then reduced,
developing a higher final current, with consequent better sensitivity. Thus in a
coulometric detector when the first electrode is in the oxidative phase and
the second reductive, sensitivity is doubled (Fig. 4), without any effect on the
noise-to-signal ratio,as it is the result of the algebraic sum of the disturbance
in the oxidation and reduction baseline; thus concentrations of the order of
0.02 pmol NA and A, and 0.05 pmol DA can be determined as a matter of
routine (Fig. 5).

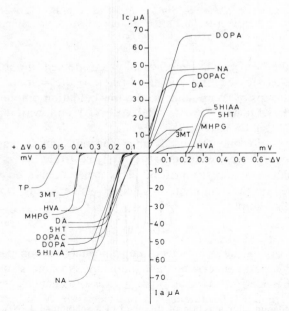

Figure 3. Voltammograms of NA, DA, 5-HT, their metabolites and precursors. Samples
flowed at the concentration of 10^{-5} M.

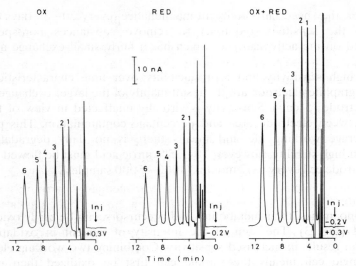

Figure 4. Chromatogram of 10 pmol of external standard: 1. L-DOPA; 2. NA; 3. A; 4. IS; 5. DOPAC; 6. DA. Samples were eluted with the following mobile phase: 85:15 (by volume) mixture of 400 ml 0.1 M COONa plus 250 ml 0.1 M citric acid, and methanol; 1.8×10^{-3} M heptasulphonic acid sodium salt and 2×10^{-4} M EDTA were added to the solution, final pH 4.1 and were detected either in oxidation (left part of the figure) with an applied potential of +0.3 V, or in reduction (central part) with an applied potential of −0.2 V, or in oxidation plus reduction (right part), with applied potentials of +0.3 V at the first electrode and −0.2 V at the second.

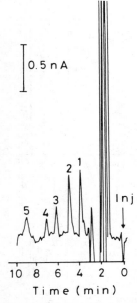

Figure 5. Chromatogram after injection of 0.1 pmol of a solution of 1. NA; 2. A; 3. IS; 4. DOPAC; 5. DA. Samples were eluted with buffer reported in Fig. 1 and were detected in oxidation +0.3 V plus reduction −0.2 V.

This method is easily applied to porous graphite coulometric detectors (Fig. 4) which, in view of the potential ratio between working and reference electrodes, are not sensitive to atmospheric oxygen, which is reduced at higher potentials than those normally used for sample reduction. Amperometric detectors, on the other hand, are highly sensitive to atmospheric oxygen dissolved in the mobile phase, and inert gas has to be bubbled through the buffer during chromatography.

A practical example of this method is shown in Fig. 6 and Table 1. In this case NA and DA neuronal functions were determined in small brain areas such as the neurohypophysis, the adenohypophysis and the pineal gland. The index for assessing neuronal function was the rate of disappearance of NA and DA after inhibition of their synthesis by α-methylparatyrosine, a specific inhibitor of tyrosine hydroxylase activity, the enzyme that regulates catecholamine synthesis. It would be inappropriate in this case to determine the rate of accumulation of the precursor L-DOPA after inhibition of decarboxylase, as this aminoacid is the direct precursor of both catecholamines under investigation (Cooper *et al.*, 1982).

In these brain areas DA disappeared faster than NA, indicating that dopaminergic neurons were functioning more than noradrenergic ones (Cooper *et al.*, 1982). In the areas examined, the sensitivity of the method was such that decreases of the order of 92% in DA levels in the pineal gland could be determined.

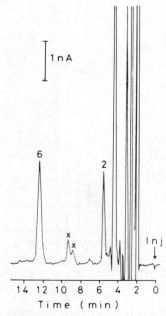

Figure 6. Twenty µl injection of an Al_2O_3 extract from rat neurohypophysis containing NA (Peak No. 2, 0.879 pmol) and DA (peak No. 6, 3.3 pmol).

Table 1. Effect of α-methylparatyrosine on NA and DA concentrations in rat neurohypophysis, adenohypophysis and pineal gland.

	Controls (pg/organ)		α-Methylparatyrosine (pg/organ)	
	NA	DA	NA	DA
Neurohypophysis	372 ± 15	1268 ± 55	145 ± 3.8[a]	286 ± 44[a]
Adenohypophysis	57 ± 1.9	209 ± 11	39 ± 5.6[a]	105 ± 18[a]
Pineal gland	1273 ± 82	331 ± 22	498 ± 59[a]	60 ± 5.5[a]

Rats received α-methylparatyrosine (250 mg/kg i.p.) and were killed 2 h thereafter. Each value represents the mean ± S.E. of six determinations. Statistical significance was analysed by Student's *t*-test.
[a] $p < 0.01$ compared to the respective control.

REFERENCES

Achilli, G., Perego, C., and Ponzio, F. (1985). Application of the dual-cell coulometric detector: a method for assaying monoamines and their metabolites. *Anal. Biochem. 148*, 1—9.

Adams, R. N. (1969). Applications of modern electroanalytical techniques to pharmaceutical chemistry. *J. Pharm. Sci. 58*, 1171—1184.

Adams, R. N., and Marsden, C. A. (1982). Electrochemical detection methods for monoamine measurements *in vitro* and *in vivo*. In: *Handbook of Psychopharmacology, Vol. 15, New Techniques in Psychopharmacology*, L. L. Iversen, S. D. Iversen, and S. H. Snyder (eds.), Plenum Press, New York, pp. 1—74.

Alonso, R., Gibson, C. J., and McGill, J. (1981). Determination of 3-methoxy-4-hydroxy-phenylglycol in urine by high-performance liquid chromatography with amperometric detection. *Life Sci. 29*, 1689—1696.

Amin, A. H., Crawford, T. B. B., and Gaddum, J.H. (1954). The distribution of substance P and 5-hydroxytryptamine in the central nervous system of the dog. *J. Physiol. 126*, 596—618.

Anderson, G. M., Batter, D. K., Young, J. G., Shaywitz, B. A., and Cohen, D. J. (1980). Simplified liquid chromatographic—electrochemical determination of norepinephrine in rat brain. *J. Chromatogr. 181*, 453—455.

Anton, A. H., and Sayre, D. F. (1962). A study of the factors affecting of the aluminum oxide—trihydroxyindole procedure for the analysis of catecholamines. *J. Pharmacol. Exp. Ther. 138*, 360—375.

Benfenati, F., Ferretti, P., Ferretti, C., Ponzio, F., and Algeri, S. (1982). Determination of L-DOPA in brain tissue using high-performance liquid chromatography with electro-chemical detection to study the activity of central dopaminergic neurons. *IRCS Med. Sci. 10*, 425—426.

Bunyagidj, C., and Girard, J. E. (1982). A comparison of coulometric detectors for catecholamine analysis by LC—EC. *Life Sci. 31*, 2627—2634.

Carlsson, A. (1959). Detection and assay of dopamine. *Pharmacol. Rev. 11*, 300—304.

Caudill, W. L., Houck, G. P., and Wightman, R. M. (1982). Determination of γ-aminobutyric acid by liquid chromatography with electrochemical detection. *J. Chromatogr. 227*, 331—339.

Chiu, A. S., Godse, D. D., and Warsh, J. J. (1981). Determination of brain regional normetanephrine levels by liquid chromatography with electrochemical detection (LC—EC). *Prog. Neuropsychopharmacol. 5*, 559—563.

Cooper, J. R., Bloom, F. E., and Roth, R. H. (1982). In: *The Biochemical Basis of Neuropharmacology*, J. R. Cooper, F. E. Bloom, and R. H. Roth(eds.), New York.

Costa, E., and Neff, N. H. (1970). Estimation of turnover rates to study the metabolic regulation of the steady-state level of nueronal monoamines. In: *Handbook of Neurochemistry, Vol. IV*, A. Lajtha (ed.), Plenum Press, New York, pp. 45—90.

Cross, A. J., and Joseph, M. H. (1981). The concurrent estimation of the major monoamine metabolites in human and nonhuman primate brain by HPLC with fluorescence and electrochemical detection. *Life Sci. 28*, 499—505.

Dahlstrom, A., and Fuxe, K. (1964). Evidence for the existence of monoamine-containing neurons in the central nervous system. I. Demonstration of monoamines in the cell bodies of brain stem neurons. *Acta Physiol. Scand. 62*, 1—55.

Dayton, M. A., Geier, G. E., and Wightman, R. M. (1979). Electrochemical measurement of release of dopamine and 5-hydroxytryptamine from synaptosomes. *Life Sci. 24*, 917—924.

Di Giulio, A. M., Groppetti, A., Cattabeni, F., Galli, C. L., Maggi, A., Algeri, S., and Ponzio, F. (1978). Significance of dopamine metabolites in evaluation of drugs acting on dopaminergic neurons. *Eur. J. Pharmacol. 52*, 201—207.

Eva, C., and Meek, J. L. (1984). Acetylcholine measurement by HPLC using an enzyme loaded post-column reactor. *Anal. Biochem. 143*, 320—324.

Falck, B., Hillarp, N. A., Thieme, G., and Torp, A. (1962). Fluorescence of catecholamines and related compounds condensed with formaldehyde. *J. Histochem. Cytochem. 10*, 348—354.

Felice, L. J., and Kissinger, P. T. (1978). Determination of homovanillic acid in urine by liquid chromatography with electrochemical detection. *Anal. Chem. 48*, 794—796.

Fenn, R. J., Siggia, S., and Curran, D. J. (1978). Liquid chromatography detector based on single and twin electrode thin-layer electrochemistry: application to the determination of catecholamines in blood plasma. *Anal. Chem. 50*, 1067—1073.

Fleisher, L. N., Simon, J. R., and Aprison, M. H. (1979). A biochemical—behavioral model for studying serotonergic supersensitivity in brain. *J. Neurochem. 32*, 1613—1619.

Frattini, P., Santagostino, G., Cucchi, M. L., Corona, G. L., and Schinelli, S. (1982). 3-Methoxy-4-hydroxyphenylglycol in human cerebrospinal fluid. *Clin. Chim. Acta 125*, 97—105.

Freed, C. R., and Asmus, P. A. (1979). Brain tissue and plasma assay of 1-DOPA and α-methyldopa metabolites by high performance liquid chromatography with electrochemical detection. *J. Neurochem. 32*, 163—168.

Glowinski, J. (1970). Storage and release of monoamines in the central nervous system. In: *Handbook of Neurochemistry, Vol. IV, Control Mechanisms in the Nervous System*, A. Lajtha (ed.), Plenum Press, New York, pp. 91—114.

Goldman, M. E., Hamm, H., and Erickson, C. K. (1980). Determination of melatonin by high-performance liquid chromatography with electrochemical detection. *J. Chromatogr. 190*, 217—220.

Hallman, H., Farnebo, L.-O., Hamberger, B., and Jonsson, G. (1978). A sensitive method for the determination of plasma catecholamines using liquid chromatography with electrochemical detection. *Life Sci. 23*, 1049—1052.

Higa, S., Suzuki, T., Hayashi, A., Tsuge, I., and Yamamura, Y. (1977). Isolation of catecholamines in biological fluids by boric acid gel. *Anal. Biochem. 77*, 18—24.

Hjemdahl, P. (1984). Inter-laboratory comparison of plasma catecholamine determinations using several different assays. *Acta Physiol. Scand. 527*, 43—54.

Hjemdahl, P., Daleskog, M., and Kahan, T. (1979). Determination of plasma catecholamines by high performance liquid chromatography with electrochemical detection: comparison with a radioenzymic method. *Life Sci. 25*, 131—138.

Hökfelt, T., Fuxe, K., Goldstein, M., and Johansson, O. (1974). Immunohistochemical evidence for the existence of adrenaline neurons in the rat brain. *Brain Res. 66*, 235—251.

Hornspenger, J.-M., Wagner, J., Hinkel, J.-P., and Jung, M. J. (1984). Measurement of 3-methoxy-4-hydroxyphenylglycol sulfate ester in brain using reversed-phase liquid chromatography and electrochemical detection. *J. Chromatogr. 306*, 364—370.

Jackman, G. P., Oddie, C. J., Skews, H., and Bobik, A. (1984). High-performance liquid chromatographic determination of plasma catecholamines during α-methyldopa therapy. *J. Chromatogr. 308*, 301—305.

Jonsson, G., Hallman, H., Mefford, I., and Adams, R. N. (1980). The use of liquid chromatography with electrochemical detection for the determination of adrenaline and other biogenic monoamines in the CNS. Wenner—Gren Center International Symposium Series. In: *Central Adrenaline Neurons, Vol. 33*, K. Fuxe (ed.), Pergamon Press, Oxford, pp. 59—71.

Keller, R., Oke, A., Mefford, I., and Adams, R. (1976). Liquid chromatographic analysis of catecholamines routine assay for regional brain mapping. *Life Sci. 19*, 995—1004.

Kilts, C. D., Breese, G. R., and Mailman R. B. (1981). Simultaneous quantification of dopamine, 5-hydroxytryptamine and four metabolically related compounds by means of reversed-phase high-performance liquid chromatography with electrochemic detection. *J. Chromatogr. 225*, 347—357.

Kissinger, P. T. (1977). Amperometric and coulometric detectors for high-performance liquid chromatography. *Anal. Chem. 49*, 447A—456A.

Koch, D. D., and Kissinger, P. T. (1980). Liquid chromatography with precolumn sample enrichment and electrochemical detection. Regional determination of serotonin and 5-hydroxyindoleacetic acid in brain tissue. *Life Sci. 26*, 1097—1107.

Krstulovic, A. M. (1981). Quantitative determination of 3-methoxy-4-hydroxyphenylethylene-glycol and its sulfate conjugate in human lumbar cerebrospinal fluid using liquid chromatography with amperometric detection. *J. Chromatogr. 223*, 305—314.

Lackovic, Z., Parenti, M., and Neff, N. H. (1981). Simultaneous determination of femtomole quantities of 5-hydroxytryptophan, serotonin and 5-hydroxyindoleacetic acid in brain using HPLC with electrochemical detection. *Eur. J. Pharmacol. 69*, 347—352.

Langlais, P. J., McEntee, W. J., and Bird, E. D. (1980). Rapid liquid-chromatographic measurement of 3-methoxy-4-hydroxyphenylglycol and other monoamine metabolites in human cerebrospinal fluid. *Clin. Chem. 26*, 786—788.

Loullis, C. C., Felton, D. L., and Shea, P. A. (1979). HPLC determination of biogenic amines in discrete brain areas in food deprived rats. *Pharmacol. Biochem. Behav. 11*, 89—93.

Lyness, W. H., Friedle, N. M., and Moore, K. E. (1980). Measurement of 5-hydroxy-indoleacetic acid in discrete brain nuclei using reverse phase liquid chromatography with electrochemical detection. *Life Sci. 26*, 1109—1114.

Maruyama, Y., Oshima, T., and Nakajima, E. (1980). Simultaneous determination of catechol-amines in rat brain by reversed-phase liquid chromatography with electrochemical detection. *Life Sci. 26*, 1115—1120.

McCaman, M. W., McCaman, R., and Lees, G. J. (1972). Liquid cation exchange a basis for sensitive radiometric assays for aromatic amino acid decarboxylase. *Anal. Biochem. 45*, 242—252.

Mefford, I. N., Gilberg, M., and Barchas, J. D., (1980). Simultaneous determination of catecholamines and unconjugated 3,4-dihydroxyphenylacetic acid in brain tissue by ion-pairing reverse-phase high-performance liquid chromatography with electrochemical detection. *Anal. Biochem. 104*, 468—472.

Mefford, I. N., Ward, M. M., Miles, L., Taylor, B., Chesney, M. A., Keegan, D. L., and Barchas, J. D. (1981). Determination of plasma catecholamines and free 3,4-dihydroxy-phenylacetic acid in continuously collected human plasma by high performance liquid chromatography with electrochemical detection. *Life Sci. 28*, 477—483.

Moleman, P., and Borstrok, J. J. M. (1982). Analysis of urinary 3-methoxy-4-hydroxy-phenylglycol by high-performance liquid chromatography and electrochemical detection. *J. Chromatogr. 227*, 391—405.

Molnar, I., and Horvath, C. (1976). Reverse-phase chromatography of polar biological substances: separation of catechol compounds by high-performance liquid chromato-graphy. *Clin. Chem. 22*, 1497—1502.

Moyer, P. T., Jiang, N.-S., Tyce, G. M., and Sheps, S. G. (1979). Analysis of urinary catecholamines by liquid chromatography with amperometric detection: methodology and clinical interpretation of results. *Clin. Chem. 25*, 256—263.

Nielsen, J. A., and Johnston, C. A. (1982). Rapid, concurrent analysis of dopamine, 5-hydroxytryptamine, their precursors and metabolites utilizing high performance liquid chromatography with electrochemical detection: analysis of brain tissue and cerebrospinal fluid. *Life Sci. 31*, 2847—2856.

Orsulak, P. J., Kizuka, P., Grab, E., and Schildkraut, J. J. (1983). Determination of urinary normetanephrine and metanephrine by radial-compression liquid chromatography and electrochemical detection. *Clin. Chem. 29*, 305—309.

Palermo, A., Ponzio, F., Sega, R., Lomuscio, G., Algeri, S., and Libretti, A. (1979). Catecholamines and sympathetic stimulation tests. In: *Radioimmunoassay of Drug and Hormones in Cardiovascular Medicine*, A. Albertini, M. Da Praga, B. A. Peskar (eds.), Elsevier/North-Holland. Biochemical Press, Amsterdam, pp. 1249—159.

Palkovits, M. (1980). Topography of chemically identified neurons in the central nervous system: progress in 1977—1979. *Med. Bio. 58*, 188—227.

Perry, B. D., Stolk, J. M., Vantini, G., Guckhait, R. B., and U'Prichard, D. C. (1983). Strain differences in rat brain epinephrine synthesis: regulation of α-adrenergic receptor number by epinephrine. *Science 221*, 1297—1299.

Ponzio, F., Achilli, G., and Algeri, S. (1981a). A rapid and simple method for the determination of picogram levels of 3-methoxytryramine in brain tissue using liquid chromatography with electrochemical detection. *J. Neurochem. 36*, 1361—1367.

Ponzio, F., Achilli, G., Perego, C., Algeri, S. (1981b). Differential effects of certain dopaminergic drugs on the striatal concentration of dopamine metabolites with special reference to 3-methoxytryptamine *Neurosci. Lett. 27*, 61—67.

Ponzio, F., and Jonsson, G. (1978). Effects of neonatal 5,7-dihydroxytryptamine treatment on the development of serotonin neurons and their transmitter metabolism. *Dev. Neurosci. 1*, 80—89.

Ponzio, F., and Jonsson, G. (1979). A rapid and simple method for the determination of picogram levels of serotonin in brain tissue using liquid chromatography with electro-chemical detection. *J. Neurochem. 32*, 129—132.

Potter, P. E., Meek, J. L., and Neff, N. H. (1983). Acetylcholine and choline in neuronal tissue measured by HPLC with electrochemical detection. *J. Neurochem. 41*, 188—194.

Racagni, G., Mocchetti, I., Renna, G., and Cuomo, V. (1982). *In vivo* studies on central-noradrenergic synaptic mechanisms after acute and chronic antidepressent drug treatment: biochemical and behavioral comparison. *J. Pharmacol. Exp. Med. 223*, 227—234.

Refshauge, C., Kissinger, P. T., Dreiling, R., Blank, L. R., Freeman, R., and Adams, R. N. (1974). New high performance liquid chromatographic analysis of brain catecholamines. *Life Sci. 14*, 311—322.

Reinhard, J. F. Jr., Moskowitz, M. A., Sved, A. F., and Fernstrom, J. D. (1980). A simple, sensitive, and reliable assay for serotonin and 5-HIAA in brain tissue using liquid chromatography with electrochemical detection. *Life Sci. 27*, 905—911.

Riggin, R. M., Alcorn, R. L., and Kissinger, P. T. (1976). Liquid chromatographic method for monitoring therapeutic concentrations of L-DOPA and dopamine in serum. *Clin. Chem. 22*, 782—784.

Riggin, R. M., and Kissinger, P. T. (1977). Determination of catecholamines in urine by reverse-phase liquid chromatography with electrochemical detection. *Anal. Chem. 49*, 2109—2111.

Sandu, R. S., and Sponsor, R. M. F. (1968). An improved method for the determination of urinary catecholamines. *Clin. Chem. 14*, 824—825.

Santagostino, G., Frattini, P., Schinelli, S., Cucchi, M.-L., and Corona, G. L. (1982). Urinary 3-methoxy-4-hydroxyphenylglycol determination using reversed-phase chromatography with amperometric detection. *J. Chromatogr. Biomed. Appl. 232*, 89—95.

Sasa, S., and Blank, C. L. (1977). Determination of serotonin and dopamine in mouse brain tissue by high performance liquid chromatography with electrochemical detection. *Anal. Chem. 49*, 354—359.

Sasa, S., and Blank, C. L. (1979). Simultaneous determination of norepinephrine, dopamine, and serotonin in brain tissue by high-pressure liquid chromatography with electrochemical detection. *Anal. Chim. Acta. 104*, 29—45.

Sasa, S., Blank, C.-L., Wenke, D. C., and Sczupak, C. (1978). Liquid-chromatographic determination of serotonin in serum and plasma. *Clin. Chem. 24*, 1509—1514.

Shoup, R. E., and Kissinger, P. T. (1977). Determination of urinary normetanephrine, metanephrine, and 3-methoxytryamine by liquid chromatography, with amperometric detection. *Clin. Chem. 23*, 1268—1274.

Smedes, F., Kraak, J. C., and Poppe, H. (1982). Simple and fast solvent extraction system for selective and quantitative isolation of adrenaline, noradrenaline and dopamine from plasma and urine. *J. Chromatogr. 231*, 25—39.

Snyder, L. R., and Kirkland, J. J. (1978). Reversed phase liquid chromatography and its application to biochemistry. *Anal. Chem. 50*, 1048A—1073A.

Tagari, P. C., Boullin, D. J., and Davies, C. L. (1984). Simplified determination of serotonin in plasma by liquid chromatography with electrochemical detection. *Clin. Chem. 30*, 131—135.

Temple, D. M., and Gillespie, R. (1966). Liquid ion-exchange extraction of some physiologically active amines. *Nature 209*, 714—715.

Voelter, W., Zech, K., Arnold, P., and Ludwing, G. (1980). Determination of selected pyrimidines, purines and their metabolites in serum and urine by reversed-phase ion-pair chromatography. *J. Chromatogr. 199*, 345—354.

Vogt, M. (1954). Concentration of sympathin in different parts of the central nervous system, under normal condition and after the administration of drugs. *J. Physiol. 123*, 451—481.

Von Euler, U. S. (1946). A specific sympathomimetic ergone in adrenergic nerve fibers (sympathin) and its relation to adrenaline and noradrenaline. *Acta Physiol. Scand. 12*, 73—97.

Wagner, J., Palfreyman, M., and Zraika, X. (1979). Determination of DOPA, dopamine, DOPAC, epinephrine, norepinephrine, α-monofluoromethyldopa and α-difluoromethyldopa in various tissue of mice and rats using reversed-phase ion-pair liquid chromatography with electrochemical detection. *J. Chromatogr. 164*, 41—45.

Ward, M. M., Mefford, I. N., Parker, S. D., Chesney, M. A., Taylor, C. B., Keegen, D. L., and Barchas, J. D. (1983). Epinephrine and norepinephrine responses in continuously collected human plasma to a series of stressors. *Psychosom. Med. 45*, 471—486.

Warnhoff, M. (1984). Simultaneous determinatiobn of norepinephrine, dopamine, 5-hydroxytryptamine and their main metabolites in rat brain using high-performance liquid chromatography with electrochemical determination. Enzymatic hydrolysis of metabolites prior to chromatography. *J. Chromatogr. 307*, 271—281.

Warsh, J. J., Chiu, A., Godse, D. G., and Coscina, D. V. (1979). Determination of picogram levels of brain serotonin by a simplified liquid chromatographic electrochemical detection assay. *Brain Res. 4*, 567—570.

Weil-Malherbe, H. (1971). The chemical estimation of catecholamines and their metabolites in body fluids and tissue extracts. *Methods Biochem. Anal. Suppl.* 119—152.

Weil-Malherbe, H., and Bone, A. D. (1952). Chemical estimation of adrenaline-like substances in blood. *J. Biochem. 51*, 311—318.

Westerink, B. H. C. (1979a). The effects of drugs on dopamine biosynthesis and metabolism in the brain. In: *The Neurobiology of Dopamine.* A. S. Horn, J. Korf, and B. H. C. Westerink (eds.), Academic Press, London, pp. 255—291.

Westerink, B. H. C. (1979b). Effects of drugs on the formation of 3-methoxytyramine, a dopamine metabolite, in the *substantia nigra*, nucleus accumbens and tuberculum olfactorium of the rat. *J. Pharm. Pharmacol. 31*, 94—99.

Westerink, B. H. C., and Korf, J. (1976). Turnover of acid dopamine metabolites in striatal and mesolimbic tissue of the rat brain. *Eur. J. Pharmacol. 37*, 249—255.

Westerink, B. H. C., and Korf, J. (1979). Rapid concurrent automated fluorimetric assay of noradrenaline, dopamine, 3-4-dihydroxyphenylacetic acid, homovanillic acid and

3-methoxytyramine in milligram amounts of nervous tissue after isolation in Sephadex GlO. *J. Neurochem. 29*, 697—706.

Westerink, B. H. C., and Mulder, T. B. A. (1981). Determination of picomole amounts of dopamine, noradrenaline, 3,4-dihydroxyphenylalanine, 3,4-dihydroxyphenylacetic acid, homovanillic acid, and 5-hydroxyindolacetic acid in nervous tissue after one-step purification on Sephadex G-10, using high-performance liquid chromatography with a novel type of electrochemical detection. *J. Neurochem. 36*, 1449—1462.

Westerink, B. H. C., and Spaan, S. J. (1982a). Estimation of the turnover of 3-methoxy-tryamine in the rat striatum by HPLC with electrochemical detection: implications for the sequence in the cerebral metabolism of dopamine. *J. Neurochem. 38*, 342—347.

Westerink, B. H. C., and Spaan, S. J. (1982b). On the significance of endogenous 3-methoxy-tryamine for the effects of centrally acting drugs on dopamine release in the rat brain. *J. Neurochem. 38*, 680—686.

Westerink, B. H. C., and Van Oene, J. C. (1980). Evaluation of the effect of drugs on dopamine metabolism in the rat superior cervical ganglion by HPLC with electrochemical detection. *Eur. J. Pharmacol. 65*, 71—79.

Wightman, R. M., Plotsky, P. M., Strope, E., Delcome, R. Jr., and Adams, R. N. (1977). Liquid chromatographic monitoring of CSF metabolites. *Brain Res. 131*, 345—349.

Wolf, F. J. (1969). Ion exchange. In: *Separation Methods in Organic Chemistry and Biochemistry*. Academic Press, London, pp. 137—181.

Progress in HPLC, Vol. 2, pp. 329—375
Parvez *et al.* (Eds)
© 1987 VNU Science Press

Determination of sulfated biogenic amines and some metabolites by HPLC with electrochemical detection: Emphasis on dopamine sulfate isomers

MARY ANN ELCHISAK

Department of Physiology and Pharmacology, School of Veterinary Medicine, Purdue University, West Lafayette, Indiana, USA

INTRODUCTION

The three catecholamines (CA) which occur endogenously in most mammalian species are epinephrine (EPI), norepinephrine (NE) and dopamine (DA). DA and NE are well-established neurotransmitters in the mammalian central nervous system, while NE is the primary neurotransmitter released from peripheral post-ganglionic sympathetic neurons. EPI is the major neurohormone released from the adrenal medulla during sympathetic nervous system activity in man and in many other species. Varying amounts of NE are also released. DA may be the inhibitory neurotransmitter released from interneurons in autonomic nervous system ganglia.

Analysis of the concentrations of these three CA in brain and peripheral tissues and body fluids has been useful for basic understanding of the normal and diseased peripheral sympathetic and central nervous systems in man and experimental animals. Quantitative analysis of the various metabolites of the CA has been at least equally important for understanding of the neurodynamics of the CA, since they are rapidly metabolized to presumably inactive products. Each of the CA can be metabolized by the enzymes catechol-*O*-methyl transferase (COMT) and monoamine oxidase (MAO). The metabolism may be sequential, in either order of COMT and MAO, or non-sequential. The major metabolites produced from the action of either or both of these enzymes on EPI are metanephrine and vanillylmandelic acid (VMA). Major NE metabolites include normetanephrine, VMA and 3-methoxy-4-hydroxyphenylglycol (MHPG). Major metabolites of DA include dihydroxyphenylacetic acid (DOPAC), homovanillic acid (HVA) and 3-methoxytyramine (3-MT). The parent CA, as well as most of the metabolites, are also conjugated as sulfates or glucuronides. There are, in addition, many quantitatively 'minor' metabolites not mentioned here.

This chapter is primarily concerned with procedures for the measurement of one probable DA metabolite, dopamine—*O*-sulfate (DAS), in biological material. DAS may occur in two isomeric forms, DA—3-O-sulfate and DA—4-O-sulfate.

Reasons for interest in DA-sulfate

There are four major reasons for our interest in quantitative detection of DAS in mammals:

    1. DAS is widely, but not homogenously, distributed in mammals.
    2. DAS is probably a DA metabolite in mammals.
    3. DAS is possibly a source for free DA, NE and EPI in mammals.
    4. The two isomers of DAS, DA—3-O-sulfate and DA—4-O sulfate, may differ in distribution and activity.

Since there are no recent reviews of the literature supporting these reasons, each of these is discussed below.

    Occurrence of DA-sulfate in mammals. Only work concerning endogenous occurrence of DAS in non-drug treated subjects is reviewed here. Conjugated DA has been found in greater concentrations than free DA in human plasma (Buu and Kuchel, 1977; Claustre *et al.*, 1983; Corneille *et al.*, 1983; Cuche *et al.*, 1982; Davidson *et al.*, 1981; de Champlain *et al.*, 1984; Johnson *et al.*, 1980; Kuchel *et al.*, 1982a; Nagel and Schumann, 1980; Scott and Elchisak, 1985; abstract; Vlachakis *et al.*, 1984; Wang *et al.*, 1983; Yoneda *et al.*, 1983); red blood cells (de Champlain *et al.*, 1984; Yoneda *et al.*, 1983), urine (Arakawa *et al.*, 1983; Buu and Kuchel, 1977; Da Prada, 1980; Elchisak and Carlson, 1982b; Hoeldke and Sloan, 1970; Kahane *et al.*, 1967; Kissinger *et al.*, 1975; Rutledge and Hoehn, 1973; Weil-Malherbe and Van Buren, 1969) and CSF (Scheinin *et al.*, 1984; Sharpless *et al.*, 1981). It is also the predominant form of DA in both lumbar and ventricular CSF from the Rhesus monkey (Elchisak *et al.*, 1982, 1983). Much of the conjugated DA in these fluids has been shown to occur as DAS.

    Conjugated DA has also been identified in non-primate species. It has been found to be the primary form of DA in dog urine (Swann and Elchisak, unpublished) and also occurs in dog plasma, adrenal medulla, liver, small intestine and kidney (Unger *et al.*, 1980). In the cat, conjugated DA has been detected in urine (Claustre and Peyrin, 1982) and in ventriculocisternal perfusates (Hammond *et al.*, 1984). In the rat, conjugated DA (mostly glucuronide, but some sulfate) has been demonstrated to occur in plasma (Alexander *et al.*, 1984; Claustre *et al.*, 1983; Wang *et al.*, 1983; Yoneda *et al.*, 1983) and urine (Peyrin *et al.*, 1978, 1982). Conjugated DA, thought to be primarily DAS, has also been detected in superfusates of rat striatum slices (Tyce and Rorie, 1982). Relatively low concentrations of DAS has been detected in the bovine adrenal cortex and medulla (Racz *et al.*, 1984).

    In rat tissues, conjugated DA has been found in spinal cord (Karoum, 1983), in some brain areas, such as hypothalamus, septum, striatum and hippocampus (Buu *et al.*, 1981a; Karoum *et al.*, 1983), and in peripheral organs, such as adrenals (Elchisak and Carlson, 1982b) and kidney (Elchisak and Carlson, 1982b; Kuchel *et al.*, 1979). Thorough studies of the distribution of conjugated DA in the mouse (Shoaf and Elchisak, 1983), rat (Elchisak and Carlson, 1982b; Shoaf and Elchisak, 1983), dog (Shoaf and Elchisak, 1983, abstract) monkey (Elchisak *et al.*, 1983) and human (Elchisak, 1983c) all indicate that conjugated DA is widely, but not homogenously, distributed in mammals. The kidney was the only tissue studied in which conjugated DA was found to account for a substantial proportion (range 20—34%) of the total DA in all species studied.

Conversion of DA to DA-sulfate in mammals. DAS can be formed from DA by the action of the enzyme phenol sulfotransferase (PST; EC 2.8.2.1). This enzyme has been partially purified from brain and liver of several species (Jenner and Rose, 1973; Renskers *et al.*, 1980). In humans, PST activity has been found in brain, liver, adrenals, kidney, jejunum (Bostrom and Wengle, 1967), platelets (Hart *et al.*, 1979) and erythrocytes (Anderson and Weinshilboum, 1979). Sulfate conjugating activity has also been demonstrated in the adrenal, intestine and kidney of several other species (Maus *et al.*, 1982; Merits, 1976; Wong, 1976).

Because of the widespread distribution of PST, coupled with the high concentrations of DAS found in the circulation, it is likely that DA-sulfate is an endogenous metabolite of DA. Surprisingly, however, this has not yet been very extensively investigated. There are several studies which report the formation of conjugated DA or DAS after administration of L-DOPA to humans (Anden *et al.*, 1970; Goodall and Alton, 1972; Imai *et al.*, 1972; Jenner and Rose, 1974; Rutledge and Hoehn, 1973; Tyce *et al.*, 1980), monkeys (Bronaugh *et al.*, 1976) and dogs (Tyce *et al.*, 1980). However, it is not known in these studies whether DAS was formed from DA or from L-DOPA-sulfate. Only a few investigators have studied DAS formation directly from DA. In humans, conjugated DA accounted for only 1.5% of the total urinary radioactivity after intravenous infusion of ^{14}C-DA (Goodall and Alton, 1968). Plasma DAS increased dramatically during infusion of unlabelled free DA into humans (Kuchel *et al.*, 1982b). This was presumably due to conversion of DA to DAS. Merits (1976) demonstrated that ^{14}C-DA—3-*O*-sulfate appeared in the urine of two dogs after oral or intravenous administration of ^{14}C-DA, and that the DA—3-*O*-sulfate accounted for 79% and 21% of the total urinary radioactivity, respectively. After oral administration of ^{14}C-DA, ^{14}C-DAS accounted for a small proportion of the total urinary radioactivity in monkey (14%), mouse (10%), rat (6%), guinea pig (2%) and rabbit (1%) (Merits *et al.*, 1973). In the rat, DAS was found to decrease, rather than increase as expected, in the striatum and plasma after intraventricular injection of free DA (Buu *et al.*, 1982). Additional studies demonstrating the conversion of DA to DA—3-*O*-sulfate or DA—4-*O*-sulfate clearly need to be done to document the assumption that DAS is, in fact, an endogenous DA metabolite.

Precursor role of DA-sulfate. There is some *in vivo* evidence to indicate that DAS is a precursor for free DA, and possibly for the other CA, in mammals. Buu *et al.* (1981b) isolated labelled DA, NE and EPI from the urine of adrenalectomized rats after injection of ^{3}H-DAS. This group has also provided indirect evidence that the dog kidney hydrolyzes DAS to free DA during surgery (Unger *et al.*, 1979). Quantitative studies determining the actual amounts of each conversion were not reported, however. Merits (1976) found labelled DA metabolites, homovanillic acid and dihydroxy-phenylacetic acid, in urine from guinea pig, rat and dog after administration of ^{14}C-DA—3-*O*-sulfate. Presumably, these DA metabolites were formed after conversion of DA—3-*O*-sulfate to free DA, although the alternate pathways of oxidation and/or methylation followed by desulfation were not ruled out. The latter explanation is plausible between DA—3-*O*-sulfate and DA—4-*O*-sulfate have recently been shown to undergo *O*-methylation by a crude COMT preparation *in vitro* (Qu *et al.*, 1983).

There is *in vitro* evidence, from one laboratory, which indicates that DA—3-*O*-sulfate and DA—4-*O*-sulfate can be converted directly (not via DA) to NE by action of the enzyme dopamine β-hydroxylase (DBH) isolated from bovine adrenals (Buu and Kuchel, 1979a, b).

Although the concept that sulfated compounds may act as metabolic intermediates is new to the field of CA synthesis and metabolism, such conversions are commonly accepted to occur in steroid biosynthesis in mammals (Baulieu *et al.*, 1964). In insects, DA—3-*O*-sulfate is converted to *N*-acetyldopamine—3-*O*-sulfate. The *N*-acetyldopamine moiety is then incorporated into the cuticle during sclerotization (Bodnaryk *et al.*, 1974). The sulfated compound thus appears to act as a storage or transport form for the eventual use of *N*-acetyldopamine in insects. The possibility that DA—sulfoconjugates could serve as convenient forms of DA for transport to tissues in mammals was suggested several years ago by Jenner and Rose (1973). However, to the author's knowledge, this concept has not been experimentally tested in mammals. It is of interest here that the presence of free and conjugated *N*-acetyldopamine has recently been demonstrated in human urine (Elchisak and Hausner, 1984). The possibility that DAS may serve as a precursor for either free or sulfated *N*-acetyldopamine in mammals has not yet been investigated.

Distinction between isomers of DA-sulfate. Investigations concerning which isomers of DAS are present in the body have been hampered, until recently, by the lack of specific and sensitive methodology for their detection at endogenous levels. Several HPLC methods have recently been developed which can distinguish between DA—3-*O*-sulfate and DA—4-*O*-sulfate in biological samples (Arakawa *et al.*, 1983; Elchisak, 1983a, b; Elchisak and Carlson, 1982a). Using these methods, DA—3-*O*-sulfate was found to account for 73% of the total conjugated DA (Elchisak and Carlson, 1982a) and for 81% of the total DAS (Arakawa *et al.*, 1983) excretion in human urine. DA—3-*O*-sulfate was also found to be the major DAS isomer excreted in human urine when qualitatively measured using a thin layer chromatography system (Elchisak and Carlson, unpublished). In urine from the Rhesus monkey, however, both DA—3-*O*-sulfate and DA—4-*O*-sulfate were detected (Carlson and Elchisak, unpublished). Earlier work by others supports these findings. Both ^{14}C-DA—3-*O*-sulfate and ^{14}C-DA—4-*O*-sulfate were found in urine from the pigtail monkey after intravenous administration of ^{14}C-1-DOPA (Bronaugh *et al.*, 1976). ^{3}H-DA—3-*O*-sulfate was the primary isomer detected in the urine of Parkinsonian patients receiving ^{3}H-1-DOPA orally (Arakawa *et al.*, 1979; Bronaugh *et al.*, 1975; Jenner and Rose, 1974).

Methods for quantitation of free CA in biological samples
It is beyond the scope of this chapter to detail methods for the determination of free CA in biological material. However, since conjugated CA are usually hydrolyzed to free CA before quantitation, these methods are briefly mentioned.

Extraction of free CA from tissues and fluids. It is usually, although not always, necessary to remove interfering substances from biological material before estimation of the CA content. The most common methods used to purify and concentrate the free CA are extraction by activated alumina (Anton and Sayre, 1962) or binding to boric acid gel (Higa *et al.*, 1977). Solvent extraction methods have also been employed (Smedes *et al.*, 1982; Weil-Malherbe and Bone, 1957). Some authors use ion-exchange chromatography as part of the sample purification scheme (Valori *et al.*, 1969), usually in addition to one of the above methods. Many modifications of these procedures have been utilized and published.

Detection methods for free CA. The detection methods for free CA measurement have been reviewed and summarized recently (Hjemdahl, 1984; Raum, 1984). These include spectrofluorometric assays, radioenzymatic assays, radioimmunoassay and HPLC coupled with fluorometric or electrochemical detection.

Methods for quantitation of conjugated CA in biological samples
Quantitation of conjugated CA in biological specimens is more difficult than quantitation of the free CA. It is usually assumed that the only two conjugated forms of CA are *O*-glucuronides and *O*-sulfates. Since the conjugated compounds are difficult to detect directly, they are usually hydrolyzed to free CA, and then extracted and detected as free CA. A detailed discussion of methodology concerning conjugates of glucuronic acid is outside the scope of this chapter, but may be found elsewhere (Marsh, 1966). Methods for direct detection of the CA-*O*-sulfates, especially DAS, are detailed below.

Conversion of conjugated CA to free CA. Most procedures which measure conjugated CA in biological samples convert the conjugated CA to free CA by acid or enzymatic hydrolysis, and then purify and quantitate the free CA. Usually, two aliquots of a sample are used. Only one aliquot is submitted to a hydrolysis procedure; total and free CA are then measured in the respective aliquots. Conjugated CA is the difference between total and free. Occasionally, only one aliquot of a sample is used. Free CA is removed from the sample, conjugated CA remaining in the sample is then hydrolyzed to free CA, and the free CA then measured corresponds to conjugated CA.

Although hydrolysis of conjugated CA to free CA is the currently accepted technique for measuring conjugated CA in biological material, there are many difficulties inherent in these procedures. For some hydrolysis methods, the distinction between sulfate and glucuronide conjugates is not maintained. Furthermore, the distinction between isomeric forms of the conjugate (e.g. DA—3-O-sulfate and DA—4-O-sulfate) is always lost unless the isomers have been separated before the hydrolysis procedure. The hydrolysis procedures are also difficult to standardize, since authentic standards of the glucuronidated or sulfated forms of the CA are not commercially available. Small amounts of some CA-O-sulfates can be relatively easily synthesized and purified, or can be otherwise obtained (see below), but the CA glucuronides are virtually unavailable as pure compounds. Another problem adding to the difficulty of standardizing hydrolysis procedures is the fact that the different CA are not all equally stable in various solutions or fluids.

Most of the rest of this chapter will discuss some relatively simple techniques for direct detection of the CA-O-sulfates which have only recently become available. Use of one of these techniques avoids most of the problems associated with conversion of the CA-O-sulfates to the free CA. Unfortunately, similar techniques for the direct detection of CA-glucuronides are not, to the author's knowledge, available.

1. Enzymatic hydrolysis. Until recently, the most common method for the enzymatic hydrolysis of conjugated CA used 'glusulase', a crude extract of *Helix pomatia* which contains both sulfatase and β-glucuronidase activity (Weil-Malherbe, 1971). This procedure presumably converts both sulfate and glucuronide conjugates of CA to free CA under appropriate conditions. Purified glucuronidase has also been widely utilized for measurement of CA glucuronides. These enzymatic hydrolysis techniques have several limitations. They often result in poor recoveries of free CA because of the prolonged incubation period required (usually 12—24 h). Furthermore, some fluids contain compounds which inhibit the activity of the enzyme(s). The utility of these techniques is further limited because the crude enzyme preparation often contains substantial amounts of DA (Anton *et al.*, 1973).

More recently, specific purified enzyme preparations and relatively short incubation times have been used for preferential hydrolysis of CA-sulfates (Buu *et al.*, 1981a; Claustre *et al.*, 1983; Johnson *et al.*, 1980; Wang *et al.*,

1983; Yoneda *et al.*, 1983) or CA-glucuronides (Claustre *et al.*, 1983; Wang *et al.*, 1983; Yoneda *et al.*, 1983). These procedures are relatively rapid and reproducible, and, when coupled with an appropriate sample preparation and detection technique, are quite useful for determination of CA-sulfates or glucuronides in large numbers of samples. Although these procedures are an improvement over the previously-utilized glusulase procedures, most of the difficulties detailed above for procedures which convert conjugated CA to free CA still apply.

2. Acid hydrolysis. An alternative to enzymatic hydrolysis to convert conjugated CA to free CA is acid hydrolysis. Until very recently, the usual procedure was heating of the sample or extract at 100° C for 20—30 min at a pH ranging from 0 (von Euler and Orwen, 1955) to 1.5 (Weil-Malherbe and Bone, 1957). Although destruction of free CA and some metabolites can occur under these conditions, exposing the samples to nitrogen during the procedure was thought to offer a protective effect (Weil-Malherbe, 1964). In this author's experience, bubbling nitrogen through the solution prior to sealing the tubes for the acid-hydrolysis procedure has been found to prevent destruction of the free CA, even during very harsh (90—120 min) heating procedures at 100° C (Elchisak *et al.*, 1982; Elchisak and Carlson, 1982b). It is likely, however, that destruction of some metabolites would occur under such conditions.

It is this author's opinion that the main difficulty with the acid hydrolysis procedure, besides possible destruction of some compounds, is the uncertainty concerning which of the CA conjugates are actually hydrolyzed. It is usually assumed that glucuronides are stable to acid hydrolysis (Weil-Malherbe, 1971), but this is probably true only for the relatively mild hydrolysis proccedures. Although optimal conditions for hydrolysis versus destruction vary depending on the composition of the fluid or extract, wide extremes of acid hydrolysis conditions producing maximal recovery of CA have recently been reported. Vlachakis *et al.* (1984) reported that heating for 7 min in boiling water (pH 0.8, approximately 0.4 M perchloric acid) produced complete deconjugation of conjugated CA in human plasma with minimal destruction of free CA (5% for DA and NE, 15% for EPI). Significant losses of some metabolites occurred, however. These authors reported greater losses with increased heating times, but nitrogen flushing was apparently not part of their procedure. It is clear from work in that laboratory that a mild acid hydrolysis procedure (up to 10 min at pH 0.8) produces complete hydrolysis of sulfate conjugates of the CA, without affecting the glucuronide conjugates (Yoneda *et al.*, 1983).

Unfortunately, it is not known with certainty if the more harsh acid hydrolysis procedures, such as that reported for human urine (Elchisak and Carlson, 1982b) which exposes the sample to 100° C for 120 min in 0.4 M perchloric acid, cleaves the CA glucuronides. Comparative data is lacking, but it is reasonable to assume that this procedure, which does not destroy free CA, is hydrolyzing some of the CA glucuronide conjugates, since DA freed by this procedure does not plateau until approximately 120 min hydrolysis time. Unfortunately, to the author's knowledge, the specificity of longer hydrolysis procedures (greater than approximately 10 min) such as this has not been carefully evaluated.

An acid lyophilization technique which apparently hydrolyzes both CA-sulfates and glucuronides has been reported (Buu and Kuchel, 1977; Wang *et al.*, 1983). However, the routine use of this technique has been questioned because destruction and inconsistent recoveries of free CA, especially NE and EPI, have been reported to occur during this procedure (Nagel and Schumann, 1980; Vlachakis *et al.*, 1984).

A tentative conclusion regarding use of acid hydrolysis procedures to preferentially hydrolyze CA-sulfates in the presence of CA-glucuronides can be reached based on the evidence given above. CA-sulfates can be preferentially hydrolyzed by very mild and short procedures. The assumption that longer procedures, such as a 20—30 min hydrolysis, hydrolyzes only CA-sulfates (as opposed to CA-glucuronides) has not been experimentally tested. Longer procedures (greater than approximately 60 min) probably hydrolyze at least some CA-glucuronides. Comparative data is needed to evaluate these tentative conclusions.

A novel method related to acid hydrolysis techniques for the determination of conjugated CA has recently been developed by Karoum (1983) for rat spinal cord. In this procedure, advantage is taken of the facts that free CA are relatively stable in acid solution while CA-sulfates are easily hydrolyzed in acid solution. CA-sulfates are relatively stable in alkaline solution, while free CA are easily destroyed. Tissues are homogenized in sodium hydroxide, allowed to set at room temperature (which destroys free CA), then made acidic, and CA measured. This CA measurement corresponds to conjugated (presumably sulfated) CA. When the tissues are homogenized in acid, total (free plus conjugated) CA are measured, since this procedure hydrolyzes the sulfated CA. Free CA is the difference between total and conjugated CA. This procedure is very simple and appears promising for routine use for many applications. However, hydrolysis conditions must be carefully evaluated for use in different tissues or fluids, since hydrolysis of the CA-sulfates under similar conditions definitely did not occur for rat adrenal or kidney homogenates (Elchisak and Carlson, 1982b).

Direct measurement of conjugated CA. This section will review work concerning DA—*O*-sulfate only. Most of the procedures described also apply to measurement of sulfated esters of NE and EPI. Routine methods for direct measurement (not after hydrolysis) of glucuronide conjugates of the CA are, to the author's knowledge, not available.

1. Extraction of DA-sulfate from tissues and fluids. Classical ion-exchange procedures for the preliminary isolation of DA—3-*O*-sulfate and DA—4-*O*-sulfate from a sample matrix have been utilized by several investigators (Goodall and Alton, 1968, 1972; Jenner and Rose, 1973; Merits *et al.*, 1973; Rutledge and Hoehn, 1973; Bronaugh *et al.*, 1975; Merits, 1976). These

procedures are tedious, time-consuming and do not usually result in quantitative recovery of small amounts of DAS such as occurs endogenously. Recently, procedures have been described which can be utilized for the processing of DAS in multiple samples (Arakawa *et al.*, 1979; Buu *et al.*, 1981a, b; Elchisak and Carlson, 1982a). These are still quite time-consuming, however, and all require use of ion-exchange columns. A new procedure utilizing pre-packed solid-phase extraction columns is currently under development in the author's laboratory, and is described below. This technique allows the very rapid processing of large numbers of samples with quantitative recoveries of the DAS isomers.

2. Detection methods for DA—*O*-sulfate. Direct, quantitative detection of non-radioactive DA—3-*O*-sulfate and DA—4-*O*-sulfate is possible by ultraviolet detection. However, this method does not offer sufficient specificity and sensitivity for the determination of endogenous concentrations of these compounds in most biological samples, except after administration of a drug, such as *L*-dopa, which increases the concentrations of DAS in various tissues and fluids. Ultraviolet detection has been utilized to measure DA—3-*O*-sulfate in urine from normal humans (Elchisak and Carlson, 1982a), but this method requires extensive sample purification and lacks the specificity and sensitivity desired for routine determinations of this type.

Two procedures utilizing HPLC coupled with post-column reactions have recently been described. Each of these allows the sensitive and relatively specific determination of DA—3-*O*-sulfate and DA—4-*O*-sulfate. Since the DAS isomers are separated by the HPLC procedure, they can be detected separately, even though the post-column reaction forms the same compound from each isomer. One procedure simply uses acid hydrolysis in-stream to convert each DAS isomer to free DA, which is then detected by single-electrode electrochemical detection (Elchisak, 1983a). This procedure is described in more detail below. The other procedure utilizes a photo-induced in-stream reaction of DAS with ethylenediamine and detection of the fluorescent compounds thus formed by a spectrofluorometer with a flow-through cell (Arakawa *et al.*, 1983). Lower detection limits are reported to be 1—2 pmol.

One other procedure has been developed for the direct detection of DAS in biological samples. This utilizes a commercially-available electrochemical detector with dual electrodes (Elchisak, 1983b). This procedure is simple, relatively specific and sensitive, and is considered by the author to be the current method of choice for determination of DAS (and other CA-sulfates) in biological samples. It is described in detail below.

Summary of introduction
Investigations concerning the specific conjugates of DA in the body have been hampered by the lack of specific and sensitive methods for their detection. Since there is increasing interest in the occurrence and measurement of sulfated metabolites of DA and the other CA, NE and EPI, reliable methods for the detection and quantitation of these compounds in biological

samples are needed. This chapter reviews work, primarily conducted in the author's laboratory, directed towards the direct detection of DAS in biological material. Appropriate data concerning other sulfated biogenic amines and metabolites is presented as appropriate.

GENERAL CHEMICAL METHODS

Synthesis of DA—O-sulfate

Radiolabelled (^{3}H) or unlabelled DA—3-O-sulfate and DA-4-O-sulfate are synthesized by a modification of the method of Jenner and Rose (1973) described elsewhere (Elchisak and Carlson, 1982a). This procedure simply requires the mixing of DA—HCl and sulfuric acid while keeping the mixture ice cold. Several unidentified products are produced in addition to DA—3-O-sulfate and DA—4-O-sulfate, so careful purification of the products is necessary.

Purification of the synthetic mixture is accomplished by reversed-phase HPLC using an octadecyl silica (C_{18}) column, a mobile phase of 2% methanol in water and ultraviolet detection at 280 nm. These HPLC procedures are described in more detail below. An analytical column (5 μm, 250 × 4 mm) is used for purification of small batches of radiolabelled compounds of high specific activity, and a semipreparative column (10 μm, 300 × 7.9 mm) is used for larger batches (100 mg) of non-radioactive DAS isomers. The appropriate fractions are collected and the methanol evaporated under a stream of nitrogen. The pure compound (DA—3-O-sulfate or DA—4-O-sulfate) in water is then standardized quantitatively by HPLC using authentic re-crystallized DAS as a standard. The advantage of this purification method is its rapidity compared to the more tedious ion-exchange and recrystallization procedures described previously (Elchisak and Carlson, 1982a). The HPLC procedure takes about one hour compared to two or three days for the 'classical' purification and recrystallization procedure. Furthermore, there are no salts or other compounds in the mobile phase to be removed from the collected fraction. This allows the compounds to be used for experiments without recrystallization.

The DAS solutions are routinely analyzed in at least two different HPLC systems (described below) to be sure that they co-chromatograph exactly with authentic DA—3-O-sulfate or DA—4-O-sulfate. Authentic compounds are supplied by Dr J. Stephen Kennedy of the Neurosciences Research Branch, NIMH. Unfortunately, the authentic compounds have been found to deteriorate when stored in crystalline form, and it has been determined that they contain up to 20% free DA after six months of storage at 0° C. This prevents the use of the 'authentic' compounds for some experiments without purification.

The amount of free DA in any given DAS solution is routinely determined by HPLC as described below. This is usually less than 0.025% of the DAS amount. If necessary (depending on the experiment), this is removed by alumina extraction (Elchisak and Carlson, 1982a). The amount of one isomer in a 'pure' solution of the other isomer is also routinely determined by HPLC. A 'pure' solution of DA—3-*O*-sulfate usually contains 2—3% DA—4-*O*-sulfate because of the 'tailing' of the earlier eluting DA—4-*O*-sulfate peak into the DA—3-*O*-sulfate peak. A 'pure' DA—4-*O*-sulfate solution usually contains less than 0.25% DA—3-*O*-sulfate when purified by the methanol—water HPLC method.

Sample clean-up procedures

The direct detection and quantitation of DA—3-*O*-sulfate and DA—4-*O*-sulfate is sometimes possible in certain tissues without preliminary sample purification procedures. However, routine quantitation of these compounds in plasma, urine, CSF and most tissues requires preliminary sample clean-up procedures to isolate the compounds of interest from the sample matrix.

Solid-phase extraction: traditional ion-exchange columns. This description is from unpublished work by Swann and Elchisak. The sample (1—2 ml, pH 2—3) is applied to a 4 × 1 cm column of strong cation-exchange resin (Dowex 50 W, H$^+$ form). The column is eluted with 23 ml 0.5 M acetic acid using a vacuum of approximately 20 mm Hg. Free CA are retained on the column and, in this procedure, are assayed by standard alumina methods in a second aliquot of the sample. The CA-sulfates, specifically DA—3-*O*-sulfate and DA—4-*O*-sulfate, are contained in the eluate.

The entire eluate (24—25 ml) is then applied to a 4 × 1 cm column of strong anion exchange resin (AG 1, acetate form). The eluate is collected and combined with a column wash of 24 ml 0.5 M acetic acid. The total eluate volume (48—49 ml) is then evaporated to dryness (overnight) in a vacuum centrifuge in tubes containing 1 ml 0.03% EDTA and 1 ml 0.03% sodium metabisulfite as preservatives. The sample is then redissolved in a small volume (usually 0.5—1 ml) of HPLC mobile phase and injected into one of the HPLC systems described below.

This procedure is somewhat unwieldly, but produces very good results. The recovery of standards run through the entire procedure is quite low, but reproducible. For three experiments, recovery averaged 46.5 ± 2.2% for DA—3-*O*-sulfate and 45.8 ± 2.1% for DA—4-*O*-sulfate. The purification of the DAS isomers from urine produces chromatograms which have no peaks within 2—3 min on either side of the DAS peaks. The total HPLC run-time is approximately 40 min per sample. Using this technique, values obtained for urinary excretion of DA—3-*O*-sulfate and DA—4-*O*-sulfate in humans are similar to those obtained previously (Arakawa *et al.*, 1983; Elchisak and Carlson, 1982a).

Solid-phase extraction: pre-packed columns. The brief description below is based on work by Swann and Elchisak and will be published in detail elsewhere (manuscript in preparation).

This is the method of choice for sample preparation in the author's laboratory. The procedure is based on the same principles which govern the 'workability' of the classical ion-exchange separation procedure described above. However, silica-based columns [Baker-10 solid phase extraction (SPE)] are utilized. The columns are eluted using a 'Baker-10' SPE vacuum manifold system. The total working time to process 10 columns is less than one hour. This is an obvious improvement over the procedure described above.

The procedure utilizes an SPE strong anion-exchange (quaternary amine) column in sequence with an SPE C-18 column. The anion-exchange column is washed with hexane or water, and eluted with acetic acid. Free DA elutes primarily with the wash and void volume from the anion-exchange column, and this portion of eluate is saved and then either injected directly into the HPLC system or further purified by standard alumina procedures. The portion of eluate containing the DAS isomers is collected in tubes containing EDTA and sodium metabisulfite, evaporated to dryness in a speed-vac concentrator and redissolved in appropriate mobile phase for injection into the HPLC system.

This procedure has been utilized less extensively in the author's laboratory than the classical ion-exchange procedure discussed above. However, the HPLC chromatograms of the DAS isomers in human urine are at least as 'clean' after this procedure as after sample clean-up with the classical ion-exchange procedure. This procedure is definitely the method of choice for use in most experiments in the author's laboratory.

Internal standard

There is currently no adequate internal standard commercially available for the DAS sample purification and HPLC procedures. ^{3}H-DA—3-O-sulfate and ^{3}H-DA—4-O-sulfate, synthesized from ^{3}H-DA, are currently being used as internal standards. For analysis of free CA, dihydroxybenzylamine (DHBA) and/or deoxyepinephrine (DOE) are commonly used as internal standards. Arakawa *et al.* (1983) have recently reported the use (and synthesis) of DHBA—3-O-sulfate and DHBA—4-O-sulfate as suitable internal standards for analyses of the CA-sulfates in human urine. These authors used the reaction of DHBA with chlorosulfonic acid to prepare the O-sulfated products. In this author's hands, however, neither the chlorosulfonic acid method nor the sulfuric acid method (Jenner and Rose, 1973) was successful for the synthesis of the DHBA—O-sulfate isomers. Similarly, synthesis of DOE—O-sulfate was unsuccessful by these methods. These difficulties are probably related to the fact that DHBA and DOE are relatively unstable compounds. This, of course, is a serious limitation when the chemical reaction of interest is performed in concentrated sulfuric acid or utilizes chlorosulfonic acid.

The author's laboratory is currently investigating the synthesis of isoproterenol—O-sulfate using methods analagous to those described in detail above for DAS synthesis. If the synthesis is successful, the behavior of isoproterenol—3-O-sulfate and isoproterenol—4-O-sulfate will then be compared to DAS in the purification schemes described above and the HPLC systems described below. It is expected that, since isoproterenol is a

CA, the biochemical and electrochemical (for detection) properties of the *O*-sulfate derivatives will be similar to the DAS isomers. Isoproterenol was chosen as the compound to investigate in detail because it is reasonably stable, is commercially available and is inexpensive.

HPLC METHODS FOR DA-SULFATE ANALYSES

This section will only describe work conducted in the author's laboratory.

Equipment and materials

Chromatography equipment. The liquid chromatographs used in experiments described in this chapter were assembled from the following components purchased from Bioanalytical Systems, Inc. (West Lafayette, IN, USA) or Waters Associates (Milford, MA, USA): Waters Models 6000-A or M-45 pumps; Rheodyne Model 7125 injection valves equipped with 20 or 200 μl loops; and Houston Instruments dual-channel B-5000 strip chart recorders. A column of air connected in parallel with the solvent flow-path served as a pulse damper. In some experiments a completely automated liquid chromatograph was used. This consisted of the following components purchased from Waters Associates: Waters M-45 or 6000A pump; WISP Model 710-B automatic sample injector; Model 720 system controller; and Model 730 data module. Each HPLC system was equipped with a stainless-steel column (250 × 4 mm ID) prepacked with 5 μm Biophase octadecyl silica (Bioanalytical Systems) and a guard column (20 × 4.5 mm ID) packed with 40 μm C_{18}/Corasil (Waters Associates). The following detectors were utilized: ultraviolet absorbance detector (Model 440, Waters Associates) equipped with either a 254 nm or a 280 nm filter; single-electrode electrochemical detector (Bioanalytical Systems, Inc.) consisting of a thin-layer carbon paste or glassy carbon working electrode, an Ag/AgCl reference electrode and an LC-4 or LC-4A controller; dual-electrode electrochemical detector (LC-4B/17(D), Bioanalytical Systems, Inc.) consisting of a thin-layer dual glassy-carbon electrode in the series configuration, an Ag/AgCl reference electrode mounted downstream from the dual electrode and 2 LC-4B amperometric controllers; flow-through scintillation counter (Model HP radioactive flow detector; Radiomatic Instruments, Tampa, FL, USA).

Sulfated compounds. The following sulfated compounds were furnished by Drs Albert A. Manian and J. Stephen Kennedy, Neurosciences, Research Branch, National Institute of Mental Health (Rockville, MD, USA): dopamine—3-*O*-sulfate; dopamine—4-*O*-sulfate; norepinephrine—3-*O*-sulfate; norepinephrine—4-*O*-sulfate; serotonin—*O*-sulfate; tyramine—*O*-sulfate; and 3-methoxy-4-(hydroxysulfonyloxy) phenylacetic acid (homovanillic acid sulfate). 3-Methoxy-4-(sulfonyloxyphenyl)-glycol (MHPG-sulfate) was obtained from Fluka AG (Switzerland). For some experiments, dopamine—3-*O*-sulfate and dopamine—4-*O*-sulfate were synthesized and purified by the modification of the method of Jenner and Rose described above (Jenner and Rose, 1973).

Columns and mobile phases

All currently-utilized HPLC techniques for these separations utilize a reversed-phase stainless steel column (250 × 4 mm) prepacked with 5 μm Biophase octadecyl silica (C_{18}; Bioanalytical Systems) and a guard column (20 × 4.5 mm ID) packed with 40 μm C_{18}/Corasil (Waters Associates). Four different mobile phases are used and these are discussed below. In all cases, the mobile phase is passed through a 0.45 μm filter, degassed by sonication, maintained at approximately 63° C and stirred continuously during use.

Determination of free and conjugated CA: reversed-phase paired ion HPLC. Reversed-phase paired-ion chromatography is used for routine determination of free CA. Conjugated CA are measured after hydrolysis to free CA. This procedure has been published in detail (Elchisak and Carlson, 1982b). The mobile phase consists of phosphate buffer (75 mM), pH 2.8, containing EDTA (1 mM), sodium octyl sulfate (30—40 mg/l) and methanol (0—10%). Typical catechol retention times, expressed as capacity factors (k'), are as follows: NE (1.6), E (2.9), 1-DOPA (3.7), DHBA (4.1), DA (7.4), DOE (9.4), DOPAC (11). Capacity factors for these and other compounds of interest are shown in Table 1. The DAS isomers are first hydrolyzed to free DA in acid (Elchisak and Carlson, 1982b) or enzymatically (Johnson *et al.*, 1980) before measurement as free DA. This method does not distinguish between the two isomers of DAS, but it has been utilized in the author's laboratory to study the distribution of conjugated DA in human, monkey, rat and mouse body fluids and tissues (Elchisak, 1983c; Elchisak and Carlson, 1982b; Elchisak *et al.*, 1983; Shoaf and Elchisak, 1983). A chromatogram of

Table 1. Catechol retention times expressed as capacity factors

Compound	Capacity factor
Norepinephrine	1.60 ± 0.041 ($n = 3$)
Epinephrine	2.91 ± 0.093
Dihydroxybenzylamine	4.05 ± 0.13
Dopamine	7.37 ± 0.29
Deoxyepinephrine	9.43 ± 0.36
3,4-Dihydroxyphenylglycol	1.19 ($n = 1$)
Uric acid	1.34
3,4-Dihydroxymandelic acid	2.02
1-DOPA	3.66
3,4-Dihydroxyphenylacetic acid	10.5
Caffeic acid	11.3
α-Methyldopa	13.2
α-Methyldopamine	13.4
N-Acetyldopamine	14.1

Capacity factor = (peak retention time − void volume time)/void volume time.
Mobile Phase: Phosphate buffer (75 mM); EDTA (1 mM); sodium octyl sulfate (30 mg/1); pH 2.80; methanol (3%).

CA from rat kidney utilizing this chromatographic system coupled with single-electrode electrochemical detection is shown in Fig. 1. Using this technique, significant amounts of conjugated DA were found in the heart, kidney and liver of the dog (Table 2). However, conjugated DA was not detected in the following areas of dog brain: striatum, frontal cortex, hindbrain, diencephalon, midbrain, cerebellum and pituitary (Shoaf and Elchisak, unpublished). Similar methods, which do not distinguish between the two isomers of DAS, are widely utilized by other investigators.

Determination of DA-sulfate isomers: reversed-phase HPLC. Reversed-phase chromatography is used for the routine determination of DA—3-*O*-sulfate and DA—4-*O*-sulfate in the author's laboratory. Three different mobile phases are currently used for different applications.

1. Paired-ion HPLC for DA-sulfate isomers. Reversed-phase paired-ion chromatography is used for the routine determination of DA—3-*O*-sulfate and DA—4-*O*-sulfate. This procedure has been published in detail (Elchisak and Carlson, 1982a). The mobile phase consists of monochloroacetic acid (25 mM) and EDTA (1 mM), and is adjusted to pH 2.8 with sodium hydroxide. The ion-pairing agent is *n*-octylamine (4 mM). This is added after pH adjustment and results in a final mobile phase pH of 3.1. Typical capacity factors are 3.3 and 3.7 for DA—3-*O*-sulfate and DA—4-*O*-sulfate, respectively. This system is also suitable for separation of sulfated esters of NE, tyramine, HVA, MHPG and serotonin. These compounds have capacity factors ranging from approximately 1 to 10 (Elchisak, 1983b, Table 3). Free NE, E and DA have capacity factors significantly less than 1 (Elchisak, unpublished), however, making quantitation very difficult in complex sample matrices. This HPLC system has been widely utilized in the author's laboratory for development of methods to directly quantitate DAS isomers in biological samples without prior hydrolysis of the sample (Elchisak, 1983a, b; Elchisak and Carlson, 1982a). A chromatogram of DAS standards using this HPLC system and ultraviolet detection is shown in Fig. 2.

2. Phosphate buffer HPLC for free catechols and DA-sulfate isomers. Two reversed-phase procedures not utilizing ion-pairing agents are routinely used in the author's laboratory for separation of the two DAS isomers. The first of these is a modification of a procedure published by Buu *et al.* (1981b). As utilized in the author's laboratory, the mobile phase consists of phosphate buffer (75 mM) and EDTA (1 mM), pH 4.8. Capacity factors are satisfactory for the determination of DAS isomers and free CA. Typical capacity factors are as follows: NE (0.59), E (1.4), l-DOPA (1.5), DHBA (1.6), DA (3.1), DA—4-*O*-sulfate (4.1), DA—3-*O*-sulfate (4.5). Capacity factors for these and other compounds are shown in Table 4. Although Buu *et al.* (1981b) reported poor resolution between the DAS isomers using this mobile phase (with less EDTA), the separation of these two compounds has been very satisfactory in this laboratory. The ratio of the *k*'s (selectivity) for the DAS isomers was found to be 1.1, which is quite acceptable for most

M. A. Elchisak

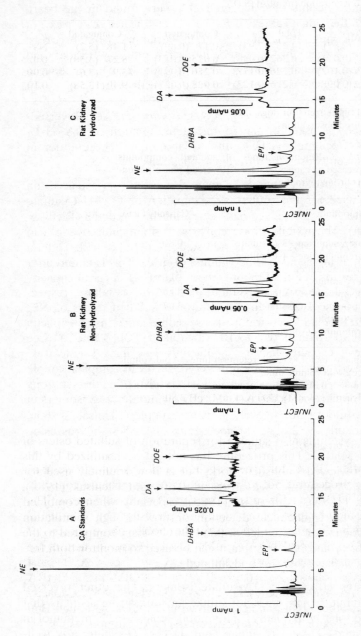

Figure 1. Chromatograms of alumina extracts of: (A) catecholamine standard solution and rat kidney homogenate supernatants (B) before and (C) after hydrolysis. There was no breakdown of any of the CA during the hydrolysis procedure. Note that total DA (hydrolyzed) is greater than free DA in this sample, indicating that substantial amounts of conjugated DA were present. The mobile phase was phosphate buffer (75 mM), pH 2.8, containing EDTA (1 mM), sodium octyl sulfate (30 mg/l) and methanol (5%). The detector was a single-electrode electrochemical detector, and a thin-layer CP-O carbon paste working electrode maintained at +0.70 V was utilized. The flow-rate was 1.1 ml/min. Injection volume was 50 µl. Other HPLC conditions were as described in the text. The standard solution contained the following amounts of each CA per injection (assuming 100% recovery): NE, 200 pmol; EPI, 10 pmol; DHBA, 100 pmol; DA, 5 pmol; DOE, 20 pmol. [From Elchisak and Carlson (1982b); reprinted with permission, copyright Elsevier Science Publishers b.v.]

Table 2. Free and conjugated DA in dog visceral tissues

	DA Concentration (nmol/g) (S.D.)				
Tissue (n)	Free	Total	Conjugated	% Conjugated	p^b
Adrenal (8)[a]	14.06 (7.87)	14.23 (7.93)	0.17 (0.89)	1.48 (5.2)	NS
Heart (4)	0.297 (0.011)	0.362 (0.048)	0.025 (0.013)	8.27 (4.58)	0.025
Kidney (8)[a]	0.476 (0.102)	0.726 (0.177)	0.250 (0.091)	33.9 (5.6)	0.005
Liver (4)	0.189 (0.098)	0.216 (0.123)	0.028 (0.027)	9.40 (13.58)	0.10

[a] Eight tissues (left and right) from four dogs.
[b] Student's *t*-test for paired observations.

Table 3. Retention times of sulfated compounds expressed as capacity factors: Reversed-phase paired-ion chromatography

Compound	Capacity factor
Norepinephrine—3-*O*-sulfate	0.80
Norepinephrine—4-*O*-sulfate	0.90
Tyramine-sulfate	1.87
HVA-sulfate	3.00
Dopamine—3-*O*-sulfate	3.27
Dopamine—4-*O*-sulfate	3.67
MHPG-sulfate	7.40
Serotonin—*O*-sulfate	10.80

Capacity factor = (peak retention time − void volume time)/void volume time.
Mobile phase: Monochloroacetic acid (25 mM); *n*-octylamine (4 mM); EDTA (1 mM); pH 2.80.

routine work. This system is also suitable for separation of sulfated esters of NE, serotonin and MHPG. This procedure has not yet been utilized by this laboratory in any reviewed published work, but is now routinely used for most applications [for example, see abstracts: Toth and Elchisak (1984); Scott and Elchisak (1985); Toth *et al.* (1985)] especially when combined with dual-electrode electrochemical detection or flow-through scintillation counting (Fig. 3). The major advantage of this HPLC system compared to the others discussed above and below is that it can be used to monitor both free and sulfated DA, as well as other free and sulfated CA.

3. Methanol/water HPLC for DA-sulfate isomers. The other reversed phase HPLC technique which does not utilize an ion-pairing agent is that published by Osikowska *et al.* (1982). The mobile phase is 2% methanol in water. In this system, DA—4-*O*-sulfate and DA—3-*O*-sulfate are well-resolved from each other (*k*'s of 4.6 vs 4.1, respectively), but the free catechols are retained indefinitely on the column. This system is also suitable

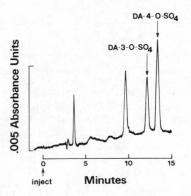

Figure 2. Chromatogram of DA—3-*O*-sulfate and DA—4-*O*-sulfate. 20 µl of a synthetic mixture containing approximately 1 nmol of each DAS isomer was injected into the HPLC system. The DAS peaks disappeared upon hydrolysis of the sample. The mobile phase was monochloroacetic acid (0.025 M), adjusted to pH 2.8, containing EDTA (1 mM) and *n*-octylamine (4 mM). Ultraviolet detection at 280 nm was used. Other HPLC conditions were as described in the text. [From Elchisak and Carlson (1982a); reprinted with permission, copyright Pergamon Press, Ltd.]

Table 4. Retention times expressed as capacity factors: Reversed-phase (non-paired-ion) chromatography

Compound	Capacity factor
3,4-Dihydroxymandelic acid	0.37
Norepinephrine	0.59
Epinephrine	1.2
1-Dopa	1.5
3,4-Dihydroxyphenylglycol	1.5
Dihydroxybenzylamine	1.6
Caffeic acid	1.7
α-Methyldopa	2.0
α-Methyldopamine	2.0
n-Acetyldopamine	2.2
Norepinephrine—4-*O*-sulfate	2.7
Norepinephrine—3-*O*-sulfate	2.8
Dopamine	3.1
Dopamine—4-*O*-sulfate	4.1
Dopamine—3-*O*-sulfate	4.5
3,4-Dihydroxyphenylacetic acid	4.8
Deoxyepinephrine	4.9
Serotonin—*O*-sulfate	16.0
MHPG-Sulfate	34.0

Capacity factor = (peak retention time − void volume time)/ void volume time.
Mobile phase: Phosphate buffer (75 mM); EDTA (1 mM); pH 2.80.

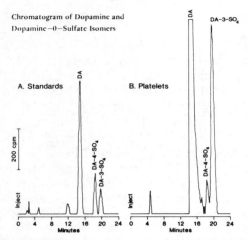

Figure 3. chromatograms of ^{3}H-DA, ^{3}H-DA—3-O-sulfate and ^{3}H-DA—4-O-sulfate (A) in a standard solution and (B) from human platelets incubated at 37° C in the presence of 1 μM ^{3}H-DA. The mobile phase was phosphate buffer (75 mM) and EDTA (1 mM), pH 4.8. The detector was a flow-through scintillation counter. The mobile phase flow rate was 0.7 ml/min and the scintillator (ACS) flow rate was 4.6 ml/min. Injection volume was 200 μl. Other HPLC conditions were as described in the text.

for separation of sulfated esters of NE and serotonin, since these compounds have capacity factors ranging from approximately 1 to 13 (Table 5). The main application in the author's laboratory for this chromatographic system is for isolation and purification of the DAS isomers from a gross synthetic mixture, since the methanol can be readily evaporated and the resulting water solution of the DAS isomer of interest can be used for most experiments. The major limitations of the system are that the free CA are retained indefinitely on the column and that the mobile phase is 'non-buffered', thus allowing for pH changes (which might modify retention time) with injections of samples from various sources.

Table 5. Retention times of sulfated compounds expressed as capacity factors: Methanol—water mobile phase

Compound	Capacity factor
Norepinephrine—4-O-sulfate	0.76
Norepinephrine—3-O-sulfate	0.85
Tyramine-sulfate	2.07
Dopamine—4-O-sulfate	4.18
Dopamine—3-O-sulfate	4.55
Serotonin—O-sulfate	12.84

Capacity factor = (peak retention time − void volume time)/void volume time.
Mobile phase: 2% methanol in water.

Detection systems

Ultraviolet detection at 280 nm. Ultraviolet (UV) detection at 280 nm was originally used for analyses of the DAS isomers (Elchisak, 1983a; Elchisak and Carlson, 1982a). Although there is a linear relationship between the UV detector response and the amount of either DAS isomer injected onto the HPLC column (Fig. 4), the lower limits of detection of this method are approximately 15—20 pmol for each of the DAS isomers. Consequently, this detection method is not sensitive enough for much routine work requiring quantitation of endogenous concentration of the DAS isomers in biological samples. Furthermore, the UV detection method also has relatively poor specificity, which makes quantitation of the DAS isomers difficult in complex sample matrices, such as urine (Fig. 5). Parameters describing the UV detection of several sulfated compounds in two different mobile phases are given in Tables 6 and 7.

Since UV detection is non-destructive to the sample of interest, it is useful for routine use in monitoring the synthesis of the DAS isomers, and for occasional use when it is desired to collect a fraction and submit it to other identification tests. Specificity can be improved by monitoring absorbance at two different wavelengths and comparing the ratio (e.g. 280/254) in samples and standards.

Single-electrode electrochemical detection. Single-electrode electrochemical detection has been extensively used in the author's laboratory, as well as in other laboratories, for the quantitation of free and conjugated CA in biological fluids and tissue samples (Elchisak, 1983c; Elchisak and Carlson 1982b; Elchisak *et al.*, 1983; Shoaf and Elchisak, 1983). The conjugated CA

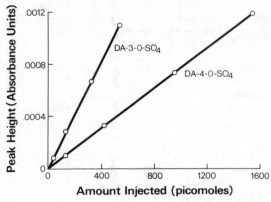

Figure 4. Standard curve for DA—3-*O*-sulfate and DA—4-*O*-sulfate. There was a linear correlation between the detector response and the amount of DA—3-*O*-sulfate or DA—4-*O*-sulfate injected into the HPLC system. The mobile phase was monochloroacetic acid (0.025 M) adjusted to pH 2.8, containing EDTA (1 mM) and *n*-octylamine (4 mM). Ultraviolet detection at 280 nm was used. Other HPLC conditions were as described in the text [From Elchisak and Carlson (1982a); reprinted with permission, copyright Pergamon Press, Ltd.]

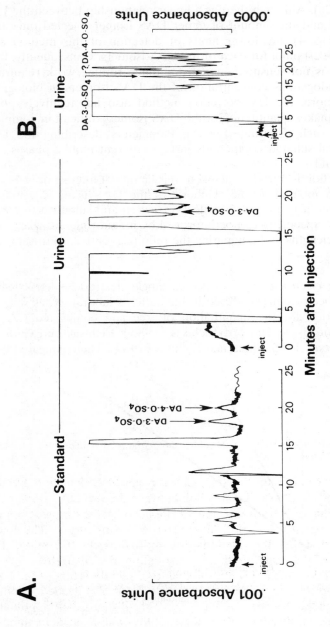

Figure 5. (A) Chromatograms of DAS standards and eluate from human urine 'purified' by passage through cation- and anion-exchange columns (Elchisak and Carlson, 1982a) before injection into the HPLC system. The mobile phase was monochloracetic acid (0.025 M), pH 2.8, containing EDTA (1 mM) and *n*-octylamine (4 mM). Ultraviolet detection at 280 nm was utilized. The flow rate was 0.5 ml/min and the pressure was approximately 700 psi. Injection volume was 200 µl. (B) Chromatogram of eluate from human urine prepared according to the method of Buu *et al.* (1981a). [From Elchisak and Carlson (1982a); reprinted with permission, copyright Pergamon Press, Ltd.]

Table 6. Ultraviolet detection of sulfated compounds: Reversed-phase paired-ion chromatography

Compound	Capacity factor	280/254	Lower limits of detection (pmol)
Norepinephrine—3-O-sulfate	1.16	4.43	14
Norepinephrine—4-O-sulfate	1.27	4.92	11
Tyramine-sulfate	2.68	12.4	10
HVA-sulfate	3.00	?	?
Dopamine—3-O-sulfate	4.52	5.11	21
Dopamine—4-O-sulfate	5.18	4.34	14
MHPG-sulfate	7.40	?	?
Serotonin—O-sulfate	13.73	2.41	12

Capacity factor = (peak retention time — void volume time)/void volume time.
Mobile phase: Monochloroacetic acid (25 mM); *n*-octylamine (4 mM); EDTA (1 mM); pH 2.80.

Table 7. Ultraviolet detection of sulfated compounds: Reversed-phase (non-paired ion) chromatography

Compound	Capacity factor	280/254	Lower limits of detection (pmol)
Norepinephrine—4-O-sulfate	0.76	4.65	5.1
Norepinephrine—3-O-sulfate	0.85	4.38	7.0
Tyramine-sulfate	2.07	13.6	9.2
HVA-sulfate		(?4.78)	18
Dopamine—4-O-sulfate	4.18	4.41	12
Dopamine—3-O-sulfate	4.55	5.59	18
MHPG-sulfate		(?4.25)	5.7
Serotonin—O-sulfate	12.8	2.55	11

Capacity factor = (peak retention time — void volume time)/void volume time.
Mobile phase: 2% Methanol in water.

must first be hydrolyzed to the free CA for sensitive detection by this method. This results in a lack of distinction between the two DAS isomers, as well as the possibility of hydrolyzing non-sulfated conjugated forms. A sample chromatogram using this detection method is shown in Fig. 1. The lower limits of detection using this method are approximately 0.2 pmol. The specificity is reasonable due to the low oxidation potential (0.5—0.7 V) utilized. This detection method is not suitable for direct detection of the DAS isomers because an oxidation potential in excess of 1.4 V is needed for these compounds (discussed below; see Fig. 19). This high oxidation potential results in unacceptable specificity for analysis of the DAS isomers in complex sample matrices, as well as technical difficulties resulting from rapid fouling of the electrode surface.

Post-column hydrolysis. This technique is one of two techniques used by the author which is practically useful for the detection of the DAS isomers in biological samples. Another recently published technique, not used by the author personally, also appears to be very promising for this purpose. Arakawa *et al.* (1983) have utilized a post-column reaction with ultraviolet light at 280 nm to cleave the *O*-sulfate bond from the CA—*O*-sulfates and then utilized an in-stream ethylenediamine condensation method to form derivatives which were then detected fluorometrically. This technique is not reviewed here.

1. General principle and description. DA—3-*O*-sulfate and DA—4-*O*-sulfate are separated by HPLC and the column effluent then flows into a mixing chamber into which a strong acid is continuously introduced. The effluent exits the mixing chamber and flows through a coil maintained at a high temperature. DA—3-*O*-sulfate and DA—4-*O*-sulfate are each hydrolyzed to free DA in the heated coil, but their temporal separation is maintained. The effluent is then passed through a cooling chamber and into a single-electrode electrochemical detector. Free DA is detected as two separate peaks corresponding to each of the original DA-sulfate isomers.

2. Post-column hydrolysis apparatus. The basic liquid chromatography system was modified for the post-column hydrolysis procedure from commercially available parts. A diagram of the entire liquid chromatography system, including the post-column hydrolysis apparatus, is shown in Fig. 6. The post-column hydrolysis apparatus is connected in series between the exit from the liquid chromatography column and electrochemical detector. Stainless steel tubing (0.01 inch ID) is used throughout the system, except between the cooling chamber exit and the working electrode of the electrochemical detector. These components are connected with polyethylene tubing (0.034 inch ID). The mixing chamber is constructed from a reference manifold assembly for a Model 6000A pump (Waters Associates) (Fig. 7). The acid used for the hydrolysis procedure is pumped into the mixing chamber by either a Waters Model M-45 pump or a Constameric I pump (Laboratory Data Control, Riviera Beach, FL). A column of air connected in parallel with the acid flow serves as a pulse damper. The effluent from the mixing chamber is then passed through a hydrolysis coil, constructed from stainless steel tubing (for most experiments, 50 ft × 0.01 inch ID), which is maintained at a constant high temperature in a standard laboratory electric oven. The hydrolysis of DA—3-*O*-sulfate and DA—4-*O*-sulfate to free DA occurs during passage through this heated coil. The hydrolysis coil is connected to a cooling coil which is immersed in a dewar flask containing an ice-water mixture. The effluent then flows into the single-electrode electrochemical detector where the hydrolysis products (and any other electrochemically-active species) are detected.

3. Chromatographic conditions. For all experiments described in this section, the column was 5 µm octadecyl silica and the mobile phase was

POST-COLUMN HYDROLYSIS SYSTEM

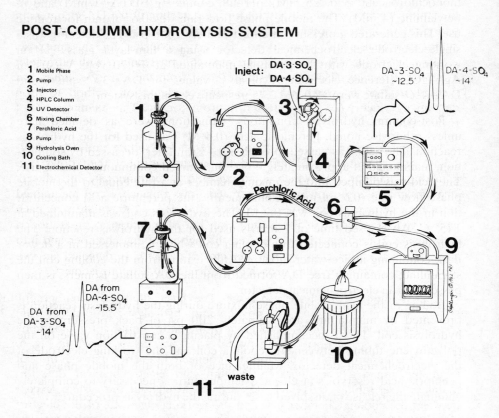

Figure 6. Schematic diagram of the liquid chromatography system, including the post-column hydrolysis apparatus. Details are given in text. [From Elchisak (1983a); reprinted with permission, copyright Elsevier Science Publishers b.v.]

MIXING CHAMBER

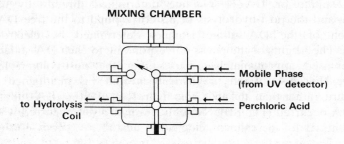

Figure 7. Schematic diagram of mixing chamber used for the post-column hydrolysis apparatus showing mobile phase and perchloric acid flow directions. The flow paths shown resulted in optimal mixing of the acid and mobile phase. [From Elchisak (1983a); reprinted with permission, copyright Elsevier Science Publishers b.v.]

monochloroacetic acid (25 mM), pH 2.8, containing EDTA (1 mM) and *n*-octylamine (4 mM). The mobile phase flow rate was 0.6 ml/min for routine use. This generated a pressure of approximately 1000 psi. The detector was a single-electrode electrochemical detector using a thin-layer glassy-carbon working electrode with the potential maintained at +0.70 volts versus an Ag/AgCl reference electrode. Typical k' values for DA—3-O-sulfate and DA—4-O-sulfate were 2.40 and 2.75, respectively, in this system.

Post-column hydrolysis conditions. Conditions were as detailed here unless otherwise noted. Perchloric acid (0.4 M) was used for the hydrolysis reaction. The acid was passed through a 0.45 μm filter, degassed by sonication, maintained at approximately 63° C and stirred continuously during use. The acid was pumped into the mixing chamber at a rate equal to the mobile phase flow rate (0.6 ml/min). Consequently, the perchloric acid concentration in the hydrolysis coil was 0.2 M. The hydrolysis coil was maintained at 145° C. An elapsed time of 1 min is used for the hydrolysis reaction. The hydrolysis coil is connected to a cooling coil which is immersed in a dewar flask containing an ice-water mixture. After exiting from the cooling coil the streamline, containing free DA formed from the DA-sulfate isomers, is then passed into an electrochemical detector.

Boiling of the acid/mobile phase mixture during the hydrolysis reaction is prevented by maintaining approximately 200 psi of back pressure in the hydrolysis coil. This is accomplished by placing a constriction valve on the polyethylene tubing between the cooling coil and the working electrode of the electrochemical detector. Maintenance of both the mobile phase and pechloric acid reservoirs at elevated temperature is necessary to completely eliminate the release of dissolved gases during the hydrolysis procedure.

5. Determination of DA-sulfate isomers pre- and post-hydrolysis. Chromatograms of DA—3-O-sulfate and DA—4-O-sulfate standards before and after the post-column hydrolysis procedure are shown in Fig. 8. Approximately 1.5 nmol DA—3-O-sulfate and 2.5 nmol DA—4-O-sulfate were injected into the HPLC system. The pre-hydrolysis peaks correspond to DA—3-O-sulfate or DA—4-O-sulfate determined directly by ultraviolet detection, and the post-hydrolysis peaks correspond to the free DA formed from each of the DA-sulfate isomers determined by electrochemical detection. The heights of the peaks corresponding to each DA-sulfate isomer were increased approximately 15-fold after the post-column hydrolysis procedure (Fig. 8A) When the hydrolysis coil was maintained at room temperature, rather than at 145° C, no hydrolysis of either DA-sulfate isomer to free DA occurred (Fig. 8B). Similarly, no hydrolysis occurred when water was substituted for perchloric acid in the heated hydrolysis coil (data not shown).

6. Determination of optimal conditions. Hydrolysis of the DA-sulfate isomers with perchloric acid produced larger post-hydrolysis free DA peaks than any of the other acids tested [sulfuric, acetic and phosphoric (Fig. 9)].

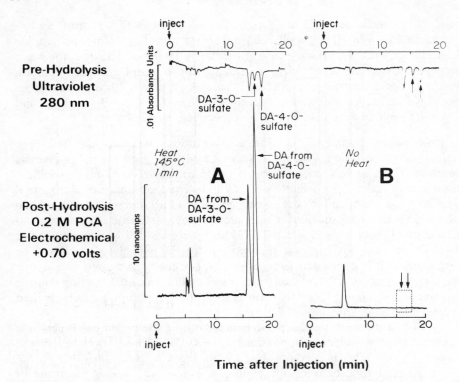

Figure 8. (A) Chromatograms of DAS isomers before and after the post-column hydrolysis procedure. The pre-hydrolysis peaks are DA—3-*O*-sulfate and DA—4-*O*-sulfate determined by ultraviolet detection at 280 nm. The free DA peaks on the post-hydrolysis chromatogram correspond to each of the original DAS isomers, and were detected electrochemically. Approximately 1.5 nmol DA—3-*O*-sulfate and 2.5 nmol DA—4-*O*-sulfate were injected onto the column. The mobile phase was monochloroacetic acid (0.025 M), adjusted to pH 2.8, containing EDTA (1 mM) and *n*-octylamine (4 mM). The single-electrode electrochemical detector utilized had a thin-layer glassy carbon working electrode with the potential maintained at +0.70 V. Other conditions were as described in the text. (B) The same sample as in (A) except that the hydrolysis coil was at room temperature. [From Elchisak (1983a); reprinted with permission, copyright Elsevier Science Publishers b.v.]

The height of the post-hydrolysis DA peaks was not appreciably affected by varying the final concentration of perchloric acid between 0.2 and 1.0 M (Fig. 10). Decreased peak heights were seen outside this range. Consequently, 0.2 M perchloric acid was chosen for routine use in the post-column hydrolysis procedure.

Experiments in which the hydrolysis temperature was systematically varied indicated that the peak heights of the free DA formed from both DA—3-*O*-sulfate and DA—4-*O*-sulfate approached plateaus between 130° and 145° C (Fig. 11). The peak heights of both DA peaks formed from the DA-sulfate isomers increased approximately 20% with increased heating times between 0.7 and 1.8 min (Fig. 12).

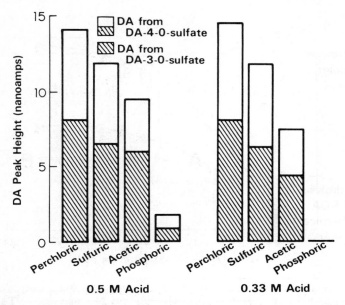

Figure 9. Effect of different acids on post-column hydrolysis efficiency. The mobile phase was monochloroacetic acid (0.025 M), adjusted to pH 2.8, containing EDTA (1 mM) and *n*-octylamine (4 mM). A single-electrode electrochemical detector was utilized with a thin-layer glassy carbon working electrode. The potential was maintained at +0.70 V. Other conditions were as described in the text.

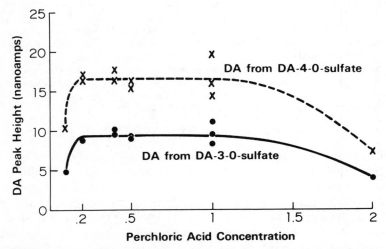

Figure 10. Effect of perchloric acid concentration on post-column hydrolysis efficiency. The mobile phase was monochloroacetic acid (0.025 M), adjusted to pH 2.8, containing EDTA (1 mM) and *n*-octylamine (4 mM). A single-electrode electrochemical detector was utilized with a thin-layer glassy carbon working electrode. The potential was maintained at +0.70 V. Other conditions were as described in the text.

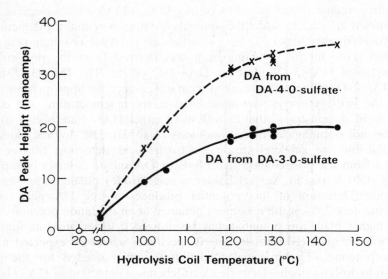

Figure 11. Effect of hydrolysis temperature on post-column hydrolysis efficiency. The mobile phase was monochloroacetic acid (0.025 M), adjusted to pH 2.8, containing EDTA (1 mM) and *n*-octylamine (4 mM). A single-electrode electrochemical detector was utilized with a thin-layer glassy carbon working electrode. The potential was maintained at +0.70 V. Other conditions were as described in the text.

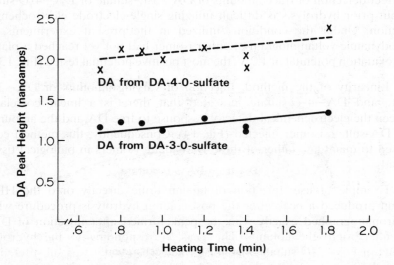

Figure 12. Effect of heating time on post-column hydrolysis efficiency. The mobile phase was monochloroacetic acid (0.025 M), adjusted to pH 2.8, containing EDTA (1 mM) and *n*-octylamine (4 mM). A single-electrode electrochemical detector was utilized with a thin-layer glassy carbon working electrode. The potential was maintained at +0.70 V. Other conditions were as described in the text.

Hydrodynamic voltammograms of free DA and DA-sulfate isomers were constructed in order to select the optimal oxidation potential for detection of the hydrolysis products of DA—3-*O*-sulfate and DA—4-*O*-sulfate (Fig. 13). The free DA on the voltammogram was derived from the post-column hydrolysis of DA—3-*O*-sulfate or DA—4-*O*-sulfate. The DA—3-*O*-sulfate and DA—4-*O*-sulfate peaks resulted from injection of the appropriate isomer while the hydrolysis coil was maintained at room temperature. An earlier experiment demonstrated that hydrolysis of either DA-sulfate isomer to free DA did not occur under these conditions (Fig. 8B). The optimal oxidation potential for the single-electrode electrochemical detection of free DA derived from the post-column hydrolysis of DA-sulfate isomers is approximately 0.80 V (vs an Ag/AgCl reference electrode) on the glassy carbon electrode. Detection at this potential produces a free DA peak height approximately 75% of the maximum obtained at an oxidation potenial of 1.0 V. A higher oxidation potential may be utilized if lower absolute limits of detection are required, but decreased specificity would be expected at the higher potential. The maximal oxidation potential needed for the post-column hydrolysis single-electrode electrochemical detection of DA is higher than that usually needed for detection of free DA (0.5—0.7 V, Elchisak and Carlson, 1982b). Although there are several factors contributing to this, the primary reason for the higher oxidation potential required is probably the low pH of the mobile phase/perchloric acid mixture crossing the electrode. Final choice of oxidation potential used depends on the requirements for specificity and lower limits of detection.

Direct detection of trace amounts of DA—3-*O*-sulfate or DA—4-*O*-sulfate without prior hydrolysis is difficult utilizing single-electrode electrochemical detection. Under the conditions utilized in the present experiments, the hydrodynamic voltammogram of either isomer had not yet reached a plateau at an oxidation potential of 1.2 V, the most positive potential tested (Fig. 13).

7. Linearity of the method. Injection of varying amounts of DA—3-*O*-sulfate and DA—4-*O*-sulfate indicated that there is a linear correlation between the electrochemical detector response to free DA and the amount of each DA-sulfate isomer injected (Fig. 14). Consequently, this method could be used to quantitate either of these DA-sulfate isomers in biological tissues or fluids.

8. Example of use. Injection of human urine directly onto the HPLC column produced a peak after the post-column hydrolysis procedure which co-chromatographed exactly with the peak formed after injection of DA—3-*O*-sulfate onto the column. The peak corresponding to the hydrolysis product of DA—3-*O*-sulfate was not formed when water was substituted for acid in the hydrolysis procedure, or when the hydrolysis coil was maintained at room temperature (Fig. 15).

9. Conclusions regarding post-column hydrolysis procedure. Use of the post-column hydrolysis technique resulted in increased lower limits of

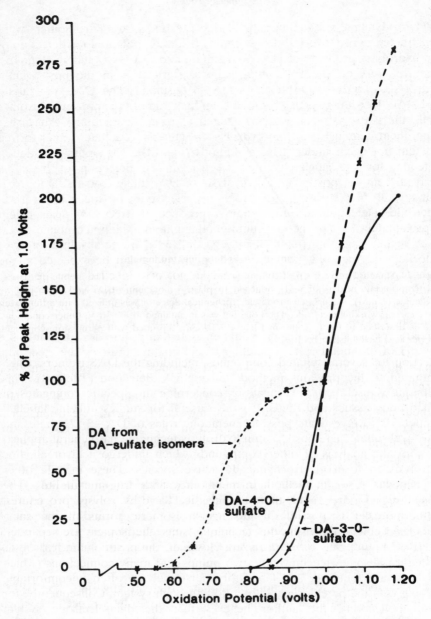

Figure 13. Hydrodynamic voltammograms of free DA and DAS isomers determined after post-column hydrolysis. The free DA was derived from the post-column hydrolysis of either DA—3-*O*-sulfate or DA—4-*O*-sulfate. The DA-sulfate peaks resulted from injection of the appropriate isomer without post-column hydrolysis, i.e. the hydrolysis coil was maintained at room temperature instead of at 145° C. The mobile phase was monochloroacetic acid (0.025 M), adjusted to pH 2.8, containing EDTA (1 mM) and *n*-octylamine (4 mM). A single-electrode electrochemical detector was utilized with a thin-layer glassy carbon working electrode. Other conditions were as described in the text.

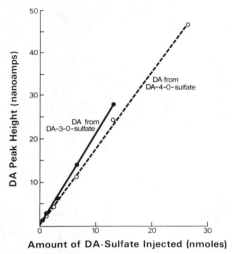

Figure 14. Standard curve for post-column hydrolysis of DAS isomers. The mobile phase was monochloroacetic acid (0.025 M), adjusted to pH 2.8, containing EDTA (1 mM) and *n*-octylamine (4 mM). A single-electrode electrochemical detector was utilized with a thin-layer glassy carbon working electrode. The potential was maintained at +0.70 V. Other conditions were as described in the text. [From Elchisak (1983a); reprinted with permission, copyright Elsevier Science Publishers b.v.]

detection for several sulfated compounds, including the DAS isomers, when compared to the previous method utilizing UV detection (Table 8). This technique is useful for investigations concerning the specific conjugates of DA in body tissues and fluids. It is also useful for studies utilizing labelled compounds to investigate possible metabolic roles of DA—3-*O*-sulfate and DA—4-*O*-sulfate. The post-column hydrolysis technique has general applications for the analysis of other compounds which undergo acid or alkaline hydrolysis to form electrochemically active species. These include other conjugated CA or indoles, their metabolites and drug metabolites. The procedure is more rapid than the classical acid-hydrolysis procedures. Furthermore, it is possible to distinguish isomeric forms of the same conjugated compound using this technique, since the isomers are separated by HPLC before they are hydrolyzed. However, the post-column hydrolysis procedure is somewhat unwieldy, two pumps and many connections (which can leak) are required, and laboratory personnel are often uncomfortable working with the perchloric acid in a flow-through system. Consequently, for most applications, the author believes that the dual-electrode system described below is preferable for the direct analysis of the DAS isomers in biological samples.

Dual-electrode electrochemical detection.

1. General principle and summary. A dual-electrode electrochemical detection procedure, with a lower limit of detection for each of the DAS isomers of slightly less than 1 pmol, has more recently been developed (Elchisak,

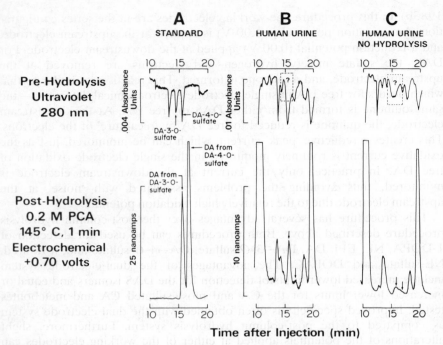

Fig. 15. Post-column hydrolysis chromatograms of human urine injected directly onto the HPLC column. The mobile phase was monochloroacetic acid (0.025 M), adjusted to pH 2.8, containing EDTA (1 mM) and *n*-octylamine (4 mM). A single-electrode electrochemical detector was utilized with a thin-layer glassy carbon working electrode. Other conditions were as described in the text. (A) Standard solution containing approximately 5 nmol of each DA-sulfate isomer. (B) Human urine (20 μl) injected directly onto the HPLC column. (C) The same urine sample as in (B) was injected directly onto the HPLC column, but the hydrolysis coil was maintained at room temperature instead of at 145° C. [From Elchisak (1983a); reprinted with permission, copyright Elsevier Science Publishers b.v.]

Table 8. Lower detection limits for sulfated compounds: Reversed-phase paired-ion chromatography

Compound	Lower detection limits ultraviolet (pmol)[a]	Post-column hydrolysis 0.70 V	0.80 V
Norepinephrine—3-*O*-sulfate	14	3.8	0.84
Norepinephrine—4-*O*-sulfate	11		
Tyramine-sulfate	10		
HVA-sulfate		34	35
Dopamine—3-*O*-sulfate	21	8.8	6.0
Dopamine—4-*O*-sulfate	14	7.9	5.7
Serotonin—*O*-sulfate	12	7.5	5.9

[a] Injected amount producing signal: noise of approximately 2.
Mobile phase: Monochloroacetic acid (25 mM); *n*-octylamine (4 mM); EDTA (1 mM); pH 2.80.

1983b). In this procedure, the working electrodes are in the series configuration. An oxidation potential (+1.00 V) is applied at the upstream electrode, and a reduction potential (0.00 V) applied at the downstream electrode. For DAS, the sulfate moiety, hydrogen and electrons are removed at the upstream electrode, and a quinone is formed. This reaction is similar to that which occurs for free DA in single electrode electrochemical detection — the same quinone is formed from either DAS or free DA. At the downstream electrode, the quinone is reduced to free DA by 'recapture' of the electrons. This creates a reductive 'peak' current, which can be monitored, just as the oxidative current is routinely monitored in the single electrode oxidation of free DA. In practice, only the current at the downstream electrode is monitored, thus avoiding the problems associated with 'noise' at the upstream electrode due to the relatively high oxidation potential.

This procedure has several advantages over the post-column hydrolysis procedure described above. Both procedures can be used for detection of 1-DOPA, NE, EPI, DA, DA—3-*O*-sulfate, DA—4-*O*-sulfate and, if desired, NE-sulfate and DOPAC. The advantages of the dual-electrode system include increased lower limits of detection for the DAS isomers and equal or increased lower limits for the CA and most sulfated CA and metabolites tested. Improved specificity is often obtained with the dual-electrode system as compared to the post-column hydrolysis system. Furthermore, slight alterations of the potentials applied at either of the working electrodes can sometimes eliminate or reduce a co-eluting or near co-eluting interfering peak while having only a slight effect on the size of the peak of interest. The dual-electrode system is technically less complex than the post-column hydrolysis procedure, and consequently it is more reliable for routine use on a day-to-day basis.

This procedure is currently used for the majority of DAS analyses in the author's laboratory. Other detection methods are used in special circumstances, such as when a non-destructive technique is needed or for verification of a tentatively identified peak.

2. Chromatographic conditions. The column was octadecyl silica and the mobile phase was monochloroacetic acid (25 mM), pH 2.8, containing EDTA (1 mM) and *n*-octylamine (4 mM). For experiments described in this section, the mobile phase flow rate was usually 0.8 ml/min. The pressure was approximately 1800 psi.

3. Detector conditions. The thin-layer glassy-carbon dual electrodes were utilized in the series configuration. Potentials were maintained versus a Ag/AgCl reference electrode located downstream. For many experiments, the potential was +1.00 V on the upstream (W1) electrode and 0.00 V on the downstream (W2) electrode. The potentials were maintained, and varied as indicated in some experiments, with LC-4B amperometric controllers.

4. Capacity factors. Capacity factors for the sulfated compounds of interest, including the DAS isomers, are given in Table 3. None of the compounds injected interfered with the detection of any of the other com-

pounds, except that NE—3-O-sulfate and NE—4-O-sulfate were not satisfactorily resolved from each other under the conditions utilized. Consequently, the chromatographic conditions would have to be changed if one wished to distinguish between NE—3-O-sulfate and NE—4-O-sulfate in samples in which a mixture of the two might occur. Since the primary interest of this laboratory is DA—3-O-sulfate and DA—4-O-sulfate, no attempt was made to optimize the separation of NE—3-O-sulfate and NE—4-O-sulfate.

5. Determination of DA-sulfate isomers. The series chromatograms for DA—3-O-sulfate and DA—4-O-sulfate, with the upstream (W1) potential maintained at +1.00 V and the downstream electrode (W2) at +0.70 V, are

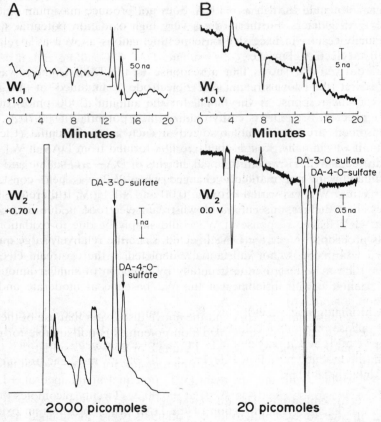

Figure 16. Series dual-electrode chromatograms for DA—3-O-sulfate and DA—4-O-sulfate. The mobile phase was monochloroacetic acid (0.025 M), adjusted to pH 2.8, containing EDTA (1 mM) and *n*-octylamine (4 mM). Injection volume was 20 μl. Other HPLC conditions were as described in the text. (A) 2000 pmol of each DAS isomer was injected. Both electrodes were maintained at oxidation potentials, W1 = +1.00 V and W2 = +0.70 V. (B) 20 pmol of each DA-sulfate isomer was injected. Upstream electrode (W1) was maintained as in (A), and the downstream electrode (W2) was maintained at a reduction potential of 0.00 V. [From Elchisak (1983a); reprinted with permission, copyright Elsevier Science Publishers b.v.]

shown in Fig. 16A. Two nanomoles (2000 pmol) of each isomer was injected. The direction of the DAS peaks indicates oxidative responses (loss of electrons) at both electrodes. Quantitation of the upstream response is limited by the oxidation of mobile phase components at the relatively high potential necessary for the oxidation of the DAS isomers. This response is essentially identical to that which occurs when DAS isomers are detected by a single-electrode electrochemical detector. Direct detection of small amounts of DA—3-*O*-sulfate and DA—4-*O*-sulfate by single-electrode electrochemical detection is usually unsuitable for routine use because the hydrodynamic voltammogram of each isomer indicates that maximal oxidation of these compounds requires an oxidation potential in excess of 1.2 V (Elchisak, 1983a). Experiments described below (Fig. 19) suggest that an oxidation potential as high as +1.4 V does not produce maximum oxidation of the DAS isomers. Furthermore, a very high oxidation potential such as this usually results in excessive baseline fluctuations as well as a relatively short-lived electrode surface.

Figure 16A also shows that a response to the DAS isomers was also obtained at the downstream (W2) electrode maintained at +0.70 V. However, the response is very small for the amount (2000 pmol) of each isomer injected. Free DA is easily oxidized at a potential of +0.70 V, while DAS isomers are not readily oxidized at such a low potential (Elchisak, 1983a). It was initially thought that free DA formed from DAS at W1 might be detected at W2. However, the peak heights of DA—3-*O*-sulfate and DA—4-*O*-sulfate were not markedly changed when W2 was held constant at +0.70 V and W1 was varied between 0.00 and +1.10 V. It therefore seems unlikely that the response at the downstream electrode is due to free DA formed at W1. The response at W2 could simply be due to oxidation of a very small proportion of the DAS injected, since the relatively large quantity injected was probably not substantially depleted at the upstream electrode. This could not be tested experimentally by injection of smaller amounts of DAS because of the instability of the W2 baseline at moderate and high oxidation potentials.

It is apparent from Fig. 16A and the above discussion that use of the dual-series electrode at two positive oxidation potentials is not suitable for detection of picomolar amounts of DA—3-*O*-sulfate or DA—4-*O*-sulfate.

Figure 16B shows the series chromatograms for DA—3-*O*-sulfate and DA—4-*O*-sulfate, with the upstream (W1) potential maintained at +1.00 V and the downstream electrode (W2) at +0.00 V. Twenty picomoles of each isomer was injected. The direction of the DAS peaks indicates an oxidative response (loss of electrons) at W1 and a reductive response (gain of electrons) at W2. Quantitation of the upstream response is limited by the oxidation of mobile phase components at the high oxidation potential. This was discussed above for Fig. 16A. A large reductive response was produced at W2 by injection of 20 pmol of each DAS isomer. Measurement of the response at W2 under these conditions appears to be more selective for DAS than measurement at W1, since there are fewer peaks present. Furthermore, after the electrode has stabilized, there is little or no fluctuation of the baseline at W2. The electrode had not yet completely stabilized in Fig. 16B.

It appears from Fig. 16B that low picomolar amounts of DA—3-O-sulfate and DA—4-O-sulfate can be detected at the downstream electrode of a dual series electrochemical detector. The upstream electrode should be maintained at a suitable oxidation potential, and the downstream electrode maintained at a suitable reduction potential. A similar oxidation—reduction coupling has been utilized for the dual-electrode detection of the catechol caffeic acid (Roston and Kissinger, 1982).

6. Mechanism of dual-electrode DA-sulfate detection. The exact mechanism of the reactions at each electrode have not been determined, but it seems likely that oxidation of DAS at W1 procedures a quinone, just as the oxidation of free DA produces a quinone. By setting W2 at an appropriate reduction potential (0.00 V), the quinone is reduced by 'recapture' of electrons.

A schematic representation of the events leading to the dual-electrode detection of DAS is shown in Fig. 17. The top frame shows the mechanism of the single-electrode electrochemical detection of free DA. Free DA is easily oxidized at low potential (+0.5 to 0.7 V) (Elchisak and Carlson, 1982b) to form a quinone. The electrons released are detected as a 'peak'. The procedure is highly sensitive and specific because of the low oxidation potential required. The middle frame of Fig. 17 shows that DA—3-O-sulfate and DA—4-O-sulfate can each be hydrolyzed to free DA and then detected as free DA by single-electrode electrochemical detection. Acid or enzymatic hydrolysis may be used. If the hydrolysis procedure is done in the traditional 'test-tube' fashion, and then DA determined, the distinction between the two isomers is lost (Elchisak and Carlson, 1982b). If an 'in-stream' post-column hydrolysis procedure is used, distinction between the two DAS isomers is maintained (Elchisak, 1983a) (discussed above). The bottom frame of Fig. 17 illustrates the proposed mechanism for the dual-electrode detection of DAS isomers. Each DAS isomer is oxidized to a quinone at the upstream (W1) electrode. A high potential is required so the procedure has poor specificity for quantitation. The quinone is reduced to free DA at the downstream (W2) electrode by 'recapture' of electrons. Quantitation at W2 is relatively specific, since oxidation—reduction reactions are sequentially required at W1 and W2. Because the reactions occur post-column, the distinction between the DAS isomers is maintained.

7. Linearity of DA-sulfate response. Injection of varying amounts of DA—3-O-sulfate and DA—4-O-sulfate between 5 and 1000 pmol indicates that there is a linear correlation between the downstream detector response and the amount of each DAS isomer injected (Fig. 18). The response was not linear above 1000 pmol. The upstream electrode was maintained at +1.00 V and the downstream electrode was maintained at 0.00 V.

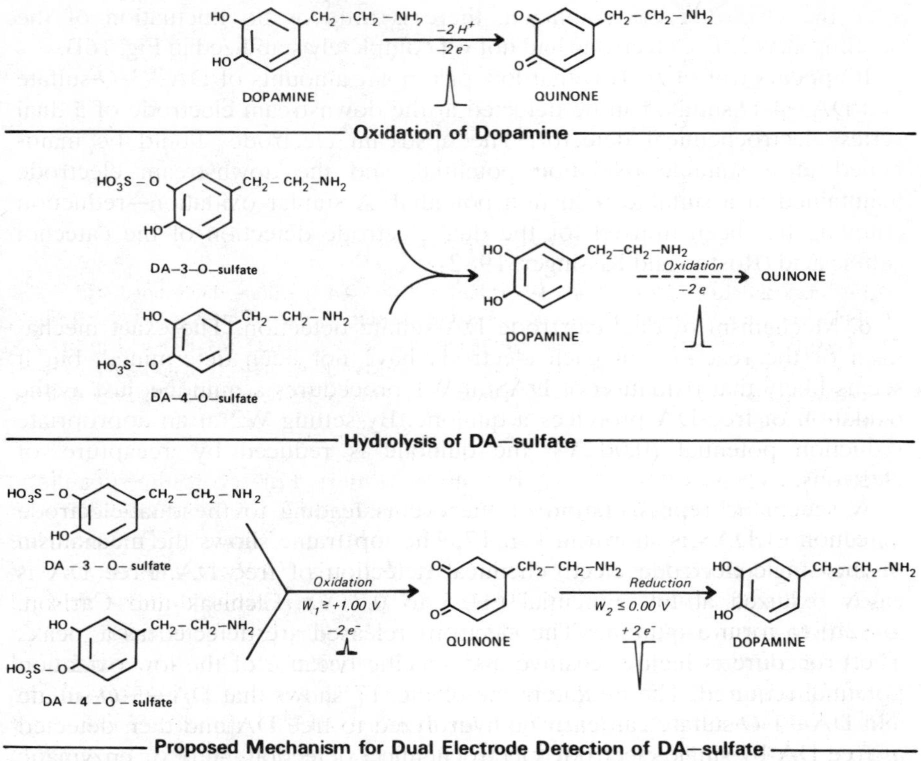

Oxidation of Dopamine

Hydrolysis of DA—sulfate

Proposed Mechanism for Dual Electrode Detection of DA—sulfate

Figure 17. Mechanisms for electrochemical detection of DA and DA—*O*-sulfate isomers. (Top) Single-electrode electrochemical detection of free DA. Free DA is easily oxidized at low potential (+0.5 to 0.7 V) to form a quinone. The electrons released are detected as a 'peak'. (Middle) DA—3-*O*-sulfate and DA—4-*O*-sulfate can be hydrolyzed to free DA and detected as free DA by single-electrode electrochemical detection. Acid or enzymatic hydrolysis may be used. If the hydrolysis procedure is done in the traditional 'test-tube' fashion, and then DA determined, the distinction between the two isomers is lost. If an 'in-stream' post-column hydrolysis procedure is used, distinction between the two isomers is maintained. (Bottom) Series dual-electrode electrochemical detection of DA—3-*O*-sulfate and DA—4-*O*-sulfate. Each DAS isomer is oxidized to a quinone at the upstream (W1) electrode. A high potential is required, so the procedure has poor specificity for quantitation. The quinone is reduced to free DA at the downstream (W2) electrode by 'recapture' of electrons. Quantitation at W2 is relatively specific, since oxidation—reduction reactions are sequentially required at W1 and W2. Because the reactions occur post-column, the temporal distinction ebtween the DAS isomers is maintained.

8. Optimal electrochemical conditions. Hydrodynamic voltammograms for DA—3-*O*-sulfate, DA—4-*O*-sulfate, NE—3-*O*-sulfate, NE—4-*O*-sulfate and serotonin—*O*-sulfate are shown in Fig. 19. Figure 19A shows the responses measured at W2, while W1 was held constant at +1.00 V and W2 was varied from −0.35 to +0.20 V. The voltammogram indicates that the optimal potential (that producing the maximum response) for W2 is at or very near

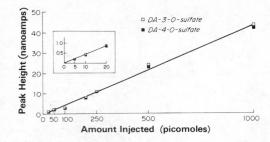

Figure 18. Standard curve for DA—3-*O*-sulfate and DA—4-*O*-sulfate determined with series dual-electrode electrochemical detection. Responses were measured at W2. W1 = +1.00 V; W2 = 0.00 V. The mobile phase was monochloroacetic acid (0.025 M), adjusted to pH 2.8, containing EDTA (1 mM) and *n*-octylamine (4 mM). Other HPLC conditions were as described in the text. [From Elchisak (1983a); reprinted with permission, Elsevier Science Publishers b.v.]

0.00 V for each of the DA- or NE-sulfate isomers. For serotonin—*O*-sulfate, however, a W2 potential more negative than −0.20 V was necessary to produce a maximal response (all data not shown). Figure 19B shows the responses measured at W2. In this case, W2 was held constant at 0.00 V while W1 was varied from +0.70 to +1.40 V. The results indicate that an oxidation potential (applied at W1) in excess of +1.40 V is necessary to produce maximal oxidation of the DAS isomers, while a potential of approximately +1.15 V appears to be optimal for the oxidation of the NE-sulfate isomers. A potential greater than +1.20 V is apparently necessary for maximal oxidation of serotonin—*O*-sulfate. The optimal oxidation (W1) and reduction (W2) potentials necessary for the detection of each of the sulfated amines listed above are summarized in Table 9.

Table 9. Lower detection limits and optimal dual electrode potentials for determination of sulfated compounds

	Signal: noise (20 pmol injected)	Lower limits of detection[a] (pmol)	Optimal potential (V)	
			W1	W2
Dopamine—3-*O*-sulfate	45	0.9[b]	+1.40	0.00
Dopamine—4-*O*-sulfate	34	1.2[b]	+1.40	0.00
Norepinephrine—3-*O*-sulfate	60	0.7	+1.15	0.00
Norepinephrine—4-*O*-sulfate	49	0.8	+1.15	0.00
Serotonin—*O*-sulfate	3	13[b]	+1.20	−0.20
Tyramine-sulfate		25[b]	+1.20	
HVA-sulfate		> 200[b]		
MHPG-sulfate		> 200[b]		

W1 = +1.00 V; W2 = 0.00 V.
[a] Injected amount producing signal: noise of approximately 2; W1 = +1.00 V; W2 = 0.00 V.
[b] See "note added in proof" at end of chapter.
Mobile phase: Monochloroacetic acid (25 mM); *n*-octylamine (4 mM); EDTA (1 mM); pH 2.80.

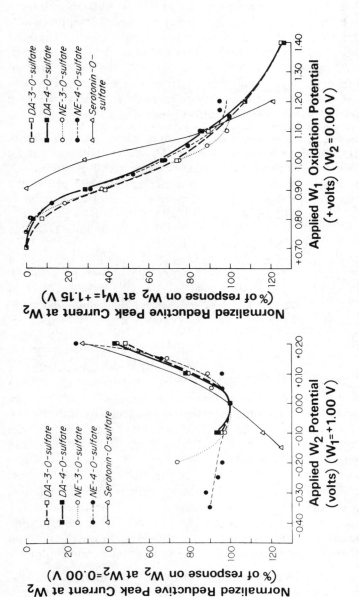

Figure 19. Hydrodynamic voltammograms of DA–3–*O*-sulfate (□), DA–4–*O*-sulfate (■), NE–3–*O*-sulfate (○), NE–4–*O*-sulfate (●) and serotonin—*O*-sulfate (△) determined with series dual-electrode detection. 200 pmol of each compound in a volume of 20 μl was injected. The mobile phase was monochloroacetic acid (0.025 M), adjusted to pH 2.8, containing EDTA (1 mM) and *n*-octylamine (4 mM). Other HPLC conditions were as described in the text. (A) W1 was held constant at +1.00 V and W2 varied from −0.35 to +0.20 V. Responses were measured at W2 and expressed as a percentage of the appropriate response at W2 = 0.00 V. (B) W2 was held constant at 0.00 V and W1 varied from +0.70 to +1.40 V. Responses were measured at W2 and expressed as a percentage of the response at W1 = +1.15 V. [From Elchisak (1983a); reprinted with permission, Elsevier Science Publishers b.v.]

9. Lower limits of detection. Table 9 also shows the signal-to-noise ratios for 20 pmol of each of the DA—, NE— and serotonin—O-sulfates. W1 was maintained at +1.00 V and W2 was maintained at 0.00 V, even though these potentials were not necessarily optimal for each compound. The approximate lower limit of detection is defined as the amount of a given compound injected which produced a signal-to-noise response ratio at W2 of approximately 2. With W1 at +1.00 V and W2 at 0.00 V, the lower limit of detection ranged from 0.7 to 1.2 pmol per injection for each of the DA- and NE-sulfate isomers, while serotonin—O-sulfate required a 13 pmol injection to produce the same response. (See "note added in proof".)

10. Other sulfated compounds. The dual-electrode detection system, under the conditions utilized here (W1 = 1.00 V, W2 = 0.00 V), was no more satisfactory than other detection systems for analysis of serotonin—O-sulfate (Tables 8 and 9). In the monochloroacetic acid mobile phase, the lower limit of detection for this compound utilizing ultraviolet detection at 280 nm was approximately 12 pmol. The post-column hydrolysis procedure, described above, resulted in lower detection limits for serotonin—O-sulfate of approximately 6—7 pmol. (See "note added in proof".)

The dual-electrode system was also utilized in an attempt to detect each of the following compounds: HVA-sulfate, MHPG-sulfate and tyramine-O-sulfate. The dual-electrode detection system, under the conditions utilized (W1 = 1.00 V, W2 = 0.00 V), was less satisfactory than other detection systems for analysis of any of these compounds (Tables 8 and 9). Peaks were not reliably produced at W2 for an injection of 200 pmol of MHPG-sulfate or HVA-sulfate. A lower detection limit of 25 pmol was obtained for tyramine—O-sulfate. When W1 was increased to 1.20 V, the lower detection limit for tyramine—O-sulfate was 8 pmol. Ultraviolet detection at 280 nm was more suitable for each of these compounds in terms of absolute lower limits of detection, but there are often problems with specificity with the UV detection mode. Lower limits of detection utilizing ultraviolet detection were 18, 6 and 10 pmol, respectively, for HVA-sulfate, MHPG-sulfate and tyramine—O-sulfate. (See "note added in proof".)

The actual lower detection limits could be substantially increased for most of these compounds by increasing the oxidation potential at W1. While this does not necessarily increase the noise generated at W2, an increased W1 oxidation potential usually requires more frequent polishing of the electrode surface. This procedure requires only a few minutes, however, plus an equilibration time which varies depending on the mobile phase composition.

11. Example of use. Injection of human urine (20 µl) directly onto the HPLC column produced a peak at the W2 electrode which co-chromatographed exactly with DA—3-O-sulfate (Fig. 20A, B). Quantitation of the amount of DA—3-O-sulfate represented by this peak, and calculation of a 24-hour urinary excretion rate for DA—3-O-sulfate, produced a value of 1.40 µmol/day in this subject. This value is within the range previously reported by this laboratory using a much more tedious method (Elchisak and

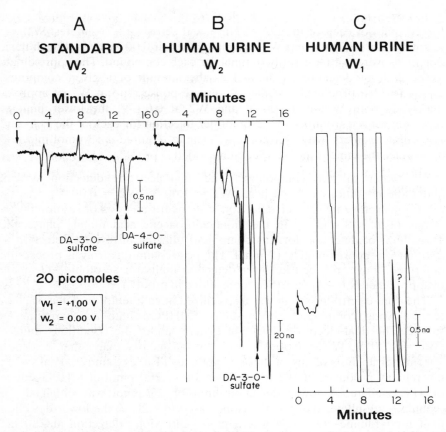

Figure 20. Series dual-electrode chromatograms for 20 μl human urine injected directly onto the HPLC column. The mobile phase was monochloroacetic acid (0.025 M), adjusted to pH 2.8, containing EDTA (1 mM) and *n*-octylamine (4 mM). W1 = +1.00 V; W2 = 0.00 V. Other conditions were as described in the text. (A) Standard solution containing 20 pmol of each DAS isomer. Responses monitored at W2. (B) and (C) Human urine sample injected and responses monitored at W2 (B) and W1 (C). [From Elchisak (1983b); reprinted with permission, Elsevier Science Publishers b.v.]

Carlson, 1982a). There were two peaks roughly corresponding to DA—3-*O*-sulfate when the detector response was monitored at the W1 electrode (Fig. 20C). If the appropriate peak were identified, quantitation would be very difficult, if not impossible, because of the very large responses seen at W1 and the accompanying difficulty of finding appropriate minima from which to measure the peak.

The dual-electrode technique for detection of DAS isomers is now in routine use in the author's laboratory and has been used in several recent investigations (for example, see abstracts of Scott and Elchisak, 1985; Toth *et al*, 1985). It is used for virtually all current investigations and other detection techniques are used primarily for verification and special purposes.

12. Conclusions regarding dual-electrode technique. The dual-series electrode elecetrochemical detection technique is suitable for the determination of DA—3-O-sulfate, DA—4-O-sulfate, NE—3-O-sulfate and NE—4-O-sulfate. Lower limits of detection for each of these compounds are approximately 1 pmol under the conditions utilized, which were not necessarily optimal. The technique is more sensitive for each of these compounds than either ultraviolet detction at 280 nm or the post-column hydrolysis technique utilizing single-electrode electrochemical detection. It is technically superior to the post-column hydrolysis technique for routine use. For serotonin—O-sulfate, dual-electrode electrochemical detection is less sensitive than either ultraviolet detection or the post column hydrolysis technique. However, the conditions utilized were not optimal for the detection of serotonin—O-sulfate. Maintenance of a significant reduction potential (more negative than −0.20 V) and W2 appears to be necessary for maximal detection of this compound; this was not pursued in depth in the present study. (See "note added in proof".)

The dual-electrode detection system, under the conditions utilized in the present study (W1 = 1.00 V, W2 = 0.00 V), was less satisfactory in terms of lower detection limits than other detection systems for analysis of tyramine—O-sulfate, HVA-sulfate and MHPG-sulfate. No attempts were made to determine optimal detection conditions for these three compounds. It must be realized that the dual-electrode technique might offer advantages, such as increased specificity, which are not related to the lower limits of detection. (See "note added in proof".)

CONCLUSIONS

Investigations concerning the specific conjugates of DA in the body have been hampered by the lack of specific and sensitive methodology for their detection. This chapter has reviewed recently developed methods for the measurement of the two sulfate conjugates of DA in the body, DA—3-O-sulfate and DA—4-O-sulfate. Emphasis was on work from the author's laboratory. The methods reviewed are sensitive enough for most, if not all, routine uses. The dual-electrode method is technically more appealing than the post-column hydrolysis method, but either method can be used routinely for many applications relating to DAS. With minor modifications, either method should be applicable to measurement of sulfate conjugates of many other electrochemically-active compounds, such as other CA or indoles, their metabolites, and some drugs are metabolites.

ACKNOWLEDGEMENTS

The author gratefully acknowledges the contributions of many persons and organizations for contributions to research described in this chapter. The following persons made scientific contributions: Joanne H. Carlson, Jeffrey D. Evanseck, Y. P. Grace Kao, Mary C. Scott, Susan E. Shoaf, Patrick G. Swann and Linda A. Toth. Mark A. Roberts and Mark L. Bucherl provided technical assistance, and Betsy K. Beeson provided editorial assistance. Drs Albert A. Manian and J. Stephen Kennedy of NIMH provided gifts of most of the sulfated compounds used. Dr. Walter W. Tourtellotte, Director of the National Neurological Research Bank, supplied the human tissues used in some studies. Some equipment was borrowed from Bioanalytical Systems Inc., and Dr. Peter Kissinger of that company made many helpful suggestions. J. T. Baker Chemical Co. provided solid phase extraction columns and a vacuum manifold for use in some experiments. Financial support was provided by NIH (NS 17514), Pfeiffer Research Foundation, Purdue Research Foundation, Pharmaceutical Manufacturers' Association and Purdue University.

REFERENCES

Alexander, N., Yoneda, S., Vlachakis, N. D., and Maronde, R. F. (1984). Role of conjugation and red blood cells for inactivation of circulating catecholamines. *Am. J. Physiol. 247*, R203—R207.

Anden, N. E., Carlsson, A., Kerstell, J., Magnusson, T., Olsson, R., Roos, B. E., Steen, B., Steg, G., Svanborg, A., Thieme, G., and Werdinius, B. (1970). Oral L-dopa treatment of Parkinsonism. *Acta Med. Scand. 187*, 247—255.

Anderson, R. J., and Weinshilboum, R. M. (1979). Phenolsulphotransferase: enzyme activity and endogenous inhibitors in the human erythrocyte. *J. Lab. Clin. Med. 94*, 158—171.

Anton, A. H., and Sayre, D. F. (1962). A study of the factors affecting the aluminium oxide—trihydroxyindole procedure for the analysis of catecholamines. *J. Pharmacol. Exp. Ther. 138*, 360—375.

Anton, A. H., Serrano, A., Tjandramaga, T. B., and Goldberg, L. I. (1973). Dopa and dopamine in glusulase: possible artefact in studies on catecholamine metabolism. *Science 182*, 60—61.

Arakawa, Y., Imai, K., and Tamura, Z. (1979). High-performance liquid chromatographic determination of dopamine sulfoconjugates in urine after L-dopa administration. *J. Chromatogr. 162*, 311—318.

Arakawa, Y., Imai, K., and Tamura, Z. (1983). Determination of catecholamine sulfoconjugate isomers in normal human urine by use of high-performance liquid chromatography with a photochemical detector. *Anal. Biochem. 132*, 389—399.

Baulieu, E.-E., Corpechot, C., Dray, F., Emiliozzi, R., Lebeau, M., Mauvais-Jarvis, P., and Robel, P. (1965). An adrenal-secreted "Androgen": Dehydroisoandrosterone sulfate. Its metabolism and a tentative generalization on the metabolism of other steroid conjugates in man. *Recent Prog. Horm. Res. 21*, 411—494.

Bodnaryk, R. P., Brunet, P. J., and Koeppe, J. K. (1974). On the metabolism of N-acetyl-dopamine in *Periplaneta americana*. *J. Insect Physiol. 20*, 911—923.

Bostrom, H., and Wengle, B. (1967). Distribution of phenol and steroid sulphokinase in adult human tissues. *Acta Endocrin. 56*, 691—703.

Bronaugh, R. L., Hattox, S. E., Hoehn, M. M., Murphy, R. C., and Rutledge, C. O. (1975). The separation and identification of dopamine-3-*O*-sulfate and dopamine-4-*O*-sulfate in urine of Parkinson's patients. *J. Pharmacol. Exp. Ther. 195*, 441—452.

Bronaugh, R. L., Wegner, G. R., Garver, D. L., and Rutledge, C. O. (1976). Effect of carbidopa on the metabolism of L-dopa in the pigtail monkey. *Biochem. Pharmacol. 25*, 1679—1681.

Buu, N. T., and Kuchel, O. (1977). A new method for the hydrolysis of conjugated catecholamines. *J. Lab. Clin. Med. 90*, 680—685.

Buu, N. T., and Kuchel, O. (1979a). The direct conversion of dopamine 3-*O*-sulfate to norepinephrine by dopamine-β-hydroxylase. *Life Sci. 24*, 783—790.

Buu, N. T., and Kuchel, O. (1979b). Dopamine-4-*O*-sulfate: A possible precursor of free norepinephrine. *Can. J. Biochem. 57*, 1159—1162.

Buu, N. T., Duhaime, J., Savard, C., Truong, L., and Kuchel, O. (1981a). Presence of conjugated catecholamines in rat brain: a new method of analysis of catecholamine sulfates. *J. Neurochem. 36*, 769—772.

Buu, N. T., Nair, G., Kuchel, O., and Genest, J. (1981b). The extra adrenal synthesis of epinephrine in rats: Possible involvement of dopamine sulfate. *J. Lab. Clin. Med. 98*, 527—535.

Buu, N. T., Duhaime, J., and Kuchel, O. (1982). The formation and removal of dopamine sulfate in rat brain. In: *Sulfate Metabolism and Sulfate Conjugation*, G. J. Mulder, J. Caldwell, G. M. J. Van Kempen, and R. J. Vonk (eds.), Taylor and Francis, London, pp. 93—98.

Claustre, J., and Peyrin, L. (1982). Free and conjugated catecholamines and metabolites in cat urine after hypoxia. *J. Applied Physiology. 52*, 304—308.

Claustre, J., Serusclat, P., and Peyrin, L. (1983). Gucuronide and sulfate catecholamine conjugates in rat and human plasma. *J. Neural Trans. 56*, 265—278.

Corneile, L., LaChance, S., Demassieux, S., and Carriére, S. (1983). Turnover of free and conjugated serum catecholamines during hemodialysis. *Clin. Invest. Med. 6*, 7—11.

Cuche, J. L., Prinseau, J., Ruget, G. Seiz, F., Tual, J. L., Baglin, A., Guedon, J., and Fritel, D. (1982). Plasma free and sulfoconjugated catecholamines in healthy men. *Eur. Heat J. 3*, 3—8.

Da Prada, M. (1980). Concentration, dynamics and functional meaning of catecholamines in urine. *Trends Pharmacol Sci. 1*, 157—159.

Davison, L., Vandongen, R., and Beilin, L. (1981). Effect of eating bananas on plasma free and sulfate-conjugated catecholamines. *Life Sci. 29*, 1773—1778.

de Champlain, J., Bouvier, M., Cléroux, J., and Farley, L. (1984). Free and conjugated catecholamines in plasma and red blood cells of normotensive and hypertensive patients. *Clin. Exp. Hypertens. A6*, 523—537.

Elchisak, M. A. (1983a). Determination of conjugated compounds by liquid chromatography with electrochemical detection using post-column hydrolysis. *J. Chromatogr. 255*, 475—482.

Elchisak, M. A. (1983b). Determination of dopamine-*O*-sulfate and norepinephrine-*O*-sulfate isomers and serotonin-*O*-sulfate by high-performance liquid chromatography using dual-electrode electrochemical detection. *J. Chromatogr. 264*, 119—127.

Elchisak, M. A. (1983c). Distribution of free and conjugated dopamine in human caudate nucleus, hypothalamus and kidney. *J. Neurochem. 41*, 893—896.

Elchisak, M. A., and Carlson, J. H. (1982a). Method for analysis of dopamine sulfate isomers by high-performance liquid chromatography. *Life Sci. 30*, 2325—2336.

Elchisak, M. A., and Carlson, J. H. (1982b). Assay of free and conjugated catecholamines by high performance liquid chromatography with electrochemical detection. *J. Chromatogr. Biomed. Appl. 233*, 79—83.

Elchisak, M. A., Cosgrove, S. E., Ebert, M. H., and Burns, R. S. (1983). Distribution of free and conjugated dopamine in monkey brain, peripheral tissues and cerebrospinal fluid determined by high-performance liquid chromatography. *Brain Res. 279*, 171—176.

Elchisak, M. A., and Hausner, E. A. (1984). Demonstration of *N*-acetyldopamine in human kidney and urine. *Life Sci. 35*, 2561—2569.

Elchisak, M. A., Powers, K. H., and Ebert, M. H. (1982). Determination of conjugated dopamine in monkey CSF by gas chromatography—mass spectrometry. *J. Neurochem. 39*, 726—728.

Goodall, McC., and Alton, H. (1968). Metabolism of 3-hydroxytyramine (dopamine) in human subjects. *Biochem. Pharmicol. 17*, 905—914.

Goodall, McC., and Alton, H. (1972). Metabolism of 3,4-dihydroxyphenylalanine (L-dopa) in human subjects. *Biochem. Pharmacol. 21*, 2401—2408.

Hammond, D. L., Yaksh, T. L., and Tyce, G. M. (1984). Conjugates of dopamine and serotonin in ventriculocisternal perfusates of the cat. *J. Neurochem. 42*, 1752—1757.

Hart, R. F., Renskers, K. J., Nelson, E. B., and Roth, J. A. (1979). Localization and characterization of phenolsulfotransferase in human platelets. *Life Sci. 24*, 125—130.

Higa, S., Suzuki, T., Hayashi, A., Tsuge, I., and Yamamura, Y. (1977). Isolation of catecholamines in biological fluids by boric acid gel. *Anal. Biochem. 77*, 18—24.

Hjemdahl, P. (1984). Catecholamine measurements by high-performance liquid chromatography. *Am. J. Physiol. 247*, E13—E20.

Hoeldtke, R. D., and Sloan, J. W. (1970). Acid hydrolysis of urinary catecholamines. *J. Lab. Clin. Med. 75*, 159—165.

Imai, K., Sugiura, M., Kubo, H., Tamura, Z., Ohya, K., Tsunakawa, N., Hirayama, K., and Narabayashi, H. (1972). Studies on the metabolism and excretion of L-3,4-dihydroxyphenylalanine (L-dopa) in human beings by gas chromatography. *Chem. Pharm. Bull. 20*, 759—764.

Jenner, W. N., and Rose, F. A. (1973). Studies on the sulphation of 3,4-dihydroxyphenylethylamine (dopamine) and related compounds by rat tissue. *Biochem. J. 135*, 109—114.

Jenner, W. N., and Rose, F. A. (1974). Dopamine-3-*O*-sulfate, an end product of L-dopa metabolism in Parkinson patients. *Nature 252*, 237—238.

Johnson, G. A., Baker, C. A., and Smith, R. T. (1980). Radioenzymatic assay of sulfate conjugates of catecholamines and DOPA in plasma. *Life Sci. 26*, 1591—1598.

Kahane, Z., Esser, A. H., Kline, N. S., and Vestergaard, P. (1967). Estimation of conjugated epinephrine and norepinephrine in urine. *J. Lab. Clin. Med. 69*, 1042—1050.

Karoum, F. (1983). Presence, distribution and pharmacology of conjugated catecholamines in the rat spinal cord. *Brain Res. 259*, 261—266.

Karoum, F., Chuang, L. W., and Wyatt, R. J. (1983). Biochemical and pharmacological characteristics of conjugated catecholamines in the rat brain. *J. Neurochem. 40*, 1735—1741.

Kissinger, P. T., Riggin, R. M., Alcorn, R. L., and Rau, L. D. (1975). Estimation of catecholamines in urine by high performance liquid chromatography with electrochemical detection. *Biochem. Med. 13*, 299—306.

Kuchel, O., Buu, N. T., and Unger, T. H. (1979). Free and conjugated dopamine: physiological and clinical implications. In: *Peripheral Dopaminergic Receptors*, J. L. Imbs and J. Schwartz (eds.), Pergamon Press, New York, pp. 15—27.

Kuchel, O., Buu, N. T., Hamet, P., Larochele, P., Bourque, M., and Genest, J. (1982a). Dopamine surges in hyperadrenergic essential hypertension. *Hypertension 4*, 845—852.

Kuchel, O., Buu, N. T., and Serri, O. (1982b). Sulfoconjugation of catecholamines, nutrition and hypertension. *Hypertension. 4, supp. III*, 93—98.

Marsh, C. A. (1966). Chemistry of *d*-glucuronic acid and its glycosides. In: *Glucuronic Acid, Free and Combined*, G. J. Dutton (ed.), Academic Press, New York, pp. 4—136.

Maus, T. P., Pearson, R. K., Anderson, R. J., Woodson, L. C., Reiter, C., and Weinshilboum, R. M. (1982). Rat phenol sulfotransferase: assay procedure, development changes and glucocorticoid regulation. *Biochem. Pharmacol. 31*, 849—856.

Merits, I. (1976). Formation and metabolism of ^{14}C dopamine-3-*O*-sulfate in dog, rat and guinea pig. *Biochem. Pharmacol. 25*, 829—833.

Merits, I., Anderson, D. J., and Sonders, R. C. (1973). Species differences in the metabolism of orally administered dopamine-^{14}C and *n*-alanyldopamine-^{14}C. *Drug Metabol. Disposition 1*, 691—697.

Nagel, M., and Schumann, H. J. (1980). A sensitive method for determination of conjugated catecholamines in blood plasma. *J. Clin. Chem. Clin. Biochem. 18*, 431—432.

Osikowska, B. A., Idle, J. R., Swinbourne, F. J., and Sever, P. S. (1982). Unequivocal synthesis and characterization of dopamine-3- and 4-*O*-sulphates. *Biochem. Pharmacol. 31*, 2279—2284.

Peyrin, L., Simon, H., Cottet-Emard, J. M., Bruneau, N., and Le Moal, M. (1982). 6-Hydroxydopamine lesions of dopaminergic A10 neurons. Long-term effets on the urinary excretion of free and conjugated catecholamines and their metabolites in the rat. *Brain Res. 235*, 363—369.

Peyrin, L., Cottet-Emard, J. M., Javoy, F., Agid, Y., Herbet, A., Glowinski, J. (1978). Long-term effects of unilateral 6-hydroxydopamine destruction of the dopaminergic nigrostriatal pathway on the urinary excretion of catecholamines (dopamine, norepinephrine, epinephrine) and their metabolites in the rat. *Brain Res. 143*, 567—572.

Qu, Y., Imai, K., Tamura, Z., Hashimoto, Y., and Miyazaki, H. (1983). *O*-methylation of dopamine-*O*-sulfates with catechol-*O*-methyltransferase. *Life Sci. 32*, 1811—1818.

Racz, K., Buu, N. T., and Kuchel, O. (1984). Regional distribution of free and sulfoconjugated catecholamines in the bovine adrenal cortex and medulla. *Can. J. Physiol. Pharmacol. 62*, 622—626.

Raum, W. J. (1984). Methods of plasma catecholamine measurement including radioimmuno-assay. *Am. J. Physiol. 247*, E4—E12.

Renskers, K. J., Feor, K. D., and Roth, J. A. (1980). Sulfation of dopamine and other biogenic amines by human brain phenol sulfotransferase. *J. Neurochem. 34*, 1362—1368.

Roston, D. A., and Kissinger, P. T. (1982). Series dual-electrode detector for liquid chromato-graphy/electrochemistry. *Anal. Chem. 54*, 429—434.

Rutledge, C. O., and Hoehn, M. M. (1973). Sulphate conjugation and L-DOPA treatment of Parkinsonian patients. *Nature 244*, 447—450.

Scheinin, M., Seppala, T., Koulu, M., and Linnoila, M. (1984). Determination of conjugated dopamine in cerebrospinal fluid from humans and non-human primates with high perfor-mance liquid chromatography using electrochemical detection. *Acta Pharmacol. Toxicol. 55*, 88—94.

Scott, M. C., and Elchisak, M. A. (1985). Determination of conjugated dopamine and dopamine sulfate in dog plasma. *Fed Proc. 44*, 1251.

Sharpless, N. S., Tyce, G. M., Thal, L. J., Waltz, J. M., Tabaddor, K., Wolfson, L. I. (1981). Free and conjugated dopamine in human ventricular fluid. *Brain Res. 217*, 107—118.

Shoaf, S. E., and Elchisak, M. A. (1983). Distribution of free and conjugated dopamine in rats and mice. *Life Sci. 33*, 625—630.

Smedes, F., Kraak, J. C., and Poppe, H. (1982). Simple and fast solvent extraction system for selective and quantitative isolation of adrenaline, noradrenaline and dopamine from plasma and urine. *J. Chromatogr. 231*, 25—39.

Toth, L. A., and Elchisak, M. A. (1984). Dopamine sulfate formation by intact human platelets. *IUPHAR 9th International Congress of Pharmacology.* Abstract No. 408 P.

Toth, L. A., Kao, G., and Elchisak, M. A. (1985). Factors influencing the recovery of dopamine sulfate in the assay of phenol sulfotransferase. *Fed. Proc. 44*, 1250.

Toth-Kennedy, L., and Elchisak, M. A. (1983). Dopamine uptake and dopamine-*O*-sulfate formation by human and dog platelets. *The Pharmacologist 25*, 197.

Tyce, G. M., and Rorie, D. K. (1982). Conjugated dopamine in superfusates of slices of rat striatum. *J. Neurochem. 39*, 1333—1339.

Tyce, G. M., Sharpless, N. S., Kerr, F. W. L., and Muenters, M. D. (1980). Dopamine conjugate in cerebrospinal fluid. *J. Neurochem. 34*, 210—212.

Unger, Th., Buu, N. T., and Kuchel, O. (1979). Renal and adrenal DA balance: implications for the role of conjugated DA. In: *Peripheral Dopaminergic Receptors*, J. L. Imbs and J. Schwartz (eds.), Pergamon Press, New York, pp. 357—367.

Unger, Th., Buu, N. T., Kuchel, O., and Schurch, W. (1980). Conjugated dopamine: peripheral origin, distribution and response to acute stress in the dog. Can. J. Physiol. Pharmacol. 58, 22—27.

Valori, C., Reanzini, V. Brunori, C. A., Porcellati, C., and Corea, L. (1969). An improved procedure for separation of catecholamines from plasma. *Ital. J. Biochem. 18*, 394—405.

Vlachakis, N. D., Kogosov, E., Yoneda, S., Alexander, N., and Maronde, R. F. (1984). Plasma levels of free and total catecholamines and two deaminated metabolites in man — rapid deconjugation by heat in acid. *Clin. Chim. Acta 137*, 199—209.

Von Euler, U. S., and Orwen, I. (1955). Preparation of extracts of urine and organs for estimation of free and conjugated noradrenaline and adrenaline. *Acta Physiol. Scand. Suppl. 118*, 1—9.

Wang, P. C., Buu, N. T., Kuchel, O., and Genest, J. (1983). Conjugation patterns of endogenous plasma catecholamines in human and rat: A new specific method for analysis of glucuronide conjugated catecholamines. *J. Lab. Clin. Med. 101*, 141—151.

Weil-Malherbe, H. (1964). The simultaneous estimation of catecholamines and their metabolites. *Zh. Klin. Chem. 2*, 161—167.

Weil-Malherbe, H. (1971). The chemical estimation of catecholamines and their metabolites in body fluids and tissue extracts. In: *Methods of Biochemical Analysis*, D. Glick (ed.), Interscience Publishers, New York, pp. 119—152.

Weil-Malherbe, H., and Bone, A. D. (1957). The estimation of catecholamines in urine by a chemical method. *J. Clin. Path. 10*, 138—147.

Weil-Malherbe, H., and Van Buren, J. M. (1969). The excretion of dopamine and dopamine metabolites in Parkinson's disease and the effect of diet thereon. *J. Lab. Clin. Med. 74*, 305—318.

Wong, K. P. (1976). Species differences in the conjugation of 4-hydroxy-3-methoxyphenyl-ethanol with glucuronic acid and sulphuric acid. *Biochem. J. 158*, 33—37.

Yoneda, S., Alexander, N., and Vlachakis, N. D. (1983). Enzymatic deconjugation of catechol-amines in human and rat plasma and red blook cell lysate. *Life Sci. 33*, 935—42.

NOTE ADDED IN PROOF:

Improved lower limits of detection using dual electrode detector: Additional pulse dampening has resulted in improved lower limits of detection (compared to those listed in Table 9) for various compounds in a mobile phase composed of phosphate buffer (75 mM) and EDTA (1 mM), pH 2.80. The improved lower limits of detection, in picomoles, are as follows: DA3S, 0.02; DA4S, 0.02; serotonin-*O*-sulfate, < 2; tyramine-sulfate, < 5; HVA-sulfate, < 2; MHPG-sulfate, < 1. Detector conditions were optimized only for the DAS isomers. NE-*O*-sulfate isomers were not evaluated, but it is likely that approximately 50-fold increases in the lower detection limits would be obtained, compared to those listed in Table 9.

Progress in HPLC, Vol. 2, pp. 377—396
Parvez *et al.* (Eds)

Automatic assay of biogenic amines and their metabolites in mice using high pressure liquid chromatography and electrochemical detection LCED

ELISABETH MORIER and RICHARD RIPS
INSERM U.98, 17 rue du Fer à Moulin, 7505 Paris, France

INTRODUCTION

Catechol and indole amine amounts in biological tissues are modified by physiological factors and influenced by pharmacological agents especially psychotropic substances. It is therefore essential to be able to detect their variations.

The first procedures which were published concerned separation on an ion-exchange resin or on alumina, followed by fluorescence assays of a limited number of compounds. Then came the gas chromatographic methods coupled with mass spectrometry, and high pressure liquid chromatography coupled with UV and fluorescence methods, or more recently with electro-chemical detection. This last method (Keller *et al.*, 1976; Kissinger, 1977; Kissinger *et al.*, 1975, 1977; Riggin and Kissinger, 1977) well-suited the catechols, methylated catechols and indoles which all have a low oxidation potential.

Most of the published LCED methods for the measurement of these compounds in biological samples used purification procedures on alumina or ion-exchange resin and extraction procedures to separate either the amines or the acid and neutral metabolites, with final separation on LCED. These purifications were time-consuming and tedious and when performed with unstable catechols or indoles the risks of degradation were increased; they also gave low yields which lowers the sensitivity limit.

Only a few investigators (Crombeen *et al.*, 1978; Scratchley *et al.*, 1979; Wagner *et al.*, 1979, 1982) have described procedures to use the HPLC at its maximum capacity and to measure simultaneously, without previous cleaning, a combination of compounds generally associated in biological samples such as noradrenaline, dopamine, serotonin and their metabolites.

Taking advantage of reverse-phase chromatography and ion-pairing, we chose a method which allowed us to separate as many compounds of this type in a standard mixture as possible in a single analysis and in isocratic mode (Morier and Rips, 1981). Then we applied this method to unpurified extracts of biological samples without previous cleaning (Morier and Rips, 1982a, b).

CHROMATOGRAPHIC METHODS

According to the literature the best choice for the separation of biogenic amines depended on reverse-phase column with ion-pair chromatography. So we started by using a new type of reverse-phase columns: Radial Pak A column (Waters). These columns are not expensive compared to steel columns and are radially compressed to avoid cavity formation, thereby permitting better separation of injected compounds. We studied the conditioning of these columns necessary to obtain reproducible results and optimum resolution of several biogenic compounds with an ion-pairing mobile phase.

Column conditioning

The results of our preliminary experiments differed depending on whether the columns were new or had already been exposed to solvents; DA retention, for example, varied from 0 to 15 min. This suggested that the more the columns were washed with a MeOH containing eluent the better they retained the samples, we therefore studied the conditioning of these columns with MeOH and Et$_3$N using different radial compression values.

MeOH. After trying several washing techniques with MeOH we concluded that to obtain good reproducible resolution of samples, both new or used ineffective columns must be washed with water for 1 h, MeOH for 1 h, water again for 1 h and then eluent for 3 h. Columns from the same or different series gave similar results. In the presence of MeOH the OH groups on the silica particles, on which the samples are retained, seem to become oriented like the 'bristles of a brush'.

Radial compression. In order to determine whether changes in radial pressure during conditioning alter the arrangement of the bonding phase, the pressure on three new dry columns was fixed at 0, 1/3 or 2/3 of the maximum (200 kg/cm^2) on a RCM 100 module, the latter being recommended by the manufacturer. The three columns were washed as stated above with water, MeOH, water and then eluent. When the radial pressure was adjusted to its usual value (about 2/3 of the maximum) identical traces were obtained with the three columns conditioned at the three different pressures. At the flow rate used (1 ml/min) the radial pressure did not affect the column conditioning.

Triethylamine. In order to determine whether the retention of biogenic amines and their derivatives is affected by blocking unbounded silanol sites the columns were washed with triethylamine. The following results were obtained:

neither the retention time nor the shape of the peaks were affected by washing with Et_3N;

less washing with eluent seemed necessary to obtain reproducible results;

distorted symmetry of the peaks obtained with amines and high pH eluent could not be corrected with this procedure.

Mobile phase

Counter-ion. As a counter-ion we used sodium heptyl sulfate (Pic B7 Waters). Two mechanisms have been proposed to explain the effects of ion-pair chromatography:

1. The amines may form ion-pairs with the heptyl sulfate of opposite charge in excess in the mobile phase and this neutral entity partitions onto the C_{18} stationary phase as a hydrophobic ion-pair.

2. In the ion-exchange model the lipophilic part of the heptyl sulfate is absorbed on the surface of the stationary organic phase, the charged part of the counter-ion being oriented toward the mobile phase; the reverse-phase column becomes a dynamic anion exchanger. The catecholamines which are positively charged in the pH range of 2—5 may then be separated by ion-exchange with the sulfate groups onto the stationary phase instead of absorption onto the C_{18} chain (Felice *et al.*, 1978). As a result the amines retention time increases with increased heptane sulfonate concentration. Ion-pair chromatography is also effective in minimizing the peak-tailing of amines which is attributed to secondary interactions between polar ionogenic compounds and unreacted silanol groups on the silica surface.

The retention of the acids is increased slightly by this addition, and that of an alcohol, such as MHPG, decreases as the heptyl sulfate concentration increases. Such behaviour was also indicated by Felice *et al.* (1978) but no explanation of these phenomena has been given until now. To favour the separation of NA and MHPG (Fig. 1) the counter-ion will therefore be used here at a concentration of 2 mM. However, to obtain the best resolution of NA from the fast eluting peaks the counter-ion concentration will be increased up to 5 mM.

Variations in pH. Varying the pH affected not only the retention time, as seen in Fig. 2, but also the shape and symmetry of the peaks. The retention times increased as pH decreased, and the larger the retention time the greater the increase. The symmetry of the peaks was acceptable at low pH, but the distortion increased with increasing pH, this was slight with the acids but greater with the amines. For the old columns (loss of C_{18} chains) the peaks of the amines became increasingly asymmetrical above pH = 3.85. Lowering the pH therefore improved the separation and the peak symmetry, but increased the analysis time for all the samples.

The retention times of the amines should be independent of the pH of the eluent as the amino groups are completely protonated in the pH range

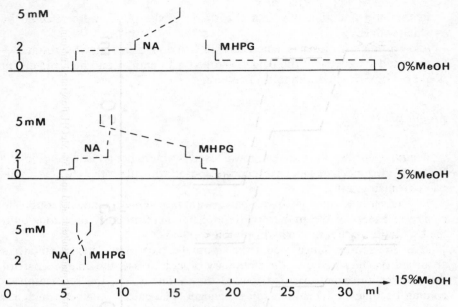

Figure 1. Effect of heptane sulfonic acid in mmol and methanol as a percentage on retention times. Radial Pak A, 0.1 M KH_2PO_4, 0.1 mM EDTA, pH = 4.1, flow rate 1 ml/min.

studied. This is so for NA, DBA and DA, but not for 5HT where the retention time slightly increases when pH decreases.

The retention times of the amino acids DOPA and DHPMA greatly increased as the pH decreased. Increased acidity of the mobile phase increased the protonation of the amino groups and subsequently increased ion-pair formation with the heptyl sulfate.

As the mobile phase pH becomes lower than the pK_a values of DOPAC, HVA and 5HIAA, they are protonated to a greater extent, which results in an increase in their retention times.

The value of the pH should be chosen so that an acid is displaced without moving the nearby amine, so giving a better separation of two neighbouring peaks, for example, DOPAC—DA.

Percentage of MeOH. Increasing the percentage of MeOH in the eluent generally decreased the sample retention times, but for some such as 5HT, the retention time is decreased more than for the others; this property allows 5HT to appear much earlier and to decrease the analysis time whilst still giving a good separation of products with the smallest retention times. This is illustrated in Fig. 3.

Optimal separation could be obtained under the following conditions:

glycols (MHPG, DHPG) and acids (DHMA, VMA, DOPA, DOPAC) were separated well with 10% MeOH at a rate of 1 ml/min;

a mixture of pure products chosen from those most frequently used could be separated with 15% MeOH at a rate of 1 ml/min (Fig. 4) within 25 min;

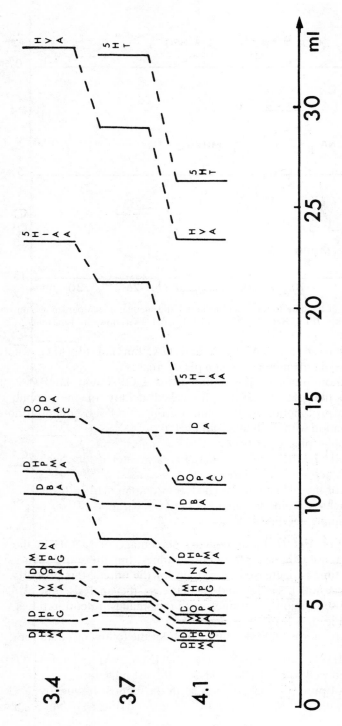

Figure 2. Effect of pH on retention times. Radial pak A, 0.1 M KH₂PO₄, 0.1 mM EDTA, 5 mM heptane sulfonic acid and 15% MeOH, flow rate 1 ml/min.

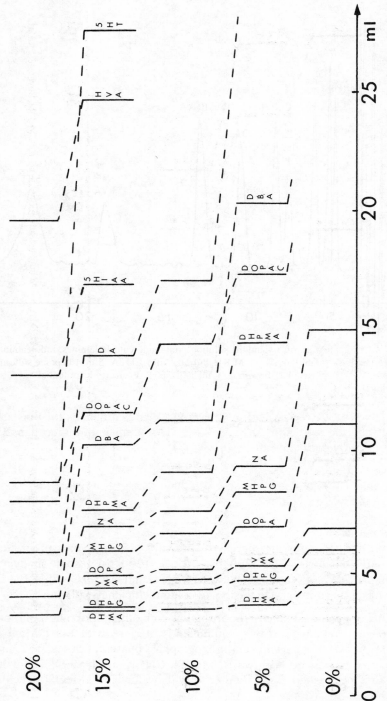

Figure 3. Effect of methanol as a percentage on retention times. Radial Pak A, 0.1 M KH$_2$PO$_4$, 0.1 mM EDTA, 5 mM heptane sulfonic acid, pH = 4.1, flow rate 1 ml/min.

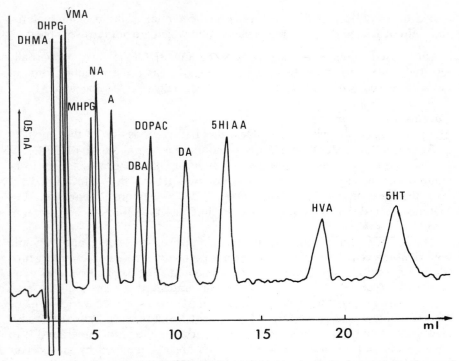

Figure 4. Chromatogram of a standard solution containing catecholamines, indoleamines and their metabolites. Radial Pak A, 0.1 M KH$_2$PO$_4$, 0.1 mM EDTA, 5 mM heptane sulfonic acid and 15% MeOH, pH = 4.1, flow rate 1 ml/min, +0.8 V electrode potential vs Ag/AgCl, sensitivity 0.5 nA.

if only NA, DBA, DA and 5 HT were to be quantified, as for the example after Amberlite extraction, then 15% MeOH at 2 ml/min, rather than 20% at 1 ml/min gave optimal results.

Several mobile phases with varying amounts of MeOH, THF and aceto-nitrile and combinations with two of them were tested, but there was no improvement of the results.

Column wear
Column wear results in increased pressure. Normal pressure with the proposed eluent was 350—700 psi. Above 1200 psi resolution decreases resulting in a loss of separation between MHPG and NA, DA and DOPAC.

Two factors contribute to the loss of column efficiency with use:

some organic susbtances (lipoproteins and other lipophilic compounds) are strongly retained by the column and interfere with the normal equi-librium between the two chromatographic phases. These are partially eliminated by washing with water, MeOH, CHCl$_3$ and DMSO. The columns must first be washed for a long time with water since MeOH precipitates heptane sulfonic acid;

hydrolysis of octadecyl molecules from the surface of the packing material and a physical loss of the column packing itself which are both irreversible.

On a used column, resolution between DA—DOPAC can be partially restored whilst increasing the pH since this decreases the retention time of the acids in a consistant manner without substantially affecting the amines.

Conclusion

The previous results will usually allow the final composition of the eluent to be fixed as follows: MeOH = 15%; heptane sulfonic acid = 5 mM; pH = 4.1.

The results of parameter-optimization using the window-diagram technique (Laub and Purnell, 1975) are as follows: pH = 4.0, % MeOH = 13.5. (We thank Professor C. Genty and J. M. Fougnion, Laboratoire des Méthodes Physico-Chimiques d'Analyse du Conservatoire National des Arts et Métiers for this work.)

The mobile phase constituents thus comprised a flexible and versatile system which could be altered as column performance decreased and which allows an optimal resolution to be maintained. The separation and peak symmetry obtained for various amines was improved by the use of a high concentration of heptane sulfonic acid (5 mM) together with a high organic solvent content (15% MeOH). The eluent employed in this study also contributes to a clearer signal at the eluent front on the chromatogram due to the rapid elution of many molecules which are not retained by an ion-pair mechanism because of the high content of MeOH.

APPLICATION TO BIOLOGICAL SAMPLES

Recordings from the brain

Three different solutions were tested for their ability to extract the compounds and to precipitate the proteins. The proteins precipitated slowly and tailing peaks appeared on the traces when using ethyl alcohol. $ZnSO_4$ could not be used because it was detected electrochemicaly with a retention time identical to that of NA. Protein precipitation and compound extraction was best with perchloric acid $HClO_4$. In a study on the conditions of extraction and separation of biogenic amines in biological media (Verbiese-Genard *et al.*, 1983), it has been observed that perchloric acid induces rapid degradation of indole structure molecules. This process is a specific proton catalyzed reaction: the higher the proton concentration, the less stable the products, whatever the acid. The results of Verbiese-Genard *et al.* (1984) have pointed out that indole derivatives are easily oxidized in two irreversible steps. This oxidation is inhibited by ascorbic acid and sodium ethylenediaminetetraacetic acid (EDTA). Whereas the role of ascorbic acid is obvious, the inhibition provided by EDTA is often attributed to its chelating power of metal ion traces (Cu^{2+}, Fe^{3+}) which act as catalysts. However, in the context of this study, the solutions used contained particularly low concentrations of metal traces, as determined by anodic stripping voltammetry. The results demonstrate that EDTA does not act as a complexing agent for metal traces but as a competitive agent interfering on the second oxidation step of the indole derivatives, thus avoiding their further oxidation.

In the perchloric acid solution we add 0.1% EDTA, but ascorbic acid could not be used as a reducing agent because it appeared with our mobile phase as a large peak which overlapped compounds with low retention times. $Na_2S_2O_5$ was used since it was eluted as a narrow peak in the void volume of the column.

Recordings were obtained quickly and easily (about 20 min per sample) and required only a limited number of preparatory steps (deproteinization) thus avoiding loss in sensitivity due to unsatisfactory recovery during the absorption procedures.

NA, DA and its metabolites DOPAC and HVA, and 5HT and its metabolite 5HIAA were readily detected in regions of the rat brain. Electrochemical detection is a specific enough detection method, since with a fixed potential only those products are detected on the brain chromatograms. Products with retention times lower than that of NA were masked by endogenous substances such as ascorbic acid.

Catechol and indole assays in regions of the mouse brain have previously been hampered by the small size of the samples. For instance three mouse brains have to be pooled to measure striatum HVA (Caccia *et al.*, 1973), four brains are required to measure striatum DOPAC (Roffler-Tarlov *et al.*, 1971) by the fluorescence method, and hypothalamus DA has not even been detected. This method described above is sensitive enough to measure catecholamines, indoleamines and their metabolites even in separate regions of the mouse brain. Typical results for several regions of the mouse brain are shown in Fig. 5. Our values agree with results reported recently by investigators using LCED (Dix Christensen and Leroy Blank, 1979; Leroy Blank *et al.*, 1979).

The use of internal standards in quantitative analytical procedures is generally considered to be superior to direct calibration because it provides an inherent correction factor. Thus DBA is used as the internal standard for injection volumes.

Required precautions
Although it was easy to compare the areas of peaks obtained for a group of control or treated mice, difficulties were encountered in accurately converting the area of each peak into ng of products for three reasons:

retention times vary with the wear of the column;
passivation of the electrode causes the sensitivity to decrease;
standard or assay solutions are not stable at ambiant temperature.

Retention times. The use of an automatic sample injector allowed the analysis of 50 tissue extracts per 24 h. Slowly the retention times decrease, but the difference is not significant within the same day. To prolong its life

E. Morier and R. Rips

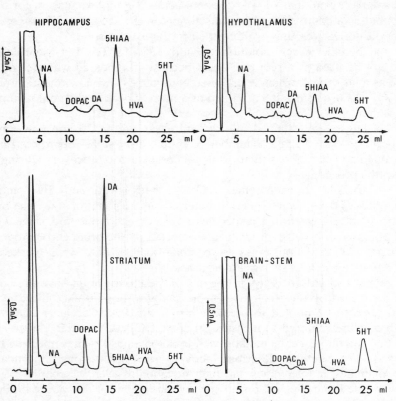

Figure 5. Chromatograms of tissue extracts from mouse brain. Conditions are the same as for Fig. 4.

the main column needed to be washed from time to time according to the procedure described previously. When DOPAC and DA are not sufficiently well separated a small increase of the pH can restore their separation. With these precautions the same column can be used for 4 months, by which time it has performed around 4000 injections (Fig. 6).

Electrode responses. The relatively high voltage applied and the large number of endogenous compounds oxidized resulted in a gradual decay in detector sensitivity within one day. To reverse the process, the recommended procedure is to alumina polish the two glassy carbon electrodes. This operation takes only a few minutes, but we have to wait several hours before obtaining again a linear base line. This cleaning operation is as far as possible done in the evening so that the electrochemical detector is ready to work the next morning.

In our hands the alumina cleaning procedure worked very well, but recently one author found that occasionally there is no improvement, or worse the responsiveness of the electrode may decrease (Anton, 1984).

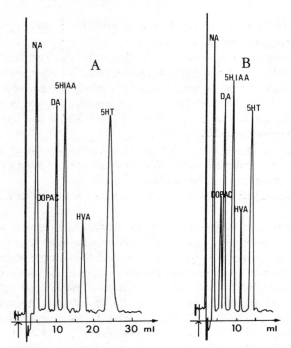

Figure 6. Retention times with (A) a new column and (B) a used one.

Various manipulations were tried to restore the original sensitivity, for instance sonification in presence of chemical substances. Many of them were ineffective, but finally sonification in a dichromate — H_2SO_4 cleaning solution not only overcame the passivation problem, but also increased the responsiveness of the electrodes over that obtained originally. This treatment also increased significantly the sensitivity beyond that produced by a successful alumina polishing trial. In the final procedure 2—3 drops of a CrO_3—H_2SO_4 solution (CrO_3, 200 mg; H_2O, 0.3 ml; H_2SO_4, 9.7 ml) were placed on the electrode; after 30 s it was thoroughly rinsed in tap water and then in distilled water. The authors used chromium trioxide rather than potassium dichromate because of its greater water solubility. An exceptionally strong oxidizing agent is necessary since other relatively strong oxidizing agents, e.g. concentrated HNO_3 are ineffective. Chromium ions and H_2SO_4 are both necessary since either one alone or alternatively one mixed with other substances is inactive. H_2SO_4, but not other acids, may catalyze an interaction between chromium and the surface of the electrode (Anton, 1984).

When we tried the CrO_3—H_2SO_4 treatment the responsiveness of the electrode was increased, but not more than with an alumina polishing procedure. The passivation of the electrode depends on several factors including the type and amount of metabolite being analyzed and the purity of the material injected. Even with crude brain extracts the passivation of our glassy carbon electrodes can be reversed as well with the alumina polishing technique as with the CrO_3—H_2SO_4 treatment.

Figure 7 shows two responses of the electrode: the first corresponds to a working electrode which has become passivated, the other corresponds to a passivated working electrode which has been newly reactivated by polishing. The decreased electrode response is not the same for every product; it is greater for NA, less for other catechols DA, DOPAC and HVA, and very weak for indoles such as 5HIAA and 5HT.

The changes in peak responses were assumed to occur linearly with respect to time, and external standard values run every 10 samples relative to a specific tissue sample were interpolated accordingly.

Electrodes are repolished when the NA reaches 3/4 of its initial value.

The importance of temperature. In the earlier assays, the samples were kept at 0° C prior to manual injection. It was not possible to use an automatic injection method at room temperature as unstable products such as catecholamines, indoleamines and their metabolites degrade in solution; the rate of degration depends on the surrounding medium, particularly pH, oxidizers, concentration and temperature. Prior to the use of an automatic injection method, we studied (Morier and Rips, 1982a, b) the degradation of six main compounds found in the mouse brain extract: i.e. NA, DA, DOPAC, HVA, 5HIAA and 5HT. We then investigated the conditions under which

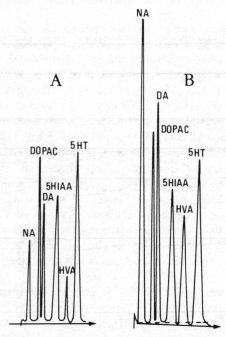

Figure 7. Response obtained with (A) a passivated electrode and (B) a newly polished electrode.

an external standard made up of a mixture of pure compounds is still a valid means of calculating the amounts of products contained in mouse brain extracts.

Perchloric acid at two different concentrations, 0.1 and 0.4 N, was used to precipitate the proteins. The standards were prepared using pure products in the same solvent and in water. The dilution gave levels close to those obtained with brain homogenates, 5 ng in a 50 μl sample. The temperatures used were identical to the temperature chosen for the thermostat regulated tray (0° to 5° C according to the position of the vials in the tray) or to the temperature obtained without coolant in an air-conditioned room at 22° C.

Degradation of the standards was compared to that for the endogenous products of mice at different time intervals (i.e. 4, 8, 12 and 20 h).

At the temperature of 22° C (dotted line on Fig. 8) three compounds were very unstable. DA decomposed in water, but was stable in acidic solution (0.1 or 0.4 N). 5HIAA and 5HT were highly unstable in both neutral and acidic solutions. Degradation was less significant in the biological medium and for all compounds stability increased when the temperature decreased. For refrigerated samples, stability was greater than 90% for all compounds (after 20 h storage) except for the standard solutions of 5HT and 5HIAA which had 80% stability. The difference between 0.1 and 0.4 N was not as large at 4° C compared to that of 22° C. We therefore used 0.4 N since the volume required to precipitate the proteins was less and this concentration increased the sensitivity of the analysis. These results demonstrate the

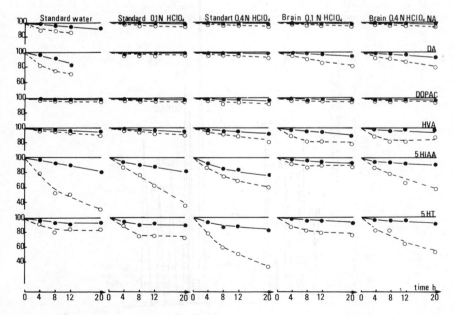

Figure 8. Stability of NA, DA, DOPAC, HVA, 5HT and 5HIAA at 0—5° C (○) and 22° C (●) in water, $HClO_4$ (0.1 N and 0.4 N), in standards or biological samples.

importance of using an automatic injector in which the samples are maintained at a low temperature (for example an injector with a cooled sample
tray) when measuring catechol and indole derivatives. They are highly
unstable whilst in the tray of the injector at room temperature, and the
standard and the biological extracts do not undergo the same rate of
degradation.

Recordings from other biological tissues or liquids

Recordings were also obtained from mouse cerebrospinal fluid (CSF)
without previous cleaning, using a technique for collecting CSF samples in
mice which was developed in our laboratory (Boschi et al., 1983). The range
of CSF volumes taken was 2 μl minimum for detection purposes and 5μl for
routine analysis in the mouse, and 5—40 μl in the rat. Some catechol or
indole metabolites such as MHPG, DOPAC, 5HIAA and HVA are clearly
seen on the chromatograms (Fig. 9).

The same technique can be applied directly to peripheral organs rich in
catecholamines such as the adrenal gland. This measurement is very easily
made as the adrenal gland NA and A concentrations are very high. The
intensity range has to be modified during the course of a recording to
measure DA since it is a hundred times less concentrated (Fig. 10).

Electrochemically active drugs or drug metabolites

All the recordings were obtained with groups of control or treated animals.
The drugs administered may also be electrochemically active, or one of their
metabolites synthesized by the animal can become so.

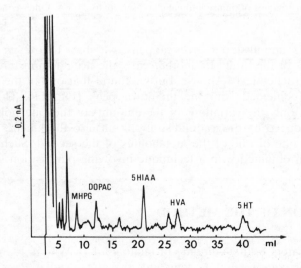

Figure 9. Chromatograms of 2 μl of mouse cerebrospinal fluid. Radial Pak 5 μ, sensitivity 0.2
nA. Other condtions are the same as for Fig. 4.

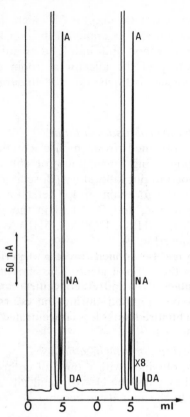

Figure 10. Chromatograms of mouse adrenal gland sensitivity 50 nA. Other conditions are the same as for Fig. 4.

If the drug administered gives a peak we may follow its kinetics in different areas of the brain. In another case, the drug administered does not give any electrochemical response, but we found that one of the metabolites of this drug coincided with the unknown peak (Fig. 11). Therefore, we measured not only the variations of the amounts of the catechol and indole derivatives induced by drug administration, but also the kinetics of the first drug and the rate of one of the metabolites of the second. Such data could not have been obtained with a technique involving purification steps before HPLC.

DESCRIPTION OF THE METHOD

Material
We used a Waters 6000 A liquid chromatograph, a Waters RCM 100 or Z module radial compression system and a Waters Radial Pack A column (C-18, 0.8 × 10 cm ID, 5 or 10 μm particle size). The automatic injector used was the Perkin-Elmer Model ISS 100 with a thermostat regulated tray

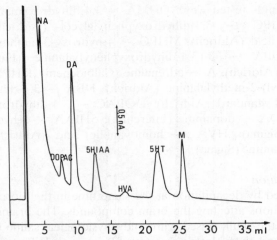

Figure 11. One of the metabolites, M, of a drug may be electrochemically active.

containing up to 100 samples. Temperature regulation of the tray between 0°
and 5° C was obtained with a Bioblock Ministat cryostat which used a
mixture of water : ethylene glycol (80 : 20). The electrochemical detector
used was the Metrohm 641 fitted with two vitrous carbon electrodes EA
286-1 and a reference electrode, Ag/AgCl, EA 442, voltage + 0.8 V and
current 5 nA. Chromatograms were recorded on a Kipp and Zonen BD 40
recorder or quantified with a Hewlett-Packard Model 3930 A integrator.

Mobile phase
The mobile phase consisted of a 0.1 M KH_2PO_4, 0.1 mM EDTA, 5mM
heptane sulfonic acid solution in demineralized water and 15% MeOH; the
pH value of 4.1 of this aqueous methanol buffer was adjusted by NaOH or
H_3PO_4 (1M) and verified before each recording. The solution was filtered
through a 0.45 µm Millipore filter and used at a flow rate of 1 ml/min.

Reagents and standard
Stock standard solutions of the various compounds consisted of 10 mg in
100 ml 0.1 N hydrochloric acid for each catecholamine and water for each of
the others. All were stored at 4° C in the dark and remained stable for at
least one month. As a matter of fact, concentrations are typically low and
both NA and DA are particularly prone to oxidative destruction when stored
in neutral solvents and they are markedly unstable when placed in basic
solution. Solutions containing 5HIAA and 5HT have to be stored in neutral
solvents (Verbiese-Genard *et al.*, 1983) and in the dark (Chow and Grushka,
1978) since chemical and photochemical alteration of 5OHindole species
has now been well established. Equal amounts of stock standards were mixed
every week and the volume adjusted with water to make a 1/10 dilution. This
was stored at 4° C and freshly diluted to 1/100 each day. All concentrations
are given as the free base or acid.

The compounds tested were: DHMA — 3,4-dihydroxymandylacetic acid (Aldrich); DHPG — 3,4-dihydroxyphenylglycol (Aldrich); VMA — vanylmandelic acid (Aldrich); MHPG — 4-hydroxy-3-methoxyphenylglycol (Aldrich); DOPA — 3-(3,4-dihydroxyphenyl)alanine (Fluka); NA — noradrenaline (Aldrich); A — adrenaline (Calbiochem); DHPMA — 3-(3,4-dihydroxyphenyl)-2-methylalanine (Aldrich); DBA — 3,4-dihydroxybenzyl-amine, internal standard (Aldrich); DOPAC — 3,4-dihydroxyphenylacetic acid (Sigma); DA — dopamine (Interchim); 5HIAA — 5-hydroxyindole-3-acetic acid (Sigma); HVA — homovanillic acid (Aldrich); 5 HT — 5-hydroxytryptamine (Sigma).

Sample preparation

Mice were killed by decapitation at the same time in the morning to prevent circadian variations affecting the brain compounds. The brains were quickly removed from the skull, rinsed with cold physiological saline to remove any excess blood, and then blotted dry with filter paper. Different regions were dissected on ice (cerebellum, brain-stem, hypothalamus, hippocampus, striatum and the 'rest' containing mainly cortex and midbrain) according to the method of Glowinski and Iversen (1966) prior to freezing on dry ice. The frozen tissue samples were then weighed before being stored at $-70°$ C until required for extraction. Small samples of < 100 mg cannot be weighed without thawing, their protein content is estimated by the method of Lowry *et al.* (1951). The tissue samples were normally processed within 3 days of dissection. However, samples could be stored at $-70°$ C for several weeks without any significant change in the amine content.

Weighed tissue samples were homogenized with an ultra turrax unit in 500 μl (1500 μl for the rest) of 0.4 N perchloric acid containing 0.1% EDTA and 0.1% $Na_2S_2O_5$. Homogenates were centrifuged (Beckman J21, rotor JA 14) at 8000 rpm and at 4° C for 10 min, and the supernatants were then injected onto the HPLC column within one day.

Calculation

The concentration of each compound was determined by comparing the respective areas of the peaks with those obtained for the standard mixture of compounds which were tested every ten samples and samples injection was checking with DBA as the internal standard. No deviation from linearity was detected over a range of 0.1—10 ng.

CONCLUSION

Over the last few years, a relatively large number of reports have appeared detailing new methods for the determination of various catecholamines, indoleamines and related compounds using HPLC. Recent reviews have appeared describing a number of them which use either fluorescence or elec-trochemical detection. Most of the currently available liquid chromatographic methods do not analyze a large number of compounds per day from a single sample injection because of the time necessary for sample clean-up. In general, extracts of tissue samples or biological fluids require a preliminary purification step using adsorption or ion-exchange chromatography or gel filtration. Some more recent methods allow for analysis without clean-up but can quantify only a few compounds.

We report a method for the determination of most of the biologically relevant substances related to catechol and indoleamine metabolism within the same tissue sample. The combination of the easy-to-handle deproteinization procedure, the high resolving power of HPLC in the reversed-phase mode and the high sensitivity of the electrochemical detection provides a fast and reliable method for the determination of the brain neurotransmitters and of some of their metabolites.

The sample preparation described offers a number of advantages: speed (due to the absence of prepurification), sensitivity (due to the absence of dilution), minimal destruction in the case of very labile compounds and absence of loss by low yields. The small volumes of $HClO_4$ used enhance the assay sensitivity by maximizing the sample amine concentration present in the injection volume. The degradation of the indoles over time is reduced to a minimum by the use of a refrigerated automatic injector and the rapidity of the analysis. Off-line isolation of compounds from brain homogenates on alumina or cation-exchange columns yielded a poor recovery percentage and destruction of the labile compounds. An average absolute recovery for NA, DA on alumina was 50—70%. The handling time, the risks of error and the degradation samples are decreased using this method.

To achieve a good separation of the various neurotransmitters from interfering peaks originating from brain samples, the chromatographic conditions have to be carefully chosen. The composition of the eluent is very versatile as retention times can be adjusted to the needs of a particular assay by varying the pH, the amount of methanol present or the concentration of counter-ion. The use of the radial compression cartridge reduces the time of the analysis by decreasing the back-pressure value.

These chromatographic conditions are well suited to rapid analysis of catechol and indole compounds of CSF, brain samples and the adrenal gland by the direct injection procedure. The possibility described here, consisting in simultaneously checking the levels of the three major monoamines in brain areas, offers a good routine method for evaluating pharmacological agents and determining the effects of drugs on monoamines. The present method also permits the examination of *in vivo* enzyme activity in small brain regions. The procedure described here shows the advantage of short time, minimal sample handling, high sensitivity, good accuracy and reproducibility for analyzing. In addition, this procedure allows detection of the administered drug or one of its metabolites if their oxidation potentials are low enough.

A number of metabolites with low retention times can be separated on standard with this method. But they tend to be obscured by peaks of unknown identity derived from the biological samples. MHPG, DHPG and VMA metabolites of noradrenaline could not be resolved. These metabolites can be determined in a separate chromatographic run after a purification step with absorption or ion-exchange chromatography or an ethyl acetate extraction of the supernatant. A modified technique with column-switching valves is a very flexible method which will allow the purification of low retention time compounds and sharpen the peaks. Even so the direct injection of serum, urine, heart and liver extracts, after deproteinization was not satisfactory.

In the brain, CSF and adrenal gland, NA, DA, DOPAC, HVA, 5HT and 5HIAA are simultaneously measured in less than 25 min (about fifty samples per 24 h). The absence of later eluting compounds and no need for column re-equilibration make possible an intersample assay interval of zero. The column life-time is enhanced by consistency of the solvent composition delivered to it and its performance is not decreased by the absence of prepurification of the samples. All these factors facilitate the automatization of this method.

REFERENCES

Anton, A. H. (1984). A simple, reliable and rapid method for increasing the responsiveness of the glassy carbon electrodes (GCE) for the analysis of biogenic amines by high performance liquid chromatography with electrochemical detection (LCED). *Life Sci. 35*, 79—85.

Boschi, G., Morier, E., and Rips, R. (1983). Comparison of biogenic amines and their metabolites in the CSF of mice and rats. *IRCS Med. Sci. 11*, 518—519.

Caccia, S., Cecchieti, G., Garattini, S., and Jori, A. (1973). Interaction of (+)amphetamine with cerebral dopaminergic neurones in two strains of mice, that show different temperature responses to this drug. *Br. J. Pharm., 59*, 400—406.

Chow, F. K., and Grushka, E. (1978). Determination of 5-hydroxyindole-3-acetic acid in urine by high performance liquid chromatography. *Anal. Chem. 50*, 1342.

Crombeen, J. P., Kraak, J. C., and Poppe, M. (1978). Reversed-phase systems for the analysis of catecholamines and related compounds. *J. Chromatogr. 167*, 219—230.

Dix Christensen, H., and Leroy Blank, C. (1979). The determination of neurochemicals in tissue samples at subpicomole levels. In: *Biological Biomedical Application of Liquid Chromatography (Chromatographic Science*, Séries 12), G. Hawk (ed.), Marcel Dekker, New York, pp. 133—165.

Felice, L. J., Felice, J. D., and Kissinger, P. T. (1978). Determination of catecholamines in rat brain parts by reverse-phase ion-pair liquid chromatography. *J. Neurochem. 31*, 1461—1465.

Glowinski, J., and Iversen, L. L. (1966). Regional studies of catecholamines in the rat brain. I The disposition of 3H-norepinephrine, 3H-dopamine and 3H-dopa in various regions of the brain. *J. Neurochem. 13*, 655.

Keller, R., Oke, A., Mefford, J., and Adams, R. N. (1976). Liquid chromatographic analysis of catecholamines. Routine assay for regional brain mapping. *Life Sci. 19*, 955—1004.

Kissinger, P. T. (1977). Amperometric and coulometric detectors for liquid chromatography *Anal. Chem. 49*, 447A—456A.

Kissinger, P. T., Bruntlett, C. S., Davis, G. C., Felice, L. J., Riggin, R. M., and Shoup, R. E. (1977). Recent developments in the clinical assessment of the metabolism of aromatics by high performance reverse-phase chromatography with amperometric detection. *Clin. Chem. 23*, 1449—1455.

Kissinger, P. T., Riggin, R. M., Alcorn, R. L., and Rau, L. D. (1975). Estimation of catecholamines in urine by high performance liquid chromatography with electrochemical detection. *Biochem. Med. 13*, 299—306.

Laub, R. J., and Purnell, J. H. (1975). Criteria for use of mixed solvents in gas liquid chromatography. *J. Chromatogr. 112*, 71—79.

Leroy Blank, C., Sasa, S. Isernhagen, R., Meyerson, L. R. Wasil, D., Wong, P., Modak, A. T., and Stavinoha, W. B. (1979). Levels of norepinephrine and dopamine in mouse brain regions following microwave inactivation — rapid post mortem degradation of striatal dopamine in decapitated animals. *J. Chromatogr. 33*, 213—219.

Lowry, O. H., Rosenbrough, N. J., Farr, A. L., and Randall, R. J. (1951). Protein measurement with Folin phenol reagent. *J. Biol. Chem. 193*, 265—275.

Magnusson, O., Nilsson, L. B., and Wersterlund, D. (1980). Simultaneous determination of dopamine, dopac and homovanillic acid. Direct injection of supernatants from brain tissue homogenates in a liquid chromatography—electrochemical detection system. *J. Chromatogr. 221*, 237—247.

Maruyama, Y., Oshima, T., and Nakajima, E. (1980). Simultaneous determination of catecholamines in rat brain by reversed-phase liquid chromatography with electrochemical detection. *Life Sci. 26*, 1115—1120.

Morier, E., and Rips, R. (1981). A new technique for simultaneous determination of biogenic amines and their metabolites. *IRCS Med. Sci. 9*, 454—455.

Morier, E., and Rips, R. (1982a). A new technique for simultaneous determination of biogenic amines and their metabolites. *J. Liq. Chromatogr. 5*, 151—164.

Morier, E., and Rips, R. (1982b). Importance of temperature for the determination of biogenic amines and their metabolites in biological material. *IRCS Med. Sci. 10*, 921—922.

Moyer, T. P., and Jiang, N. S. (1978). Optimized isocratic conditions for analysis of catecholamines by high-performance reversed-phase paired-ion chromatography with amperometric detection. *J. Chromatogr. 153*, 365—372.

Riggin, R. M., and Kissinger, P. T. (1977). Determination of catecholamines in urine by reverse-phase ion-pair chromatography. *Anal. Chem. 49*, 2109—2111.

Roffler-Tarlov, S., Sharman, D. F., and Tegerdine, P. (1971). 3,4-Dihydroxyphenylacetic acid and 4-hydroxy-3-methoxyphenylacetic acid in the mouse striatum: a refletion of intra and extra-neuronal metabolism of dopamine? *Br. J. Pharmacol. 42*, 343—351.

Sasa, S., and Blank, C. L. (1977). Determination of serotonin and dopamine in mouse brain tissue by high performance liquid chromatography with electrochemical detection. *Anal. Chem. 49*, 354—359.

Scratchley, G. A., Masoud, A. N., Stroms, J. J., and Wingard, D. W. (1979). High-performance liquid chromatographic separation and detection of catecholamines and related compounds. *J. Chromatogr. 169*, 313—319.

Verbiese-Genard, N., Hanocq, M., Alvoet, C., and Molle, L. (1983). Degradation study of catecholamines, indole amines and some of their metabolites in different extraction media by chromatography and electrochemical detection. *Anal. Biochem. 134*, 170—175.

Verbiese-Genard, N., Kauffmann, J. M., Hanocq, M., and Molle, L. (1984). Study of the electrooxidative behaviour of 5-hydroxyindole-3-acetic acid, 5-hydroxytryptophan and serotonin in the presence of sodium ethylenediaminetetraacetic acid. *J. Electroanal. Chem. 170*, 243—254.

Wagner, J., Palfreyman, M., and Zraika, M. (1979). Determination of dopa, dopamine, dopac, epinephrine, norepinephrine, α-monofluoromethyldopa and α-difluoromethyldopa in various tissues of mice and rats using reversed-phase ion-pair liquid chromatography with electrochemical detection. *J. Chromatogr. 164*, 41—54.

Wagner, J., Vitali, P., Palfreyman, M. G. Zraika, M., and Huot, S. (1982). Simultaneous determination of 3,4-dihydroxyphenylalanine, 5-hydroxytryptophan, dopamine, 4-hydroxy-3-methoxyphenylalanine, norepinephrine, 3,4-dihydroxyphenylacetic acid in rat cerebrospinal fluid and brain by high-performance liquid chromatography with electrochemical detection. *J. Neurochem.* 1241—1254.

Progress in HPLC, Vol. 2, pp. 397—426
Parvez *et al.* (Eds)
© 1987 VNU Science Press

Measurement of catecholamines in blood and urine by liquid chromatography with amperometric detection

NAI-SIANG JIANG and DWAINE MACHACEK

Department of Laboratory Medicine, Mayo Clinic and Mayo Foundation,
Rochester, MN 55905, USA

INTRODUCTION

The measurement of catecholamines in biological material, including human body fluids, has evolved continuously since their discovery (Oliver and Schäfer, 1895). This chapter will begin with a brief history of the evolution of methods for the determination of catecholamines and the practicality of doing this in a clinical diagnostic laboratory. The major portion of the chapter will be devoted to describing a liquid chromatographic method with amperometric detection developed and used in our laboratory for the quantitation of the three principal catecholamines — norepinephrine, epinephrine and dopamine — in human blood and urine. For extensive reviews of method developments, the reader is referred to Mefford (1981), Krstulović (1982) and Holly and Makin (1983).

Epinephrine (or adrenaline) was first isolated, and characterized, in 1897 (Abel and Crawford), from an adrenal extract shown 2 years earlier (Oliver and Schäfer, 1895) to contain a substance or substances able to increase blood pressure in animals. Almost 50 years later, von Euler (1946) isolated a sympathomimetic compound from nerve tissue that was different from epinephrine but identical with norepinephrine which was a known compound at that time. Dopamine had been found to be a precursor in the biosynthesis of epinephrine and norepinephrine in 1939 (Blaschko), but its physiological significance only began to be appreciated recently (Cotzias *et al.*, 1970; Dorris and Shore, 1971).

Catecholamine biosynthesis takes place in the brain, sympathetic nerve endings and sites of chromaffin tissue including the adrenal medulla and neural crest tissues (Udenfriend and Wyngaarden, 1956; Musacchio and Goldstein, 1963; Udenfriend and Zaltzman-Nirenberg, 1963). Catecholamines are the few hormones whose biosynthetic and metabolic pathways are completely elucidated. The enzymes involved in each metabolic step have been isolated and characterized. Inhibitors of these enzymes have been synthesized. A complete review of catecholamine metabolism is beyond the scope of this chapter. A chart of catecholamine metabolism is presented in Fig. 1. Measurement of the three principal catecholamines and their metabolites in blood and urine have been used widely as diagnostic tools for many diseases, including: hypertension (de Champlain *et al.*, 1976), chromaffin and neural crest tumors (Sjoerdsma *et al.*, 1966; Gitlow *et al.*, 1971; Moyer *et al.*, 1979b), hyperthyroidism and hypothyroidism (Stoffer *et al.*, 1973; Coulombe *et al.*, 1976) and orthostatic hypotension (Ziegler *et al.*, 1977).

BACKGROUND: EVOLUTION OF METHODS

Bioassay

The first method developed for the estimation of catecholamines in tissue extracts and body fluids was based on the physiological response of a whole animal or of isolated tissue *in vitro* to catecholamine. Measurement of blood pressure change (Holzbauer and Vogt, 1956) was used as an end-point when a whole animal was used. The degree of contraction was measured when isolated tissue was used. Various tissues have been used including rat uterus (Gaddum and Lembeck, 1949) and blood vessels of rabbits' ears *in situ* (Armin and Grant, 1955). Although the bioassay technique is very sensitive, it lacks specificity. Many substances present in tissue extracts or body fluids could either enhance or inhibit the action of catecholamines. Other disadvantages include the lack of precision, the large number of animals required and

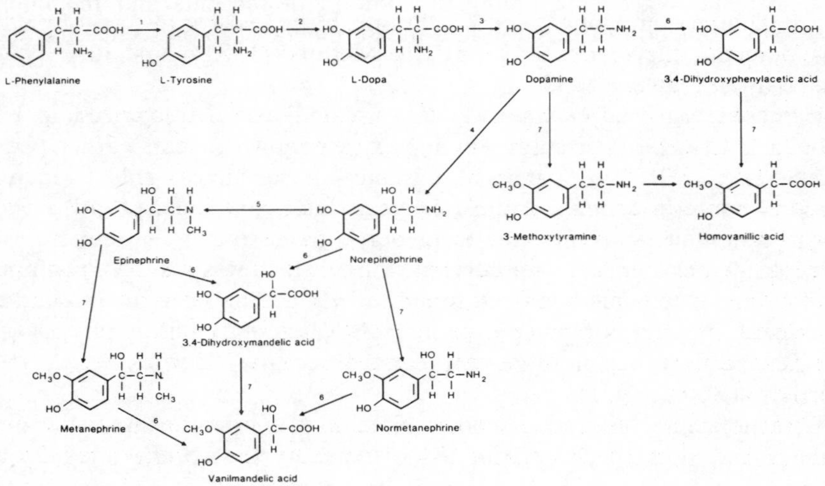

Figure 1. Pathways of catecholamine metabolism. Enzymes: 1. phenylalanine 4-hydroxylase (phenylalanine 4-monooxygenase; EC 1.14.16.1); 2. tyrosine 3-hydroxylase (tyrosine 3-monooxygenase; EC 1.14.16.2); 3. dopa decarboxylase (aromatic-L-amino-acid decarboxylase; EC 4.1.1.28); 4. dopamine β-hydroxylase (dopamine β-monooxygenase; EC 1.14.17.1); 5. phenylethanolamine N-methyltransferase (noradrenalin N-methyltransferase; EC 2.1.1.28); 6. monoamine oxidase (amine oxidase [flavin-containing]; EC 1.4.3.4); 7. catechol O-methyltransferase (EC 2.1.1.6).

the need for repeated tests by experienced workers. The gradual decline of its use was predicted (Gaddum, 1959).

Colorimetry and fluorometry

The first attempt at chemical measurement of catecholamines was the development of a colorimetric method. Catecholamines give a colored product when exposed to iodine or other oxidizing agents. However, the majority of the colorimetric methods developed lacked the sensitivity, specificity and reproducibility needed to measure catecholamines in body fluids (Bloor and Bullen, 1941).

Fluorometry generally is 3 orders of magnitude more sensitive than colorimetry and thus would be a logical choice for the estimation of catecholamines in body fluids. The two most popular fluorescence reactions used for this purpose are:

1. the oxidation of catecholamines to fluorescent trihydroxyindole (THI); and

2. condensation of the catecholamines with ethylenediamine (EDA) to form fluorescent products.

In the THI reaction, epinephrine and norepinephrine are oxidized by air or other oxidizing agents. Under alkaline conditions the oxidation product is rearranged to 1-methyl-3,5,6-trihydroxyindole which is the fluorescent agent. Because THI is not stable under alkaline conditions, the final reaction mixture usually is acidified. But even under acidic conditions the fluorescent product continues to fade. Because of this general instability of the reactants, products and blanks, the timing of addition of reagents and the interval between completion of reaction and measurement of fluorescence has to be under exact control.

Purification of the catecholamines in a biologic sample by adsorption on alumina (Anton and Sayre, 1962) or binding to an ion-exchange resin (Bergström and Hansson, 1951) is necessary for high selectivity of the assay. Even with purification, the specificity of the assay has been questioned seriously because sample readings often approximate sample blank values (Crout, 1959; Udenfriend, 1962). The introduction of a two-step purification using alumina adsorption followed by binding on a cation-exchange resin and boric acid elution provided an important improvement in sensitivity and specificity of the method (Valori *et al.*, 1969, 1970; Renzini *et al.*, 1970). These and other modifications (Jiang *et al.*, 1976) resulted in a lower detection limit of 50 pg and made the method more suitable for use in a clinical diagnostic laboratory. Separate measurement of epinephrine and norepinephrine could be accomplished by the use of differential stabilization reagents (Merrills, 1963; Viktora *et al.*, 1968; Wood and Mainwaring-Burton, 1975) and measurement of fluorescence at different wavelengths (Martin and Harrison, 1968).

Fluorescence of the polycyclic compounds obtained from condensation of oxidized catecholamines and ethylenediamine forms the basis of the EDA assay (Natelson *et al.*, 1949). The EDA assay is more sensitive than the THI method but lacks specificity. Values derived from the EDA assay are often much higher than those obtained by THI or bioassay (Price *et al.*, 1960). When the EDA method is applied to extracts obtained by alumina adsorption and ion-exchange purification, the values obtained agree closely with those from the THI method (Weil-Malherbe, 1961).

It can be concluded that either the THI or the EDA fluorometric method, with proper sample purification, can yield a valid estimation of catecholamines. Alumina adsorption followed by ion-exchange purification is almost essential to ensure the specificity needed to measure catecholamines in body fluid. More is known about the THI assay, and it is used widely in clinical diagnostic laboratories. The technique has yielded much clinically useful information in our laboratory (Jiang *et al.*, 1973; Stoffer *et al.*, 1973). However, separate estimation of epinephrine and norepinephrine, especially in blood, can only be made with difficulty by the THI method.

Radiochemical techniques

Numerous techniques involving utilization of radioactive compounds for the measurement of catecholamines have been reported. For example, a double isotope-dilution method has been tried (Rentzhog, 1972). After oxidation of catecholamine by ^{131}I (Berod *et al.*, 1978) or [^{51}Cr]-dichromate (Bois *et al.*, 1978), the reduced radioactive compounds were extracted and the radioactivity of the stoichiometrically formed products was determined. Radioimmunoassay (Raum and Swerdloff, 1981) was also developed. But, because of lack of sensitivity, lack of specificity or insufficient development, none of these methods enjoyed wide use.

The radiochemical method that caught investigators' imagination and enjoys wide application today is the radioenzymatic method (REA) which utilizes enzymes to catalyze the labelling of catecholamines. The specificity of the assay depends on the enzyme used. There are two enzymes used for this purpose: phenylethanolamine *N*-methyltransferase (PNMT) and catechol *O*-methyltransferase (COMT) (Fig. 1). The enzyme PNMT transfers the ^{3}H- or ^{14}C-labelled methyl group of *S*-adenosylmethiomine (SAMe) to the primary amino nitrogen in norepinephrine (Axelrod, 1962). The specificity of this method depends on the specificity of the enzyme and on the selectivity of alumina to adsorb the substrate and product of the enzymatic reaction (Henry *et al.*, 1975). Other potential substrates, when present, either are not methylated in the presence of norepinephrine or are not adsorbed by alumina. This method has a lower detection limit of 25 pg. One disadvantage of it is that it can only measure norepinephrine.

To extend the technique for the measurement of other catecholamines and their metabolites, the enzyme COMT was examined (Engelman *et al.*, 1968). COMT is less specific than PNMT in that it transfers a methyl group from the methyl group donor to catechol compounds. In addition to the three principal catecholamines, dihydroxyphenylacetic acid (DOPAC), dihydroxymandelic acid (DOMA), dihydroxyphenylethylene glycol (DOPEG) and other metabolites also can be determined by this technique in a single 25 μl sample of plasma (Thiede and Kehr, 1981). The COMT method (Peuler and

Johnson, 1977), which was later adapted into a kit form and enjoys wide application, involved methylation with [³H]-SAMe and the enzyme, extraction of the methylated products with organic solvent, and isolation of individual derivatives by thin-layer chromatography.

Because of its specificity and sensitivity, the REA has gradually gained favor over the fluorescence assay. This is especially true in clinical laboratories when catecholamine determinations are required on pediatric specimens. The disadvantages of the technique are that it is laborious, with multiple handling steps, and the inhibition of the enzymatic reaction by compounds in the sample causes erroneous results.

Gas—liquid chromatography
The application of gas-liquid chromatography to the measurement of catecholamines is limited by the sensitivity of the detectors. The popular flame ionization detector (Kawai and Tamura, 1968) lacks sensitivity. Most methods developed for gas chromatographic determination of catecholamines use an electron-capture detector. Halogenated derivatives of catecholamines, such as trifluoroacetate (Imai *et al.*, 1973), pentafluoropropionate (Horning *et al.*, 1964; Gelpi *et al.*, 1974), heptafluorobutyrate (Änggård and Sedvall, 1969; Karoum *et al.*, 1972) and perfluorobenzoate (Moffat *et al.*, 1972), have been examined for this purpose. In general, the electron capture detector has high absolute sensitivity but the usable sensitivity is usually decreased because of high background caused by reagents and solvents. The technique also lacks specificity. Many compounds present in the sample could form halogenated derivatives with retention times similar or identical to those of catecholamine derivatives.

Mass spectrometry is a highly sensitive and specific detector. When used in tandem with gas—liquid chromatography, it can eliminate the problems associated with other kinds of detectors. Both halogenated (Wang *et al.*, 1975) and trimethylsilated (Donike, 1975) derivatives of catecholamines yield excellent fragmentation patterns with prominent molecular ions. Gas—liquid chromatography with mass spectrometry detection (GL—MS) certainly is a method of great potential in the measurement of catecholamines.

Liquid chromatography
With its speed and the availability of versatile column packings that are particularly suitable for thermolabile compounds without derivatization, high-performance liquid chromatography (HPLC) provides an attractive alternative for the quantitation of catecholamines. Ion exchange (Zambotti *et al.*, 1975), normal-phase with ion-pair (Persson and Karger, 1974) and reverse-phase with ion-pair (Wittmer *et al.*, 1975; Crombeen, *et al.*, 1978; Molnár and Horváth, 1978) chromatography have been studied. Because of its high resolution, column stability, reproducibility and ability to separate large numbers of catecholamines and their metabolites, the reverse-phase ion-pair procedure enjoys wide popularity. Only the fluorescent and electrochemical detectors possess the sensitivity required for the detection and quantitation of catecholamines in HPLC column effluents.

For fluorescence detection, both pre-column and post-column derivatization have been tried. The derivatives used include THI (Mori *et al.*, 1976; Hamaji and Seki, 1979), EDA (Viktora *et al.*, 1968), and fluorescamine (Udenfriend *et al.*, 1972; Weigele *et al.*, 1972). The reaction between o-phthalaldehyde and catecholamines in the presence of 2-mercaptoethanol (Benson and Hare, 1975; Lindeberg, 1976), and that among ninhydrin, aldehyde and amines to generate fluorescence products (Samejima *et al.*, 1971) were also studied. Among these methods, the THI technique has proven to be the most reliable and useful (Yui *et al.*, 1980). A fully automated procedure has been developed based on the THI method (Yamatodani and Wada, 1981).

The first electrochemical device applied to chromatographic detection was constructed in 1952 (Kemula, 1952). It was later refined and applied to catecholamine analysis (Kissinger *et al.*, 1973; Refshauge *et al.*, 1974). The amperometric and coulometric detectors are the two most commonly used with liquid chromatographic systems. The amperometric detector oxidizes only a portion of the compound present in the mobile phase as it passes through the electrode cell. The coulometric detector attempts to achieve 100% oxidation efficiency. The coulometric detector should offer greater sensitivity than the amperometric detector because of the difference in their efficiencies but, in practice, the two detectors are about equally sensitive. This is due to the higher background current and noise caused by the larger electrode surface area in the coulometric detector. A detailed review of the theory and detector design for electrochemical detection is beyond the scope of this chapter. For that purpose, the reader is referred to Kissinger (1977) and Krstulović (1982).

Reverse-phase ion-pair HPLC with amperometric detection has been applied to the determination of urinary catecholamines (Riggin and Kissinger, 1977; Moyer *et al.*, 1979b) and plasma catecholamines (Freed and Asmus, 1979) with a detection limit of 10 pg. Recently, a comparison study (Hjemdahl, 1984) showed that HPLC with electrochemical detection yields reasonable agreement in results among laboratories. The study also showed that results obtained by HPLC with electrochemical detection are similar to those obtained by a radioenzymatic method. Similar results were obtained between HPLC and GC—MS technology (Warsh *et al.*, 1980).

HPLC with electrochemical detection represents the latest and most promising technique for the quantitation of catecholamines. It possesses the sensitivity and specificity required. The procedure is simple and convenient to perform. The instrumentation is of relatively low cost and easy to maintain. It is particularly well suited for use in a clinical diagnostic laboratory. The disadvantage of the technique is that it does not provide absolute chemical identification, whereas GC—MS does.

DETERMINATION OF CATECHOLAMINES IN BLOOD

Material and equipment
The chemicals used were norepinephrine, epinephrine, dopamine, cyanogen

bromide and tris(hydroxymethyl)aminomethane (Tris; Trizma base) purchased from Sigma Chemical Company. 3,4-Dihydroxybenzylamine hydro-
bromide (DHBA), 3-(3,4-dihydroxyphenyl)-2-methyl-L-alanine (methyldopa;
Aldomet) and *m*-aminophenylboronic acid hemisulfate were obtained from
Aldrich Chemical Company. Sodium 1-heptanesulfonate was from Eastman
Kodak Company. The aluminum oxide (Brockman activity II), used in
the plasma purification, was ordered from BDH Chemicals Ltd. (Poole,
England); the aluminum oxide (Catalog number A-540) used in the urine
purification was from Fisher Scientific Company. Amberlite CG-50, 100—
200 mesh, was from Mallinckrodt. Avicel microcrystalline cellulose (Catalog
number 66-00-550-0) was from Brinkman Instruments. Perchloric, acetic,
hydrochloric and phosphoric acids were obtained from J. T. Baker Chemical
Company. The chromatographic grade solvents including methanol, acetonitrile and ethyl acetate were from Burdick and Jackson Laboratories, Inc.
All other chemicals were of ACS grade and purchased from Fisher Scientific
Company.

The HPLC system included a Model 6000A solvent delivery system, a
710B automatic injecting system (WISP) and a model 730 integrator (data
module) from Waters Associates, Inc. (Milford, Massachusetts 01757). The
reverse-phase analytical column packed with Ultrasphere-IP, 5 μm particle
size, was 4.6 × 250 mm (Part number 235335) and was purchased from
Beckman Instruments, Inc. (1780 Fourth Street, Berkeley, California 94710).
The column inlet filter, Model 7302, with a 2 μm filter was from Rheodyne,
Inc. (2809 Tenth Street, Berkeley, California 94710). An LC-4 amperometric controller and TL-5 glassy carbon thin-layer detector cell (transducer)
with T6-2M cell gaskets were purchased from Bioanalytical Systems, Inc.
(West Lafayette, Indiana 47906). An OmniScribe, two-pen strip chart
recorder from Houston Instruments (Austin, Texas 78753) was used. The
vitreous carbon (GC-30S) used in the transducer modification was obtained
from IMC Industry Group, Inc. (New York, New York). The silicon carbide
abrasives used to resurface the vitreous carbon were obtained from Ultralap,
Moyco Industries, Inc. (Philadelphia, Pennsylvania).

Modification of the transducer to accommodate a larger vitreous carbon
surface has been an important contribution to catecholamine analysis (Moyer
et al., 1979a). The modification consists of removing the carbon paste
electrode in the TL-3 or the glassy carbon electrode in the TL-5 and milling
a cavity in the Kel-F block large enough to accommodate a $2 \times 10 \times 22$ mm
vitreous carbon plate and an underlying $22 \times 10 \times 0.27$ mm platinum
contact plate. The $2 \times 10 \times 22$ mm electrode is cut from a larger plate of
vitreous carbon (GC-305, IMC Industry Group, Inc.) by using a silicon
carbide saw. The upper surface of the plate is polished with successively finer
grades of Ultralap silicon carbide abrasive, particle sizes range from 63 to 1
μm. This polishing produces a surface that has a mirror-like finish.

Because individual differences in polishing will result in uniquely shaped
electrodes, we have found it desirabe to cut and polish the glassy carbon
electrode before milling the Kel-F block to accommodate it. When fitted into

the assembled cell, the glassy carbon electrode should extend approximately 0.2 mm above the block surface to ensure a leak-proof seal. A plate not extending this high may allow the transducer to leak, and one extending higher than 0.2 mm could result in the fracture of the glassy carbon plate when the transducer block is reassembled. The platinum plate is connected by a multi-stranded wire to a gold-plated pin to provide the electrical contact for the glassy carbon electrode.

Procedures

Chromatographic adsorbents. Aluminum oxide is prepared for use by a series of acid and water washes with decanting after each wash to remove suspended fine particles. Four hundred grams of alumina is placed in a 2 l beaker containing 1 l of 2 M HCl. The mixture is boiled on a steam bath for 4 h with stirring every 15 min. The recommended stirring consists of a swirling motion in the liquid phase induced by using a glass paddle. This has the effect of suspending the particles yet does not fracture them. After 4 h, the acid is decanted and 1.5 l of preheated (80° C) 2 M HCl is added to the alumina for an additional 4 h of heating and stirring. The acid is removed and the alumina is washed three times with 0.5 l portions of 2 M HCl; 15 min is allowed for the alumina to settle before decanting. The final washing consists of repetitive washing with 8 l of distilled water in 0.5 l portions. The pH of the final alumina—water slurry is adjusted to 7.0 with 10 M NaOH.

The alumina is then filtered under vacuum on a Büchner funnel lined with Whatman No. 1 filter paper. When the filtering is completed, the alumina is spread in a thin layer in several large evaporating dishes and dried in a 300° C oven for 4 h. After drying and cooling, it is spread over the surface of a large watch glass and stored in a water-saturated container for at least 4 days and preferably 1—2 weeks. When the moisture content reaches about 20 wt %, the alumina is ready for use and can be stored in a tightly sealed bottle.

This rehydration of the aluminum oxide is essential to maintain adequate recoveries. Omitting the rehydration will cause a drastic reduction in recovery efficiency (Jiang *et al.*, 1976).

Amberlite CG-50 is prepared in 100 g lots by the procedure of Valori *et al.* (1969). The resin is soaked in 1 l of 2 M HCl overnight. The following morning, the HCl is removed and the resin is washed with glass-distilled water three times or until the supernatant appears clear. The resin then is slowly stirred for 30 min with 1 l of 2 M NaOH followed by several water washes. The resin is suspended in 1 l of 2 M HCl for 30 min and the acid—water—base—water wash cycle is repeated twice more. The final water wash is removed and 500 ml of 0.1 M sodium acetate buffer, pH 6.0, is added to the resin. The resin—buffer slurry is adjusted to pH 6.0 and stored in a glass reagent bottle at 4° C. One milliliter of the Amberlite—buffer mixture is pipetted into the purification column when required.

Plasma collection. Considerable care and attention must be given to each step in the entire procedure, beginning with sample collection. Plasma catecholamines are present in picogram quantities and are unstable, which makes accurate measurements difficult. Care should be taken to maintain the integrity of the sample.

To ensure that a representative sample is collected, we have found it desirable to instruct the patient a day prior to sample collection to refrain from smoking and from drinking coffee or tea. On the morning of the collection, each patient is told what the collection procedure will be, and an attempt is made to make the patient feel comfortable to reduce the level of anxiety that the patient might feel from new surroundings and the impending venipuncture. The patient is placed in a reclining position and an intermittent infusion set, PRN-21 filled with heparin (to prevent clotting), is inserted into an antecubital vein. The patient remains resting in the reclined position for 30 min; then the infusion set is cleared of heparin by drawing 2 ml of blood. Two 100×16 mm (10 ml) stoppered Vacutainers containing 0.07 ml of 0.2 M EDTA and 0.1 ml of 0.52 M sodium metabisulfite that have been chilled on ice are filled with blood, mixed and returned to a crushed ice container. The infusion needle and tubing are again filled with sterile heparin to prevent clotting. With the infusion set still in place, the patient is requested to stand and remain standing for 10 min. After the 10 min has elapsed, the infusion set is again cleared, and 14 ml of blood is collected into two 10 ml chilled Vacutainers containing EDTA and sodium metabisulfite (standing sample). The four Vacutainers are centrifuged at $1000 \times g$ for 10 min. The plasma from the supine and the standing position are combined into respective collection tubes marked 'Supine' and 'Standing' and forwarded to the laboratory on ice. The sample could be stored in a $-70°$ C freezer for at least 4 weeks without deterioration. Samples are slowly thawed on the day of analysis. Rapid thawing of the frozen plasma by warming in a water bath should be avoided.

Plasma purification. The use of alumina in the first step of the plasma catecholamine purification takes advantage of the attraction of the catechol function to the alumina particles. Instead of placing the alumina in a column and passing the serum directly through it, a batch extraction procedure is used. The serum is thawed and centrifuged to remove fibrin. Seven milliliters of each plasma sample is added to a 25 ml beaker containing 5 ml of 0.13 M EDTA, 7.8 ml of glass-distilled water, 0.150 ml of DHBA (working internal standard) and 0.2 ml of 0.5 M sodium metabisulfite.

Pretreated alumina is added to each sample with a measuring spoon that delivers 0.5 g with continuous stirring on a stirring plate (3×10 mm stirring bar). The pH is adjusted to 8.4—8.5 by dropwise addition of 0.5 M Na_2CO_3. Each sample is mixed for 10 min to ensure good recovery. We have found that five samples can be processed within 15 min by using five stirring plates. The stirring motion should be gentle enough to keep the alumina suspended without causing excessive fragmentation of the particles. After the sample is removed from the stirring plate, the alumina is allowed to settle for 5 min and then the supernatant is carefully decanted. The alumina is then transferred to a 6×140 mm glass column with a 5 ml reservoir (a squeeze bottle filled with distilled water is used to flush the alumina into the column as a slurry). In the column, the alumina is washed with 4 ml of glass-distilled water and then 0.8 ml of 0.05 M $HClO_4$. These fractions are discarded and the columns are eluted with 5 ml of 0.5 M $HClO_4$. The eluate is collected in a 10 ml beaker containing 0.1 ml of 0.13 M EDTA. The beakers are stored on crushed ice. While the beakers remain on ice, the pH is adjusted to 6.0–6.2 by dropwise addition of 0.5 M and 0.2 M Na_2CO_3.

After the pH has been adjusted in all of the samples, each eluate is transferred to a 5×130 mm glass column (with a 4 ml reservoir) containing 1 ml of the Amberlite slurry. Each resin then is washed with 3 ml of glass-distilled water. One hundred and twenty microliters of 0.66 M boric acid is added to the column, allowed to run through and discarded. The column is eluted with two 150 µl and one 100 µl portions of 0.66 M boric acid. These fractions are collected at 2 min intervals in a single 10×75 mm glass tube stored on ice. Each boric acid eluate is immediately frozen and stored at $-20°$ C until ready for analysis. Samples that are stored in this manner are stable for 4 days.

Quantitation. A 60 µl portion of the boric acid eluate from the Amberlite column is injected onto a reverse-phase column. The individual components are resolved by using reverse-phase ion-pair chromatography, and the eluted catechols are oxidized with an amperometric detector that is operated at a constant potential. The current resulting from the oxidation of catechols to o-quinones is converted to a voltage, amplified and then measured by using a strip chart recorder or digital integrator. The heights of the unknown component peaks are quantitated by comparison to known amounts of catecholamine standard. A fixed quantity of internal standard, DHBA, is added to both the calibration and analysis samples to correct for losses during purification.

In the chromatographic separation, the mobile phase consists of 5 vol of acetonitrile and 95 vol of 70 mM sodium phosphate buffer containing 0.1 mmol of Na_2EDTA and 0.5 mmol of sodium 1-heptanesulfonate per liter. The mobile phase is adjusted to pH 3.2 with concentrated phosphoric acid, filtered through a 0.45 µm filter and degassed under reduced pressure. The column is equilibrated with each new mobile phase preparation for about 40 min at a flow rate of 1 ml/min.

The electrochemical detector, LC-4 with modified transducer, is operated at an oxidation potential of $+0.63$ V relative to a Ag/AgCl reference electrode at a detector sensitivity of 5 nA. Freshly prepared calibration standards are injected, the resulting peak heights are measured and correction factors for each component are calculated (peak height divided by the amount of the component in the calibration standard). The individual correction factors are used to calculate the amount of each catechol in the unknown samples.

A 15 μl aliquot of a standard mixture, containing norepinephrine at 1000 ng/ml, epinephrine at 500 ng/ml, and dopamine at 400 ng/ml in 0.01 M HCl, is added to 3.97 ml of 0.66 M boric acid. Also, 15 μl of working internal standard (DHBA at 1500 ng/ml) is added to give a total volume of 4.00 ml. Sixty microliters of this standard calibration mixture is injected into the system to give an on-column injection of 225 pg of norepinephrine, 113 pg of epinephrine, 90 pg of dopamine and 338 pg of DHBA.

The pH of the sample after it has been eluted from the Amberlite ion-exchange column is slightly more basic than the pH of the mobile phase. Therefore, 1—5 μl of 5 M sodium acetate buffer (pH 3.2) is added to the samples to make the pH more compatible with the mobile phase. At high detector sensitivity, less than 10 nA, pH differences are important in that too great a difference in pH between mobile phase and injected sample will cause a temporary baseline instability that can interfere with the early-eluting chromatographic peaks. We have found that a small amount of sodium acetate will prevent or minimize this problem and make the analysis less subject to these problems.

It is advantageous if the system can be dedicated exclusively to use for plasma catecholamine analysis. When the chromatographic system is not in use, the mobile phase is continuously pumped through the column and detector at 0.1 ml/min, and the detector and reference cell are left in the ON position. We have found this greatly facilitates rapid analysis and minimizes start-up time that would otherwise be needed for the detector to reach a stable baseline.

Our assay protocol calls for preparation of new mobile phase daily. One hundred milliliters of freshly prepared mobile phase is pumped through the column, followed by two or three calibration mixture samples before sample analysis begins. A sample blank consisting of 0.66 M boric acid is injected to check for background contamination.

With the above protocol and an automatic sample injector, a single technician can extract and quantitate 20—25 plasma catecholamine samples per day.

Results and discussion

Assay condition, reproducibility and recovery. The chromatographic conditions for the separation and quantitation of norepinephrine, epinephrine and dopamine have been published (Moyer and Jiang, 1978). In the current assay, acetonitrile is used in place of methanol in the mobile phase, and a 5 μm particle reverse-phase column packing is used instead of a 10 μm particle. These changes decrease retention times and improve resolution with use of less organic solvent. Figure 2 shows a chromatogram of the three principal catecholamines, internal standard and some of the commonly used

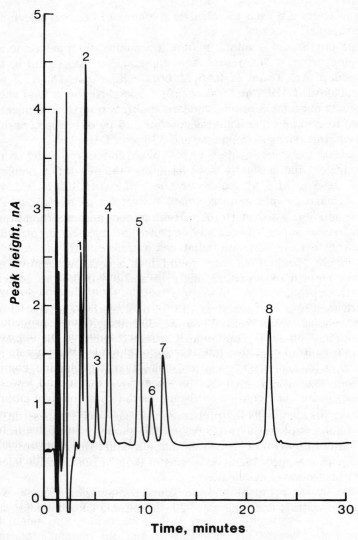

Figure 2. Chromatogram of catecholamines and some common drugs. Peaks: 1. norepine-phrine; 2. L-dopa; 3. epinephrine; 4. DHBA (internal standard); 5. methyldopa; 6. dopamine; 7. carbidopa; 8. L-isoproterenol D-bitartrate.

antihypertensive medications. These drugs are well separated from the peaks of interest and make the determination of catecholamines possible in blood from patients being treated with these drugs.

The linearity of the amperometric detector response for the catechol-amines is shown in Fig. 3. Ten picograms of norepinephrine or epinephrine injected onto the column can generate reproducible responses at a detector sensitivity of 5 nA. Using the modified transducer for plasma quantitation over a perod of 14 months (6300 injections), we experienced only a 13.6%

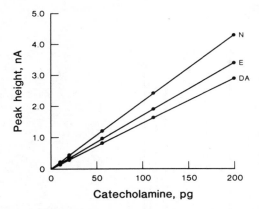

Figure 3. Linearity of amperometric detector response for norepinephrine (N), epinephrine (E) and dopamine (DA).

decrease in sensitivity before resurfacing of the glassy carbon was required. No loss of linearity occurred throughout the 14-month period.

Purification of plasma samples by adsorption on alumina followed by Amberlite CG-50 chromatography prior to HPLC is essential. Chromatograms of sample extracts prepared by the two-column procedure are much cleaner than those prepared by either alumina or ion-exchange purification alone. There are interfering compounds that migrate in and around the catecholamine peaks. In a plasma catecholamine comparison study (Hjemdahl, 1984), our laboratory generated the second lowest basal norepinephrine and epinephrine levels among the group of eight laboratories that performed HPLC analysis. The lower result agreed well with that obtained by the PNMT radioenzymatic method. We believe this is because other laboratories used a one-column sample purification instead of a two-column one and therefore had interfering substances in their samples that caused the higher values.

On rare occasions, a patient's chromatogram will show an elevated norepinephrine and/or epinephrine peak with a very high DHBA peak, resulting in a recovery greater than 100%. We believe this might be due to an unknown catecholamine metabolite in the patient's plasma that co-elutes with DHBA. An example is shown in Fig. 4. We have seen three patients with similar patterns in the past 4 years. Each of the three patients had elevated norepinephrine levels and all were found, at operation, to have pheochromocytoma.

The cellulose aminophenylboronate gel that is used in the purification of urinary catecholamines (see later) cannot be used in place of Amberlite CG-50 in the plasma assay. In the plasma catecholamine assay, the detector sensitivity is at 5 nA and the small amount of aminophenylboronate that is constantly leaching from this gel will produce a significant peak. The retention time of aminophenylboronate is identical to that of the DHBA (internal standard). In the urine catecholamine assay, the detector sensitivity

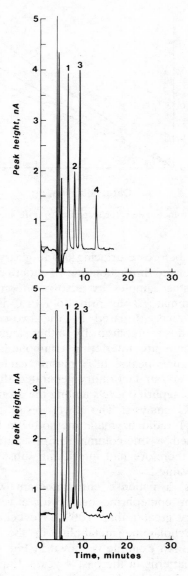

Figure 4. Chromatograms of calibration mixture (upper) and plasma from patient with pheochromocytoma (lower). 1. Norepinephrine; 2. epinephrine; 3. DHBA (internal standard); 4. dopamine.

is 1—2 orders of magnitude less than it is in the plasma assay, and the small amount of aminophenylboronate present cannot be detected.

The assay reproducibility is shown in Fig. 5 and Table 1. The mean coefficients of variation for norepinephrine, epinephrine and dopamine for 1983 were 4.3, 7.2 and 9.5%, respectively.

Table 1. Reproducibility of plasma catecholamine assays in 1983

Month	No. of observations	Norepinephrine Mean ± S.D. (pg/ml)	CV (%)	Epinephrine Mean ± S.D. (pg/ml)	CV (%)	Dopamine Mean ± S.D. (pg/ml)	CV (%)
January	24	201 ± 8.6	4.3	100 ± 4.3	4.3	64 ± 5.3	9.0
February	24	199 ± 7.1	3.5	96 ± 6.6	7.0	68 ± 4.6	6.9
March	30	203 ± 10.6	5.2	98 ± 9.2	9.4	69 ± 7.2	10.5
April	25	217 ± 8.5	3.9	110 ± 8.3	7.6	80 ± 6.0	7.0
May	25	215 ± 10.0	4.7	110 ± 9.0	8.0	76 ± 7.0	9.2
June	26	217 ± 6.1	2.8	115 ± 6.7	5.9	77 ± 5.6	7.3
July	27	218 ± 7.2	3.3	115 ± 7.4	6.5	79 ± 8.0	10.2
August	22	217 ± 10.4	4.8	114 ± 7.1	6.3	77 ± 7.1	9.2
September	18	221 ± 8.2	3.7	118 ± 5.1	4.3	76 ± 9.3	12.2
October	23	222 ± 9.7	4.4	117 ± 10.7	8.9	78 ± 7.4	9.4
November	25	215 ± 11.8	5.5	112 ± 11.1	9.9	80 ± 8.4	10.4
December	23	205 ± 12.3	5.9	105 ± 8.5	8.1	78 ± 10.3	13.2

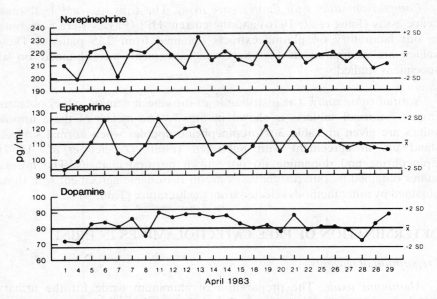

Figure 5. Day-to-day variations in plasma catecholamine assays in April 1983. Appearance of the same date twice in the figure indicates that two sets of assays are performed on that date.

The mean recovery of DHBA added to plasma samples at the beginning of the extraction is shown in Table 2. The recovery of DHBA varied from 44 to 56%. The overall mean for the year was 50 ± 3%. In a blind interlaboratory comparison study (Hjemdahl, 1984) the recoveries of norepinephrine and epinephrine added to human plasma obtained by our laboratory after correction according to DHBA recovery were 90 and 112%, respectively.

Table 2. Recovery of DHBA added to plasma samples

Month	No. of observations	Recovery (%) (Mean ± S.D.)
January	340	51.4 ± 6.1
February	342	53.1 ± 6.7
March	349	49.3 ± 7.7
April	341	50.6 ± 6.1
May	306	44.0 ± 7.5
June	320	48.6 ± 7.6
July	270	48.5 ± 7.7
August	322	48.6 ± 6.6
September	241	49.7 ± 6.1
October	322	48.5 ± 7.3
November	374	50.1 ± 6.9
December	282	56.0 ± 7.1

Comparison study with fluorescence assay. The trihydroxyindole fluorescence assay (Jiang *et al.*, 1976) and the current HPLC assay were performed in our laboratory on plasma extracts obtained from 235 patients. These values are compared in Fig. 6. Good agreement was obtained on all specimens studied.

Normal value study. The distribution of the catecholamine values obtained from 67 normal subjects is shown in Fig. 7. The means of these normal values are given in Table 3. Norepinephrine doubles when normal subjects stand up, in agreement with published results (Ziegler *et al.*, 1977). Epinephrine and dopamine do not shown postural changes. The normal values obtained by our current assay are in reasonable agreement with those obtained by other methods selected from the literature (Table 4).

DETERMINATION OF FREE CATECHOLAMINES IN URINE

Preparation of reagents

Aluminum oxide. The preparation of aluminum oxide for the urinary catecholamine purification was similar to that used for the plasma catecholamine assay with the exception that the rehydration step is not required.

Borate gel. The cellulose aminophenylboronate (CAPB) (Porath, 1974; Higa *et al.*, 1977) is prepared by washing 20 g of Avicel microcrystalline cellulose with 200 ml of 1 M HCl. The cellulose is allowed to settle for 10 min before it is centrifuged at 750 × g for 6 min (it is convenient to split the lot and do the washings and centrifugings in 250 ml plastic centrifuge bottles). The acid is poured off and the gel is washed twice with 150 ml portions of distilled water, centrifuged and decanted. Distilled water is added

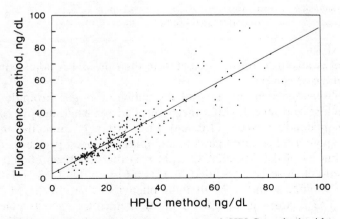

Figure 6. Comparison of THI fluorescence assay and HPLC method with amperometric detection. $y = 3.4 \pm 0.89\, x\, (r = 0.92)$.

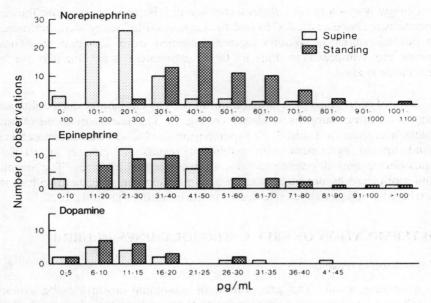

Figure 7. Distributions of catecholamine values in normal plasma.

Table 3. Plasma catecholamine values in normal subjects

| | Catecholamines (pg/ml) | | | | | |
| | Norepinephrine | | Epinephrine | | Dopamine | |
	Supine	Standing	Supine	Standing	Supine	Standing
n	67	67	44	49	16	29
Mean	254	542	30	42	15	13
S.D.	122	217	20	23	12	7
95% confidence range	10—499	109—975	0—69	0—89	0—38	0—26
Observed range	74—750	201—1473	7—71	11—140	3—45	3—30

and the gel is mixed to form a suspension; then the pH is adjusted to 11 by titration with 4 M NaOH.

In a well-ventilated hood, 20 g of crystalline cyanogen bromide is added to the cellulose and mixed with a mechanical stirrer while the pH is adjusted to and maintained between 11.0 and 11.5. After 12 min, the mixture is filtered through a 3 l sintered glass funnel under reduced pressure. The gel is washed with 2 l of 0.1 M NaHCO$_3$ (pH 9.5) that has been cooled to 4° C. The washed cellulose is suspended in 130 ml of cold 0.1 M NaHCO$_3$ (pH 9.5) and combined with 0.1 M *m*-aminophenylboronate in 0.1 M NaHCO$_3$ (pH 9.5). The mixture in a 250 ml plastic centrifuge tube is placed on a rotating mixer where it is allowed to react (in the dark) for 24 h at 4° C. Then, the product, CAPB, is washed on a sintered glass funnel with 1 l of 0.1

Table 4. Normal plasma catecholamine values by various methods

Method	Mean plasma concentration (pg/ml)			
	Norepinephrine	Epinephrine	Dopamine	Reference
REA, PNMT	254	—	—	Henry *et al.* (1975)
REA, COMT	200	50	—	Engelman and Portnoy (1970)
RIA	240	22	—	Raum and Swerdloff (1981)
GC—MS	200	58	48	Ehrhardt and Schwartz (1978)
HPLC, Fluorimetric	185	32	—	Yui *et al.* (1980)
HPLC, EC	519	40	—	Hallman *et al.* (1978)
HPLC, EC[a]	254	30	15	Present method
HPLC, EC[b]	542	42	13	Present method

[a] Supine.
[b] Standing.

M HCl followed by 10 l of distilled water. The CAPB is again suspended in water, the pH is adjusted to 8.5 with NaOH, and the CAPB is washed with 1 l of distilled water. The gel is harvested, suspended in 500 ml of water and stored at 4° C. When needed, the CAPB is resuspended by shaking, and 1.0 ml volumes are pipetted into the purification columns.

Urine collection

A 24 h urine collection is required for the catecholamine analysis. Twenty milliliters of glacial acetic acid, as a preservative, is added to a 2 l collection bottle prior to dispensing the collection container to the patient. At this time the patient receives instructions about the proper collection procedure. When the urine collection is returned, the total volume is recorded and the pH is adjusted to between 3 and 4 with 6 M HCl. If a urine collection is received with a pH less than 1, it should be rejected. Low pH levels cause hydrolysis of the conjugated catecholamines present in the urine, resulting in falsely high urinary free catecholamine levels. An aliquot of urine is removed from the 24 h collection and is stored at 4° C until analysis.

Urine purification

The urinary catecholamine purification procedure consists of an extraction with ethyl acetate to remove interfering acid metabolites followed by column chromatography on alumina. The catecholamine extract is further purified by taking advantage of the attraction of the *cis*-diol function to a boronate gel. This purification system ultimately produces an extract that is free of nearly all interfering susbtances.

Four milliliters of urine from a 24 h collection is extracted twice with 2 ml of ethyl acetate. The urine is extracted in a 16 × 25 mm glass tube, with vortex mixing. The ethyl acetate is removed by allowing the layers to separate for 5 min and then aspirating the organic phase.

One milliliter of the extracted urine is transferred to a 10 ml beaker containing 1 ml of 200 μg/l DHBA (internal standard), 1 ml of 0.13 M EDTA, and 1 ml of 1 M NaHCO$_3$. The pH is adjusted to 8.4—8.5 with 0.3 M NaOH. We have found it helpful, especially when processing large numbers of urine samples, to use a circular wooden board (20 × 250 mm in diameter) with circular holes drilled out to accommodate 24 10 ml beakers. By using this holder and a magnetic stirring plate, the pH electrode can be positioned for titrating each sample with a minimum of individual handling. Each urine extract is then transferred to a 10 × 80 mm glass column containing 0.5 g of alumina that has been washed with 20 ml of distilled water. The urine extract is allowed to pass through the alumina column. The column is washed with 7 ml of distilled water. Five milliliters of 50 mM acetic acid containing 5 mM sodium metabisulfite is used to elute the catecholamines from the column into a beaker containing 0.1 ml of 0.13 M EDTA and 0.2 ml of 1 M Tris. The eluted fraction is transferred to a 15 ml screw-capped conical centrifuge tube containing 7 ml of distilled water and 1 ml of CAPB gel. The total volume in the centrifuge tube is adjusted to 14 ml with distilled water, and the tube is capped and gently tumbled on a rotating mixer for 10 min.

The extraction tube is centrifuged at 700 × g for 5 min and the supernatant is removed by aspiration. The CAPB gel is washed by resuspending it in 10 ml of distilled water, shaking it for 10 s and centrifuging it for 5 min at 700 × g. The supernatant is carefully removed by aspiration to remove all of the liquid without disturbing the CAPB gel. The catecholamines are removed from the gel by adding 0.8 ml of 0.67 M boric acid, and gently vortexing the centrifuge tube for 1 min. This mixture is centrifuged for 7 min at 1000 × g and the supernatant is removed with a transfer pipette and stored in a 12 × 75 mm glass tube at 4° C until analysis. If the catecholamine determination cannot be made on the same day as the extraction, the extract can be stored frozen at −20° C for up to 4 days.

Quantitation
The quantitative analysis of urinary free catecholamines by HPLC consists of injecting 30 μl of the boric acid fraction that has been eluted from the CAPB gel onto an analytical column. The individual components are resolved with reverse-phase, ion-pair chromatography, and the eluted catecholamines are oxidized with an amperometric detector. A working calibration standard containing norepinephrine, epinephrine, dopamine, methyldopa and DHBA is prepared just prior to analysis by making a 1:100 dilution of the catecholamine stock standard with 0.66 M boric acid. This dilution will give a working calibration standard containing norepinephrine at 25 ng/ml, epinephrine at 8 ng/ml, DHBA at 200 ng/ml, dopamine at 164 ng/ml and methyldopa at 300 ng/ml. Thirty microliters of the working calibration standard is injected on column. Methyldopa is included because it is one of the few hypertensive drugs that can interfere. The retention time of methyldopa is made to be between that of the internal standard and that of

dopamine by adjusting the pH of the mobile phase to 3.2. By including methyldopa in the calibration standard and monitoring the retention time, we can avoid peak interference and can determine qualitatively which samples contain methyldopa.

The extracted unknown samples are injected and assayed, and the internal standard is used to correct for purification losses. The chromatographic configuration and mobile phase composition used for the urine analysis are identical to those listed for the plasma analysis. The oxidation potential across the glassy carbon electrode is +0.63 V with a detector setting of 200 nA. With an autosampler and a data reduction system, the unattended quantitative analysis of up to 48 samples is possible in less than 13 h.

Results and discussion

Assay condition, reproducibility and recovery. The procedure for urinary catecholamines assay is that reported previously from our laboratory (Moyer *et al.*, 1979b) with two modifications. As in the plasma assay, acetonitrile replaced methanol in the eluting buffer and a 5 μm particle was used in place of the 10 μm particle.

The initial urinary extraction with ethyl acetate is used to remove acidic metabolites. If not removed, these metabolites will cause overlapping with the peaks of interest. Amberlite CG-50 has been tried in place of CAPB gel for the purification of the urine but, for reasons not entirely understood, the recovery of catecholamine was drasticaly decreased when the ion exchanger was used. The constant leaching of small amounts of aminophenylboronate from the gel, which caused difficulty in the plasma assay, did not affect the urine assay.

Samples collected from patients being treated with the antihypertensive drug methyldopa produce chromatograms that contain not only methyldopa but also three additional chromatographic peaks not normally seen in urine obtained from subjects who are not taking methyldopa (Fig. 8). The peak (peak 7) immediately following the internal standard (peak 3) is methyldopa itself. One peak (peak 6) migrates in the vicinity of epinephrine (peak 2), sometimes overlapping it. By adjusting the composition of the mobile phase, pH and organic strength, it can generally be separated from the peak of interest. The other two unknown peaks (peaks 8 and 9) have retention times of approximately 40 min. Because the elution time is normally only 16—20 min for each sample, these two late peaks will interfere with subsequent samples. We advise that when a methyldopa peak appears, the catecholamine values of the subsequent two samples should be subjected to close examination for possible interfering peaks.

Figure 9, the quality control chart of urinary free catecholamines for the month of April 1983, shows the day-to-day variations of the results. Table 5 gives the monthly means of the quality control pool. The yearly mean coefficient variations of norepinephrine, epinephrine and dopamine are 4.2, 12.3 and 4.0%, respectively.

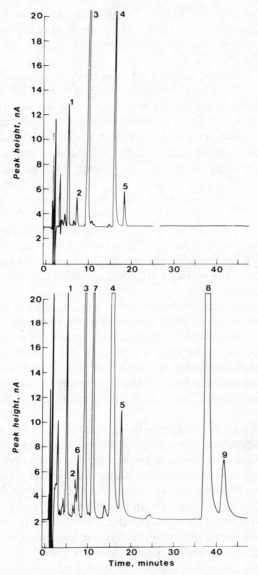

Figure 8. Chromatograms of urinary free catecholamine from subject not taking methyldopa (upper) and subject taking methyldopa (lower). 1. Norepinephrine; 2. epinephrine; 3. DHBA (internal standard); 4. dopamine; 5. unknown; 6. methyldopa metabolite; 7. methyldopa; 8 and 9. methyldopa metabolites.

The mean recovery of DHBA added to 377 consecutive urine samples at the beginning of extraction in our clinical diagnostic laboratory was 50.8% ± 5.3%, which is not significantly different from the value 59% ± 4% reported previously (Moyer *et al.*, 1979b).

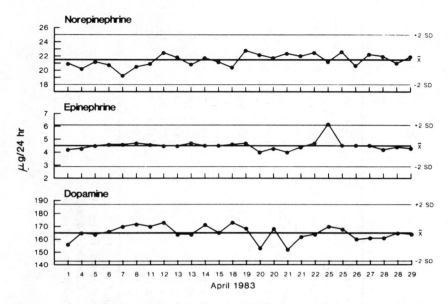

Figure 9. Day-to-day variations of urinary free catecholamine assays in April 1983. Appearance of the same date twice in the figure indicates that two sets of assays are performed on that date.

*Comparison study.** The extracts of 60 urine samples were subjected to both THI and current HPLC analysis. Good agreement was found for all three catecholamines (Fig. 10).

Population studies and diagnostic efficacy of urinary free catecholamines have been published (Moyer *et al.*, 1979b). It is our experience that the urinary catecholamine assay is the preferred test for diagnosis of pheochromocytoma. The plasma assay is reserved as an ancillary test when urine metanephrine or catecholamine values are normal (Jiang *et al.*, 1973).

CONCLUSION

HPLC with amperometric detection represents the latest technical development in a long evolution of catecholamine estimation in biological fluids. Among the methods available today, HPLC with electrochemical detection is the simplest and most elegant way of measuring catecholamines. The technique is ideally suited for operation in a clinical diagnostic laboratory. The instrumentation required is relatively inexpensive and easy to maintain. Although the method is quite specific, it does not provide absolute identification. For this reason, we suggest that the HPLC support the daily function of the laboratory and GC—MS be used as a reference or standard method for catecholamine determination.

---

* We are indebted to Dr G. M. Tyce and Dr T. P. Moyer for carrying out this study.

Table 5. Reproducibility of urinary catecholamine assays in 1983

Month	No. of observations	Norepinephrine		Epinephrine		Dopamine	
		Mean ± S.D. (µg/24 h)	CV (%)	Mean ± S.D. (µg/24 h)	CV (%)	Mean ± S.D. (µg/24 h)	CV (%)
January[a]	25	28.1 ± 1.3	5.0	4.0 ± 0.8	20	191 ± 13.6	7.1
February	23	28.7 ± 0.6	2.2	3.7 ± 0.7	18.9	199 ± 10.4	5.2
March[b]	31	21.1 ± 0.9	4.2	4.6 ± 0.4	8.7	169 ± 6.5	3.8
April	26	21.5 ± 0.9	4.2	4.5 ± 0.4	8.7	165 ± 5.6	3.4
May	24	22.2 ± 0.7	3.0	4.3 ± 0.5	11.5	160 ± 5.5	3.4
June	23	21.4 ± 1.4	6.4	4.6 ± 1.0	21	162 ± 6.4	3.9
July	21	21.0 ± 1.1	5.4	4.4 ± 0.6	15	159 ± 6.2	3.9
August	21	20.2 ± 1.5	6.6	4.0 ± 0.8	20	160 ± 5.6	3.5
September[c]	21	37.4 ± 1.5	3.9	17.1 ± 0.6	3.4	247 ± 7.9	3.2
October	20	37.9 ± 1.5	4.0	16.3 ± 1.5	9.0	240 ± 10.7	4.4
November[d]	15	58.8 ± 1.8	3.1	9.6 ± 0.6	6.0	241 ± 7.2	3.0
December	21	57.6 ± 1.0	1.8	9.1 ± 0.5	5.0	232 ± 6.0	2.6

a Established target value: norepinephrine, 28.0 µg/24 h; epinephrine, 3.8 µg/24 h; dopamine, 180 µg/24 h.
b New pool established target value: norepinephrine, 21.5 µg/24 h; epinephrine, 4.5 µg/24 h; dopamine, 165 µg/24 h.
c New pool established target value: norepinephrine, 38.0 µg/24 h; epinephrine, 17.5 µg/24 h; dopamine, 248 µg/24 h.
d New pool established target value: norepinephrine, 58.2 µg/24 h; epinephrine, 9.5 µg/24 h; dopamine, 240 µg/24 h.

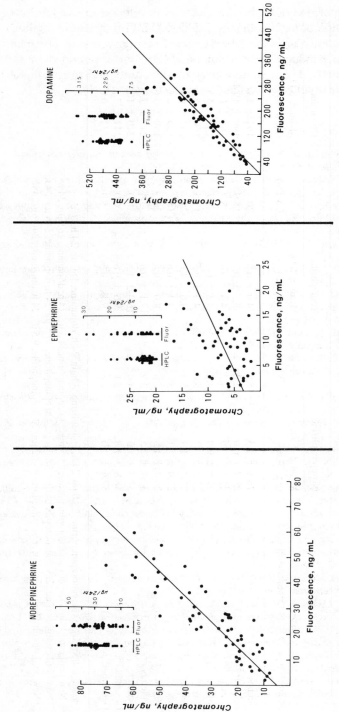

Figure 10. Comparisons of urinary free catecholamines norepinephrine by THI fluorescence assay and HPLC with amperometric detection. Insets show normal values obtained by the two methods. Left, norepinephrine: $y = 4.9 \pm 0.98\ x\ (r = 0.90)$. Center, epinephrine: $y = 3.5 \pm 0.45\ x\ (r = 0.54)$. Right, dopamine: $y = 1.9 \pm 0.95\ x\ (r = 0.91)$.

In this chapter, we describe the methods developed in our laboratory for the determination of catecholamines in blood and urine. These methods have been tested in our diagnostic laboratory for several years. In our hands, the HPLC method produces results comparable to those obtained by the THI fluorescence method. The method has good stability and precision, and it is easily adaptable to other biological fluids and tissue extracts.

REFERENCES

Abel, J. J., and Crawford, A. C. (1897). On the blood-pressure-raising constituent of the suprarenal capsule. *Johns Hopkins Hosp. Bull. 8*, 151—157.

Änggård, E., and Sedvall, G. (1969). Gas chromatography of catecholamine metabolites using electron capture detection and mass spectrometry. *Anal. Chem. 41*, 1250—1256.

Anton, A. H., and Sayre, D. F. (1962). A study of the factors affecting the aluminum oxide—trihydroxyindole procedure for the anaysis of catecholamines. *J. Pharmacol. Exp. Ther. 138*, 360—375.

Armin, J., and Grant, R. T. (1955). Vasoconstrictor activity in the rabbit's blood and plasma. *J. Physiol. 128*, 511—540.

Axelrod, J. (1962). Purification and properties of phenylethanolamine-*N*-methyl transferase. *J. Biol. Chem. 237*, 1657—1660.

Benson, J. R., and Hare, P. E. (1975). *o*-Phthalaldehyde: fluorogenic detection of primary amines in the picomole range; comparison with fluorescamine and ninhydrin. *Proc. Natl. Acad. Sci. USA 72*, 619—622.

Bergström, S., and Hansson, G. (1951). The use of Amberlite IRC-50 for the purification of adrenaline and histamine. *Acta Physiol. Scand. 22*, 87—92.

Berod, A., Bois, F., Chermette, H., Pujol, J. F., and Blond, P. (1978). Radiochemical micro-determination of catecholamines. I. Oxidation with labelled iodine. *Mikrochim. Acta 2*, 213—220.

Blaschko, H. (1939). The specific action of L-dopa decarboxylase (abstract). *J. Physiol. 96*, 50P—51P.

Bloor, W. R., and Bullen S. S. (1941). The determination of adrenaline in blood. *J. Biol. Chem. 138*, 727—739.

Bois, F., Berod, A., Chermette, H., Pujol, J. F., and Blond, P. (1978). Radiochemical micro-determination of catecholamines. II. Oxidation with labelled potassium dichromate. *Mikrochim. Acta 2*, 221—227.

Cotzias, G. C., Papavasiliou, P. S., Fehling, C., Kaufman, B. and Mena, I. (1970). Similarities between neurologic effects of L-dopa and of apomorphine. *N. Engl. J. Med. 282*, 31—33.

Coulombe, P., Dussault, J. H., and Walker, P. (1976). Plasma catecholamine concentrations in hyperthyroidism and hypothyroidism. *Metabolism 25*, 973—979.

Crombeen, J. P., Kraak, J. C., and Poppe, H. (1978). Reversed-phase systems for the analysis of catecholamines and related compounds by high-performance liquid chromatography. *J. Chromatogr. 167*, 219—230.

Crout, J. R. (1959). Some spectrophotofluorimetric observations on blood and urine catechol-amine assays. *Pharmacol. Rev. 11*, 296—299.

de Champlain, J., Farley, L., Cousineau, D., and van Ameringen, M.-R. (1976). Circulating catecholamine levels in human and experimental hypertension. *Circ. Res. 38*, 109—114.

Donike, M. (1975). *N*-Trifluoracetyl-*O*-trimethylsilylphenolalkylamine: Darstellung und mas-senspezifischer gaschromatographischer Nachweis. *J. Chromatogr. 103*, 91—112.

Dorris, R. L., and Shore, P. A. (1971). Amine uptake and storage mechanisms in the *corpus striatum* of rat and rabbit. *J. Pharmacol. Exp. Ther. 179*, 15—19.

Ehrhardt, J.-D., and Schwartz, J. (1978). A gas chromatography—mass spectrometry assay of human plasma catecholamines. *Clin. Chim. Acta 88*, 71—79.

Engelman, K., Portnoy, B., and Lovenberg, W. (1968). A sensitive and specific double-isotope derivative method for the determination of catecholamines in biological specimens. *Am. J. Med. Sci. 255*, 259—268.

Engelman, K., and Portnoy, B. (1970). A sensitive double-isotope derivative assay for norepinephrine and epinephrine: normal resting human plasma levels. *Circ. Res. 26*, 53—57.

Freed, C. R., and Asmus, P. A. (1979). Brain tissue and plasma assay of L-dopa and α-methyldopa metabolites by high performance liquid chromatography with electrochemical detection. *J. Neurochem. 32*, 163—168.

Gaddum, J. H. (1959). Bioassay procedures. *Pharmacol. Rev. 11*, 241—249.

Gaddum, J. H., and Lembeck, F. (1949). The assay of substances from the adrenal medulla. *Br. J. Pharmcol. Chemother. 4*, 401—408.

Gelpi, E., Peralta, E., and Segura, J. (1974). Gas chromatography—mass spectrometry of catecholamines and tryptamines: determination of gas chromatographic profiles of the amines, their precursors and their metabolites. *J. Chromatogr. Sci. 12*, 701—709.

Gitlow, S. E., Pertsemlidis, D., and Bertani, L. M. (1971). Management of patients with pheochromocytoma. *Am. Heart J. 82*, 557—567.

Hallman, H., Farnebo, L.-O., Hamberger, B., and Jonsson, G. (1978). A sensitive method for the determination of plasma catecholamines using liquid chromatography with electro-chemical detection. *Life Sci. 23*, 1049—1052.

Hamaji, M., and Seki, T. (1979). Estimation of catecholamines in human plasma by ion-exchange chromatography coupled with fluorimetry. *J. Chromatogr. 163*, 329—336.

Henry, D. P., Starman, B. J., Johnson, D. G., and Williams, R. H. (1975). A sensitive radioenzymatic assay for norepinephrine in tissues and plasma. *Life Sci. 16*, 375—384.

Higa, S., Suzuki, T., Hayashi, A., Tsuge, I., and Yamamura, Y. (1977). Isolation of catechol-amines in biological fluids by boric acid gel. *Anal. Biochem. 77*, 18—24.

Hjemdahl, P. (1984). Inter-laboratory comparison of plasma catecholamine determinations using several different assays. *Acta Physiol. Scand. Suppl. 527*, 43—54.

Holly, J. M. P., and Makin, H. L. J. (1983). The estimation of catecholamines in human plasma. *Anal. Biochem. 128*, 257—274.

Holzbauer, M., and Vogt, M. (1956). Depression by reserpine of the noradrenaline concentra-tion in the hypothalamus of the cat. *J. Neurochem. 1*, 8—11.

Horning, E. C., Horning, M. G., Vanden Heuvel, W. J. A., Knox, K. L., Holmstedt, B., and Brooks, C. J. W. (1964). Part II. Tryptamine-related amines and catecholamines. *Anal. Chem. 36*, 1546—1549.

Imai, K., Wang, M.-T., Yoshiue, S., and Tamura, Z. (1973). Determination of catecholamines in the plasma of patients with essential hypertension and of normal persons. *Clin. Chim. Acta 43*, 145—149.

Jiang, N.-S., Machacek, D., and Wadel, O. P. (1976). Further study on the two-column plasma catecholamine assay. *Mayo Clin. Proc. 51*, 112—116.

Jiang, N.-S., Stoffer, S. S., Pikler, G. M., Wadel, O., and Sheps, S. G. (1973). Laboratory and clinical observations with a two-column plasma catecholamine assay. *Mayo Clin. Proc. 48*, 47—49.

Karoum, F., Cattabeni, F., Costa, E., Ruthven, C. R. J., and Sandler, M. (1972). Gas chromatographic assay of picomole concentrations of biogenic amines. *Anal. Biochem. 47*, 550—561.

Kawai, S., and Tamura, Z. (1968). Gas chromatography of catecholamines as their trifluoro-acetates. *Chem. Pharm. Bull. (Tokyo) 16*, 699—701.

Kemula, W. (1952). Chromato-polarographic studies I. General considerations and descrip-tion of the set of apparatus. *Rocz. Chem. 26*, 281.

Kissinger, P. T. (1977). Amperometric and coulometric detectors for high-performance liquid chromatography. *Anal. Chem. 49*, 447A—456A.

Kissinger, P. T., Refshauge, C., Dreiling, R., and Adams, R. N. (1973). An electrochemical detector for liquid chromatography with picogram sensitivity. *Anal. Lett. 6*, 465—477.

Krstulović, A. M. (1982). Investigations of catecholamine metabolism using high-performance liquid chromatography: analytical methodology and clinical applications. *J. Chromatogr.* *229*, 1—34.

Lindeberg, E. G. G. (1976). Use of *o*-phthalaldehyde for detection of amino acids and peptides on thin-layer chromatograms. *J. Chromatogr. 117*, 439—441.

Martin, L. E., and Harrison, C. (1968). An automated method for determination of noradrenaline and adrenaline in tissues and biological fluids. *Anal. Biochem. 23*, 529—545.

Mefford, I. N. (1981). Application of high performance liquid chromatography with electrochemical detection to neurochemical analysis: measurement of catecholamines, serotonin and metabolites in rat brain. *J. Neurosci. Methods 3*, 207—224.

Merrills, R. J. (1963). A semiautomatic method for determination of catecholamines. *Anal. Biochem. 6*, 272—282.

Moffat, A. C., Horning, E. C., Matin, S. B., and Rowland, M. (1972). Perfluorobenzene derivatives as derivatising agents for the gas chromatography of primary and secondary amines using electron capture detection. *J. Chromatogr. 66*, 255—260.

Molnár, I., and Horváth, C. (1979). Catecholamines and related compounds: effect of substituents on retention in reversed-phase chromatography. *J. Chromatogr. 145*, 371—381.

Mori, K., Saito, S., Tak, Y., Suzuki, H., and Sato, S. (1976). Determination of urinary catecholamines by a completely automated high speed liquid chromatography system (abstract). *Nippon Naibunpi Gakkai Zasshi 52*, 523.

Moyer, T. P., and Jiang, N.-S. (1978). Optimized isocratic conditions for analysis of catecholamines by high-performance reversed-phase paired-ion chromatography with amperometric detection. *J. Chromatogr. 153*, 365—372.

Moyer, T. P., Jiang, N.-S., and Machacek, D. (1979a). *Biological and Biochemical Applications of Liquid Chromatography*, Vol. II. Marcel Dekker, New York, pp. 75—92.

Moyer, T. P., Jiang, N.-S., Tyce, G. M., and Sheps, S. G. (1979b). Analysis for urinary catecholamines by liquid chromatography with amperometric detection: methodology and clinical interpretation of results. *Clin. Chem. 25*, 256—263.

Musacchio, J. M., and Goldstein, M. (1963). Biosynthesis of norepinephrine and norsynephrine in the perfused rabbit heart. *Biochem. Pharmacol. 12*, 1061—1063.

Natelson, S., Lugovoy, J. K., and Pincus, J. B. (1949). A new fluorometric method for the determination of epinephrine (letter to the editor). *Arch. Biochem. 23*, 157—158.

Oliver, G., and Schäfer, E. A. (1895). The physiological effeects of extracts of the suprarenal capsules. *J. Physiol. 18*, 230—276.

Persson, B.-A., and Karger, B. L. (1974). High performance ion pair partition chromatography: the separation of biogenic amines and their metabolites. *J. Chromatogr. Sci. 12*, 521—528.

Peuler, J. D., and Johnson, G. A. (1977). Simultaneous single isotope radioenzymatic assay of plasma norepinephrine, epinephrine and dopamine. *Life Sci. 21*, 625—636.

Porath, J. (1974). General methods and coupling procedures. *Methods Enzymol. 34*, 13—30.

Price, H. L., Linde, H. W., and Price, M. L. (1960). Failure of ethylenediamine condensation method to detect increased plasma norepinephrine concentrations during general anesthesia in man. *Clin. Pharmacol. Ther. 1*, 298—302.

Raum, W. J., and Swerdloff, R. S. (1981). A radioimmunoassay for epinephrine and norepinephrine in tissues and plasma. *Life Sci. 28*, 2819—2827.

Refshauge, C., Kissinger, P. T., Dreiling, R., Blank, L., Freeman, R., and Adams, R. N. (1974). New high performance liquid chromatographic analysis of brain catecholamines. *Life Sci. 14*, 311—322.

Rentzhog, L. (1972). Double isotope dilution derivative technique for measurement of catecholamines. *Acta Physiol. Scand. Suppl. 377*, 1—101.

Renzini, V., Brunori, C. A., and Valori, C. (1970). A sensitive and specific fluorimetric method for the determination of noradrenalin and adrenalin in human plasma. *Clin. Chim. Acta 30*, 587—594.

Riggin, R. M., and Kissinger, P. T. (1977). Determination of catecholamines in urine by reverse-phase liquid chromatography with electrochemical detection (letter to the editor). *Anal. Chem. 49*, 2109—2111.

Samejima, K., Dairman, W., Stone, J., and Udenfriend, S. (1971). Condensation of ninhydrin with aldehydes and primary amines to yield highly fluorescent ternary products. II. Application to the detection and assay of peptides, amino acids, amines, and amino sugars. *Anal. Biochem. 42*, 237—247.

Sjoerdsma, A., Engelman, K., Waldmann, T. A., Cooperman, L. H., and Hammond, W. G. (1966). Pheochromocytoma: current concepts of diagnosis and treatment: combined clinical staff conference at the National Institutes of Health. *Ann. Intern. Med. 65*, 1302—1326.

Stoffer, S. S., Jiang, N.-S., Gorman, C. A., and Pikler, G. M. (1973). Plasma catecholamines in hypothyroidism and hyperthyroidism. *J. Clin. Endocrinol. Metab. 36*, 587—589.

Thiede, H. M., and Kehr, W. (1981). Conjoint radioenzymatic measurement of catecholamines, their catechol metabolites and DOPA in biological samples. *Naunyn Schmiedebergs Arch. Pharmacol. 318*, 19—28.

Udenfriend, S. (1962). *Fluorescence Assay in Biology and Medicine*, Vol. 1. Academic Press, New York, p. 141.

Udenfriend, S., Stein, S., Böhlen, P., Dairman, W., Leimgruber, W., and Weigele, M. (1972). Fluorescamine: a reagent for assay of amino acids, peptides, proteins, and primary amines in the picomole range. *Science 178*, 871—872.

Udenfriend, S., and Wyngaarden, J. B. (1956). Precursors of adrenal epinephrine and norepinephrine *in vivo. Biochim. Biophys. Acta 20*, 48—52.

Udenfriend, S., and Zaltzman-Nirenberg, P. (1963). Norepinephrine and 3,4-dihydroxyphenethylamine turnover in guinea pig brain *in vivo. Science 142*, 394—396.

Valori, C., Brunori, C. A., Renzini, V., and Corea, L. (1970). Improved procedure for formation of epinephrine and norepinephrine fluorophors by the trihydroxyindole reaction. *Anal. Biochem. 33*, 158—167.

Valori, C., Renzini, V., Brunori, C. A., Porcellati, C., and Corea, L. (1969). An improved procedure for separation of catecholamines from plasma. *Ital. J. Biochem. 18*, 394—405.

Viktora, J. K., Baukal, A., and Wolff, F. W. (1968). New automated fluorometric methods for estimation of small amounts of adrenaline and noradrenaline. *Anal. Biochem. 23*, 513—528.

Von Euler, U. S. (1946). A specific sympathomimetic ergone in adrenergic nerve fibres (sympathin) and its relations to adrenaline and nor-adrenaline. *Acta Physiol. Scand. 12*, 73—97.

Wang, M.-T., Imai, K., Yoshioka, M., and Tamura, Z. (1975). Gas—liquid chromatographic and mass fragmentographic determination of catecholamines in human plasma. *Clin. Chim. Acta 63*, 13—19.

Warsh, J. J., Chiu, A., Li, P. P., and Godse, D. D. (1980). Comparison of liquid chromatography—electrochemical and gas chromatography—mass spectrometry methods for brain dopamine and serotonin. *J. Chromatogr. 183*, 483—486.

Weigele, M., DeBernardo, S. L., Tengi, J. P., and Leimgruber, W. (1972). A novel reagent for the fluorometric assay of primary amines (letter to the editor). *J. Am. Chem. Soc. 94*, 5927—5928.

Weil-Malherbe, H. (1961). II. The fluorimetric estimation of catecholamines. *Methods Med. Res. 9*, 130—146.

Wittmer, D. P., Nuessle, N. O., and Haney, W. G. Jr. (1975). Simultaneous analysis of tartrazine and its intermediates by reversed phase liquid chromatography. *Anal. Chem. 47*, 1422—1423.

Wood, W. G., and Mainwaring-Burton, R. W. (1975). The development and evaluation of a semi-automated assay for catecholamines suitable for plasma and urine. *Clin. Chim. Acta 61*, 297—308.

Yamatodani, A., and Wada, H. (1981). Automated analysis for plasma epinephrine and norepinephrine by liquid chromatography, including a sample cleanup procedure. *Clin. Chem. 27*, 1983—1987.

Yui, Y., Fujita, T., Yamamoto, T., Itokawa, Y., and Kawai, C. (1980). Liquid-chromatographic determination of norepinephrine and epinephrine in human plasma. *Clin. Chem. 26*, 194—196.

Zambotti, F., Blau, K., King, G. S., Campbell, S., and Sandler, M. (1975). Monoamine metabolites and related compounds in human amniotic fluid: assay by gas chromatography and gas chromatography—mass spectrometry. *Clin. Chim. Acta 61*, 247—256.

Ziegler, M. G., Lake, C. R., and Kopin, I. J. (1977). The sympathetic-nervous-system defect in primary orthostatic hypotension. *N. Engl. J. Med. 296*, 293—297.

Progress in HPLC, Vol. 2, pp. 427—466
Parvez *et al.* (Eds)
© 1987 VNU Science Press

Recent advances in electrochemical detector for highly selective and sensitive analysis of biogenic compounds

YUTAKA HASHIMOTO and YUJI MARUYAMA

Research Laboratories, Pharmaceutical Division, Nippon Kayaku Co., 3-31-12 Shimo, Kitaku, Tokyo 115, Japan

INTRODUCTION

Developments in mass spectrometry have been advanced through its connection with gas chromatography (GC—MS) which is a highly selective and sensitive analytical technique (Hashimoto and Miyazaki, 1979; Waller, 1971). High performance liquid chromatography (HPLC) has been widely used as a simple and inexpensive analytical technique, which also allows determination of low volatile compounds which are difficult to analyze by GC—MS. Recently, fast atom bombardment—mass spectrometry (FAB—MS) has appeared (Barber *et al.*, 1981) and the mass spectra of high molecular weight compounds, such as human proinsuline (mol. wt 9388.6), can be obtained by this technique. Thus development of a more sensitive and selective detector of HPLC comparable to GC—MS is desirable. The enhancement of the detection limit in HPLC is primary in selecting sensitive methods of detection. Another consideration must be simplifying the whole HPLC system, especially using a microcolumn (less than 1 mm ID). The electrochemical detector (ECD) is one of the most highly sensitive detectors (Kissinger *et al.*, 1979a, b; Maruyama and Kusaka, 1978). This type of detector has recently become commercially available.

In this chapter we will describe several different methods used for the enhancement of sensitivity and selectivity in ECD. One aspect of this development has been hardware (cell design, amperometry or coulometry, dual electrodes, microelectrodes, etc.) and another aspect has been software (derivatization, combination of enzyme reaction). We will also examine possible future developments of ECD.

DESIGN AND FUNCTION OF THE ELECTROCHEMICAL DETECTOR

Fundamental aspects of electrochemical cell design have been discussed in several recent reviews (Stulik and Pacakova, 1981; Kissinger *et al.*, 1979a, b;

Weber and Purdy, 1978; Kissinger, 1977). ECD could be divided into two categories according to cell design and electrolytic efficiencies: amperometric and coulometric. There are no fundamental differences in the principles of current change associated with electrochemical redox reactions. A schematic diagram of the electrochemical cell in an ECD is presented in Fig. 1. This is a typical, thin-layer type amperometric detector. The electrodes usually consist of three types: working, reference and auxillary. In the thin-layer type cell, two plastic blocks are tightly pressed around a slotted Teflon gasket (50 μm thick) to form a rectangular channel. The working electrode(s) is (are) embedded in the wall of one block along this channel. The dead volume of such a cell is typically between 0.1 and 1 μl, and therefore, the chromatographic profile is not disturbed even for the most efficient commercially available HPLC. The surface area of the working electrode is relatively small (-0.5 cm) and their electrolytic efficiency (= coulometric yield) is low, usually in the range of 3—30% (a coulometric detector converts 100%).

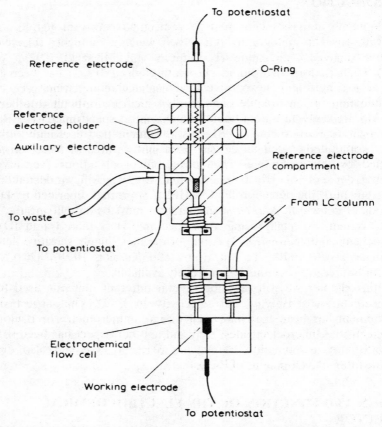

Figure 1. Structure of an electrochemical detector flow cell. (Reprinted from LC 4 amperometric controller operations/maintenance manual, Bioanalytical Systems Inc., West Lafayette, Indiana).

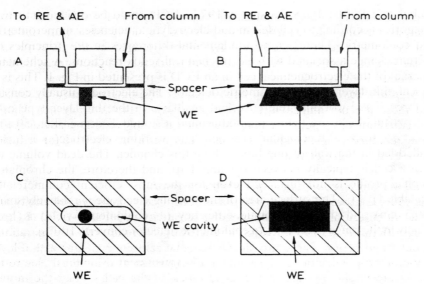

Figure 2. Cross section of (A) and (C) an amperometric detector cell and (B) and (D) a coulometric detector cell. AE = auxiliary electrode, RE = reference electrode, WE = working electrode.

In contrast to their low electrolytic efficiencies, compensation of background current, which flows steadily through the circuit, is feasible because the background current is very low. As the surface area is small, it is possible to place the working electrode where the effluent flow is continuous and steady. In a coulometric detector, the large surface area of the electrode is necessary to enhance a coulometric yield. Cross sections (vertical and horizontal) of representative amperometric and coulometric detector are provided in Fig. 2. Electrochemically active components separated by HPLC continuously enter the EC cell, are oxidized or reduced on the surface of the working electrodes, and are excreted out of the cell after contact (direct or electrical) with the reference and auxillary electrodes.

The most commonly used working electrode materials for HPLC are glassy carbon, carbon paste and mercury/gold. Glassy carbon and carbon paste make the best electrodes. Besides being chemically resistant and highly conductive, glassy carbon is also mechanically rigid. This glass-like material can be brought to a high luster by metallographic polishing procedures, and then can be firmly embedded in a plastic block to form a detector cell. Because of their mechanical ruggedness and chemical resistance, the glassy carbon cells can be used under high fluid velocity and/or totally non-aqueous solvent conditions. Due to their excellent mechanical stability, these cells are ideally suited for routine determination where carbon paste is not required. The sensitivity of glassy carbon is approximately comparable to that of carbon paste electrodes and there is generally less noise on glassy carbon. The paste, prepared by kneading carbon powder with an oil such as liquid parafin or silicone grease, is highly sensitive and has a low background current when carefully prepared, packed polished and conditioned before analysis. Unfortunately, carbon paste electrodes quickly deteriorate and cannot be applied for extended use. Air bubbles lodged in the detector cell, cause sudden changes in the flow-rate and other related phenomena damage the electrode.

Mercury has conventionally been used as dropping mercury electrodes, but these are not amenable to the dead volume thin-layer design. Thus a mercury film on a polished gold substance is employed. The advantage of using mercury as an electrode material is that it is more effective for reducing noise than carbon electrodes.

Dual working electrodes in amperometric detector
Recently, dual working electrodes have been used to improve selectivity and peak identification, and to enhance the sensitivity in HPLC—ECD (LCEC) system (Krull *et al.*, 1983; Roston *et al.*, 1982). There are two configurations; parallel and series. These are shown schematically in Fig. 3 as a thin-layer amperometric detector. In the parallel arrangement, two identical electrodes are arranged side-by-side in the same block of the cell so that the mobile phase meets both sensors simultaneously. The series configuration places one electrode upstream of the other.

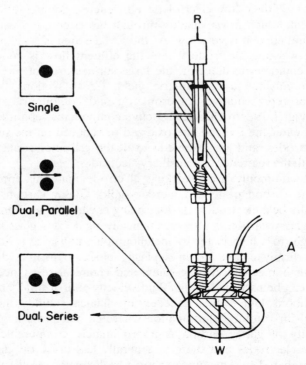

Figure 3. Single, dual parallel and dual series electrodes in the standard thin-layer transducer. R = reference electrode, W = working electrode, A = amperometric detector. From Shoup (1982).

Shoup (1982) conveniently compared the functions of these detectors in terms of well-known optical techniques.

Parallel multiple electrodes are analogous to dual wavelength ultraviolet absorbance detectors. In both cases, the LC eluent is monitored at two different energies — whether optical or redox — and simultaneous chromatograms may be obtained at each wavelength or potential. For example, in LC—UV, it is now commonplace to monitor at 254 and 280 nm and to use the ratio of the two absorbance values during the elution of a peak to ascertain peak purity. These schema are shown in Fig. 4.

The series electrode is analogous to fluorescence, since series electrochemistry is a similar two-step process consisting of both an excitation and a reversion reaction. With series detection the eluant is modified by the upstream detector (the excitation step in fluorescence) so that a more useful chromatogram (the emission spectrum in fluorescence) can be recorded at the downstream detector. These schema are shown in Fig. 5.

Kissinger (1984) summarizes the following useful applications in the parallel mode.

1. The ratio of current monitored at each electrode can provide confirmation of peak identity and purity.

2. Oxidations and reductions can be carried out simultaneously. This saves time and enhances selectivity. This can be ideal for compounds in several different redox states.

3. Signals from low and high potential reactions can be recorded simultaneously, providing both greater selectivity and wider applicability in a single experiment.

4. A different signal can be plotted to substract out 'common mode' information while enhancing detection of the desired compound.

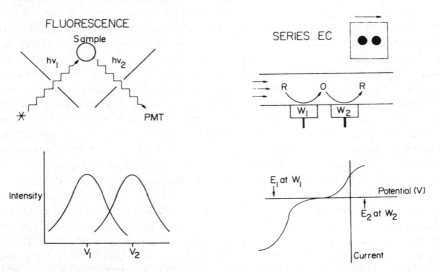

Figure 4. Analogies between parallel dual electrodes and dual wavelength UV detection in liquid chromatography. R = reduced form, O = oxidized form. From Shoup (1982).

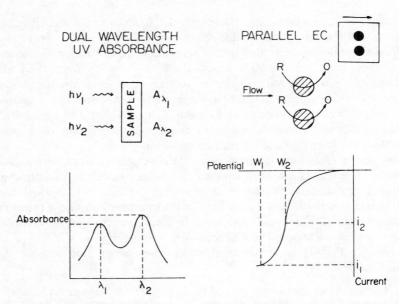

Figure 5. Analogies between series dual electrodes and fluorescence detection in HPLC. Left: Fluorescence (PMT = photomultiple tube); Right: Series EC W_1 and W_2 = working electrodes, R = reduced form, O = oxidized form. From Shoup (1982).

Applications of parallel dual electrode are shown as follows. Peak identification by the ratio of two electrodes' currents (Lunte and Kissinger, 1983; Roston and Kissinger, 1983; Shoup and Mayer, 1982). The electrodes are placed side-by-side and operated at two different potentials along the hydrodynamic voltammograms of the solutes of interest. Simultaneous chromatograms are generated, and the ratios of the peak currents (i_2/i_1) may be calculated to assess peak purity in real extracts by comparison to standard injections. Mayer and Shoup (1982) used this mode for improved selectivity of biogenic amines and metabolites. For these compounds, the potentials were +800 (W_1) and +700 mV (W_2) vs Ag/AgCl/3 M NaCl. Two chromatograms were obtained simultaneously (Fig. 6) and peak height ratios were calculated (see inset of Fig. 6). Samples were injected in the same manner and the ratios for each peak in the sample were then compared to those obtained when callibrating the standard. Table 1 shows good agreement between the values, event at fairly low tissue homogenate concentrations, thereby confirming peak identities by two independent parameters: chromatographic retention times and the ratios of two electrodes' currents.

Another interesting approach to the enhancement of the coulometric yield is the application of redox couples by a parallel-opposed dual-electrode detection. This will be described in *Sensitivity and selectivity considerations.* In the series mode, Kissinger and other scientists (Allison *et al.*, 1983; Bratin *et al.*, 1981; Eggli and Asper, 1978; Goto *et al.*, 1981, 1982; King and Kissinger, 1980; Lunte and Kissinger, 1984; MacCrehan and Durst, 1981;

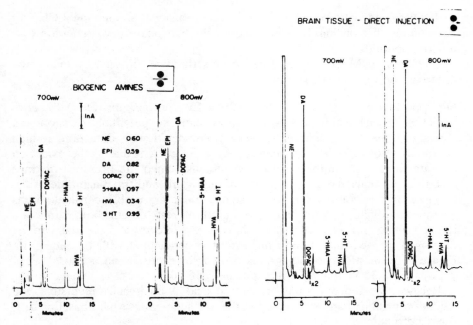

BRAIN TISSUE - DIRECT INJECTION

Figure 6. Left: dual parallel chromatograms of standard at +700 and +800 mV. Ratios of the current at 700 mV to the current at 800 mV are calculated. Right: chromatogram of brain tissue homogenate under the same conditions. NE = norepinephrine; Epi = epinephrine; DA = dopamine; DOPAC = 3,4-dihydroxyphenylacetic acid; 5-HIAA = 5-hydroxyindoleacetic acid; HVA = homovanillic acid; 5-HT = 5-hydroxytryptamine. From Mayer and Shoup (1982).

Table 1. Comparison of ratios for parallel outputs at 700 and 800 mV for both standards and crude homogenate injections (Mayer and Shoup, 1982)

Compounds	NE	Epi	DOPAC	DA	5-HIAA	HVA	5-HT
Standard*	0.60	0.59	0.82	0.87	0.97	0.34	0.95
Sample*	0.59	—	0.83	0.92	0.92	0.33	0.94
Concentration (ng/ml)	38.6	—	33	2.7	6.5	4.7	7.5

* 700 mV/800 mV.

Mayer and Shoup, 1982; Roston and Kissinger, 1982; Schieffer, 1980, 1981) also describe the following common applications.

1. The ratio of currents monitored at each electrode can provide confirmation of peak identity and purity.

2. Selectivity is enhanced at the downstream electrode because compounds with chemically irreversible reactions upstream are discriminated against.

3. The upstream electrode can derivatize compounds to enhance detectability at the downstream electrode. Overall selectivity and detection limits can be greatly improved.

4. Dissolved oxygen can be discriminated against, simplifying LCEC of compounds that ordinarily would require mobile phase deoxygenation (e.g. nitro compounds).

5. 'Common Mode' currents can be discriminated against by taking the difference between the two signals.

One problem of baseline drift in which LCEC can be improved with series detection is high gain detection at more extreme potentials (Roston and Kissinger, 1982). An inherent problem associated with detection at high potentials (1.0 V) is the significant background current due to the oxidation of water and other mobile phase constituents. When the presence of trace quantities of analyte requires the use of high gains and the magnitude of the baseline drift is comparable to the signal, obtaining usable chromatograms becomes a difficult task. If the compound of interest contains both forms of a redox couple, for example, then downstream electrode potential can be employed at low, thus minimizing problems associated with baseline drift. Figure 7 details the chromatograms of 2 ng of gentisic acid achieved when W_1 was at +1.35 V, and W_2 was maintained at 0.20 V.

Because the redox couple of gentisic acid is chemically reversible, the oxidation product can be detected at the lower potential where baseline drift is not a problem.

It is also possible to use the upstream electrode as a 'generator' or 'reactor' electrode. In essence, a thin-layer cell acts as a very low dead volume post-column reactor. Using this mode, Allison *et al.* (1983) reported a simultaneous assay for thiols and disulfides. They used a dual Hg/Au electrode thin layer cell to perform both the reduction and detection functions. Two mercury/gold electrodes were utilized in a series arrangement (as in Fig. 3) with reduction of disulfide to thiol at the upstream (generator) electrode, followed by conventional thiol detection at the downstream (detector). The upstream electrode behaves as a kind of on-line post-column reactor (Fig. 8).

Detection of cysteine (CSH), glutathione (GSH) and the pure disulfides of these molecules (CSSC and GSSG) is demonstrated by Fig. 9. Both the thiols and the disulfide are readily quantitated in this standard solution. It is interesting to recall that each disulfide is actually being detected as the corresponding free thiol, e.g. GSSG as GSH and CSSC as CSH. No confusion occurs in measurement, however, because the thiols are chromatographically separated from the disulfides.

Coulometric detector

We have already explained that two types of electrodes, amperometric and coulometric, are used in the HPLC detector. The coulometric electrode has a larger surface area than that of the amperometric electrode because of the increase in coulometric yield. However, when obtaining high electrode efficiency, background current also increases in this mode. Compared to the coulometric detector, the electric efficiency of an amperometric electrode is very low. Background current maintains low levels as well. Bunyagidji and

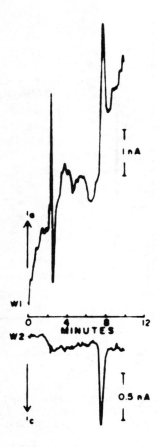

Figure 7. Series dual-electrode chromatogram of gentisic acid. Conditions: 2 ng injected, W_1 = +1.35 V, W_2 = +0.20 V; 25 cm Biophase C_{18} column; flow rate 1 ml/min. From Roston and Kissinger (1982).

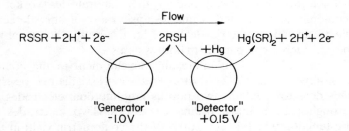

Figure 8. Schematic diagram of series dual Hg/Au detector and reactions which occur at each electrode. (RSH = thiol compounds; RSSR = disulfide forms of thiol compound.) From Allison *et al.* (1982).

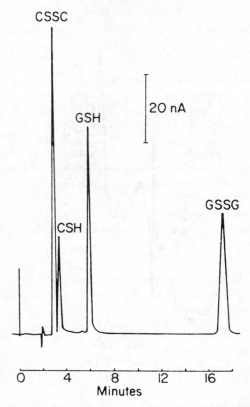

Figure 9. Chromatogram of a standard solution containing cysteine (CSH), glutathione (GSH) and the disulfides CSSC and GSSG. Column: Biophase ODS 5 μm; Mobile phase: 94% 0.5 M monochloroacetate, pH 3.0/6% methanol, 150 mg/l sodium octyl sulfate. From Allison *et al.* (1982).

Girard (1982) have compared coulometric detector with amperometric one for catecholamine analysis, and the resulting sensitivity of the coulometric detector was comparable to the conventional amperometric detector. Kurahashi and Nishino (1984) have also reported that the coulometric detector using a flow through cell made in their laboratories worked at only 80% of coulometric yield and signal-to-noise ratio of this detector was comparable to an amperometric detector. So generally, the signal-to-noise ratio can be considered comparable in both systems.

There has recently been an interesting development in the coulometric detector (Andrews *et al.*, 1982; Maston *et al.*, 1981). This can be shown in Fig. 10. This detector contains two porous graphite dual electrodes' as well as flow-through type in-line working electrodes. Two electrodes have a coulometric function which keeps 100% of the coulometric yield in the flow rate for up to 4 ml/min. As each electrode has this coulometric function, many interesting applications are possible which are not possible with an amperometric detector.

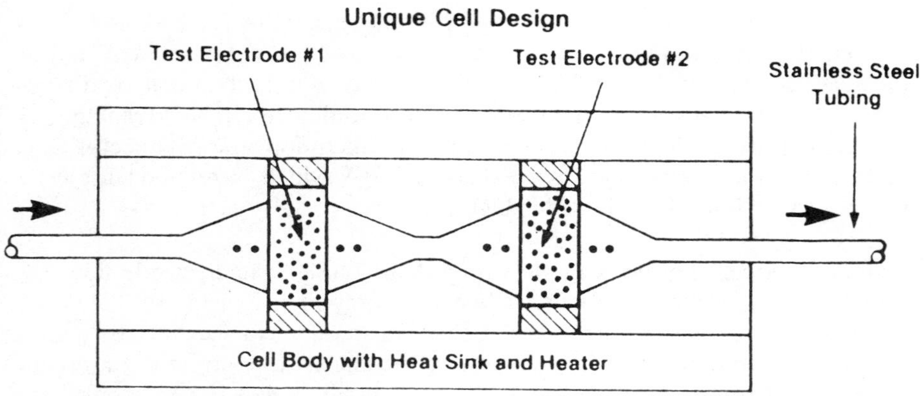

Unique Cell Design

Figure 10. The design of the flow through type coulometric detector (Reprinted from Coulochem 5100A instruction manual of Environmental Sciences Associates, Inc., Bedford, Mass).

Screen mode. In this screen mode the potential of the first (upstream) detector is set at a value which is a few millivolts less than the rising portion of the $C-V$ curve (typically 0.05—0.10 V less than the $E_{1/2}$ for the analyte is adequate). For example, a compound is set in the following mode and potential:

Mode	Detector 1	Detector 2
Oxidation	E_1	E_2
Reduction	E_3	E_4

This will result in the electrolysis of all solutes contained in the mobile phase which have potentials less than that of the analyte. Following elimination of interferences by the first detector, the second detector can be used to detect the analyte at its appropriate potentials. This selective detection can be used as a result of the quantitative (coulometric yield is 100%) oxidation or reduction of a component at each detector electrode. By selection of the proper potentials at each electrode, it is possible to resolve and detect each component. With a single electrode detector, the two compounds (hydroquinone and acetaminophen) are not completely resolved. If the potential of the first detector is set to oxidize hydroquinone at +0.15 V; the second detector is set at a higher potential (+0.50 V) which oxidizes acetaminophen, the two compounds are completely separated by this screen mode.

Octopamine, a monohydroxyphenolamine, has been implicated in certain neurological disorders. By this screen mode octopamine can be detected selectively in the presence of other biogenic amines as shown in Fig. 11 (Martin *et al.*, 1983). This figure shows the detection of norepinephrine (NE), epinephrine (E), dihydroxybenzylamine (DHBA), dopamine (DA), serotonin (5-HT) and tryptamine (TP) at the first detector (set at 0.5 V) whereas tryptamine (TA), octopamine (OA), 5-HT and TP are detected at the second detector (set at 0.75 V). Thus, many compounds that oxidize at the lower potential are screened from the second detector. The advantages of this system are evident by reference to OA which appears at Detector 2 at 5.80 min without any interference from DHBA which is retained for 6.10 min but which is totally oxidized at Detector 1.

Redox mode. The redox mode with a dual coulometric electrode detector permits more selective detection when the components of interest undergo reversible oxidation—reduction reactions. This mode also seems to be a kind of reaction chromatography, which allows the analyst to convert a compound from one chemical form to another, thus further enhancing the sensitivity and/or selectivity of detection. Potentials are set to oxidize all sample components at the first detector and selectively only reduce (or oxidize) only two reversible components at the second detector. Figure 12 is a chromatogram of a mixture of three compounds along with other impurities. The oxidation potential was set at +0.80 V. This figure shows the second detector set to reduce the oxidation product of acetaminophen at −0.50 V. The applied potentials were optimized from the $C-V$ curves shown in this figure. This selectivity is only practical when the first electrode is 100% efficient. Haroon *et al.* (1984) used this redox mode to determine vitamin K compounds. This combination could be possible to reduce the background current. From the result of the hydrodynamic voltammogram for repetitive injection of vitamin K_1 when the detector electrode (Detector 2) potential was held constant at +0.05 V and the generator electrode (Detector 1) potential was varied from −0.1 to −1.7 V, the limiting current plateau was reached at −1.2 V. The optimum conditions chosen for the reduction and oxidation of vitamin K_1 were −1.20 and +0.05 V, respectively. Figure 13 shows the schematic diagram of the reactions which occur at each electrode. An advantage of the redox mode over direct reduction techniques for the measurement of K vitamins is the high gain detection that can be achieved at extreme potentials as in Fig. 14. A comparison can be made between results obtained by direct reduction of K vitamins and those obtained by the dual electrode, redox mode. Because the redox couple for vitamin K is reversible, the reduction products can be detected at a considerably lower potential, where baseline drift is not a problem. Usable chromatograms can easily be obtained.

The pure standard of vitamin K_1 can be detected down to approximately 100 pg (Fig. 15). This represents an improvement of sensitivity at least five times greater than that obtained with the UV photometric detector.

In general, it is very difficult to keep a driftless baseline in a reducing reaction because of the interference of dissolved oxygen and traces of metal ions leached by the eluent from the stainless-steel parts of the chromatograph. The redox mode compensates very well for the high background currents caused by these problems. Thus it was not necessary to remove

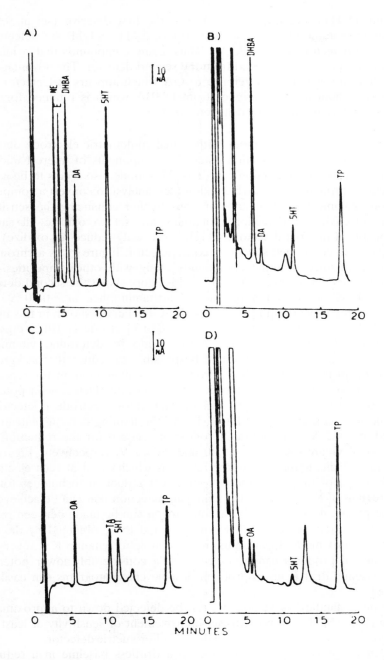

Figure 11. Chromatograms of 1 ng of standards (A) and (C) and cockroach nerve cord extracts (B) and (D) with Detector 1 set at 0.5 V (A) and (B) and Detector 2 at 0.75 V (C) and (D). Flow-rate = 0.8 ml/min. NE = norepinephrine; E = epinephrine; DHBA = 3,4-dihydroxy-benzylamine; DA = dopamine; 5-HT = 5-hydroxytryptamine; TP = tryptamine; OA = octopamine. From Martin *et al.* (1983).

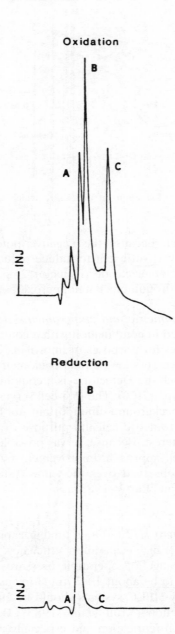

Figure 12. Top is a chromatogram of a mixture of three compounds along with other impurities. The oxidation potential is set at 0.80 V. Bottom shows the second detector set to reduce the oxidation product of acetaminophen at −50 V. A = pyridoxine; B = acetaminophen; C = theophylline. Reprinted from technical report of Environmental Sciences Associates, Inc., Bedford, Mass.

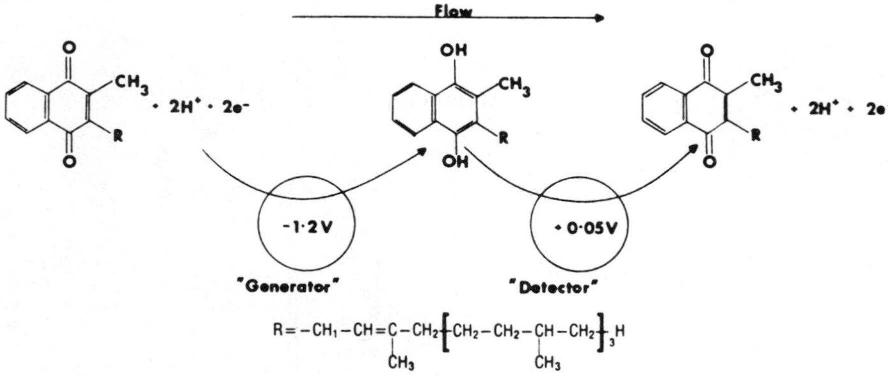

Figure 13. Schematic diagram of reactions of Vitamin K_1 which occur at each electrode. From Haroon *et al.* (1984).

oxygen from the system because the negative potential at the generator electrode (Detector 1) was sufficient to induce reduction of vitamin K_1 and oxygen, while the detector electrode (Detector 2) potential was sufficiently positive to oxidize K_1 hydroquinone but not peroxide or water.

Combination of coulometric and amperometric detector. The dual detectors have been developed in combination with a coulometric and an amperometric electrode in series for purposes of improving detection limits (Hepler and Purdy, 1983). Figure 16 shows a dual coulometric—amperometric flow-through type cell in which the Detector 1 is a coulometric type and Detector 2 is an amperometric type (70%). The first cell serves as a 'scavenger' cell to remove electroactive background unimportant to the analysis, while the second cell is an amperometric cell the purpose of which is to detect and assay the analytes of interest. Because, it was possible to reduce background current using a dual cell approach, improvement by a factor of two in the limit of detection was observed over the same system using only amperometric detector (Hepler and Purdy, 1983).

Other advantages in cell design

Small size cell for micro HPLC. Open tubular capillary columns for liquid chromatography offer a high separation efficiency. The optimum column diameter for a wide range of operation pressures and analysis times is calculated theoretically to be about 1—2 μm (Jorgenson and Guthrie, 1983). Microbore and capillary HPLC systems should be employed to enhance the sensitivity of EC detection (Krull *et al.*, 1983). But certain problems of sample injection, column fabrication and especially detector design must be overcome for the open-tubular capillary HPLC to become a useful and practical technique. The prime requirements of a detector for open-tubular HPLC are low dead volume (requiring essentially 'on-column' detection) and high sensitivity. Knecht *et al.* (1984) designed a simple yet very sensitive on-

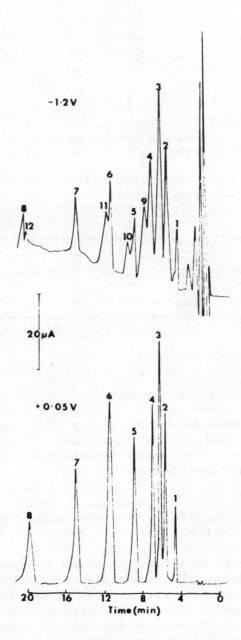

Figure 14. Reductive (top) and oxidative (bottom) detection of vitamin K_2 compounds on Zorbax ODS column. Peaks: 1 and 2 = MK-4 and MK-5; 3 = K_1; 4—8 = MK-6—MK-10; 9—12 = unknowns. MK = monaquinone. From Haroon *et al.* (1984).

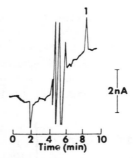

Figure 15. Detection of 115 pg Vitamin K_1 on μ-Bondapak ODS column. Mobile phase = solvent A : dichloromethane : methanol in 10:30:70 and Solvent A = methanol : 14.7 M phosphoric acid : 1.45 M sodium phosphate monobasic (95:9:5). Flow rate = 1 ml/min. From Haroon *et al.* (1984).

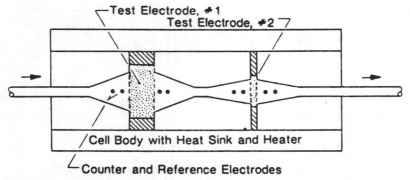

Figure 16. Schematic diagram of Model 5011 electrochemical cell (Environmental Sciences Associates, Bedford, Mass.). Approximate test electrode areas; Detector 1, 5 cm²; Detector 2, 0.6 cm². Approximate cell volume (Detectors 1 and 2), 5 µl. From Haroon *et al.* (1984).

column electrochemical detector for capillary liquid chromatography. Its main feature is an indicator electrode constructed from a single graphite fiber (9 µm diameter, *c.* 0.7 mm length) inserted into the outlet end of the capillary column (15 µm diameter). The response of the graphite fiber detector was more nearly coulometric and detection limits for ascorbic acid, catechol and 4-methylcatechol are on the order of 1 femto mol.

The use of graphite fibers as microvoltammetric electrodes has been reviewed by Wightman (1981).

Small size electrode for in vivo *voltammetry.* The small size electrode technique allows the direct determination of the neurotransmitter in the brain where electrode is inserted (Butcher *et al.*, 1984; Conti *et al.*, 1978; Kissinger *et al.*, 1973). Kissinger *et al.* (1973) proposed the direct assay method using *in vivo* voltammetry. Neurotransmitters and their metabolites such as dopamine, serotonine, 3,4-dihydroxyphenylacetic acid (DOPAC) and 5-hydroxyindoleacetic acid (5-HIAA) in rats' brains can be detected while watching rats' behavior (Suzuki and Taguchi, in press).

Figure 17 shows the design of the microelectrode and Fig. 18 illustrates the electrode implanted in the rat brain. This technique is applicable in behavior pharmacological experiments (Dayton *et al.*, 1981; Gonon *et al.*, 1984). It is difficult, however, to determine selectively one biogenic amine in a rat brain because this technique does not include a chromatographic separation step. Enhancing the selectivity of the electrode is a future problem.

Wall-jet cell. Hanekamp and de Jong (1982) reported the best design for achieving high sensitivity at solid and polarographic electrodes, which has the fluid stream impinging normally to the electrode surface. The most obvious application of this principle occurs in EC detectors of the so-called 'wall-jet' type. Gunasingham *et al.* (1984) designed a large volume wall-jet type cell and the cell was applied to the detection in normal-phase HPLC. Figure 19 shows the schematic diagram of the large-volume wall-jet cell. This cell has a capacity of more than 35 ml. This can be compared to the wall-jet cell used in the earlier work which had a cell volume of less than 100 μl. The chromatographic separation of estron, estradiol and estriol in a normal-phase HPLC mode has been applied. The mobile phase was composed of 80% hexane and 20% ethanol (consisting 0.01 M $(C_4H_9)_4BF_4$ and 0.5% ammonia solution). With these non-aqueous systems low background currents can be expected. The chromatograms obtained from the wall-jet detector and

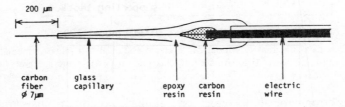

Figure 17. Schematic diagram of carbon fiber electrode for *in vivo* voltammetry. From Suzuki and Taguchi (in press).

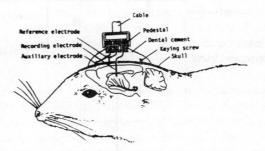

Figure 18. Placement of carbon fiber recording electrode, Ag/AgCl reference electrode and silver wire auxiliary electrode placed on the dura surface. Electrical connection from these electrodes was made via pins in the pedestal to the electronic equipment. (Reprinted from technical report of Bioanalytical Systems Inc. West Lafayette, Indiana).

fluorescence detector can be compared to the slight decrease in resolution caused by spreading which occurs as the sample passes through the fluorescence detector.

Furthermore, they provide an illustration of the analysis of all three estrogens without supporting electrolyte in the mobile phase. Conductance in the wall-jet cell was maintained by an external solvent delivery system which delivered 0.05 M $(C_4H_9)_4BF_4$ at a flow rate of 1 ml/min. Figure 20 shows a block diagram of a post-column addition set-up (Gunasingham *et al.*, 1984). They also described that this system enabled detection in gradient HPLC where a significant part of the gradient run was completely nonaqueous.

In spite of these efforts, the sensitivity could not be increased in the wall-jet type cells. Thus, the following comment has appeared, 'We are not yet convinced that the energy dissipated in producing these detectors results in a significant improvement over the simplest of thin-layer detectors with laminar flow' (Johnson *et al.*, 1984).

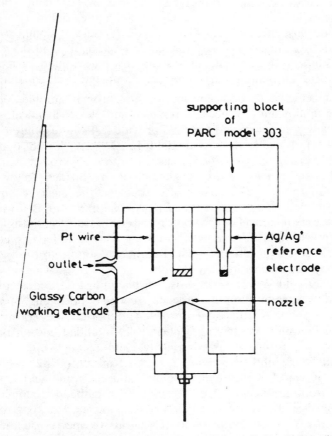

Figure 19. Schematic diagram of the large-volume wall-jet cell. From Gunasingham *et al.* (1984).

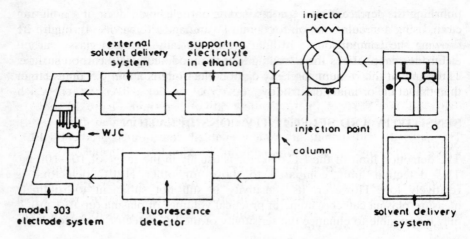

Figure 20. Schematic diagram of post-column set-up using large-volume wall-jet cell. WJC = wall-jet cell. From Gunasingham *et al.* (1984).

Surface modified electrode for enhancement of sensitivity. There have been many efforts regarding surface modification of electrodes in order to enhance the reversibility of electrochemical redox reaction by permitting the lowering of the detector operating potential. These modifications result in high sensitivity and specificity. To be useful for routine chromatographic applications, the surface modification, especially physical and electrochemical, should be experimentally convenient and fast, and provide a stable response (Wang and Freiha, 1984). Ravichandran and Baldwin (1983) showed that electrochemically pre-treated (modified) glassy carbon electrodes can enhance significantly the amperometric detection of hydrazines following their chromatographic separation. They also reported the electrochemical pre-treatment of carbon paste electrodes (Ravichandran and Baldwin, 1984). Although a pre-treated glassy carbon electrode is sufficiently long-lived to provide reproducible current levels over at least a full day's continuous usage, an electrochemically pre-treated carbon paste electrode was shorter-lived than any obtained with glassy carbon.

Sittampalam and Wilson (1983) used a platinum LCEC detector, coated with cellulose acetate film, to prevent electrode poisoning arising from protein adsorption. Korfhage *et al.* (1984) reported the construction of cobalt phthalocyanine containing a chemically modified carbon paste electrode which was prepared by thoroughly hand-mixing Nujol oil, graphite powder and cobalt phthalocyanine. Zak and Kuwana (1982) reported that oxidations of catechol, 1,4-hydroquinone and ascorbic acid at a glassy carbon electrode are catalyzed by the present of embedded alumina on the electrode surface. It has been suggested that the electrocatalysis at these electrodes involves adsorption of the electroactive species on the alumina and electrolysis of the surface species that then undergoes catalytic reaction with the solution species. The glassy carbon electrode was modified by polishing the surface with 1 μm α-alumina particles on a deck of a polishing cloth, using a circular motion for 1 min (Wang and Freiha, 1984). Figure 21 showing the comparison to ordinary (ultrasonically cleaned) glassy carbon electrodes, emphasizes the advantages of the modified glassy carbon surface. The signal at the α-alumina glassy carbon electrode is about 3.5 times larger than that at the ordinary electrode.

SENSITIVITY AND SELECTIVITY CONSIDERATION

Sensitivity

The detection limit of the EC detector is usually in the range of 10—100 pg. This detection limit, compared to those for other HPLC detectors, is relatively low. However, its sensitivity is still not sufficient to routinely measure plasma catecholamine in practical volumes of plasma (up to 1 ml). It may be possible to enhance the sensitivity of the EC detector in the following ways:

1. improvement in coulometric yield;
2. effective compensation for background current and stable amplification;
3. development and application of highly efficient columns.

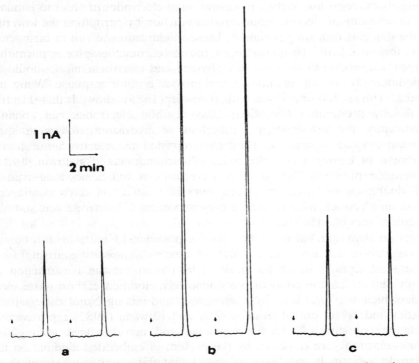

Figure 21. Chromatograms for injections of 10 ng of ascorbic acid at glassy carbon electrodes with (b) and without (a) and (c) alumina. Conditions: applied potential, +0.2 V; flow rate, 1.0 ml/min; mobile phase, 50:50 methanol:0.02 M acetic acid buffer (pH 4.5). From Wang and Freiha (1984).

The importance of coulometric yield has been described, but should be re-emphasized here. The coulometric yield of the amperometric detector is in the range of 3—30%, which is significantly lower than that of the coulometric detector. This is largely due to the small surface area of the working electrode available for electrochemical reaction. In contrast, the background current of the amperometric detector remains at lower levels than that of the coulometric detector. Improvement in coulometric yield is directly related to the enhancement of sensitivity, however, expansion of the electrode surface area results in other complications, i.e. difficulty in compensating for background current and in maintaining the electrode. If high background current can be compensated for by modifying the electrical circuitry, the coulometric detector will no doubt be superior to other methods of detection in sensitivity.

Another approach to enhancement of coulometric yield is based on the oxidation reaction mechanism of CA (Young *et al.*, 1980). It is well known that electrochemical reactions of most organic compounds are pH dependent. The primary pH effect is the alteration of half-wave potentials and changes in reaction rates (Fike and Curran, 1977; Young *et al*, 1980). Secondary effects are not directly associated with electrochemical reactions; however, they are useful for the enhancement of coulometric yield in some cases. For example, the oxidation reaction mechanism for CA varies with pH. Oxidation of CA at low pH values proceeds with losses of two electrons and two protons to form corresponding *o*-quinones. These *o*-quinones are rather stable in acidic solution since the amine groups in these compounds are completely protonated and are weak nucleophiles. In contrast, at higher pH values (neutral or basic), CA undergo two-step oxidation as illustrated in Fig. 22 (Fike and Curran, 1977; Kissinger *et al.*, 1979a, b; Young *et al.*, 1980). The *o*-quinones produced by the first electrochemical oxidation undergo cyclization to form indolines, which are further oxidized on the electrode surface with losses of two electrons and two protons. Therefore, the net electron transfer is twice that of the one-step reaction at lower pH values. This double electron transfer is the result of the rapid rate of cyclization and the completion of the reacting during the time CA are present on the electrode surface. The rate of the cyclization reaction is fastest for EN, moderate for NE and slowest for DA. The post-column addition of a buffering agent resulted in a considerable increase in the peak height of EN: a slight increase for NE, but the peak height of DA remained unchanged, as shown in Fig. 23. Increases in reaction current for NE, EN and DA were approximately 20, 0 and 70%, respectively, when the pH value of the reaction media was raised from 3.75 to 7.28. This method suggests similar but separate methods in which stronger nucleophiles are intentionally added to the reaction media. As can be seen from Fig. 22, indoline formation reactions are simply the result of the nucleophilic addition of amines to benzene nuclei. Differences in reaction rates can be ascribed to differences in the nucleophilicity of CA amine groups. When a nucleophilic stronger than the amines is added to the reaction media after separation (post-column

A) At low pH (acidic conditions)

B) At high pH (neutral or basic conditions)

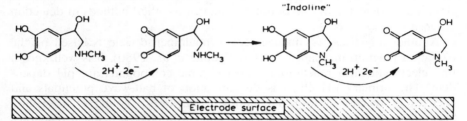

Figure 22. Electrochemical reactions of epinephrine at different pH values. From Kissinger *et al.* (1979a).

addition), nearly equal enhancement is expected for all CA. It is well known that sulfhydryl compounds (-SH containing compounds) are strong nucleophiles. Post-column addition of cysteine to the column eluate approximately doubled the coulometric yield for each CA as expected (Kissinger *et al.*, 1979a). This doubly effective coulometric yield may be partially ascribable to redox cross reaction. Redox cross reaction converts *o*-quinones into catechols and added sulfhydryl compounds to disulfides.

One additional, interesting approach to the enhancement of coulomeric yield is the application of redox couples (Fenn *et al.*, 1978; Kissinger *et al.*, 1979a). This approach is effective only when the redox reaction is reversible and rapid because this approach requires regeneration of the original electrochemically active compound. This is illustrated in Fig. 24. The electrochemical detector is composed of two working electrodes positioned opposite each other. They are separated by a thin spacer. If the potentials of these two electrodes are properly chosen and the flow rate of the stream in the EC cell is sufficiently slow to allow diffusion of the electrochemical product from one electrode to the other, a solute molecule oxidized at the anode is regenerated at the cathode and oxidized again at the anode. A single electrochemically active molecule is repeatedly cycled between the anode and the cathode, and gives (or takes up) numerous electrons to (or from) the electrodes. This suggests increased coulometric yield. Note, however, that the flow rate of the column effluent must be sufficiently slow and the length of these two electrodes must be sufficiently long to allow repeated utilization of a single molecule during its presence in the EC cell.

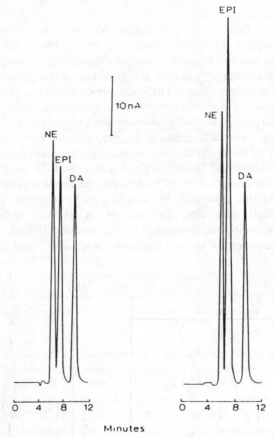

Figure 23. Effects of pH on coulometric yields (left: pH = 3.75, post column addition of the mobile phase; right: pH = 7.2, Na$_2$HPO$_4$). From Armentrout *et al.* (1979).

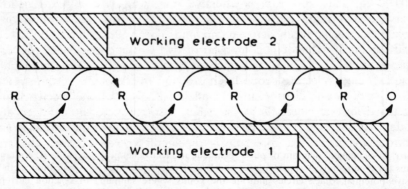

Figure 24. Enhancement of coulometric yield in a twin electrode cell. Working electrode 1 is set at a positive potential and working electrode 2 is set at a negative potential. O = oxidized form, R = reduced form. From Kissinger *et al.* (1979a).

Selectivity

The EC detection method is so selective that only electrochemically active species are detected. However, there are many electrochemically active components contained in biological samples, especially in body fluid samples. If an interfering substance exhibits a higher half-wave potential than target compounds, no problem occurs. Difficulty arises when the target compounds have higher half-wave potentials than the interfering substances. The screen mode in the coulometric detector is one of the previously described solutions to this problem. Another solution is the use of the differential pulse technique. Refer to the hypothetical voltammograms shown in Fig. 25. In this figure, 'A' represents the interfering substance and 'B' is the target compound. If sensitive detection of 'B' is desired, the applied potential should be E_2 or higher. At this potential, 'A' reaches its limiting current and is also detected with high sensitivity. The use of an EC cell with two working electrodes is obviously one approach to the solution (Blank, 1979; Fenn *et al.*, 1978). If the potential of one electrode is set at E_1 and the other at E_2,

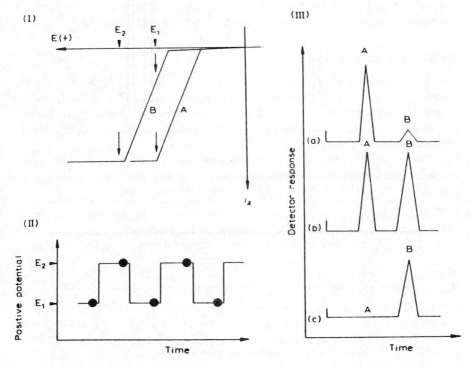

Figure 25. Schematic diagram of pulse technique in electrochemical detection. (I) Voltammograms for compounds A (interference) and B (object compound). (II) Application method of pulsive potential. Black circles indicate the points where sampling of reaction currents is performed. (III) Chromatograms obtained by conventional potential application and pulsive potential application: (a) Potential is set at E_1; (b) Potential is set at E_2; (c) Potential is applied as pulses. From Hashimoto and Maruyama (1983).

elimination of peak 'A' from the chromatogram is possible by the subtraction of the retention current. This is not difficult with current electronic technology. The results of several studies suggest that this can also be done with one electrode. In fact, application of the pulse technique can solve this problem (Brunt and Bruins, 1979; Kissinger, 1977; Schwartzfager, 1976). Two potentials, E_1 and E_2, are applied alternately, and the reaction current at these two potentials is sampled and subtracted. At potential E_1, 'A' reaches its limiting current and gives its maximum response while 'B' gives only a slight response. At increased potential E_2, both 'A' and 'B' give maximum resposnes. Thus, the current associated with the electrochemical reaction of 'A' remains nearly constant. In contrast, considerable current change is observed for 'B'. Subtraction of currents at these two potentials gives the 'B' signal, but not that for 'A'. Hypothetical chromatograms are also presented in Fig. 25 to illustrate the effects of this technique. This technique is applicable when 'A' and 'B' are the target compound and the interfering substance, respectively, and have close half-wave potentials. Large potential differences will result in good responses, but the peak of the interfering substance may appear on the chromatogram. Small potential differences will result in excellent selectivity, though sensitivity will be significantly reduced.

DERIVATIZATION FOR LCEC

The EC method is very sensitive, but not universally detectable. One of the solutions to this problem is to increase electrochemical sensitivity in the analytes (Kissinger et al., 1979a, b). It has been an usual technique as derivatization in gas chromatographic analysis (electron capture) and HPLC with fluorometric detection.

In selecting the reagent for the electrochemical detecting derivatization, the following factors are significant (Ikenoya et al., 1980):

1. the reagent for the derivatization should be oxidized at low potential;
2. the derivatization reaction should be specific, quantitative, free from side reaction and completed in a short time under mild conditions;
3. it should be easy to eliminate excess reagent from the reaction solution.

Nitroaromatic compounds

A compound containing −CO or −CHO is easily formed to hydrazone by using hydrazine. This reaction is applied to an optical identification of these compounds by a 2,4-dinitrophenylhydrazine and a 2,4,6-trinitrobenzenesulfonic acid. Fortunately, this same group is ideal as a derivatization approach for LCEC. The nitro group is easily reduced to a hydroxylamine and then to amine (Equations 1 and 2). The hydroxyamine is easily oxidized to nitroso compound (Equation 3).

$$R-NO_2 + 4e + 4H^+ \longrightarrow R-NHOH \qquad (1)$$

$$R-NHOH + 2e + 2H^+ \longrightarrow R-NH_2 \qquad (2)$$

$$R-NHOH \longrightarrow R-NO + 2e + 2H^+ \qquad (3)$$

In the reducing mode, however, dissolved oxygen exerts significant interferences, thus a careful procedure is required for the removal of oxygen
dissolved in the mobile phase. A similar approach was applied to the
determination of 17-ketosteroid sulfates in human serum using pre-column
derivatization with *p*-nitrophenylhydrazine. Shimada *et al.* (1984) have
attempted to develop a new method for the direct determination of 17-
ketosteroid sulfates without solvolysis. They found suitable conditions for derivatization using dehydroepiandrosterone sulfate as a model compound.
Condensation of dehydroepiandrosterone sulfate with *p*-nitrophenylhydra-
zine proceeded quantitatively without fission of the sulfate bond, when
heated together at 60° C for 20 min in 0.3% trichloroacetic acid—benzene
solution. The *p*-nitrophenylhydrazones formed from dehydroepiandrosterone,
epiandrosterone, etiocholanolone and androsterone sulfates, and 2-hydroxy-
estrone 3-methylether (internal standard) were satisfactorily resolved on a
μ-Bondapak C_{18} column.

The most significant problems in the derivatization procedure is how to
remove the excess reagent as mentioned above. In this case, they have
overcome the problem by disconnecting the ECD from the HPLC system for
about 5 min after injection. The detection limit of the *p*-nitrophenyl-
hydrazone derivatives was 360 pg at 4 nA full scale.

Nitroaromatics (Fig. 26) 2,4-dinitrofluorobenzene (DNFB), 2-chloro-3,5-
dinitropyridine (DNCP) and 2,4,6-trinitrobenzene sulfonic acid (TNBS) also
have been and continue to be widely used as reagents for derivatizing amino
compounds, especially amino acids (Edwards, 1977; Knapp, 1979). These
methods normally utilize TLC or LCUV for identification or quantification
of derivatives.

Jacobs and Kissinger (1982) have evaluated these three reagents for use in
conjunction with LCEC. Table 2 shows the detection limits of alanine derivatives by EC or UV detector as an *S/N* ratio of 3. The lowest detection limit is
obtained for the TNP-derivative by both UV and EC but in all cases EC
provides lower detection limits than UV by about one order of magnitude.

It is important to consider not only the sensitivity of the resulting
derivatives but also the ease with which interferences of by-products and
excess reagent can be removed as mentioned above. In this study, the reagents are detectable by ECD and furthermore, each reagent hydrolyzes to
the same extent under the derivatization conditions employed, resulting in

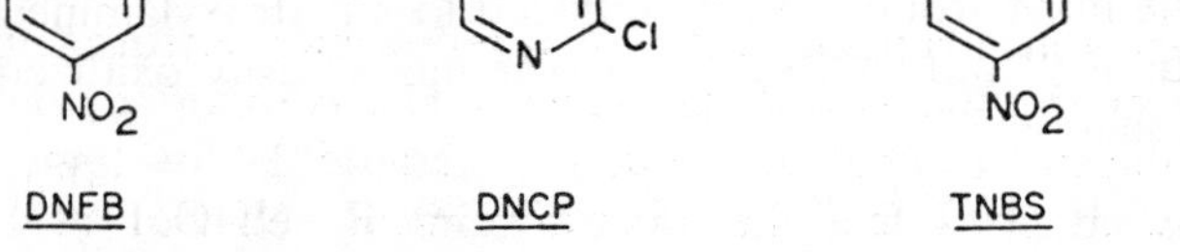

Figure 26. Structures of nitroaromatics as derivatization reagents for electrochemical detection.
DNFB = 2,4-dinitrofluorobenzene; DNCP = 2-chloro-3,5-dinitropyridine; TNBS = 2,4,6-
trinitrobenzene sulfonic acid. From Jacobs and Kissinger (1982).

Table 2. Detection limits for alanyl derivatives
by LCUV and LCEC (pmol, $S/N = 3$). (Jacobs
and Kissinger, 1982)

Derivatives	EC	UV
DNP—Ala	1.0	10
DNPy—Ala	0.60	12
TNP—Ala	0.31	5.7

DNP = Dinitrophenyl; DNPy = Dinitropyridine; TNP = Trinitrophenyl.

phenolic by-products which also constitute potential interferences. Procedures to remove the excess reagents and interferences from by-produts were
investigated. The resulting chromatograms are shown in Fig. 27. They
concluded that TNBS is best suited for application as a pre-column
derivatization reagent of the three reagents for LCEC.

O-Phthalaldehyde (for amines)

O-Phthalaldehyde (OPA) is commonly used as a derivatizing reagent for a
fluorometric determination of alkylamines, amino acids and small peptides.
The reaction of OPA with amine compounds is to produce fluorescent
products in the presence of a thiol reducing agent (Roth, 1971). These
fluorescent products were determined to be 1-(alkylthio)-2-alkylisoindoles
(Simons and Johnson, 1976) as in Fig. 28.

Recently, Joseph and Davies (1983) reported that the isoindole product
was readily oxidized electrochemically, permitting the use of LCEC for their
determination. Allison et al. (1984) showed the gradient separation of 22
amino acids with OPA and β-mercaptoethanol on short 3 µm reversed-
phase column (4.6 × 100 mm, Fig. 29). The detection limit of these derivatives was less than 500 fmol.

The isoindoles, however, have a significant problem in that these are not
stable during the short time of their determination by the pre-column
derivatization procedure. Allison et al. (1984) speculated that the electrochemical properties of the isoindoles would be far less susceptible than fluorescence to changes in the derivatives' structure. Hence, eight thiol
compounds were investigated to search for a more stable isoindole derivative
which would be suitable to a simple, pre-column LCEC method without
stability limitations. The use of OPA/tert-butylthiol derivatization produced a
derivative of a GABA as a model compound that was completely stable for
30 min, a gross improvement over β-meraptoethanol performance (Table 3).
The stability of the methionine (another model compound) derivative also
showed enhancement. The substitution of tert-BuSH for β-meraptoethanol
did not substantially affect the electrochemical behavior of the resulting
isoindole derivatives in either their half-wave oxidation potential or their
absolute response. This derivative was applied to determine neurotransmitter

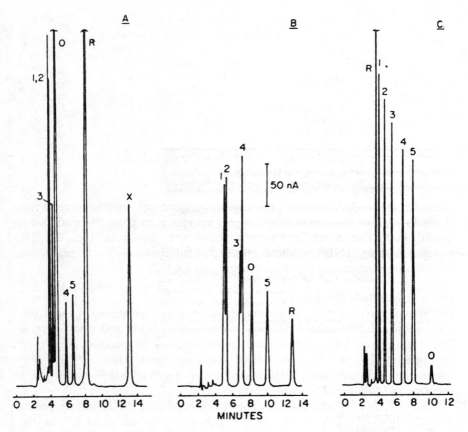

Figure 27. Chromatograms of synthetic incubations for (A) DNFB, (B) DNCP, (C) TNBS peaks: 1. glutamic acid; 2. serine; 3. threonine; 4. glycine; 5. alanine. R = reagent, O = predicted major hydrolysis product. All amino acids present in incubation mixture at 10^{-4} M. Column: Biophase C_8, 5 μm, 25 × 0.46 cm. Other abbreviations are the same as in Fig. 26. From Jacobs and Kissinger (1982).

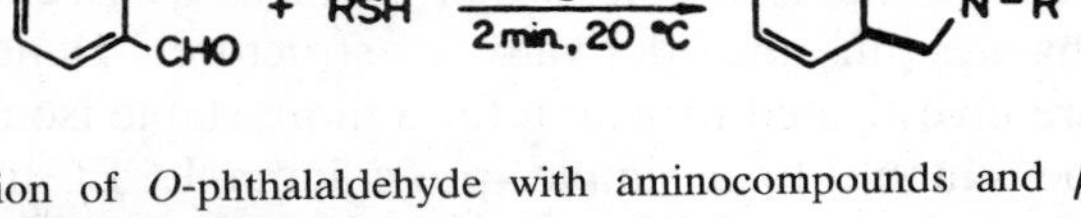

Figure 28. Reaction of *O*-phthalaldehyde with aminocompounds and β-mercaptoethanol. From Simons and Johnson (1976).

amino acids (Fig. 30) in rats' brains. A successful quantitative method using *OPA*/β-mercaptoethanol for these amino acids is very difficult to obtain, since glycine and GABA form extremely unstable *OPA*/β-mercaptoethanol derivatives. The enhanced stability of the *OPA*/tert-BuSH derivatives is thus an important advantage in this application.

Y. Hashimoto and Y. Maruyama

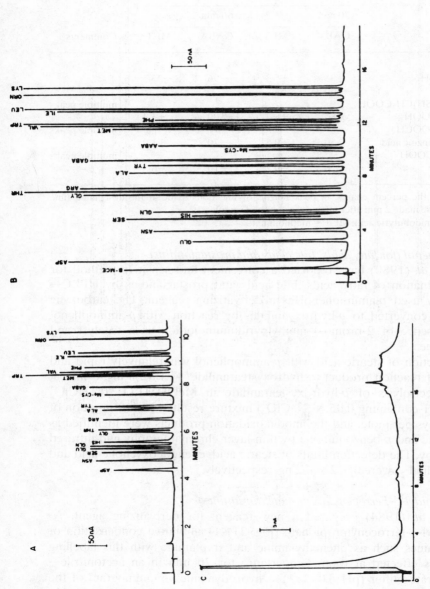

Figure 29. (A) High-speed chromatographic separation of OPA/β-MCE derivatives of 21 amino acids, 167 pmol of each amino acid derivative injected. Gradient conditions were as follows: (A) 20% MeOH, 80% 0.05 M NaClO₄/0.005 M sodium citrate (pH 5.0); (B) 76% MeOH, 19% buffer, 5% THF; flow rate 1.5 ml/min. Program: 0 min, 100% A/0% B; 8.0 min, 15/85; 9.0 min, 0/100; 9.5 min, 0/100; 9.6 min, 100/0. Stationary phase was Perkin-Elmer HS-3 3-μm C₁₈ column, 100 × 4.6 mm. (B) Chromatogram of 22 amino acids under alternate gradient, 113—258 pmol of each derivative injected. Gradient conditions were as follows: (A) 10% MeOH/90% (0.05 M NaClO₄/0.05 M sodium citrate (pH 5.80)); (B) 80% MeOH/20% buffer (pH 5.50), (C) 100% THF; flow rate 1.3 ml/min. Program: 0 min, 90% A/8% B/2% C; 4 min, 70/28/2; 5 min, 55/41.7/3.3; 7 min, 55/41.7/3.3; 13.5 min, 30/50/20; 15 min, 20/60/20; 18 min, 20/60/20. (C) Water 'blank' was derivatized and chromatographed under conditions of Fig. 3B, except at 10-fold higher detector gain. From Allison *et al.* (1984).

Table 3. Effect of thiol structure on stability (Allison *et al.*, 1984)

| | % Difference in response* | | | | |
| | 30 min | | 60 min | | |
	GABA	MET	GABA	MET	Comments
Thiol					
$HSCH_2CH_2OH$	−95	+5.3	−96	−23	
$HSCH_2CH_3$	−100				
$CH(COOH)(SH)CH_2COOH$					multiple peaks
$HSCH_2CH_2COOH$	−86	+8.2	−100	+3.4	
$HSCH_2CH_2COOCH_3$					multiple peaks
2-Mercaptobenzoic acid	−50	+0.5	−70	+1.0	
$CH_3CH(SH)COOH$					multiple peaks
$(CH_3)_3CSH$	0	−0.7	−10	−6.3	

* Defined as the percent change in peak height for injections made at the designated time relative to those made 2 min after mixing reagents and sample.
GABA = γ-aminobutyric acid; MET = methionine.

p-Aminophenol (for fatty acids, bile acids and prostaglandins)

Ikenoya *et al.* (1980) have reported a convenient and sensitive method for the determination of fatty acids, bile acids and prostaglandins by HPLC—ECD. They used *p*-aminophenol as a derivatizing reagent. The carboxylic acids were converted to *p*-hydroxyanilides by reaction with *p*-aminophenol in the presence of 2-bromo-1-methylpyridinium iodide as a catalyst and triethylamine.

The reaction of stearic acid with *p*-aminophenol was relatively rapid and the yield of reaction product, *p*-hydroxystearanilide, was 82% in 30 min at 60° C. Electrolysis of *p*-hydroxystearanilide in $MeOH:H_2O:70\%HClO_4$ (900:100:1 containing 0.05 M $NaClO_4$) mixture resulted in consumption of two Faradays per mole, and the anodic oxidation products were identified as stearylamide and *p*-benzoquinone by thin-layer chromatography and infrared spectroscopy. The detection limits of stearic acid, chenodeoxycholic acid and prostaglandin $F_{2\alpha}$ were 0.5, 2 and 2 ng, respectively.

N-Succinimidyl-3-Ferrocenylpropionate (for amines)

Tanaka *et al.* (1984) prepared a new reagent for derivatizing agent, *N*-succinimidyl-3-ferrocenylpropionate (Fig. 31). Quantitative condensation of arylalkylamines such as phenethylamine and tryptamine with this labelling reagent was effected at room tempertaure for 20 min in an acetonitrile—0.05 M borate buffer (pH 8.0, 1:1). a hydrodynamic voltammogram of the resulting derivatives indicated that the highest sensitivity would be obtainable at 0.04 V vs Ag/AgCl reference electrode with a detection limit of 0.2 pmol. The resulting ferrocene derivatives could be oxidized much more readily ($E_{1/2}$ = 0.28 V) than a catechol, 2-hydroxyestrone ($E_{1/2}$ = 0.44 V). Furthermore, from the experiment of the redox reaction by series dual electrodes the

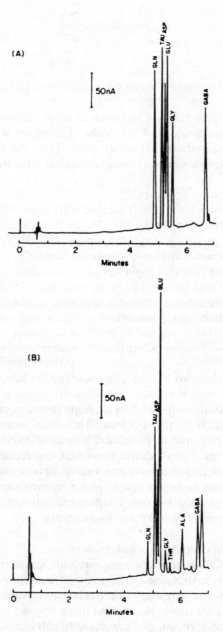

Figure 30. (A) Gradient chromatogram of *OPA/t*-BuSH derivatives of standard neurotransmitter amino acids (167 pmol of each amino acid injected). (B) Chromatogram of neurotransmitter amino acids in rat brain homogenate. Concentrations in homogenate were as follows:
glutamine, 0.216 mM; taurine, 0.808 mM; aspartic acid, 0.726 mM; glutamic acid, 1.63 mM;
glycine, 0.241 mM; and γ-aminobutyric acid, 0.501 mM. The working electrode potential was
+0.80 V vs. Ag/AgCl. From Allison *et al.* (1984).

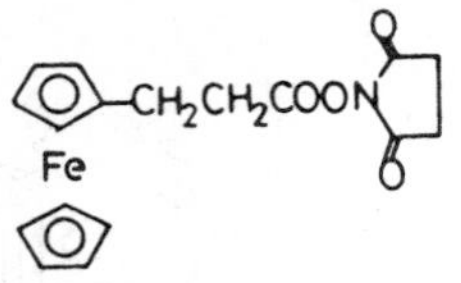

Figure 31. Structure of *N*-succinimidyl-3-ferrocenylpropionate. From Tanaka *et al.* (1984).

collection efficiency (the ratio of the current at the downstream detector to that at the upstream detector) of ferrocene derivatives was 0.289, which is higher than that of 2-hydroxyestrone (0.259). Thus the ferrocenyl group is more easily reversible in the redox reaction than the catechol group.

Using mediator (for reducing sugars)
Reducing carbohydrates have been detected with refractive index (Schmidt *et al.*, 1981), colorimetric (Simatupang, 1979) or fluoromeric methods (Honda *et al.*, 1980; Katz *et al.*, 1974; Mopper *et al.*, 1980) in HPLC. These techniques are often not sufficiently sensitive. Recently, a sensitive detection method for reducing sugars using an amperometric detector has been reported (Watanabe and Inoue, 1983). A coupling of the redox reaction of metal complex as a mediator with the reducing ability of sugars offers an essential scheme of detection as follows:

$$Cu(phen)_2^{2+} \xrightarrow[\text{chemically}]{\text{reducing sugar}} Cu(phen)_2^{+} \xrightarrow{\text{electrochemically}} Cu(phen)_2^{2+}$$

where copper bis(phenanthroline) complex (CBP) was employed as the mediator. Reducing sugars reduce the divalent complex of CBP to monovalent in alkaline solution (pH 11.0) at high temperatures (85° C) in the reaction coil placed after the column. The monovalent CBP formed is reoxidized at a low potential (+75 mV). This potential is very low compared with that commonly used in oxidative detection. the reduced form of CBP is easily reoxidizable by dissolved oxygen, thus it is very important to exclude dissolved oxygen from both the eluent and reagent solution. The detection limit of glucose is 0.2 ng by this CBP mediated method and the result represents a ten fold improvement over fluorometry.

Pre- and post-column chemistry using enzyme
Certain molecules of clinical interest (e.g. ethanol, lactate and malate) are not electroactive. However, they can be oxidized by enzymes classified as oxidoreductases. This class of enzymes can oxidize a substrate and transfer the electrons to an acceptor molecule such as nicotinamide adenine dinucleotide (NAD). The reduced form of this cofactor, NADH, has been shown to be easily oxidized at the surface of a glassy carbon electrode (Blaedal and Jenkins, 1975, 1976; Thomas and Christian, 1975). The oxidation proceeds in a 2*e* reaction (Fig. 32).

Consequently, one can indirectly quantitate the original amount of monoelectroactive compound by the oxidation of the reduced cofactor

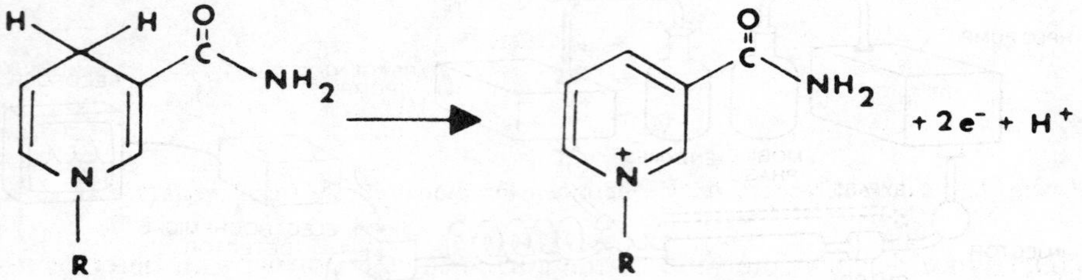

Figure 32. Electrochemical oxidation of NADH to NAD. From Aizawa *et al.* (1975).

produced from the enzymatic reaction. An assay for ethanol in small volumes of human blood (100 μl) based on this idea has been developed by Kissinger *et al.* (1977). The blood ethanol assay relies on the following reaction;

$$CH_3CH_2OH + NAD^+ \xrightarrow{\text{ADH}} NADH + CH_3CHO + H^+$$

where alcohol dehydrogenase (ADH) oxidizes ethanol and produces NADH. The amounts of NADH is determined selectivity by reverse phase HPLC with an EC detector. Kamada *et al.* (1982) have determined bile acids by a similar method using 3α-hydroxysteroid dehydrogenase coupled to amino glass beads.

Determination of choline (Ch) and acetylcholine (ACh) is relatively difficult due to the lack of a simple assay method for these compounds. The assay method for Ch and ACh using HPLC with an EC detector were presented (Ikarashi *et al.*, 1984; Maruyama and Ikarashi, in press; Potter *et al.*, 1983). The method is based on the separation of Ch and ACh by reverse phase HPLC, mixing the effluent in a post-column reaction coil with acetylcholinesterase and Ch oxidase, followed by the electrochemical detection of hydrogen peroxide enzymatically produced from both endogenous Ch and Ch generated from ACh in the reaction coil.

Figure 33 shows the design of HPLC equipment for this assay method. A post-column reaction coil composed of 10—30 m of Teflon tubing and a EC detector is equipped with a platinum electrode. The electrode potential is set to +0.5 V against a Ag/AgCl reference electrode for the detection of hydrogen peroxide. The limits of sensitivity were 1 pmol for Ch and 2 pmol for ACh. Figure 34 shows the chromatograms of Ch and ACh extracted from brain tissue samples.

One problem involved with the use of this method is that it is expensive because the enzyme can not be recovered. The connection of the immobilized enzyme column to this assay system is now under development.

In the future this may become more practicable in an 'electrochemical immunoassay' when, for example, an enzyme immunoassay using immobilized antibody could be developed by an electrochemical instrument (Wehmeyer *et al.*, 1983).

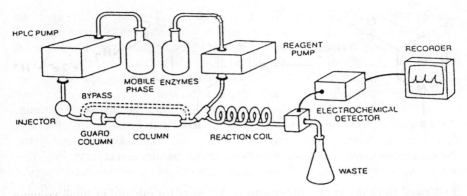

Figure 33. Design of HPLC equipment for ACh and Ch. Mobile phase is pumped through sample injector and column and is joined by enzyme solution, flow into reaction coil, and is monitored in electrochemical detector. From Potter *et al.* (1983).

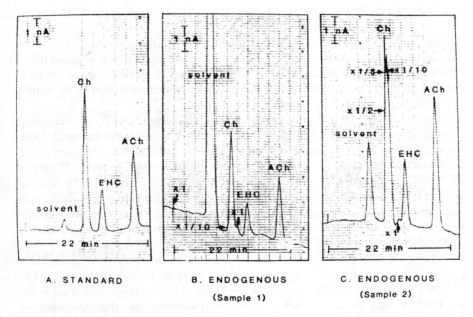

Figure 34. Chromatograms of Ch and ACh extracted from rat brain using the post-column reaction by acetylcholine esterase and choline esterase. From Maruyama and Ikarashi (in press).

CONCLUSION

We have described many projects for the enhancement of sensitivity and higher selectivity of electrochemical detector for HPLC. Dual working electrodes are quite useful for the selective determination of biological materials including many interferences. Improvement in coulometric yield is directly related to the enhancement of sensitivity. Though the sensitivity of amperometric and coulometric detector is now reported to be comparable; if high background current can be compensated for by modifying the electrical circuity, the coulometric detector will no doubt be superior to other methods of detection in sensitivity. A combination of dual electrodes, which are constructed of coulometric and amperometric cell, and a micro- or semimicro-HPLC may permit an analytical system of higher sensitivity. Derivatization techniques are also useful to expand the utility the EC detection and enhance sensitivity, and many other reagents will develop for these purposes. *In vivo* voltammetry is an interesting technique, but now selectivity is not sufficient for the specific determination of biological samples *in vivo*. We are looking forward to developing a selective electrode. In the future ECD will be a powerful tool capable of solving many biological problems.

REFERENCES

Adams, R. N., and Marsden, C. A. (1983). In: *Handbook in Psychopharmacology.* L. L. Iversen, S. D. Iversen, and S. H. Snyder (eds.), Plenum Press, New York, pp. 1—74.

Aizawa, M., Coughlin, R. W., and Charles, M. (1975). Electrochemical regeneration of nicotinamide adenine dinucleotide. *Biochim. Biophys. Acta 385,* 362—370.

Allison, L. A., Keddington, J., and Shoup, R. E. (1983). Liquid chromatographic behavior of biological thiols and the corresponding disulfides. *J. Liq. Chromatogr. 6,* 1785—1798.

Allison, L. A., Mayer, G. S., and Shoup, R. E. (1984). *O*-Phthalaldehyde derivatives of amines for high-speed liquid chromatography/electrochemistry. *Anal. Chem. 56,* 1089—1096.

Andrews, R. W., Schubert, C., Morrison, J., Zink, E. W., and Matson, W. R. (1982). Dual electrode cells for LCEC: Recent developments. *Am. Lab. 14,* 140—155.

Armentrout, D. N., Mclean, J. D., and Long, M. W. (1979). Trace determination of phenolic compounds in water by reversed-phase liquid chromatography with electrochemical detection using a carbonpolyethylene tubular anode. *Anal. Chem. 51,* 1039—1045.

Barber, M., Bordoli, R. S., Sedgwick, R. D., and Tylen, A. N. (1981). Fast atom bombardment quadrupole mass spectrometry. *J. Chem. Soc. Chem. Commun.,* 325—327.

Blaedel, W. J., and Jenkins, R. A. (1975). Study of the electrochemical oxidation of reduced nicotinamide adenine dinucleotide. *Anal. Chem. 47,* 1337—1343.

Blaedel, W. J., and Jenkins, R. A. (1976). Study of a reagent-less lactate electrode. *Anal. Chem. 48,* 1240—1247.

Blank, C. L. (1979). Dual electrochemical detector for liquid chromatography. *J. Chromatogr. 117,* 35—46.

Bratin, K., Kissinger, P. T., and Bruntlett, C. S. (1981). Reductive mode thin-layer amperometric detector for liquid chromatography. *J. Liq. Chromatogr. 4,* 1777—1795.

Brunt, K., and Bruins, H. P. (1979). Evaluation of the characteristics of the differential amperometric detector in combination with anion exchange chromatography, using 1-ascorbic acid as test compound. *J. Chromatogr. 172,* 37—47.

Bunyagidji, C., and Girard, J. E. (1982). A comparison of coulometric detectors for catecholamine analysis by LC—EC. *Life Sci. 31,* 2627—2634.

Butcher, H., Blaha, C., and Line, A. (1984). A comparison of the dopamine releasing ability of CNS stimulants, phencyclidine, using *in vivo* voltammetry. *Brain Res. Bull. 13,* 497—501.

Conti, J. C., Strope, E., Adams, R. N., and Marsden, C. A. (1978). Voltammetry in brain tissue: Chronic recording of simulated dopamine and 5-hydroxytryptamine release. *Life Sci. 23,* 2705—2716.

Dayton, M. A., Ewing, A. G., and Wightman, R. M. (1981). Evaluation of amphetamine-induced *in vivo* electrochemical response. *Eur. J. Pharmacol. 75,* 141—144.

Edwards, D. J. (1977). In: *Handbook of derivatives for chromatography*, K. Blau, and G. S. King (eds.), Heyden and Sons Ltd, London, Chap. 10.

Eggli, R., and Asper, R. (1978). Electrochemical flow-through detector for the determination of cystine and related compounds. *Anal. Chim. Acta. 101*, 253—259.

Fenn, R. T., Siggia, S., and Curran, D. J. (1978). Liquid chromatography detector based on single and twin electrode thin-layer electrochemistry: Application to the determination of catecholamines in blood plasma. *Anal. Chem. 50*, 1067—1073.

Fike, R. R., and Curran, D. J. (1977). Determination of catecholamine by thin-layer linear sweep voltammetry. *Anal. Chem. 49*, 1205—1210.

Gonon, F., Cespuglio, R., Buda, M., and Pujol, J.-F. (1983). *In vivo* electrochemical detection of monoamine derivatives. In: *Methods in Biogenic Amine Research*, S. Parvez, T. Nagatsu, I. Nagatsu, and H. Parvez (eds.), Elsevier, Amsterdam, pp. 165—188.

Gonon, F., Navarre, F., and Buda, M. J. (1984). *In vivo* monitoring of dopamine release in the rat brain with differential normal pulse voltammetry. *Anal. Chem. 56*, 573—575.

Goto, M., Nakamura, T., and Ishii, D. (1981). Micro high-performance liquid chromatographic system with micro precolumn and dual electrochemical detector for direct injection analysis of catecholamines in body fluids. *J. Chromatogr. 226*, 33—42.

Goto, M., Sakurai, E., and Ishii, D. (1982). Dual electrochemical detector for micro high-performance liquid chromatography and its application to the selective detection of catecholamines. *J. Chromatogr. 238*, 357—366.

Gunasingham, H., Tay, B. T., and Ang, K. P. (1984). Amperometric detection in normal-phase high-performance liquid chromatography with a large volume wall jet cell and Ag/Ag^+ reference system. *Anal. Chem. 56*, 2422—2426.

Hanekamp, H. B., and de Jong, H. G. (1982). Theoretical comparison of the performance of electrochemical flow-through detectors. *Anal. Chim. Acta 135*, 351—354.

Haroon, Y., Schubert, C. A. W., and Hauschka, P. V. (1984). Liquid chromatographic dual electrode detection system for vitamin K compounds. *J. Chromatogr. Sci. 22*, 89—93.

Hashimoto, H., and Maruyama, Y. (1983). High performance liquid chromatography with electrochemical detection. In: *Methods in Biogenic Amine Research*, S. Parvez, T. Nagatsu, I. Nagatsu, and H. Parvez (eds.), Elsevier, Amsterdam, pp. 35—74.

Hashimoto, Y., and Miyazaki, H. (1979). Simultaneous determinatin of endogeous norepinephrine and dopamine-β-hydroxylase activity in biological materials by chemical ionization mass fragmentography. *J. Chromatogr. 168*, 59—68.

Hepler, B. R., and Purdy, W. C. (1983). Use of a dual coulometric—amperometric detection cell approach in thyroid hormone assay. *J. Liq. Chromatogr. 6*, 2275—2310.

Honda, S., Matsuda, Y., Tanaka, M., Kakehi, K., and Ganno, S. (1980). Fluorimetric determination of reducing carbohydrates with 2-cyanoacetamide and application to automated analysis of carbohydrates as borate complexes. *Anal. Chem. 52*, 1079—1082.

Ikarashi, Y., and Maruyama, Y. (1984). High-performance liquid chromatographic analysis of regional catecholamines and 3,4-dihydroxyphenylacetic acid in rat brain following microwave irradiation. *Biogenic Amines 1*, 341—357.

Ikarashi, Y., and Maruyama, Y. (in press). Determination of catecholamines, indolamines, and related metabolites in rat brain with liquid chromatography with electrochemical detection. *Biogenic Amines 2*.

Ikarashi, Y., Sasahara, T., and Maruyama, Y. (1984). A simple method for determination of choline (Ch) and acetylcholine (ACh) in rat brain regions using high-performance liquid chromatography with electrochemical detection (HPLC—ED). *Folia Pharmacol. Jpn 84*, 529—536.

Ikeda, M., Miyazaki, H., Mugitani, N., and Matsushita, A. (1984). Simultaneous monitoring of 3,4-dihydroxyphenylacetic acid (DOPAC) and 5-hydroxyindoleacetic acid (5-HIAA) levels in the brain of freely moving rat by differential pulse voltammetry technique. *Neurosci. Res. 1*, 171—184.

Ikenoya, S., Hiroshima, O., Ohmae, M., and Kawabe, K. (1980). Electrochemical detector for high performance liquid chromatography. IV. Analysis of fatty acids, bile acids and prosagrandins by derivatization to an electrochemically active form. *Chem. Pharm. Bull. 28*, 2941—2947.

Jacobs, W. A., and Kissinger, P. T. (1982). Nitroaromatic reagents for determination of amines and amino acids by liquid chromatography/electrochemistry. *J. Chromatogr. 5*, 881—895.

Joseph, M. H., and Davies, P. (1983). Electrochemical activity of *o*-phthalaldehyde—mercaptoethanol derivatives of amino acids; Application to high-performance liquid chromatographic determination of amino acids in plasma and other biological materials. *J. Chromatogr. 277*, 125—136.

Johnson, D. C., Ryan, M. D., and Wilson, G. S. (1984). Dynamic electrochemistry: Methodology and applications. *Anal. Chem. 56*, 7R—20R.

Jorgenson, J. W., and Guthrie, E. J. (1983). Liquid chromatography in open-tubular columns; Theory of column optimization with limited pressure and analysis time. *J. Chromatogr. 255*, 335—348.

Kamada, S., Maeda, M., Tsuji, A., Umezawa, Y., and Kurahashi, T. (1982). Separation and determination of bile acids by high-performance liquid chromatography using immobilized 3α-hydroxysteroid dehydrogenase and an electrochemical detector. *J. Chromatogr. 255*, 335—348.

Katz, S., Pitt, W. W., Mrochek, J. E., and Dinsmore, S. (1974). Sensitive fluorescence monitoring of carbohydrates eluted by a borate mobile phase from an anion exchange column. *J. Chromatogr. 101*, 193—197.

King, W. P., and Kissinger, P. T. (1980). Liquid chromatography with amperometric reaction detection involving electrogenerated reagents: Applications with *in-situ* generated bromine. *Clin. Chem. 26*, 1484—1491.

Kissinger, P. T. (1977). Amperometric and coulometric detectors for high-performance liquid chromatography. *Anal. Chem. 49*, 447A—456A.

Kissinger, P. T. (1984). Liquid chromatography/electrochemistry, fundamentals and future directions. Abstract of LCEC Symposium on the Application to Life Sciences, Tokyo, pp. 3—18.

Kissinger, P. T., Bratin, K., Davis, G. C., and Pachla, L. A. (1979a). The potential utility of pre- and post-column chemical reactions with electrochemical detection in liquid chromatography. *J. Chromatogr. Sci. 17*, 137—146.

Kissinger, P. T., Bruntlett, C. S., Bratin, K., and Rice, J. R. (1979b). *Trace Organic Analysis: A New Frontier in Analytical Chemistry*, S. N. Chesler, and H. S. Hertz (eds.), National Bureau of Standards US Special publication 519, pp. 705—712.

Kissinger, P. T., Hart, J. B., and Adams, R. N. (1973). Voltammetry in brain tissue—a new neurophysiological measurement. *Brain Res. 55*, 209—213.

Knapp, D. R. (1979). *Handbook of Analytical Derivatization Reactions*, John Wiley and Sons Inc., New York.

Knecht, L. A., Guthrie, E. J., and Jorgenson, J. W. (1984). On-column electrochemical detector with a single graphite fiber electrode for open-tubular liquid chromatography. *Anal. Chem. 56*, 479—482.

Korfhage, K. M., Ravichandran, K., and Baldwin, R. P. (1984). Phthalocyanine-containing chemically modified electrodes for electrochemical detection in liquid chromatography/flow injection systems. *Anal. Chem. 56*, 1514—1517.

Krull, I. S., Bratin, K., Shoup, R. E., Kissinger, P. T., and Blank, C. L. (1983). LCEC for trace analysis: recent advances in instrumentation, method, and applications. *Am. Lab. 15*, 57—65.

Kurahashi, T., and Nishino, H. (1984). Application of electrolytic flow cell. The 27th Meeting on Liquid Chromatography. Tokyo, pp. 23—24.

Lunte, C. E., and Kissinger, P. T. (1983). Determination of pterins in biological samples by liquid chromatography/electrochemistry with a dual-electrode detector. *Anal. Chem. 55*, 1456—1462.

Lunte, C. E., and Kissinger, P. T. (1984). Investigation of 6-methylpterin electrochemistry by dual-electrode liquid chromatography/electrochemistry. *Anal. Chem. 56*, 658—663.

MacCrehan, W. A., and Durst, R. A. (1981). Dual-electrode, liquid chromatographic detector for the determination of analytes with high redox potentials. *Anal. Chem. 53*, 1700—1704.

Martin, R. J., Bailey, B. A., and Downer, R. G. H. (1983). Rapid estimation of catecholamines, octopamine and 5-hydroxytryptamine from biological tissues using high performance liquid chromatography with coulometric detection. *J. Chromatogr. 278*, 265—274.

Maruyama, Y., and Ikarashi, Y. (in press). Determination of choline (Ch) and acetylcholine (ACh) levels in rat brain regions by liquid chromatography with electrochemical detection (LC—EC). *J. Chromatogr.*

Maruyama, Y., and Kusaka, M. (1978). Assay of norepinephrine and dopamine in the rat brain after microwave irradiation. *Life Sci. 23*, 1603—1608.

Matson, W. R., Anderson, R. W., Ball, J., Skinner, D., Vitukevich, R., and Zuck, E. W. (1981). A new electrochemical HPLC detector. Pittsburgh Conference on Analytical Chemistry and Applied Spectroscopy, Atlantic City, p. 565.

Mayer, G. S., and Shoup, R. E. (1982). Improving selectivity for biogenic amines and metabolites in physiological samples using dual parallel and series electrodes. *Curr. Separ. (Bioanal. Syst. W. Lafayette) 4*, 40—42.

Mopper, K., Dawson, R., Liebezeit, G., and Hansen, H. P. (1980). Borate complex ion exchange chromatography with fluorimetric detection for determination of saccharides. *Anal. Chem. 52*, 2018—2022.

Potter, P. E., Meek, J. L., and Neff, N. H. (1983). Acetylcholine and choline in nueronal tissue measured by HPLC with electrochemical detection. *J. Neurochem. 41*, 188—194.

Ravichandran, K., and Baldwin, R. P. (1983). Liquid chromatographic determination of hydrazines with electrochemically pretreated glassy carbon electrodes. *Anal. Chem. 55*, 1782—1786.

Ravichandran, K., and Baldwin, R. P. (1984). Enhanced voltammetric response by electrochemical pretreatment of carbon paste electrodes. *Anal. Chem. 56*, 1744—1747.

Roston, D. A., and Kissinger, P. T. (1983). Series dual-electrode detector for liquid chromatography/electrochemistry. *Anal. Chem. 54*, 429—434.

Roston, D. A., Shoup, R. E., and Kissinger, P. T. (1982). Liquid chromatography/electrochemistry: Thin-layer multiple electrode detection. *Anal. Chem. 54*, 1417A—1434A.

Roth, M. (1971). Fluorescence reaction for amino acids. *Anal. Chem. 43*, 880—882.

Schieffer, G. W. (1980). Dual coulometric—amperometric cells for increasing the selectivity of electrochemical detection in high-performance liquid chromatography. *Anal. Chem. 52*, 1994—1998.

Schieffer, G. W. (1981). Precolumn coulometric cell for high-performance liquid chromatography. *Anal. Chem. 53*, 126—127.

Schmidt, J., John, M., and Wandrey, C. (1981). Rapid separation of malto-, xylo- and cello-oligosaccharides (DP2-9) on cation-exchange resin using water as eluent. *J. Chromatogr. 213*, 151—155.

Schwartzfager, D. G. (1976). Amperometric and differential pulse voltammetric detection in high performance liquid chromatography. *Anal. Chem. 48*, 2189—2192.

Shimada, K., Tanaka, M., and Nambara, T. (1984). Studies on steroids. CC. Determination of 17-ketosteroid sulphates in serum by high-performance liquid chromatography with electrochemical detection using pre-column derivatization. *J. Chromatogr. Biomed. Applic. 307*, 23—28.

Shoup, R. E. (1982). *Curr. Separ. (Bioanal. Syst., W. Lafayette), 4*, 36—37.

Shoup, R. E., and Mayer, G. S. (1982). Determination of environmental phenols by liquid chromatography/electrochemistry. *Anal. Chem. 54*, 1164—1168.

Simatupang, M. H. (1979). Ion-exchange chromatography of some neutral monosaccharides and uronic acids. *J. Chromatogr. 178*, 588—591.

Simons, S. S. Jr., and Johnson, D. F. (1976). The structure of the fluorescent adduct formed in the reaction of *O*-phthalaldehyde and thiols with amines. *J. Am. Chem. Soc. 98*, 7098—7099.

Sittampalam, G., and Wilson, G. S. (1983). Surface modified electrochemical detector for liquid chromatography. *Anal. Chem. 55*, 1608—1610.

Stulik, K., and Pacakova, V. (1981). Electrochemical detection techniques in high-performance liquid chromatography. *J. Electroanal. Chem. 129*, 1—24.

Suzuki, Y., and Taguchi, K. (in press). Effects of morphine on the serotonergic system in the cat spinal cord. *Brain Res.*

Tanaka, M., Shimada, K., and Nambara, T. (1984). Novel ferrocene reagent for pre-column labelling of amines in high-performance liquid chromatography with electrochemical detection. *J. Chromatogr. 292,* 410—411.

Thomas, L. C., and Christian, G. D. (1975). Voltammetric measurement of reduced nicotinamide—adenine nucleotides and application to amperometric measurement of enzyme reactions. *Anal. Chim. Acta 78,* 271—276.

Waller, G. R. (1971). *Biomedical application of mass spectrometry,* Wiley Interscience, New York.

Wang, J., and Freiha, B. (1984). Liquid chromatography with detection by α-alumina modified glassy carbon electrodes. *Anal. Chem. 56,* 2266—2269.

Watanabe, N., and Inoue, M. (1983). Amperometric detection of reducing carbohydrates in liquid chromatography. *Anal. Chem. 55,* 1016—1019.

Weber, S. G., and Purdy, W. C. (1978). The behaviour of an electrochemical detector used in liquid chromatography and continuous flow voltammetry. Part 1. Mass transport-limited current. *Anal. Chim. Acta 100,* 531—544.

Wehmeyer, K. R., Doyle, M. J., Wright, D. S., Eggersz, H. M., Halsall, H. B., and Heineman, W. R. (1983). Liquid chromatography with electrochemical detection of phenol and NADH for enzyme immunoassay. *J. Liq. Chromatogr. 6,* 2141—2156.

Wightman, R. M. (1981). Microvoltammetric electrodes. *Anal. Chem. 53,* 1125A—1134A.

Young, T. E., Babbitt, B. W., and Wolfe, L. A. (1980). Melanin, 2. Electrochemical study of the oxidation of α-methyldopa and 5,6-dihydroxy-2-methylindole. *J. Org. Chem. 45,* 2899—2902.

Zak, J., and Kuwana, T. (1982). Electrooxidative catalysis using dispersed alumina on glassy carbon surfaces. *J. Am. Chem. Soc. 104,* 5514—5515.

Author Index

Subject Index